TRANSLATIONS OF MATHEMATICAL MONOGRAPHS

VOLUME 84

Minimal Surfaces, Stratified Multivarifolds, and the Plateau Problem

ĐÀO TRỌNG THI

A. T. FOMENKO

American Mathematical Society · Providence · Rhode Island

ДАО ЧОНГ ТХИ
А. Т. ФОМЕНКО

МИНИМАЛЬНЫЕ ПОВЕРХНОСТИ И ПРОБЛЕМА ПЛАТО

«НАУКА», МОСКВА, 1987

Translated from the Russian by E. J. F. Primrose
Translation edited by Ben Silver

1980 *Mathematics Subject Classification* (1985 *Revision*). Primary 49F10, 53A10; Secondary 58E12, 58E20.

ABSTRACT. The book is an account of the current state of the theory of minimal surfaces and of one of the most important chapters of this theory—the Plateau problem, i.e. the problem of the existence of a minimal surface with boundary prescribed in advance. The authors exhibit deep connections of minimal surface theory with differential equations, Lie groups and Lie algebras, topology, and multidimensional variational calculus. The presentation is simplified to a large extent; the book is furnished with a wealth of illustrative material. Bibliography: 471 titles. 153 figures, 2 tables.

Library of Congress Cataloging-in-Publication Data

Dao, Trong Thi.
[Minimal′nye poverkhnosti i problema Plato. English]
Minimal surfaces, stratified multivarifolds, and the Plateau problem/Đào Trọng Thi, A. T. Fomenko.
p. cm. — (Translations of mathematical monographs; v. 84)
Translation of: Minimal′nye poverkhnosti i problema Plato.
Includes bibliographical references and index.
ISBN 0-8218-4536-5
1. Surfaces, Minimal. 2. Plateau's problem. I. Fomenko, A. T. II. Title. III. Series.
QA644.D2813 1991 90-22932
516.3′62—dc20 CIP

Printed in the United States of America

Information on Copying and Reprinting can be found at the back of this volume.
The paper used in this book is acid-free and falls within the guidelines
established to ensure permanence and durability. ♾
This publication was typeset using $\mathcal{A}_\mathcal{M}\mathcal{S}$-TEX,
the American Mathematical Society's TEX macro system.

10 9 8 7 6 5 4 3 2 1 96 95 94 93 92 91

Table of Contents

Preface

Plateau's problem is a scientific trend in modern mathematics that unites several different problems connected with the study of *minimal surfaces*, that is, surfaces of least area or volume. In the simplest version, we are concerned with the following problem: how to find a surface of least area that spans a given fixed wire contour in three-dimensional space. A physical model of such surfaces consists of soap films hanging on wire contours after dipping them in a soap solution. From the mathematical point of view, such films are described as solutions of a second-order partial differential equation, so their behavior is quite complicated and has still not finally been studied. Soap films or, more generally, *interfaces between physical media* in equilibrium, arise in many applied problems, in chemistry, physics, and also in nature. A well-known example is that of marine organisms, Radiolaria, whose skeletons enable us to see clearly the characteristic singularities of interfaces between media and soap films that span complicated boundary contours. In applications there arise not only two-dimensional but also multidimensional minimal surfaces that span fixed closed "contours" in some multidimensional Riemannian space (manifold). It is convenient to regard such surfaces as extremals of the functional of multidimensional volume, which enables us when studying them to employ powerful methods of modern analysis and topology. We should mention that an exact mathematical statement of the problem of finding a surface of least area (volume) requires a suitable definition of such fundamental concepts as a surface, its boundary, minimality of a surface, and so on. It turns out that there are several natural definitions of all these concepts, which enable us to study minimal surfaces by different methods, which complement one another.

In the framework of a comparatively small book it is practically impossible to cover all aspects of the modern problem of Plateau, to which a vast literature has been devoted. The authors have therefore tried to construct the book in accordance with the following principle: a maximum of clarity and a minimum of formalization. Of course, it is possible to satisfy such a requirement only approximately, so in certain cases (this applies mainly to the last chapters of the book) we dwell on certain nontrivial mathematical constructions that are necessary for a concrete investigation of minimal surfaces.

The book can be conventionally split into three parts: (a) Chapter 1, which contains historical information about Plateau's problem, referring to the period preceding the 1930s, and a description of its connections with the natural sciences; (b) Chapters 2–5, which give a fairly complete survey of various modern trends in Plateau's problem; (c) Chapters 6–11, in which we give a detailed exposition of one of these trends (the homotopic version of Plateau's problem in terms of stratified multivarifolds) and the Plateau problem in homogeneous symplectic spaces (Chapter 11).

The first part is intended for a very wide circle of readers and is accessible, for example, to first-year students. The second part, accessible to second- and third-year students specializing in physics and mathematics, relies mainly on information from a standard course in geometry and topology. Here we use the elements of Riemannian geometry, differential forms, homology, and the elements of complex analysis. We recall the main concepts but without going into details. The third part is intended for specialists interested in the modern theory of minimal surfaces and can be used for special courses. Here we assume that the reader has a command of the concepts of functional analysis.

At the beginning of the book we give a brief historical survey of the sources of the modern problem of Plateau. We start our account with earlier work of the 18th century, and then we dwell in more detail on the work of the 19th century, in which the fundamental properties of minimal surfaces were discovered. We pay special attention to the famous physical experiments of Plateau (1801–1883), in which he systematized various observations about the behavior of the interface between two media. One of the results of this series of experiments was a precise formulation of the so-called *principles of Plateau*, which control both the local and the global topological behavior of interfaces between media. Together with an exposition of the mathematical and physical aspects of Plateau's problem we give information about mathematicians whose work was most closely connected with the questions under consideration. We also try to characterize the concrete historical situation that brings to life various mathematical, mechanical, and physical aspects of Plateau's problem.

In Chapter 1 we present the elements of the classical theory of minimal surfaces, mainly for the two-dimensional case. Here we try to avoid complicated calculations, referring the interested reader to the more specialized literature, of which quite a large (though not complete) list is given at the end of the book.

Chapter 2 enables the reader to become acquainted quickly with those topological concepts without which the modern theory of minimal surfaces is inconceivable.

Beginning with Chapter 3 we introduce the reader directly to the very rich world of modern ideas about minimal surfaces and their role in mechanics, physics, and mathematics. Our aim is to put into the hands of the reader a guide that will enable him to orient himself quickly in the diverse information concentrated at the forefront of modern research. We shall pay a great deal of attention to the *methods* of studying minimal surfaces and to the main results obtained by means of them.

In particular, we present the solution due to A. T. Fomenko of Plateau's spectral problem in the class of spectra of manifolds with a fixed boundary, and also the solution due to Dào Trọng Thi of Plateau's problem in the homotopy class of multivarifolds with a given boundary.

The idea of creating a book of this kind and the plan of it is due to A. T. Fomenko. Chapters 1–5 were written by A. T. Fomenko, and Chapters 6–10 by Dào Trọng Thi; Chapter 11 is based on the recent results of Le Hong Van.

The structure of the book was formed as a result of A. T. Fomenko giving special courses on the theory of minimal surfaces in the Mechanics–Mathematics Department at Moscow State University. Also, the book was created under the influence of a programme of study of the connections between the topology of manifolds and the global properties of minimal surfaces developed in the research seminars "Modern Geometric Methods" and "Computer Geometry" under the supervision of A. T. Fomenko at Moscow State University. The most important results of some participants of the seminar are reflected in the book.

The book is intended for a wide circle of students, research students and mathematicians specializing in calculus of variations, topology, functional analysis, the theory of differential equations, and Lie groups and Lie algebras.

The authors thank S. P. Novikov, whose valuable support and interest stimulated the development of this scientific trend, and also the reviewer D. V. Anosov, who made a number of useful remarks and additions. The authors thank V. P. Maslov for his support.

The authors are very grateful to Dr. E. Primrose for his excellent translation and for his remarks, which have helped to improve the book.

Introduction

We have decided to begin our book with an independent topic, which by convention we can call the "one-dimensional Plateau problem." Despite the fact that here we minimize only a one-dimensional functional, namely the functional of arc length, we see quite a rich and meaningful Plateau problem even in this very simple case.

Roughly speaking, the one-dimensional Plateau problem is this: to find a one-dimensional "curve" (generally speaking, with branchings) of least length in the class of "curves" with fixed ends.

This problem can be formulated like the transport problem. Suppose, for example, that we are given N points (towns) in a plane. Find a network of roads of least length joining all these towns. Such a network is naturally called a *minimal network*.

The problem whose solution is given briefly in the Introduction (namely, the classification of minimal one-dimensional networks *with a convex boundary*), was posed in the seminar "Computer geometry", held in the Department of Mechanics and Mathematics of Moscow State University. All the figures in the Introduction (except Figure II) were produced on a computer, more precisely, they are the result of a computer analysis.

The theorems given in the Introduction are proved "in the usual way", that is, without the use of a computer. A computer has been used to check certain conjectures and models.

Many papers have been devoted to the investigation of minimal networks. First of all we should mention the classical survey of Gilbert and Pollak [467]. The authors devote their main attention to a search for an absolute minimum among all networks that span a fixed finite set M of points of the plane. There are two approaches. In the first case, the search for an absolutely minimal network is carried out in the class of networks whose vertices all belong to M. In this case, a minimal network is a tree (it does not have cycles), which is called a *minimal tree.* This approach is developed in the papers [468], [469] by Du, Hwang, and others.

In the second case, the set of vertices of the network may be larger than the set M. The vertices of a minimal network that do not belong to M are called *Steiner points.* An absolutely minimal network in this wider class (it is called a *minimal Steiner tree*) may, generally speaking, have a smaller length than a minimal tree. Thus, if M consists of the vertices of a triangle with angles less than 120°, an absolutely minimal Steiner tree necessarily has one Steiner point. This point is uniquely defined by the condition that the angles between the segments of the network that meet at this point are equal.

There is just as much interest in the investigation of locally minimal networks, that is, those for which any small fragment has least length. This means that any admissible variation of the network that has a sufficiently small support increases the length of the network. (Compare with the variational definition of minimal surfaces.)

Clearly, the problem of describing locally minimal networks (like that of describing absolutely minimal networks) can be stated for general Riemannian manifolds of arbitrary dimension. What is the local structure of such networks in the general case? The answer to this question essentially depends on what class of admissible variations we choose.

It is natural to restrict the class of admissible variations to those that leave fixed the initial (boundary) points spanned by the network. Henceforth we shall call these points the *fixed points* of the network.

It is easy to see that every locally minimal network consists of segments of geodesics that are joined in some way at vertices of the network. We recall that the number of geodesic segments that meet at a given vertex of the network is called the *degree* of this vertex. It remains to investigate how many geodesic segments meet at each vertex and how they meet. Two possibilities arise.

The first possibility: under a deformation of the network its vertices do not split, that is, variations of the type shown in Figure I(a) are prohibited. In this case it can be shown (see [470], for example) that a network is locally minimal if and only if for every fixed point the sum of the unit vectors in the directions of the geodesic segments issuing from it is zero. With this understanding the network given in Figure I(a) is minimal. The example given in Figure I(b) is not a minimal network in this sense.

The second possibility: it is permitted to split the vertices of the network. In this case, under a deformation of the network that tends to decrease its length, its vertices split into points of degree at most three. Moreover, at each vertex the angle between any pair of geodesic segments issuing from it is at least 120°. In particular, if three geodesics meet at a

FIGURE I(a)

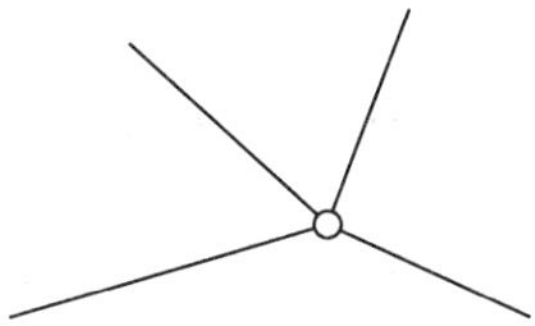

FIGURE I(b)

vertex, then they meet at angles of 120°, and so their tangent vectors at such a vertex lie in one two-dimensional plane.

All these effects can be observed in the following simple experiment. We take a flat sheet of plexiglass and drill n small holes in it (Figure II(a)). These holes will correspond to the fixed points of the network. From a string we cut a set of $n-1$ segments. At one end of $n-2$ of the segments we form a small untightened loop. We take the segment without a loop and thread it through an arbitrary number of loops. We can again thread the ends of the resulting configuration through a certain number of loops, and so on, continuing this process until all the segments are affected. We note that the number of ends of the resulting configuration is equal to the number of holes.

We place our sheet horizontally and from above we pass all the ends of the resulting configuration through the holes so that one end passes through each hole. To each end we fasten a load, the loads being equal in mass.

When the system arrives at an equilibrium position the network of string takes the form of a minimal network, in one of the senses described above (Figure II(b)). More specifically, if all the loops disperse, without hindering one another, we obtain a minimal network in the second sense. If at least one pair of loops is coupled, then there is an unresolved vertex in the resulting network, and the network is minimal in the first sense only.

We now state some recent results of Ivanov and Tuzhilin [471], who studied locally minimal networks in the second sense (henceforth simply "minimal") in a two-dimensional Euclidean plane $\mathbf{R}^2$. They obtained a *complete classification* of nondegenerate (not having vertices of degree two) minimal networks, without cycles and having a convex boundary, up to planar equivalence. In addition, they found some infinite series

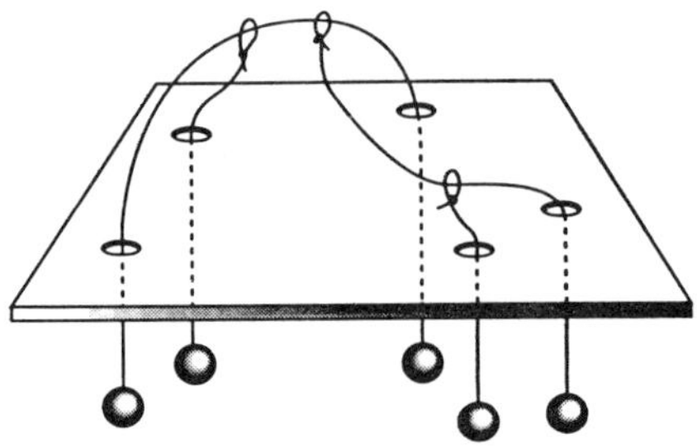

FIGURE II(a)

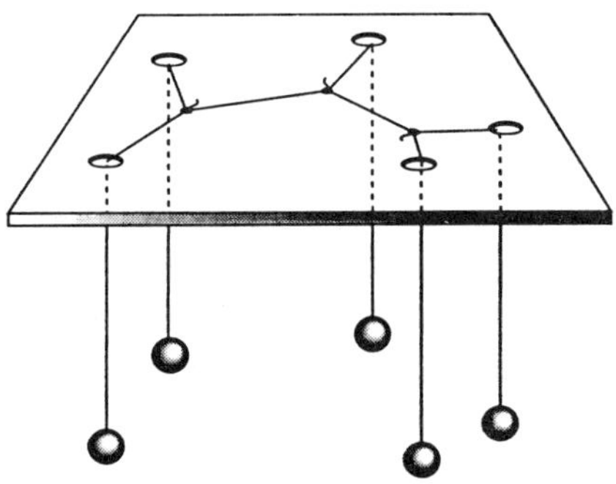

FIGURE II(b)

of minimal networks with a regular boundary, and isolated interesting examples of such networks not contained in them. In order to state these results more precisely, we need the following definitions.

A *Steiner network* is defined as an arbitrary connected planar graph whose vertices all have degree at most three. A Steiner network without vertices of degree two is said to be *nondegenerate*. Henceforth we shall study only acyclic nondegenerate minimal Steiner networks—minimal 2-trees—and we shall assume that the set of boundary points of such networks coincides with the set of vertices of degree 1.

In the classification of minimal 2-trees, it is useful to introduce the definition of the twisting number of a planar 2-tree. For an ordered pair (a, b) of edges of a planar 2-tree the *twisting number* $\mathrm{tw}(a, b)$ is the difference between the number of rotations "to the left" and "to the right" on going from a to b along the unique path in the tree joining them (the minimal subtree containing a and b). We take $\mathrm{tw}(a, a)$ to be zero. We note that $\mathrm{tw}(a, b) = -\mathrm{tw}(b, a)$.

DEFINITION. The *twisting number* of a planar 2-tree is the maximum twisting number $\mathrm{tw}(a, b)$, taken over all pairs of edges of the tree.

Proposition. *The twisting number of a minimal 2-tree with a convex boundary is not greater than five.*

It is convenient to state the classification theorem in the language of tilings. Consider a partition of the plane into congruent equilateral triangles. We call an arbitrary collection of these triangles (cells) a tiling. Consider the dual graph Γ of some tiling P. If Γ is connected, then we call P connected; if Γ is a tree, then P is a tree tiling. Henceforth a tiling will be understood as a connected tiling. The twisting number of a tree tiling is defined as the twisting number of its dual graph.

Proposition. *Any planar 2-tree with twisting number at most five can be realized as the dual graph of some tree tiling.*

Thus the problem of describing minimal 2-trees with a convex boundary reduces to the description of tree tilings with twisting number at most five. We denote the class of such tilings by $WP5$.

We now turn to the classification of tilings $WP5$. Omitting trivial cases, we assume that a tiling consists of at least three cells. We represent each tree tiling as the union of a skeleton and growths (generally speaking, this representation is not unique). We define an extreme cell of a tiling as a cell of which two sides do not lie inside the tiling; a cell whose sides all lie inside the tiling is called an interior cell.

Definition. An extreme cell of a tiling is called a *growth* if it adjoins an interior cell.

In order to obtain a skeleton, for each interior cell of the tiling we discard one of the growths adjoining it (if there are any). We note that a skeleton does not have growths.

Every skeleton of a tree tiling can be split into branch points and linear parts.

Definition. Connected components consisting of interior (resp., not interior) cells of a skeleton are called *branch points* (*linear parts*).

We note that branch points of skeletons of tree tilings can contain no more than four cells (five types of branch points are possible; they are shown in Figure III).

In each nonextreme cell of a skeleton we join by segments the midpoints of its sides lying inside the tiling. In each extreme cell we draw a midline parallel to the midline of the adjacent cell already constructed.

Definition. The *spine* (*vertebra*) of a linear part (cell) is the part of the graph constructed above that is contained in this linear part (cell).

If the twisting number of a tree tiling does not exceed five, then the spine of any linear part of it is uniquely projected onto some line l parallel to

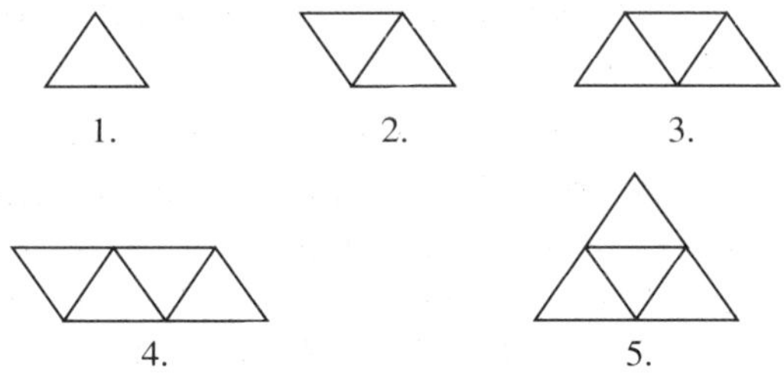

FIGURE III

one side of an arbitrary cell of the tiling. The class of lines parallel to l is called the directrix of the linear part. All such linear parts and their spines can be split into three classes, which we call a snake, stairs, or a broken snake, depending on whether the given linear part has one, two or three directrices respectively. Thus, we have described all branch points of skeletons of tree tilings and all linear parts of skeletons of tilings of $WP5$. Figure IV shows all possible types of linear parts of tilings of $WP5$.

Next we define the operation of reducing skeletons of tilings of $WP5$, which consists in cutting out some parts of the skeleton. Firstly, we can cut out from the skeleton any linear part containing an extreme cell. Secondly, inside the skeleton we can discard any snake consisting of an even number of cells and going into some linear part. It is easy to verify that the operation of reduction is well defined and does not increase the twisting number.

THEOREM (Ivanov and Tuzhilin). *All skeletons of tilings of* $WP5$ *are obtained by reduction from the three canonical types of skeletons shown in Figure* V.

A broken snake is represented by three line segments, which are parallel to its directrices. Stairs are represented by two intersecting line segments, which are parallel to its two directrices. A snake is represented by one line segment, which is parallel to the spine of the snake. To points there correspond the branch points consisting of one cell.

We now describe the possible positions of growths on a skeleton. For this we need the concept of the profile of a skeleton.

The boundary of a tiling regarded as a closed subregion of the plane is called the contour of the tiling. Consider some extreme cell of the skeleton and discard from the contour of the skeleton the edge that intersects the vertebra of this cell. Carry out the same operation for all the extreme cells of the skeleton. The contour of the skeleton splits into broken lines, which we call the profiles of the skeleton. An external side of a profile is

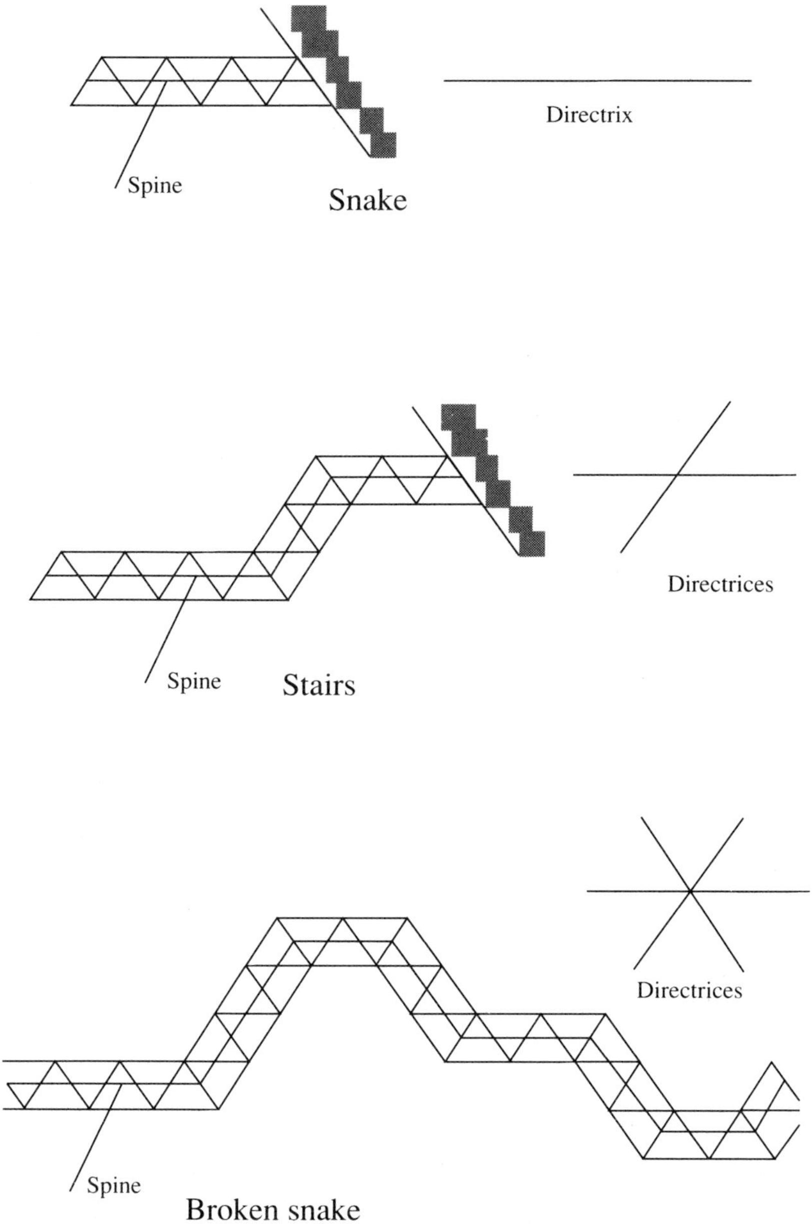
Directrix
Spine
Snake
Directrices
Spine
Stairs
Directrices
Spine
Broken snake

FIGURE IV

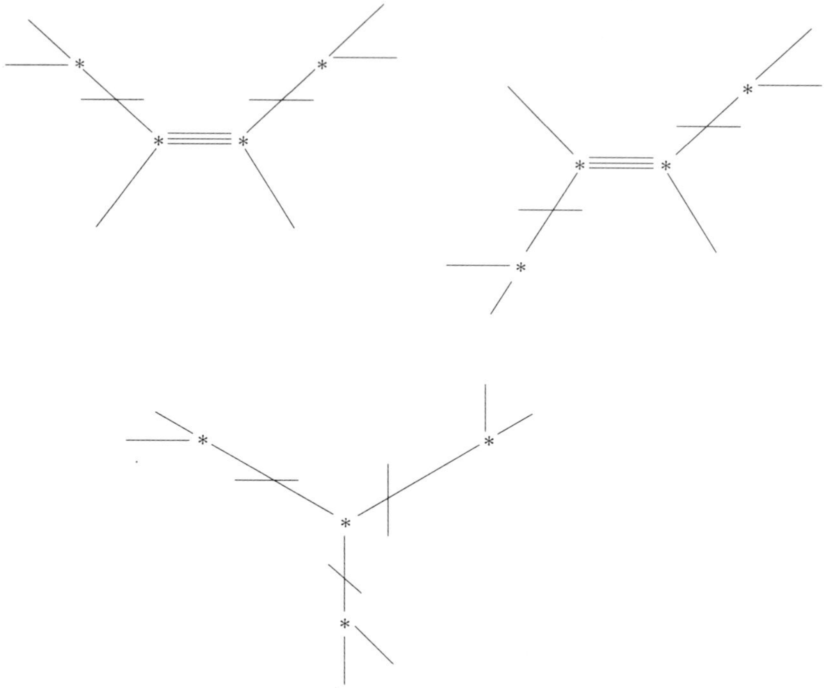

FIGURE V

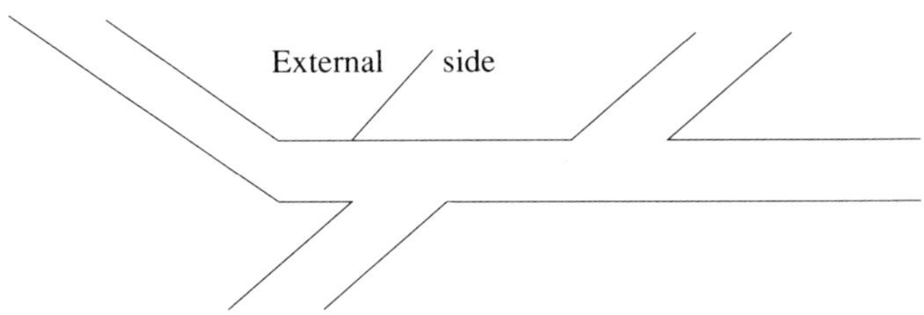

FIGURE VI

an external side with respect to the skeleton (Figure VI).

We note that the profiles of the skeleton of a tiling with twisting number not greater than five have the same properties as the spines of the linear parts of such tilings. We therefore keep the name of snake, stairs, and broken snake for the corresponding profiles.

THEOREM (Ivanov and Tuzhilin). 1. *On a profile that is a snake we can plant any number of growths* (*Figure* VII(a)).

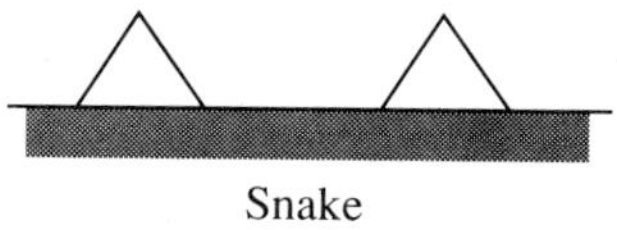

FIGURE VII(a)

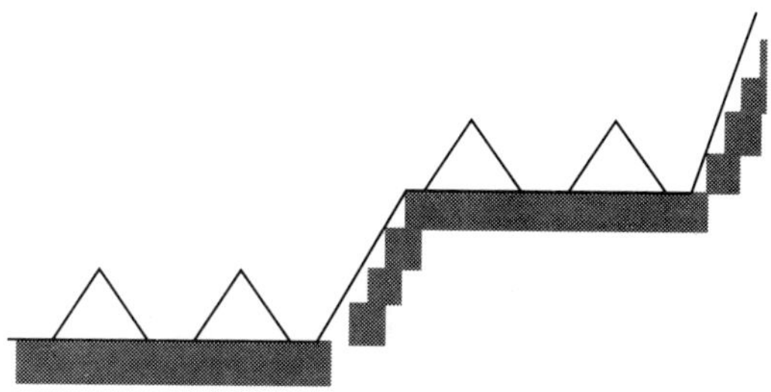

FIGURE VII(b)

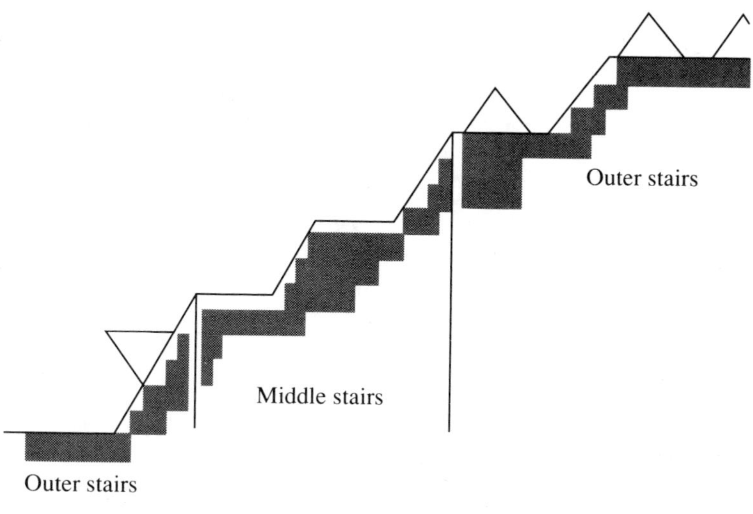

FIGURE VII(c)

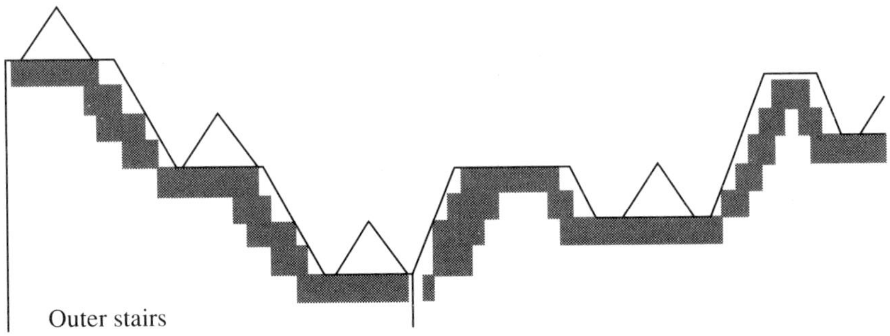

FIGURE VII(d)

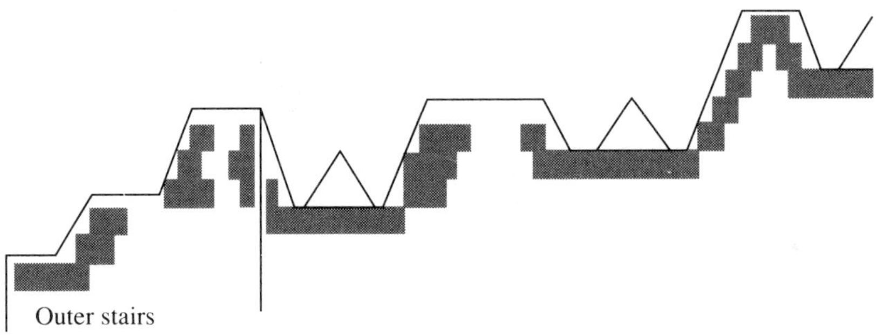

FIGURE VII(e)

2. *For stairs profiles there are two possibilities.*

(a) *The growths are distributed arbitrarily only on segments of one direction* (*Figure* VII(b)).

(b) *There is defined a partition of the stairs into three successive broken lines, the middle one of which may be empty. The middle broken line consists of an even number of links and the angle between the first pair of links, measured from the outer side, is equal to* 120°.

There are no growths on the middle broken line. On the first broken line, the growths can be situated arbitrarily on segments having the direction of its last link, and on the last broken line, they can be situated on segments having the direction of its first link (*Figure* VII(c)).

3. *We represent a profile that is a broken snake as the union of three parts, where the outer parts are maximal possible stairs, and the middle one* (*which may be empty*) *consists of all the rest. On the middle part, we can plant arbitrarily many growths only on segments parallel to the directrix of the profile. On the outer stairs we can plant growths as follows.*

Consider a segment of the profile adjacent to outer stairs. If the angle between it and the neighboring segment a *of the stairs, measured from the outer side of the profile, is equal to* 120°, *then the growths can only latch onto the segments of the stairs parallel to* a. *If this angle is equal to* $240°$, *then we can plant growths onto the stairs according to Rule* 2 (*Figures* VII(d), (e)).

We can show that any planar 2-tree that is the dual graph of the tilings described in the theorems can be realized as a minimal tree with a convex set of boundary points. Our classification is thus complete.

We now give some results of research into minimal networks spanning the vertices of regular polygons. We begin with a description of a simple algorithm that makes it possible, for a given finite set M of points of the plane, to construct by means of compasses and straight edge the minimal network spanning it (see aiso [467]). For this it is sufficient to know the structure of this minimal network as a planar 2-tree and the correspondence between the endpoints of this 2-tree and points of M. We illustrate the idea behind this algorithm by an example of the construction of a minimal network for the set M of vertices of a triangle ABC for which no angle is greater than $120°$.

We choose any pair of vertices of the triangle, say A and B, and construct on AB an equilateral triangle ABD so that C and D lie on opposite sides of AB. We draw the circle ABD.

Clearly, the only branch point V of the minimal network lies on the smaller arc d of this circle connecting the points A and B. Moreover, V lies on the ray DC (prove it). Joining this point V to the vertices of the triangle ABC, we obtain our minimal network (Figure VIII(a)).

If the triangle ABC has an angle greater than or equal to $120°$, then the corresponding minimal network is not a 2-tree. In this case we can carry out the same construction, but the angles between the segments joining the intersection point V of the ray DC and the circle to the vertices of the triangle are not equal.

For a quadrangle $ABCD$ the construction consists of two similar steps. Figure VIII(b) shows the minimal network spanning the vertices of a square. We split the vertices of the square into pairs consisting of the boundary vertices of the network, for which the edges of the network issuing from them meet at a branch point, and choose one of these pairs, say A and B. We denote by V the branch point at which the edges issuing from A and B meet, and the other branch point by W.

On the side AB we construct an equilateral triangle ABE. We place the vertex E of this triangle so that E and the branch point V lie on opposite sides of AB (Figure VIII(c)).

We now consider the triangle CDE and by the method described above we construct a minimal network for it. The branch point of this network coincides with W.

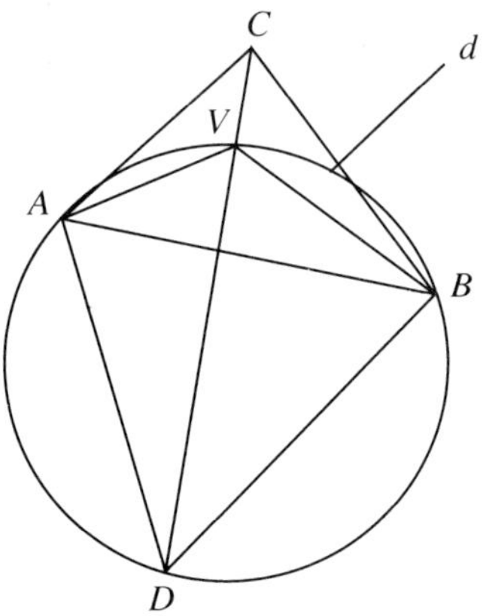

FIGURE VIII(a)

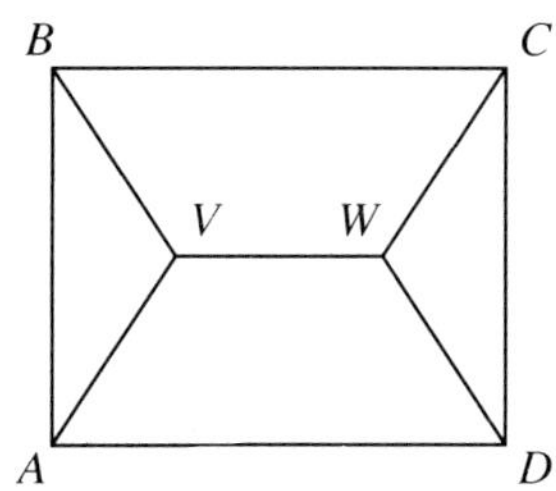

FIGURE VIII(b)

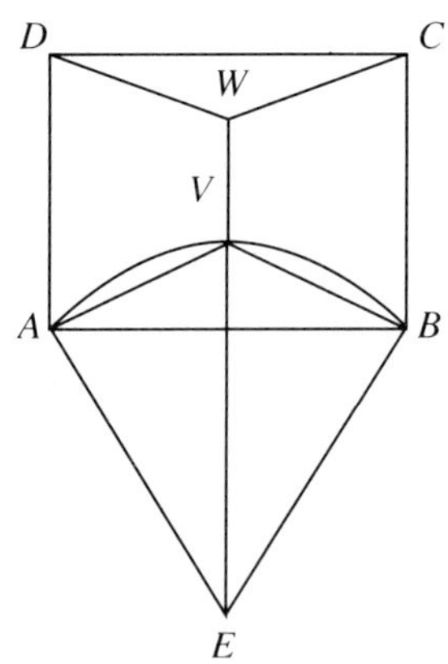

FIGURE VIII(c)

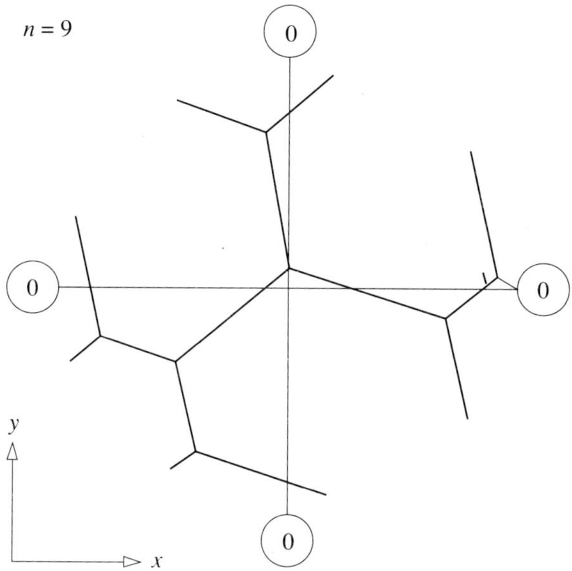

FIGURE IX(a)

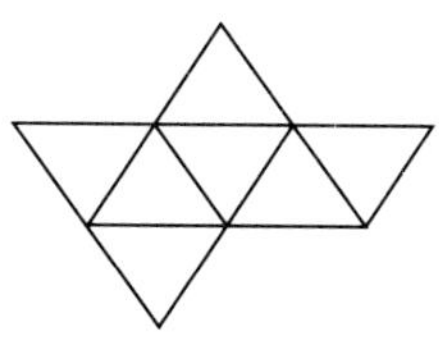

FIGURE IX(a′)

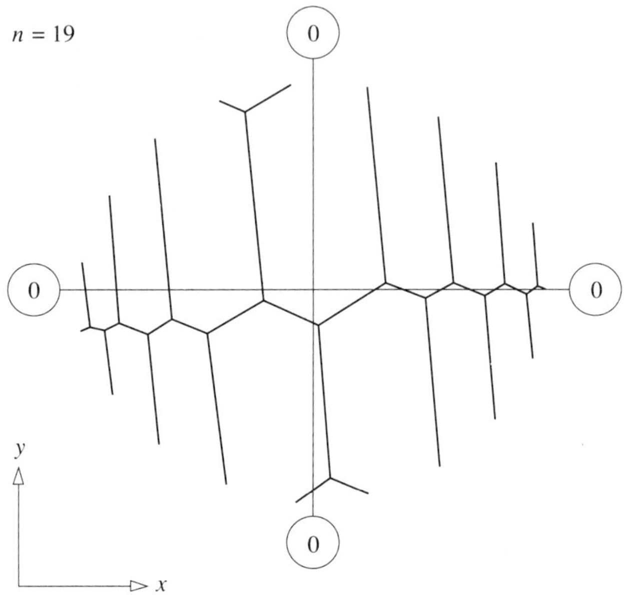

FIGURE IX(b)

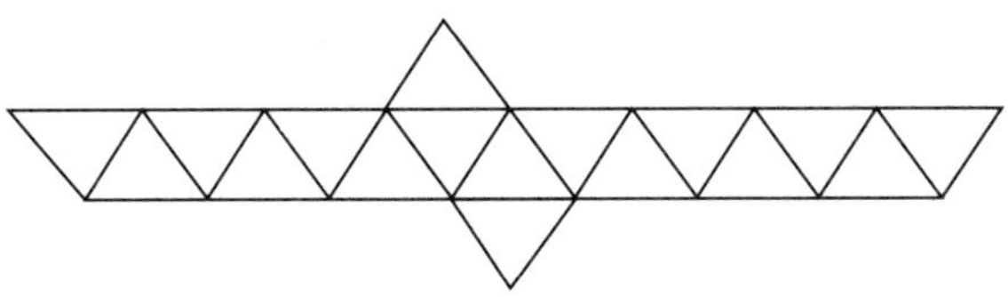

FIGURE IX(b$'$)

A minimal network for the vertices of the square $ABCD$ is obtained as follows. We draw the circle ABE. The intersection point of this circle with the minimal network we have constructed is the branch point V of the required network (prove it). It just remains to join V to A and B.

These ideas are the basis of an algorithm for constructing minimal networks with a given set of boundary points. The algorithm was realized on a computer. Because of shortage of space, we do not give a detailed description of this algorithm here. Figures IX(a), (a$'$), (b), and (b$'$) show the minimal trees constructed on the computer.

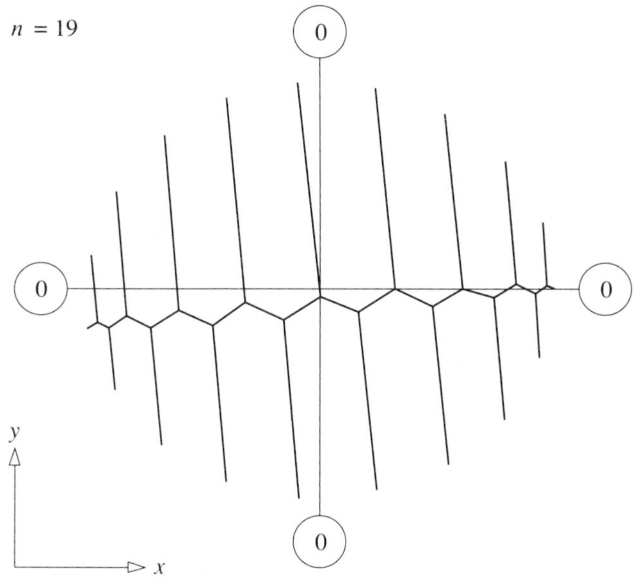

FIGURE X

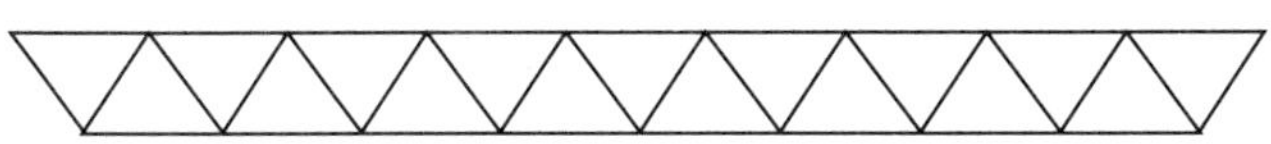

FIGURE X′

A computer experiment has enabled us to formulate a number of conjectures about the structure of minimal 2-trees spanning the vertices of regular n-gons. Some of these conjectures have been proved. Here we give a small part of the results obtained.

PROPOSITION (Ivanov and Tuzhilin). *For any* n, *we can join the vertices of a regular* n*-gon by a minimal* 2*-tree of snake type uniquely up to a motion* (*Figures* X *and* X′).

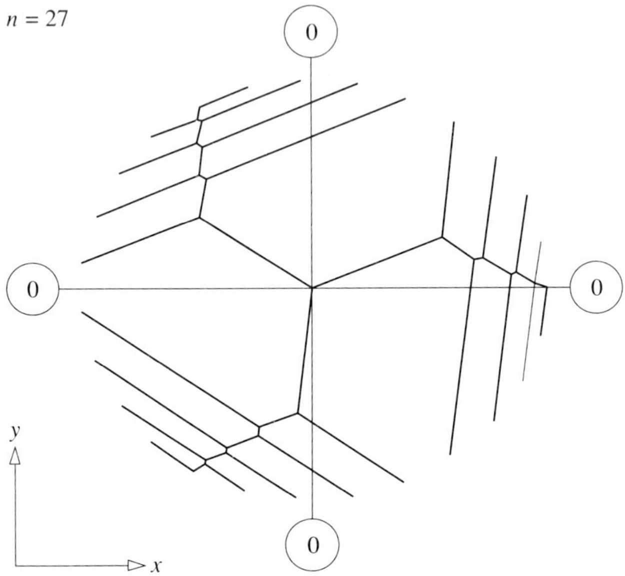

FIGURE XI

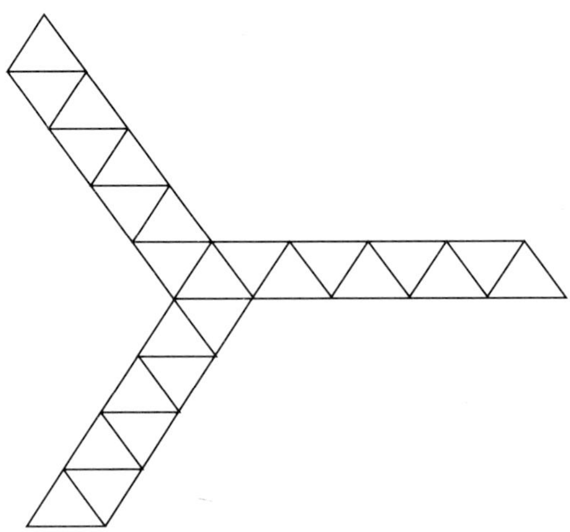

FIGURE XI′

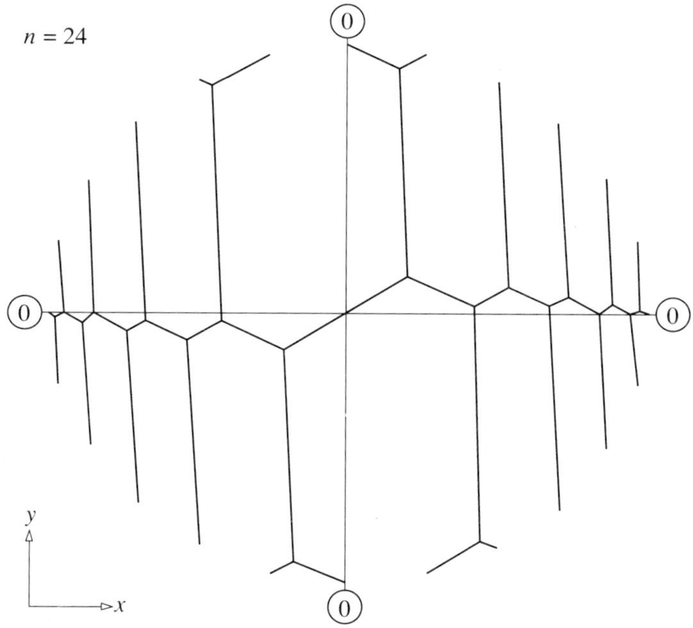

FIGURE XII

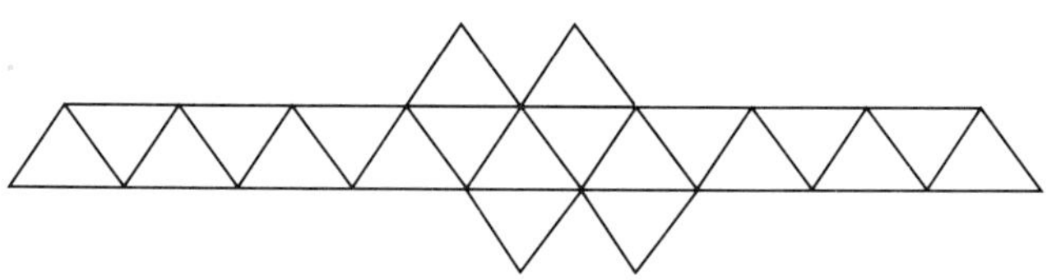

FIGURE XII′

PROPOSITION (Ivanov and Tuzhilin). *For any* $n = 6k + 3$ $(k > 0)$, *we can join the vertices of a regular* n*-gon by a minimal* 2*-tree of triple type, uniquely up to a motion* (*Figures* XI *and* XI′). *This network is invariant under rotation around the center of the* n*-gon through an angle of* 120°.

There is at least one more infinite series of minimal trees—this is a snake with pairs of symmetric growths situated close to the center of the snake (Figures XII and XII′). We have also obtained estimates for the possible position of these growths, which we do not give because of shortage of space.

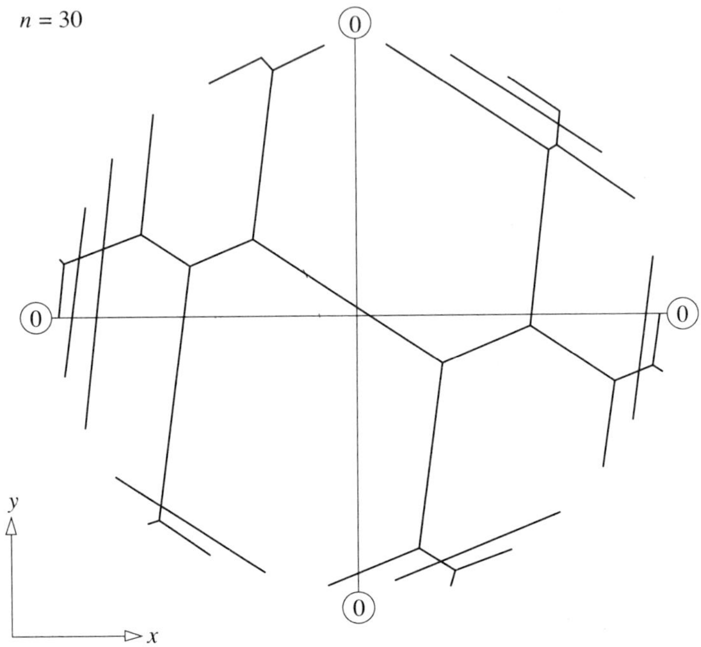

FIGURE XIII(a)

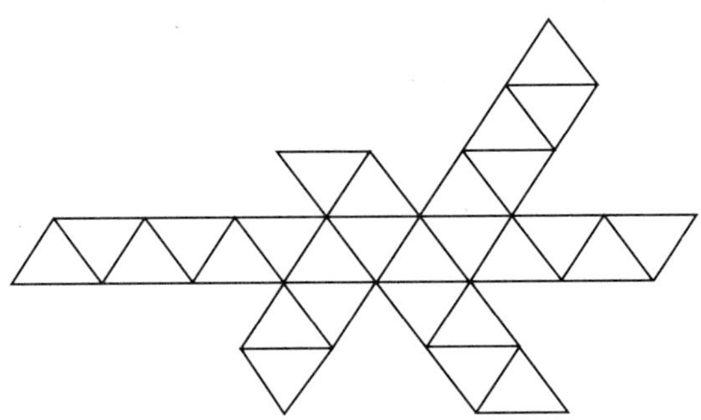

FIGURE XIII(a′)

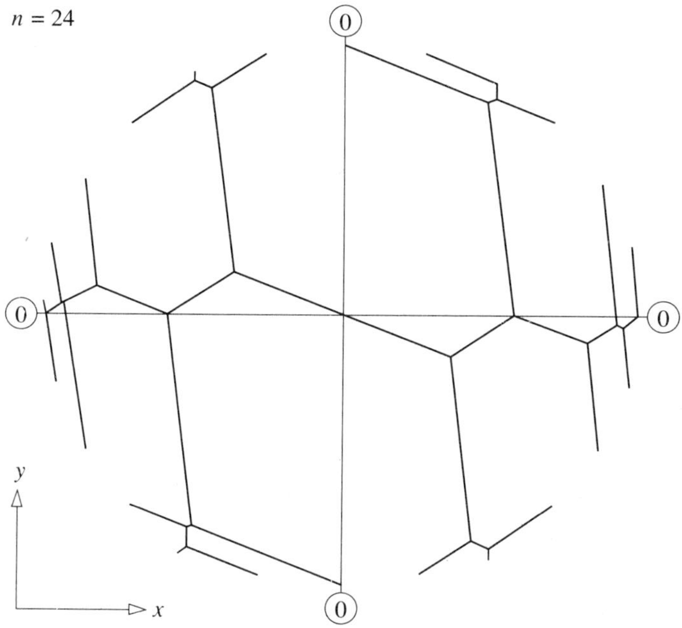

FIGURE XIII(b)

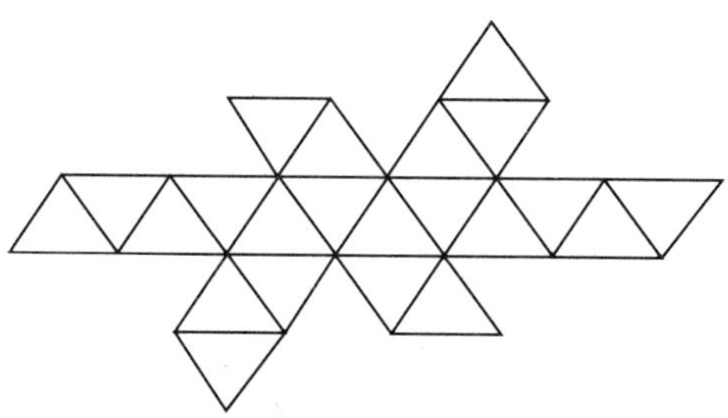

FIGURE XIII(b′)

Figures XIII (a), (a′), (b), and (b′) show representatives of an apparently finite series (as a computer experiment shows) of minimal trees realized on n-gons with $n = 24$, 30, 36, 42. We note that since the corresponding tilings have one branch point and six ends, there cannot be growths on these networks.

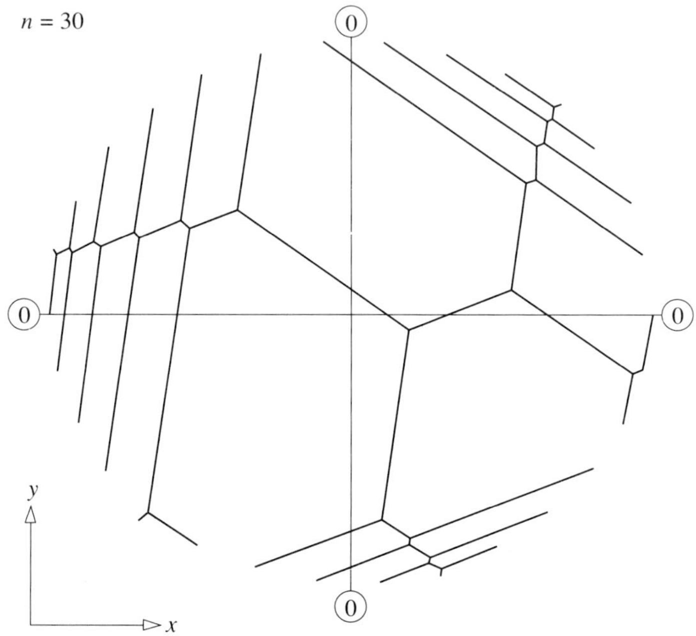

FIGURE XIV

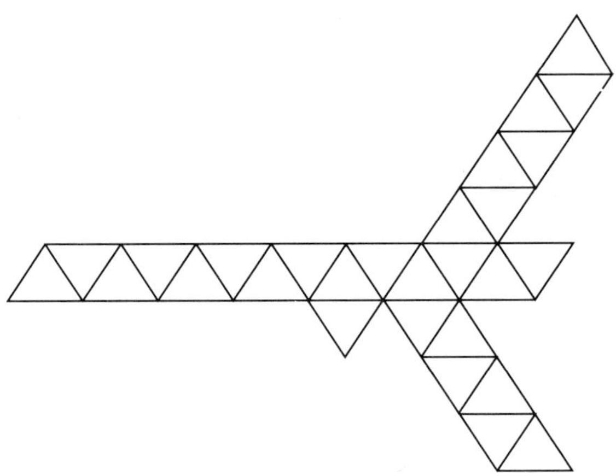

FIGURE XIV′

Figures XIV and XIV′ show an example of a network whose corresponding tiling has one branch point, four ends, and one growth.

These examples show that the problem of classifying minimal 2-trees whose sets of boundary points are the vertices of regular polyhedra is nontrivial.

CHAPTER I

Historical Survey and Introduction to the Classical Theory of Minimal Surfaces

§1. The sources of multidimensional calculus of variations

In this section we try to introduce the reader to the atmosphere of the coming into being of one of the most important branches of modern mathematics—the multidimensional calculus of variations, within which there is now developing the theory of minimal surfaces, which was first formed in the remarkable work of mathematicians and kinematicians in the 18th–19th centuries. The original development of this theory is inseparable from the personalities of the experts engaged with this question, and from the concrete historical situation under the pressure of which there began the rapid development of the theory which has found concrete applications in applied problems of mechanics and physics.

In modern calculus of variations, it is usual to distinguish the so-called *one-dimensional* and *multidimensional variational problems*. By one-dimensional problems we mean the investigation of functionals defined, for example, on a space of piecewise smooth curves $\gamma(t)$ in a Riemannian manifold. Classical examples of such functionals are the functional of length of a curve $\int |\dot\gamma|\, dt$ and the action functional $\int |\dot\gamma|^2 dt$. The extremals of such functionals are certain curves in the manifold. For example, *the extremals of the length functional are geodesics parametrized by an arbitrary continuous parameter, and the extremals of the action functional are geodesics parametrized by the natural parameter.*

However, in many questions of physics and mechanics there appear important *functionals defined on multidimensional objects and surfaces*, for example on the space of two-dimensional surfaces with a fixed boundary. An important example is the *area functional*, which associates an area with each such surface.

Another example, closely connected with the previous one, is the *Dirichlet functional*. The connection between these functionals is in many respects analogous to the well-known connection between the length functional and the action functional. In this terminology the area functional and the Dirichlet functional can be called *two-dimensional functionals.*

The main objects of study in this book are *multidimensional functionals,* including dimension three and above. In the modern terminology there occurs a certain displacement of concepts. Since the case of two-dimensional functionals is quite well known today, it is already regarded as "classical", and now under the term "multidimensional functionals" it is often understood as functionals defined on surfaces of dimension three and above. However, at the beginning of the 20th century two-dimensional functionals were numbered among multidimensional functionals.

1.1. On the work of Euler. Individual one-dimensional variational problems were studied, of course, before Euler. For example, Leibniz successfully applied his methods, taking part in competitions for solving difficult problems, in particular the problem of Galileo on the catenary and that of Johann Bernoulli on the brachistochrone. The foundations of the calculus of variations (theory of geodesics and so on) were laid by Jakob and Johann Bernoulli. Here, an important role was played by the solution of the brachistochrone problem and the isoperimetric problem advanced by Jakob Bernoulli. However, a systematic study of one-dimensional functionals is usually associated with the name of Leonhard Euler (1707–1783), who was born in Basel in Switzerland. His father, Paul Euler, intended his son for a career in the church, but at the University of Basel the young Leonhard Euler became keen on mathematics under the influence of Johann Bernoulli, whose lectures he followed and who was interested in him. We should mention that Paul Euler also studied mathematics under the supervision of Jacob Bernoulli (see [96], [108]). In 1725 Johann Bernoulli's sons Daniel and Nicolai went to St. Petersburg, and soon, on their recommendation, the young Euler received an invitation to work at the St. Petersburg Academy of Sciences in the department of physiology (see [96]). He arrived in St. Petersburg in 1727 and worked in Russia for 14 years (until 1974), gaining the reputation of the most distinguished mathematician of his epoch.

Euler's interests were very wide, and his scientific legacy was enormous. He wrote more than 850 works and an enormous quantity of letters, many of which were individual pieces of mathematical research. In St. Petersburg Euler accomplished many state tasks, was occupied with the making of geographical maps, published works on analysis, number theory, differential equations, and astronomy. As a result of overstrain in 1738 he became blind in one eye (in 1766 he lost his sight completely), but this handicap did not weaken his creative activity. Around Euler there was created a school of talented scholars: Kotel′nikov, Rumovskiĭ, Fuss, Golovin, Safronov and others (see [96]). However, he was uneasy working in St. Petersburg at that time, and in 1741 Euler received an invitation to go to Berlin to work at the Academy of Sciences. From 1766 to 1783 Euler again worked in St. Petersburg (see [108]). In the second period of his St. Petersburg scientific activity Euler presented to the Academy 416 more books and articles, dictating them to his students. Many problems solved or posed by Euler arose at the time of his research into applied problems. This applies, in particular, to his work on the theory of calculation of the lunar orbit, and the theory of boat-building and navigation. About 40% of his work was devoted to applied mathematics, physics, mechanics, hydromechanics, elasticity theory, ballistics, the theory of machines, optics, and so on. Many of Euler's discoveries were revealed after his death, in particular in the theory of differential equations (see [96]).

In the history of differential geometry it is usually assumed (see, for example, [396]) that in 1760 Euler discovered a new branch of geometry, which combined both purely geometrical and differential variational methods. In this respect there is special interest in a well-known work of Euler *Methodus inveniendi lineas curvas maximi minimive proprietate gaudentes* (1774), in which he gave methods for solving isoperimetric problems and investigated the geometrical properties of certain remarkable curves, in particular the *catenary*. This was the first account of the calculus of variations; it contained the *Euler equations* and many applications. Incidentally, Euler discovered that the *catenoid* (the surface obtained by rotating a catenary, that is, the curve formed by a sagging wire chain with fixed ends) is a two-dimensional *minimal surface*.

1.2. On the work of Lagrange and Monge. In 1762, soon after the publication of Euler's work *Recherches sur la courbure des surfaces* (Histoire de l'Académie des Sciences, Berlin, 1760, pp. 119–141) there appeared a well-known paper [278] on the calculus of variations of Joseph Louis Lagrange (1736–1813), then a young professor at Turin. Along with many remarkable results presented in [278], in an appendix to this work Lagrange reduces the *equation of minimal surfaces* (that is, extremals of the area functional) to a form in which the functions p and q are found from the following condition: the two differential 1-forms $pdx + qdy$ and $(pdy - qdx)/\sqrt{1+p^2+q^2}$ must be total differentials. These functions specify a minimal surface.

Nevertheless, we must mention Lagrange's rather sceptical attitude to the estimation of problems of differential geometry. Thus, for example, the following pronouncement of his in a letter to d'Alembert in 1772 is widely known: "Don't you think that higher geometry leads to a certain decline?" This is possibly explained by the fact that Lagrange considered these geometrical problems merely as *illustrations* for the application of the remarkable analytical methods that he developed. Thus, for example, Lagrange obtained the equation of minimal surfaces as a result of applying general methods of the calculus of variations; he did not give a geometrical interpretation of the results obtained. Moreover, Lagrange did not pose the question of finding nontrivial examples of minimal surfaces in three-dimensional space.

The next important step in the direction of the development of the theory of minimal surfaces (that is, surfaces of locally minimal area) was taken by Gaspard Monge (1746–1818). Many achievements in the field of calculus of variations are associated with the name of this distinguished mathematician. Therefore, before proceeding to expound them, we shall dwell briefly on the biography of Monge and on a characterization of the epoch that formed his unique talent.

Monge was born in 1746 into the family of a small trader who had gained a fairly firm position as head of a corporation of traders. This enabled Monge's father to give his three sons a good primary education, as a result of which all three later became professors of mathematics.

Monge received his secondary education at a college in the French town of Beaune (in the Côte d'Or department). After finishing at college Monge entered the high school of the Order of Orators in Lyons, finishing there at the age of 18. Declining an offer to take up holy orders, he soon obtained a place in the Mézières Military Engineering School, which aimed to train engineers of high qualifications.

In 1769 Monge obtained the independent position of professor at the Mézières school, which enabled him to pursue his scientific work very actively. The circle of his interests was very wide: mathematics, physics, chemistry. His first discovery, which has become widely known, was the creation of descriptive geometry. Of course, the idea of projecting an object under investigation onto different planes and the methods of restoring it from projections known beforehand was not new; many authors, starting in the 15th–16th centuries, had used it widely, for example, Dürer (see [78]). However, in Monge's work these methods were filled out with a real content of engineering problems, which enabled Struik [396], for example, to note that "his remarkable geometrical intuition went hand in hand with practical engineering applications to which his whole manner of thinking was always inclined." In particular, the development of descriptive geometry was stimulated by enquiries of the theory and practice of fortifications.

The creation of descriptive geometry was recognized as a sufficient basis for electing Monge as a member of the Academy of Sciences (Académie des Sciences). In 1780 Monge was elected junior scientific assistant in the geometry class in place of Vandermonde (see [21]). Descriptive geometry for Monge was considerably wider than the modern understanding of this subject and included elements of what are now independent disciplines, such as differential geometry, projective geometry, and the theory of machines and mechanisms. This reflected the return to synthetic methods after a period of one-sided enthusiasm for analytical methods. Although scholars had been occupied with questions of an engineering character from time immemorial (for example, Archimedes), we can say that a systematic development of these questions was begun by Monge.

At this time Monge's universalism began to show itself particularly vividly. For example, between February and May 1780 Monge submitted the following accounts: "on a model of a carriage, on two mathematical memoirs of Legendre, on wind pressure, on a machine for cleaning harbours, on the possibility of flying like birds, on a machine for threshing grain, on a system for raising water. In December of the same year he drew up information about the motion of rivers and about the calculus of probabilities. Between January and May 1781 a report on different types of pumps, on the exceptional frost of 1776, on a machine for raising ships, on windmills, on methods for preventing destruction caused by mountain streams, on axles for carriages, on the use of the waters of Béziers" (see [161], p. 38). However, most of this research has not left a noticeable trace in the history of science; it was linked with the need to prepare teaching material for the French administration.

In 1781 there was undertaken a new edition of the *Encyclopaedia*, and Monge took part in the editing of this edition as a physicist. In 1793 the *Physical dictionary* began to appear, and Monge was not only editor of this edition, but also the author of a number of articles.

A biography of Monge's scientific activity is probably one of the clearest examples of how the urgent requirements of production and technology brought forth outstanding scholars.

From 1783 Monge showed an interest in chemistry and metallurgy. He discovered (still before Lavoisier's experiments) that water is a combination of oxygen and hydrogen. Lavoisier recognized Monge's priority in this question in a memoir (jointly with Laplace). Monge was engaged in aeronautics. He tried to use hydrogen and carbon monoxide to fill a balloon.

In 1785 Monge was engaged with statistics. Examining naval cadets, he found that they had insufficient knowledge of mechanics, and in 1786 he wrote *Traité élémentaire de statique à l'usage des collèges de la Marine* (Paris, 1788). In particular, the tract contains the theory of machines and conditions for equilibrium of levers of arbitrary shape that move with respect to some point of the given lever. He considered in detail the case of a rectilinear lever, with and without weights, the case of a block, a pulley block, a winch, a cogwheel, a jack, and all these questions were investigated not only from the theoretical but also from the practical point of view.

Such an engineering approach is generally characteristic of the work of Monge and many other experts of this epoch. The mathematicians Monge and Vandermonde and the engineer Périer inspected blast furnaces and machines for metallurgical production. In 1786 Monge read to the Academy of Sciences a memoir, written jointly with Vandermonde and Périer, in which they collected the results of experimental research into the change in the properties of iron, steel and cast iron in a metallurgical process.

"Thus—mathematics, physics, chemistry, metallurgy, mechanics. In the 80's Monge was one of France's greatest scholars of the pre-war years. In 1790 he was 45, and at the height of his scientific fame: he had already laid the foundations of most of his scientific theories ... Monge was a universal scholar of the 18th century, and completely belonged to that century. There was then no sharp boundary between mathematics, mechanics, and physics, they were usually taught by the same person. Chemistry was still in a state of coming into being, it took up an intermediate position between alchemy and physics. The boundaries of physics were very indeterminate: information about mineralogy, botany and zoology was often included in it. Nearly all academicians had to be concerned with questions of technology. Therefore if Monge was distinguished among his contemporaries, it was only for his brilliant talent, his energetic activity and his constant enthusiasm." (See [21], p. 23)

On 14 July 1789 the uprising people captured the state prison—the Bastille. The revolution had begun. Monge took an active part in the events of that time: he joined the Patriotic Society of 1789, but did not give up his work at the Academy of Sciences nor teaching and research. In the most important ports of France Monge organized 12 schools for training hydrographers.

In 1791 he joined the National Society of Luxemburg, and then the famous Jacobin Club. In 1793 the Comité de salut public commissioned Monge, Vandermonde and Berthollet to draw up a textbook for armourers on steel production. Monge was engaged at the same time on questions of the manufacture of guns and cannons, and under his guidance forges and iron foundries were reorganized, and he regulated the production of the interconnecting components. In particular, Monge worked on the new technology, and at his suggestion earthenware moulds were replaced by moulds of sand, which considerably simplified and accelerated the technological process.

In 1794 Monge published the textbook *A description of the art of manufacturing cannons*. One direction of his defence activity was aerostatistics: he took part in the Commission for Aeronautics, created on the orders of the Convention in 1793. The revolutionary epoch could not fail to affect Monge's biography: he had to overcome many difficulties connected with the destiny of the Jacobin movement in France (see [160]). In November 1794 the Jacobin Club was closed down, and in May 1795 a decree was signed for the arrest of Monge. For about two months he hid in Paris and only on 22 July did he succeed in getting the charge dropped (see [21]).

Monge again actively resumed his scientific and public life. On 21 January 1795 the Normal School opened, in the organization of which Monge took part. However, this school was not long-lived: it closed down in May 1795. But the Central School of Public Works, to which Monge gave a great deal of attention, survived, and in September 1795 it was renamed the Polytechnic School (École Polytechnique). In Monge's opinion it should be preparing not professors of mathematics but engineers of various specialisms with serious scientific and practical training. This school turned into a distinguished scientific center and managed to retain a leading position throughout the whole of the 19th century.

Practically all the great mathematicians of France in the 19th century either graduated from the Polytechnic School or belonged to the teaching body. The French Revolution turned out to be a powerful influence on the whole range of knowledge.

The rapid development of industry stimulated mathematics, mechanics, physics, and chemistry. Monge was in the forefront of the new scientific trend. At the time of Napoleon's government he took part in an expedition to Egypt. Monge's main activity at that period was the leadership of the Polytechnic School: he was its director from the day of its foundation to the fall of the Empire. For more details on the history of this school see [276].

Monge's main work in differential geometry, in which are generalized not only his work but also the research of other geometers at the end of the 18th century, was the book *Application de l'analyse à la géométrie* [313].

The first version of this course was the account of Monge's lectures that he gave at the Polytechnic School and printed in separate parts. At the beginning of the book Monge develops the general theory of curves and two-dimensional surfaces; in particular, he analyzes the properties of the two principal curvatures at an arbitrary point of the surface. In §20 he considers separately the case where the two radii of curvature (at each point of the surface) are equal in magnitude but have opposite signs. From this it follows that the *mean curvature* of the surface, that is, the sum of the quantities inverse to the radii of curvature, is identically zero at each point of the surface. In other words, this is a minimal surface: the area functional of the surface has a local minimum.

In particular, Monge directly stated the following important property of this surface: *if we surround part of it with a continuous closed contour, then of all surfaces passing through this contour its area inside the contour is smallest.*

The question of the integration of the equation of a minimal surface was considered by Monge in [312].

Further progress in this direction was achieved by Legendre in [289]. At that time Monge was apparently not interested in the problem of describing specific examples of minimal surfaces in three-dimensional space; in particular, he gave no new examples beyond those already known (the catenoid and the helicoid).

1.3. On the work of Meusnier and Poisson. At this time some questions in the theory of minimal surfaces were advanced by Meusnier, a pupil of Monge.

Jean-Baptiste-Marie-Charles Meusnier (1754–1793), having graduated from the Mézières Military Academy, joined the Engineer Corps of the French army and before the Revolution he commanded a division. Also before the Revolution he was elected a member of the Academy of Sciences. After the Revolution, together with his teacher Monge he worked at the Commission of Weights and Measures. After this Meusnier was a general and fought at the front, becoming famous as a brave military leader and fighting in the ranks of the revolutionary army. During the heroic defence of Mainz against the Prussian army Meusnier received a serious wound and died soon after (in 1793). Meusnier was 22 when in 1776 he presented to the Academy his *Mémoire sur la courbure des surfaces.* Meusnier paid a great deal of attention to minimal surfaces. We

have mentioned above that Lagrange obtained the equation of minimal surfaces analytically. Meusnier proposed a geometrical derivation of this equation and derived certain new corollaries from it. He studied various properties of the catenoid and the helicoid.

One of the most prominent students of Monge associated with the Polytechnic School was Simeon Poisson (1781–1840). Like his teacher, he was interested in the applications of mathematics to mechanics and physics. In particular, his research on the theory of liquids and capillary effects served as a stimulus for the development of the mathematical theory of the interface between two media [360], [361]. See also the survey in an article by Minkowski in [214]. Poisson left a clear trace in the theory of minimal surfaces, of which we shall say something in §2 of this chapter.

Poisson subjected the following branches of the earlier mathematical physics to a fruitful reworking: capillarity, the bending of plates, electrostatistics, magnetostatistics, and heat conduction. The exceptional many-sidedness of Poisson's work is evident from the fact that his name is firmly associated with many concepts that occur in modern mathematics (the Poisson brackets in Hamiltonian mechanics, the Poisson equation, and so on).

Particularly with Poisson (and with certain pupils of Laplace) there is clearly associated the pronounced "applied stream" in the history of the development of mathematics in France during the 19th century. Poisson was the author of more than 300 papers. In his scientific outlook he was guided to a significant extent by the ideas of "atomistics in the spirit of Laplace" (see [276]). Moreover, he often preferred to assume that the derivatives and integrals required in physics are only the abbreviated notation of relations of finite increments and sums.

As a lecturer at the Polytechnic School he wrote a well-known work *Traité de Méchanique*, which had a significant influence on the development of the mathematical aspects of mechanics and physics.

§2. The 19th century—the epoch of discovery of the main properties of minimal surfaces

2.1. Plateau's experiments with soap films. Physical experiments with wire contours. Films with boundary and films without boundary (bubbles). Before starting on a mathematical study of minimal surfaces, we give a short description of their main properties that can be demonstrated in simple physical experiments and in the language of descriptive geometry.

When the Belgian physicist, professor of physics and anatomy Joseph Plateau (1801–1883), began his experiments on the study of the configuration of soap films, he could hardly have supposed that they would be a stimulus for the beginning of a significant scientific trend, which has rapidly developed to the present time and is known today under the general title of *Plateau's problem*. Apparently *the problem of finding a surface of least area with a given boundary* was called "Plateau's problem" by Lebesgue in his well-known paper [290]. At present many separate problems are united under this title, and they are often characterized by nonequivalent approaches even in the statement itself in the problem of finding surfaces of minimal area.

Some physical experiments that Plateau once carried out are very simple and well known to the reader, since there is hardly anyone who has not amused himself at some time by blowing soap bubbles through a tube or spanning a wire contour with soap films. It is well known that if a closed wire contour is dipped into soapy water and then carefully taken out of it, then on the contour there hangs an iridescent soap film bounded by the contour. The size of the film can be quite significant, but the larger the film the easier and quicker it breaks up under the action of the force of gravity. If the size of the contour is small, then we can neglect the force of gravity.

Minimal surfaces are mathematical objects that model physical soap films quite well. Conversely, many deep properties of mathematical minimal surfaces appear in experiments with soap films. There are many mathematical trends that arise from specific physical and applied problems, but by no means all of them, like Plateau's problem, are so closely associated with such a wealth of various mathematical theories. Within the bounds of the theory of minimal surfaces there are interlaced such modern theories as differential equations, Lie groups, and algebras, homology and cohomology, bordisms, and so on.

Let us illustrate by simple examples some basic concepts, techniques and results worked out in various periods of development of Plateau's problem. Some methodical discoveries in the presentation of this material are due to Poston [362]. Since exact mathematical constructions in this field sometimes require very diverse and refined techniques, we shall for now confine ourselves to descriptive constructions, omitting cumbersome definitions and calculations.

A great deal of attention was paid to the mathematical study of the properties of soap films in the 18th century by Euler and Lagrange (the problem of finding a surface of least area with a given boundary). Exact

FIGURE 1.1

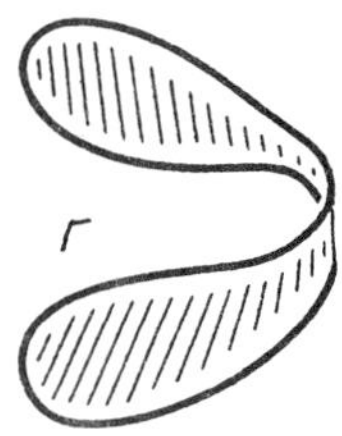

FIGURE 1.2

solutions for certain special boundary contours were found by Riemann, Schwarz and Weierstrass in the 19th century.

The theory of minimal surfaces arose from a consideration of *two sorts of soap film*: soap bubbles (without boundary) and soap films (with boundary) spanning a wire contour (the boundary).

The method of obtaining soap bubbles consists in blowing them through a tube that has first been dipped in soapy water (Figure 1.1).

To stabilize the bubbles, glycerine is added to the water. Such soap bubbles are formed and remain in equilibrium by the pressure of the gas (air) inside the bubble.

Another kind of soap film is obtained when we take a wire contour out of soapy water and fix it in space, avoiding any sudden movements. Then on the contour there arises a stable soap film, usually not having soap bubbles in its structure. This film has the original wire contour as boundary (Figure 1.2).

2.2. A mechanism for forming soap films and the physical principles that determine their properties. A soap film as a surface of minimal area. The physical principle that forms soap films and regulates their behavior and local and global properties is very simple: *a physical system preserves a definite configuration only when it cannot easily change it by taking up a position with a smaller value of the energy.*

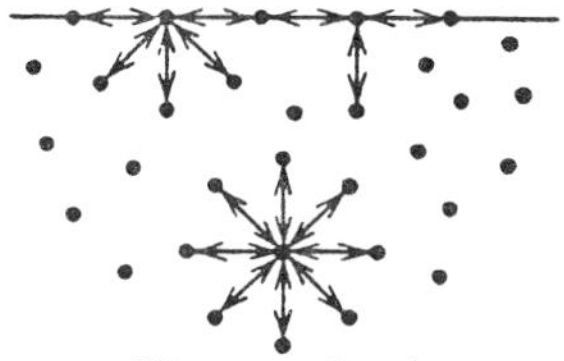

FIGURE 1.3

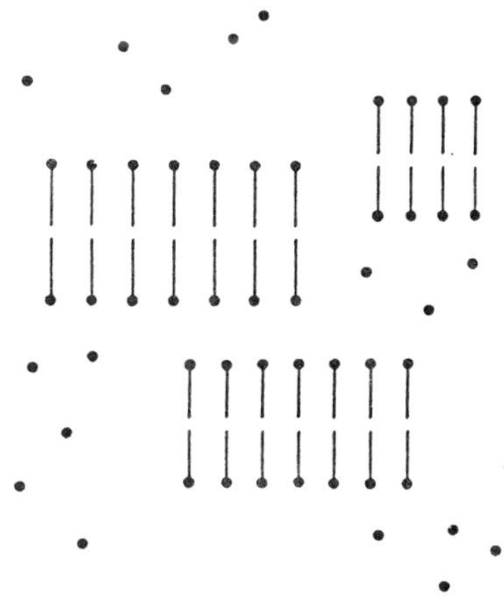

FIGURE 1.4

In our case, the energy of the surface (the soap film) is often described in terms of the *surface tension of a liquid*, conditioned by the presence of the forces of attraction between separate molecules and the imbalance of these forces on the boundary of the surface.

The presence of such forces has an interesting effect: *a liquid film turns into an elastic surface that tends to minimize its own area, and consequently to minimize the energy of the surface tension taken over a unit of area.* We neglect the force of gravity.

Let us consider the change in the properties of the surface film when soap is added to the water. Figure 1.3 shows conventionally the *interface* of two media, water and air. Molecules of water are shown conventionally by black arrows. Some of them are in the air, since the liquid evaporates, but we neglect this effect. Arrows with two ends denote the forces of mutual attraction that act between polar molecules of water characterized, as we know, by the asymmetric position of the electric charge. These forces of mutual attraction cause the appearance of *surface tension on the interface between the media.*

In contrast to water molecules, soap molecules are formed by long thin apolar hydrocarbon chains with a polar oxygen group on one end of the chain.

A soap molecule is stretched: one end of it (the polar end) is $COO^{-}Na^{+}$, and the other end (the apolar end) consists of 12–18 carbon atoms. The polar end is soluble in water (hydrophilic), while the apolar end, by contrast, is repelled by water (hydrophobic).

X-ray analysis of a concentrated soap solution shows that anions in the main mass join together, forming colloidal particles. Because of this

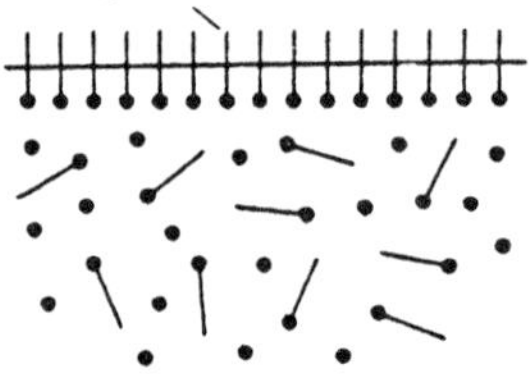

FIGURE 1.5

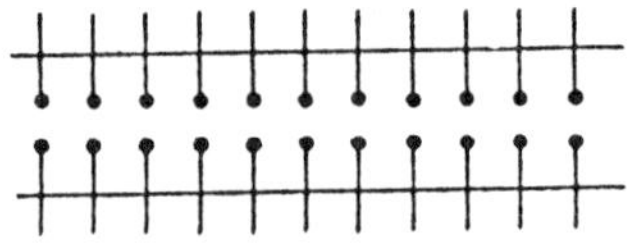

FIGURE 1.6

the soap solution acquires a fibrous structure, and the soap molecules lie parallel to each other in double layers (Figure 1.4); see [148], [5], [136].

When soap molecules are added to water, they approach its surface and fill the surface of the liquid with a uniform layer (Figure 1.5).

Each soap molecule is directed to the surface with its apolar end on the outside. From what we said above it follows that the finest soap film has thickness roughly equal to *twice the length of a soap molecule*, since such a film has the form of a layer consisting of a pair of soap molecules (Figure 1.6).

Forcing the water molecules inside the liquid, *the soap molecules decrease the surface tension on the interface between the media.* Thus, as Plateau found out, the force of surface tension of a soap solution (on the surface of the solution) is *less* than that of pure water.

It turns out that this fact gives the surface film *additional elasticity*, which is apparent especially at the moment when we dip in and then take out of the soapy water a thin wire contour (Figure 1.7). When the wire, raised from the liquid, reaches its surface, then the surface distends, covering the contour (a section of it is shown in the figure). This happens because around the wire the number of soap molecules on the surface *decreases* for a time, more precisely, the number of them by a unit of area of the film decreases, and so *at this place the surface tension increases.* We obtain a soap film spanning the contour; see the article by Almgren and Taylor [148].

2.3. Surface tension and the form of drops of liquid. The physical experiments of Boyle. Thus, *a mathematical model of a soap film is a so-called minimal surface, that is, a surface that has the least possible area (locally) among all surfaces having the same boundary.* Clearly, the form of the surface and its properties are determined to a significant extent by the configuration of the bounding contour—the boundary of the surface.

The first important step in the development of these ideas was taken by Boyle, who was interested in the form of drops of liquid in 1676. Turning his attention to the fact that raindrops have approximately circular

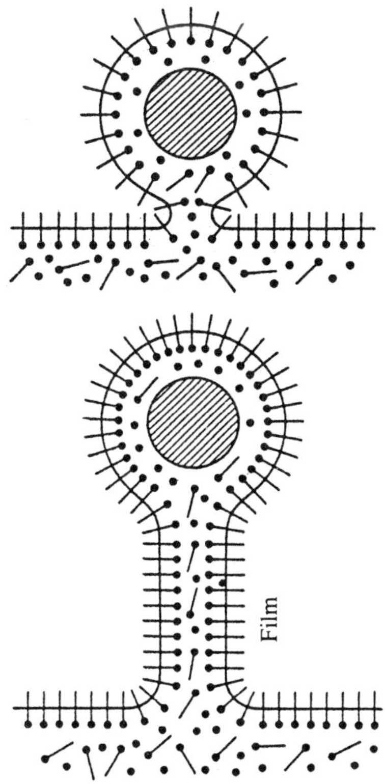

FIGURE 1.7

form, Boyle decided to find the dependence between the size of a drop and its form. He needed to set up what was for those times a rather subtle experiment. Boyle poured into a glass vessel two liquids: a solution of K_2CO_3 (a comparatively heavy dense liquid—a concentrated solution of potash) and alcohol (a light liquid). When the liquids settle, between them is formed a well distinguished interface between two media (Figure 1.8). Then on this interface between the media there is carefully located a drop of a third liquid of intermediate density and not mixed with the two liquids already present in the vessel. This drop is in equilibrium, being immersed in alcohol and touching the surface of the K_2CO_3 at its lowest point. This third liquid (butter) is picked up in such a way that it does not wet the interface between the media, that is, so that the drop does not spread over the interface. Then on the interface there are situated drops of different sizes, beginning with small drops, after which their size gradually increases. It turns out that with the growth of size of a drop its form begins to change (Figure 1.9).

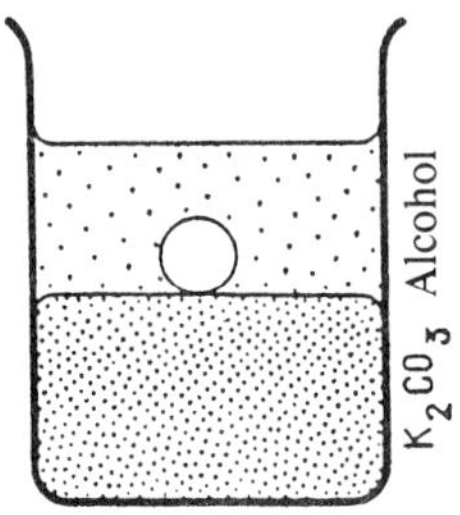

FIGURE 1.8

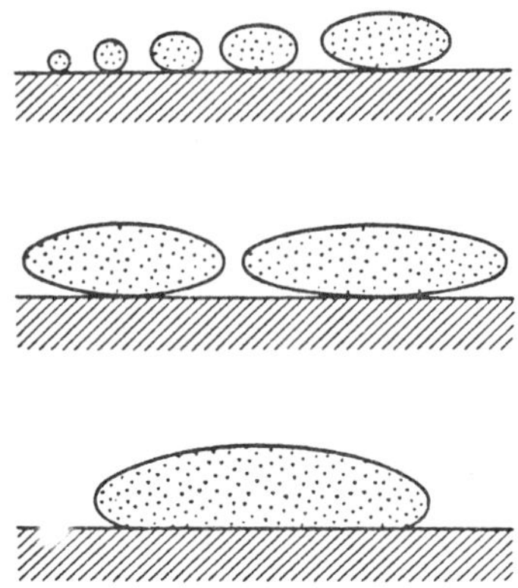

FIGURE 1.9

In 1751 Segner realized that, for an accurate study of surface tension, we should try to remove, as far as possible, the influence of the force of gravity. He partially achieved this by studying freely falling drops, and also drops immersed in a liquid of the same density. He earlier ensured that the liquid of a drop did not mix with the surrounding liquid. In particular, he was able to show that *a sphere has the least area among all closed surfaces bounding a fixed volume.* Segner was apparently the first to realize the true role of surface tension in all these processes; see [362], [357], and [356].

2.4. The second quadratic form of a surface and the mean curvature. The interface between two media in equilibrium is a surface of constant mean curvature. An important step in understanding the intrinsic geometry of the interface between media was taken by Laplace and Poisson. The first quantitative results in this field were obtained by Laplace (the study of the rise of water in capillaries). In 1828 Poisson showed that *the interface between two media in equilibrium (under the condition that we neglect the force of gravity) is a surface of constant mean curvature.* Here it may be useful for the reader to have an acquaintance with the main properties of the *second quadratic form* of a two-dimensional surface embedded in three-dimensional Euclidean space $\mathbf{R}^3$, see [48], [118].

Let $M^2 \subset \mathbf{R}^3$ be a smooth two-dimensional surface, where $\mathbf{R}^3$ is referred to Cartesian coordinates x, y, z. Suppose that the surface is specified by the radius vector $r = r(u, v) = (x(u, v), y(u, v), z(u, v))$, where the parameters u and v vary in some domain in the Euclidean plane, and they determine *regular coordinates* in a neighborhood of a point P on the surface (Figure 1.10).

Let $n(u, v)$ be the unit normal vector to the surface at P. Consider the

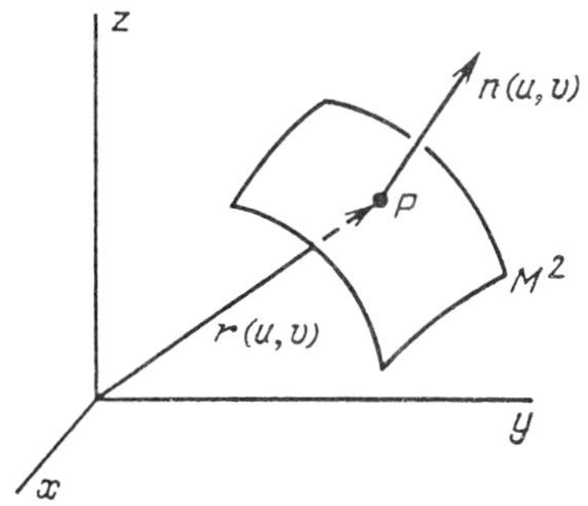

FIGURE 1.10

symmetric matrix Q formed from the numbers $\langle r_{u_i, u_j}, n\rangle = q_{ij}$, where r_{u_i, u_j} are the second partial derivatives of the radius vector with respect to the variables u and v (denoted here by u_1 and u_2 respectively). We denote by $\langle\ ,\ \rangle$ the Euclidean scalar product in $\mathbf{R}^3$.

If a and b are tangent vectors to the surface at P, then the expression $\sum_{i,j} q_{ij}a_ib_j = Q(a, b)$, where $a = (a_1, a_2)$, $b = (b_1, b_2)$, is a bilinear form, defined on the tangent plane to the surface and called the *second quadratic form of the surface.*

In the two-dimensional case the second quadratic form is usually written in the form $Ldu^2+2Mdudv+Ndv^2$, where $L = \langle r_{uu}, n\rangle$, $M = \langle r_{uv}, n\rangle$, $N = \langle r_{vv}, n\rangle$.

Here L, M, N are smooth functions of the coordinates on the surface.

We note that in some textbooks the coefficients L, M, N are taken with the opposite sign.

Together with the second quadratic form we also consider the *first quadratic form* on the surface, which defines the *Riemannian metric,* induced on the surface by the ambient Euclidean metric. This form is usually denoted as follows: $Edu^2 + 2Fdudv + Gdv^2$, where $E = \langle r_u, r_u\rangle$, $F = \langle r_u, r_v\rangle$, $G = \langle r_v, r_v\rangle$.

We denote the matrices of these forms respectively by $Q = \left(\begin{smallmatrix} L & M \\ M & N\end{smallmatrix}\right)$ and $A = \left(\begin{smallmatrix} E & F \\ F & G\end{smallmatrix}\right)$. Then (see [36], [74]) the two scalar functions $K = \det A^{-1}Q$ and $H = \operatorname{tr} A^{-1}Q$ are important geometrical characteristics of the surface and are called the *Gaussian* and *mean curvatures* respectively. At present the main interest for us is the *mean curvature.* By definition,

$$H = \frac{GL - 2MF + EN}{EG - F^2}.$$

It turns out that this expression can be simplified by a choice of suitable local coordinates, which enables us to *connect the mean curvature with the surface tension of the interface between media.* We recall that regular

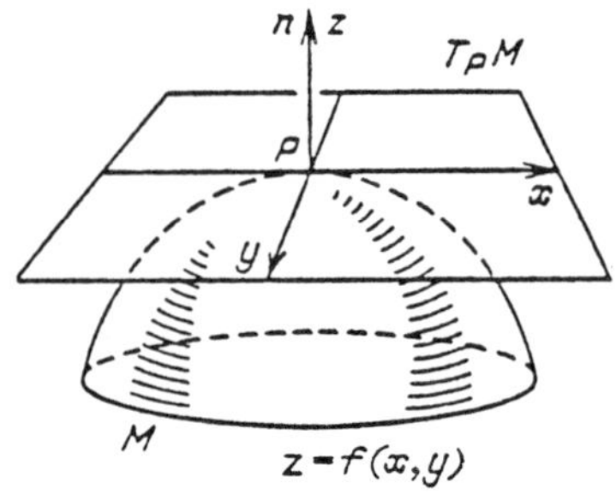

FIGURE 1.11

curvilinear coordinates u, v on a surface M are said to be *conformal* (or *isothermal*) if the induced Riemannian metric $ds^2 = Edu^2 + 2Fdudv + Gdv^2$ becomes diagonal in them, that is, $E = G$, $F = 0$ and $ds^2 = E(du^2+dv^2)$. It turns out that if the metric on the surface is real analytic, then *in a neighborhood of each point on the surface there are always local conformal coordinates* (see [48], [118]).

We may assume that surfaces of interface between media are real analytic at all interior nonsingular points, so without loss of generality we can assume the existence of conformal coordinates in a neighborhood of each regular point.

Thus, let u and v be conformal coordinates in some neighborhood of a point P on a surface M^2 in $\mathbf{R}^3$. We may assume that *the matrix of the first form is the unit matrix at* P. At one point, we can always achieve this by suitable orthogonal transformation of the conformal coordinates. In particular, at P we have $EG - F^2 = 1$. Since the coordinates are conformal in a neighborhood of P we obtain $H = (L + N)/E$. The value of the mean curvature at P becomes simpler: $H(P) = L+N$, since $E(P) = 1$. Let us recall the explicit form of the coefficients L and N. We obtain $H = E^{-1}\langle \Delta r, n\rangle$, where $\Delta = \partial^2/\partial u^2 + \partial^2/\partial v^2$ is the *Laplace operator*. The expression for the mean curvature is even simpler. Consider the tangent plane to the surface at P, choose Cartesian coordinates x, y in it (Figure 1.11), and direct a third coordinate z along the normal to the surface. Then locally the surface around P is described by means of the radius vector $r(x, y) = (x, y, f(x, y))$, where f is a smooth function whose graph specifies the surface. We then obtain $H(P) = \langle r_{xx} + r_{yy}, n\rangle$. We note that the coordinates x, y do not have to be conformal on the surface, but they are orthogonal at P and so $E = G = 1$, $F = 0$. Since in the coordinates x, y, z the normal vector $n(P)$ is written in the form $n(P) = (0, 0, 1)$, we have $H(P) = f_{xx} + f_{yy} = \Delta f$.

2.5. Poisson's theorem.

THEOREM 1.2.1 (Poisson). *Suppose that a smooth two-dimensional surface in* $\mathbf{R}^3$ *is the interface between two media in equilibrium. Let* p_1 *and* p_2 *be the pressures in the media. Then the mean curvature* H *of the surface is constant and equal to* $H = h(p_1 - p_2)$, *where the constant* $\lambda = 1/h$ *is called the coefficient of surface tension, and* $p_1 - p_2$ *is the difference between the pressures in the media.*

The formula in Theorem 1.2.1 is sometimes called *Laplace's formula* (see [5], for example).

This explains the spherical form of soap bubbles, which they acquire at the time of free fall. In this case we can neglect the force of gravity. Here the pressure of the gas inside the sphere exceeds the external pressure and equilibrium of the film is attained as a result of the action of the forces of surface tension, which stabilize the spherical film. Thus, $H = h(p_1 - p_2) = \text{const} > 0$ (under the regular choice of normal).

The main interest is in the case of a soap film stretched on a wire contour. Here there is no difference between the pressures on the two sides of the film. Consequently, the pressure on one side of the film coincides with the pressure on the other side in a neighborhood of each point on the surface. Here $H = h(p_1 - p_2) = \text{const} = 0$. Thus, the resultant of the forces is zero (we again neglect the force of gravity), and so the mean curvature of the film is zero (see [214]).

We give a proof of Poisson's theorem (see details in [5], for example). From a course on mechanics (see [10], [62], for example) it is known that the distribution of pressure in a continuous medium is described by the so-called *ellipsoid of pressure*. It is also known that in a liquid the ellipsoid of pressure is a sphere. This means that the pressures acting on each element of area in a liquid are equal in magnitude and normal to the element of area (Figure 1.12). But close to the free surface of the liquid (that is, close to the interface between the two media) this is false. Here the ellipsoid of pressure is deformed and is a real ellipsoid of revolution. Close to the interface we pick out a thin layer of constant thickness h adjacent to the interface (Figure 1.13). Consider an element of area in this layer passing through the normal to the surface. The pressure acting on this element of area (normal to it) can obviously be interpreted as the *surface tension*. Its value depends on the distance ε from the element of area to the interface (the free surface of the liquid), and also depends on the position of the base of the element of area given by the line element ds (Figure 1.13). We denote the vertical side of the element of area by $d\varepsilon$. In the thin layer

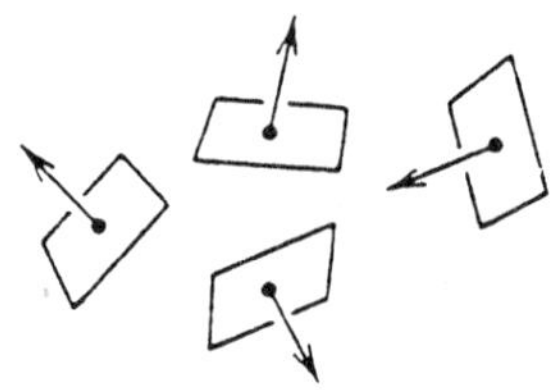

FIGURE 1.12

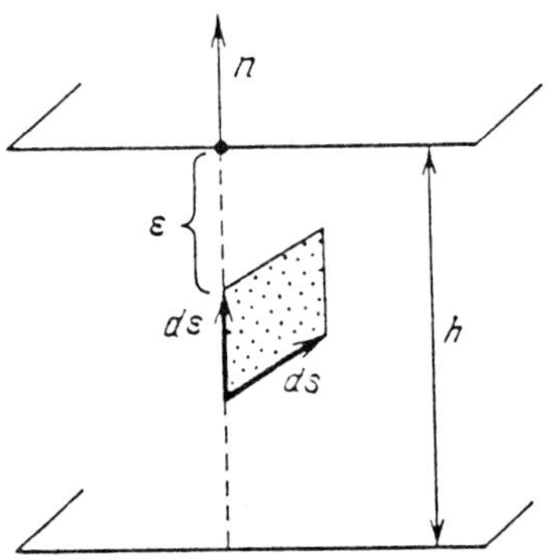

FIGURE 1.13

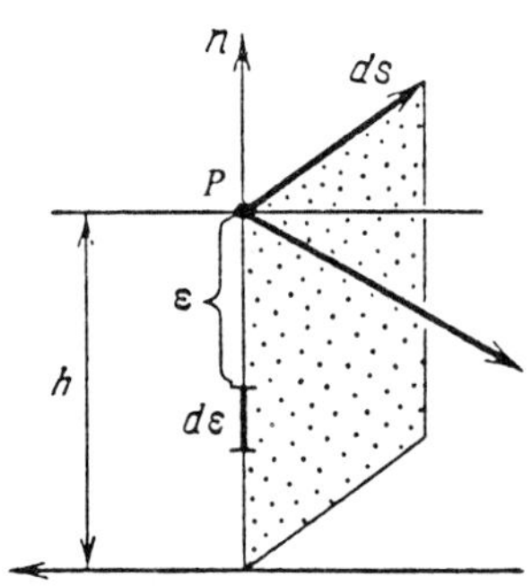

FIGURE 1.14

of thickness h under consideration the ellipsoids of pressure are ellipsoids of revolution with axis directed along the normal to the free surface. The value of the surface tension on an element of area with vertical element $d\varepsilon$ and horizontal line element of unit length is denoted by $a(\varepsilon)$. Consider an element of area passing through the whole layer of thickness h and having line element ds on the free surface (Figure 1.14). Then the pressure undergone by this element of area is made up of the pressures undergone by narrow elements of area of thickness $d\varepsilon$ (see Figure 1.14), that is, as a result the following integral is formed: $ds \cdot \int_0^h a(\varepsilon)d\varepsilon$. The vector representing the pressure on the element of area is orthogonal to it and lies in the tangent plane to the free surface at P (see Figure 1.14). It is convenient to consider the following quantity: $q = \int_0^h a(\varepsilon)d\varepsilon$. From what we said above it follows that it gives us the pressure undergone by the cross-section of the surface layer (of thickness h), referred to the unit of length. From the form of the integral we should expect that close to the free surface there is a dependence between the following three objects: *the external pressure, the internal pressure* (in the liquid), and *the curvature*

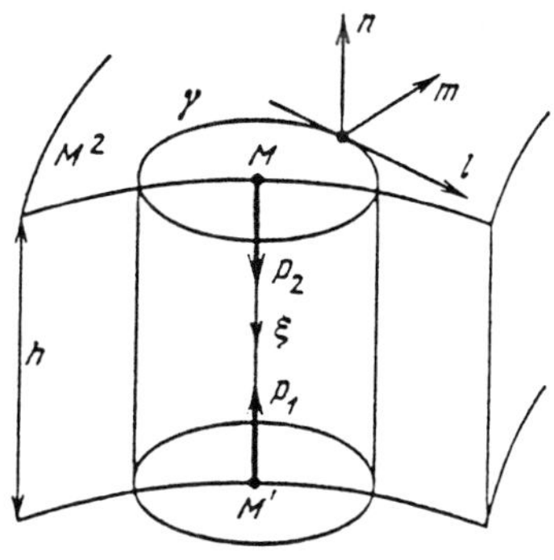

FIGURE 1.15

of the free surface (the interface). On the interface between the media we consider a smooth closed curve γ through which we draw a cylindrical surface orthogonal to the interface.

Inside the contour γ we take a point M, and let ξ be the normal to the surface directed inside the liquid. Let p_2 be the external pressure at M. We extend the normal ξ to meet the opposite boundary of the surface layer (of thickness h) at some point M'. We denote by p_1 the pressure on the layer with the opposite direction (that is, upwards in Figure 1.15). Since the thickness h of the layer is small, we may assume that the normals to its boundary at M and M' are directed towards each other.

Let m be the unit normal to γ in the tangent plane to the surface M^2 (the interface) (Figure 1.15). Let ξ_i and m_i be the Cartesian components of the vectors ξ and m (in Cartesian coordinates). Then for the part S of the free surface M^2 bounded by the curve γ we obtain the following equilibrium conditions:

$$\iint_S (p_1 - p_2)\xi_i \, dS = \int_\gamma q m_i \, ds, \qquad 1 \le i \le 3.$$

These equilibrium conditions for the part S of the surface M give us at the same time the equilibrium condition for the cylindrical element of volume whose base is S—see Figure 1.15. Let l be the tangent vector to γ lying in M. Clearly, $m = [\xi, l]$, where $[\ ,\ ]$ denotes the vector product. Writing this relation in coordinates, we obtain for the component m_3 the equation $m_3 = \xi_1 l_2 - \xi_2 l_1$. If ds is the element of displacement along γ, then it is connected with the displacements dx_i of the Cartesian coordinates of a point on the curve by the relations $dx_i = l_1 ds$.

From the Stokes formula we obtain

$$\begin{aligned} A &= \int_\gamma q m_3\, ds = \int_\gamma q(\xi_1\, dx_2 - \xi_2\, dx_1) \\ &= \iint_S \begin{bmatrix} \xi_1 & \xi_2 & \xi_3 \\ \partial/\partial x_1 & \partial/\partial x_2 & \partial/\partial x_3 \\ -q\xi_2 & q\xi_1 & 0 \end{bmatrix} dS. \end{aligned}$$

Since the length of the vector ξ is equal to one, we have

$$\begin{aligned} A &= \iint_S \{-\xi_1(q\xi_1)_{x_3} - \xi_2(q\xi_2)_{x_3} + \xi_3[(q\xi_1)_{x_1} + (q\xi_2)_{x_2}]\}\, dS \\ &= \iint_S (-q_{x_3} + \xi_3 \operatorname{div} q\xi)\, dS. \end{aligned}$$

For the mean curvature H of the surface S we have the formula $2H = \operatorname{div}\xi$ (see [48]). Hence we obtain

$$\iint_S (p_1 - p_2)\xi_3\, dS = \iint_S (-q_{x_3} + \xi_3\langle \operatorname{grad} q\,,\, \xi\rangle + q\xi_3 \cdot 2H)\, dS.$$

Since this equation holds for any arbitrarily small domain on S, there follows the equality of integrands:

$$(p_1 - p_2)\xi_3 - 2Hq\xi_3 + q_{x_3} - \xi_3\langle \operatorname{grad} q\,,\, \xi\rangle = 0.$$

We now choose the axes of a system of Cartesian coordinates in such a way that at some point $M \in S$ the normal vector ξ is directed along the x_3-axis. Then we have $p_1 - p_2 = 2Hq$. Consequently, the previous equality becomes the Laplace equality (see the statement of the theorem). Since it implies equality of two scalar functions, it does not depend on the choice of Cartesian coordinates. Therefore, in an arbitrary Cartesian coordinate system we have $q_{x_i} = \xi_i\langle \operatorname{grad} q\,,\, \xi\rangle$, $1 \le i \le 3$. This means that $\operatorname{grad} q = \xi\langle \operatorname{grad} q\,,\, \xi\rangle$, that is, the vector $\operatorname{grad} q$ is directed along the vector ξ. Consequently, the free surface is given by $q = \text{const}$. This proves the theorem.

2.6. Two types of soap films. Closed films of constant positive curvature. Films with boundary that have constant zero curvature. Having discovered that soap films—"liquid surfaces"—are surfaces of constant mean curvature, Poisson naturally posed the question of a complete description of such surfaces. Since the problem of the possible form of drops of liquid or bubbles was still not finally solved, the attention of researchers was originally concentrated on the case of positive mean curvature.

Clearly, the standard Euclidean sphere is such a surface, and therefore freely falling soap bubbles actually take up the form of a sphere.

However, it was still not clear whether there were other closed (that is, without boundary and without singularities) surfaces of constant positive mean curvature. Of course, many physical experiments with soap bubbles quickly convince us that the sphere is the only possible surface of constant positive curvature in the class of smooth closed surfaces, but a mathematical proof of this fact requires some effort. Poisson proved that in the class of spheroids sufficiently close to a sphere and being surfaces of revolution (that is, slightly compressed at the poles or slightly stretched along one axis), the sphere is the only surface with constant mean curvature. This information was practically all there was until 1853, when Jellet proved that among closed two-dimensional surfaces having star-shaped form (that is, each point of which can be seen from some interior point, the "center" of the surface), the sphere is the only surface with constant positive mean curvature. Finally, in the present century it was proved that a compact closed two-dimensional surface of constant positive curvature without self-intersections (that is, embedded in $\mathbf{R}^3$) is a standard sphere. This is a theorem of A. D. Aleksandrov. If we consider a manifold *homeomorphic to a sphere* and *immersed* in $\mathbf{R}^3$ (that is, self-intersections are allowed), then Hopf proved that from the condition that the curvature is constant and positive it follows that this submanifold is a standard (embedded) sphere. Recently Wente (Pacific J. Math. **121** (1986), 193–243) constructed an immersion of a torus in $\mathbf{R}^3$ with constant mean curvature $H \neq 0$.

We have thus distinguished two main cases.

1. If a soap film bounds a closed volume, the pressure in which exceeds the external pressure, then the surface has *constant positive mean curvature.*

2. If a soap film does not contain closed volumes and is stretched on a wire contour (that is, it has a boundary), then there is no difference between the pressures, and *the mean curvature is zero.*

The first main experiments carried out by Plateau were devoted to the study of the interface between liquids. In what follows, we concentrate attention on the case of soap films with boundary. These films admit the following two equivalent descriptions: (a) surfaces of zero mean curvature, (b) surfaces of (locally) minimal area.

§3. Topological and physical properties of two-dimensional minimal surfaces

3.1. Stable and unstable minimal surfaces. The simplest experiments show that different interfaces between two liquids react differently to small perturbations. Some of them resist destruction and are stable. Others are destroyed immediately, and are unstable. A soap film is said to be *stable*

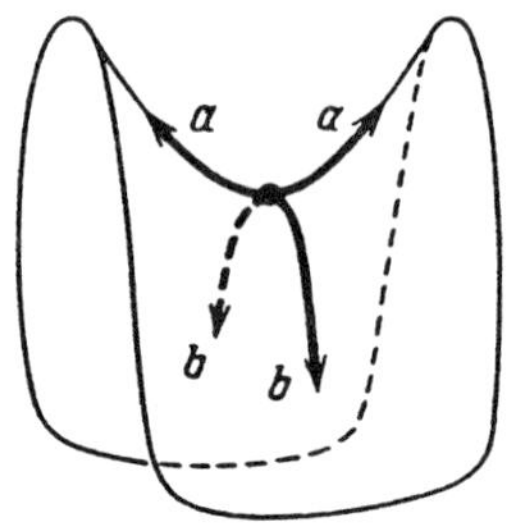

FIGURE 1.16

if small perturbations do not remove it from the equilibrium position.

Here there appears an important effect, visually illustrated even by the scalar functions. We recall that if f is a smooth function on a finite-dimensional manifold M, then a major role in the study of the behavior of a function is played by *critical points*, that is, such that $\operatorname{grad} f = 0$. These points can be of various kinds. Firstly, points of *local minima* and *maxima* of a function are critical. However, apart from these there are the so-called *saddle points* (Figure 1.16). They are characterized by the presence of two kinds of directions: along the directions a the function strictly increases, and along the orthogonal directions b the function strictly decreases.

The number of independent directions along which the function strictly decreases is called the *index of the critical point*. Points of *nondegenerate local maxima* have maximal index. A completely analogous picture is observed in the case where a function (functional) is defined on an infinite-dimensional space. Such a space is (for the right choice of formulation, which we shall not make more precise here) the set of all two-dimensional surfaces that span wire contours. Then the area functional associates an area with each point of the surface. A deformation of the surface changes the value of the area functional. If a two-dimensional film is in equilibrium, then it is critical for the area functional. This equilibrium position can be of various kinds: *local minimum, local maximum*, and *saddle*. In the last case there are sufficiently small perturbations of the film that decrease its area. In the case of a local minimum of the area functional the film is *stable*, that is, it resists small perturbations and tends to revert to its original position. In the case of a saddle the film is *unstable*, that is, it does not resist small perturbations. This means that there are perturbations arbitrarily small in amplitude that decrease the area of the film, and it begins rapidly and spontaneously to deform, taking up a new configuration corresponding to a lower value of the energy. The new configuration can differ essentially from the original, even from the topological point of view (see the examples below). The "infinite-dimensionality" of the space of all surfaces also means that there are infinitely many distinct independent directions of perturbations of the film.

3.2. Physical realization of a helicoid as a soap film. In 1842 Catalan proved that *the only complete ruled surfaces of zero mean curvature are the plane and the helicoid*. A helicoid is obtained as a result of combining two motions of a straight line: a translation with constant velocity and a rotation with constant angular velocity in a plane orthogonal to the translation vector (Figure 1.17). In other words, we need to take a line intersecting a second line at right angles and uniformly move the first line along the second, at the same time uniformly rotating it. This surface is noncompact. If we restrict ourselves to the rotation of a line segment, we obtain the surface shown in Figure 1.18. In coordinates x, y, z and r, t it is given by the following radius vector: $x = r\cos\omega t$, $y = r\sin\omega t$, $z = at$, where a and ω are the velocities of translation and rotation respectively. Plateau realized "half" of a helicoid by using a spiral of wire winding round an axial straight line (Figure 1.19). However, an attempt to realize a complete helicoid by means of a film stretched on a wire contour is fraught with difficulty. Of course, we can obtain part of a complete helicoid by putting together two copies of the film shown in Figure 1.19. However, in this approach, apart from the bounding contour consisting of two spirals with a common axis, to realize a soap film we must add this axis itself, for example as a thin thread (Figure 1.20). This thread stabilizes the film. If we remove the axial thread and keep only the bounding spirals, then in the case where the pitch of the spirals is not very large, the helicoid turns into the surface shown in Figure 1.21. Clearly, this surface is not ruled,

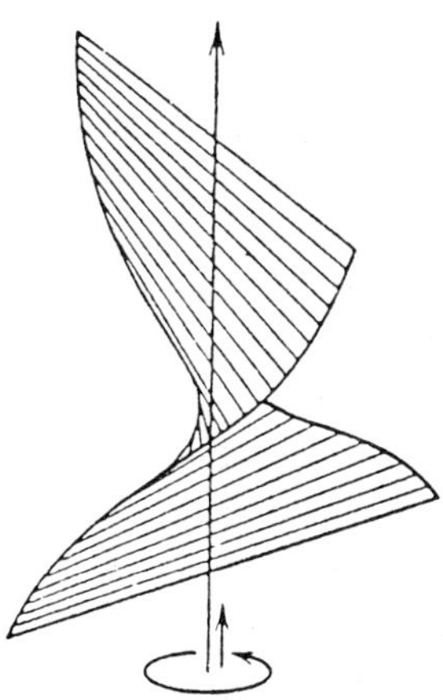

FIGURE 1.17

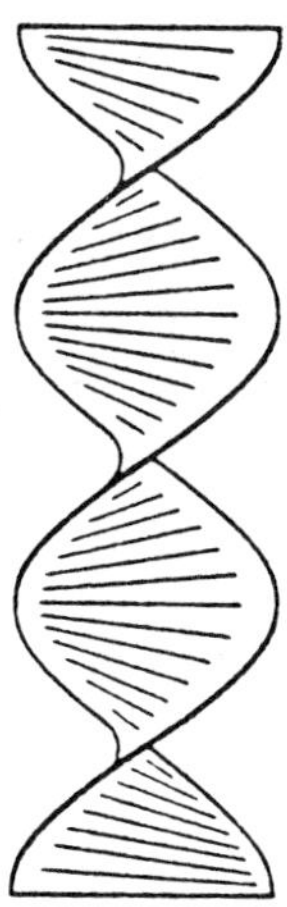

FIGURE 1.18

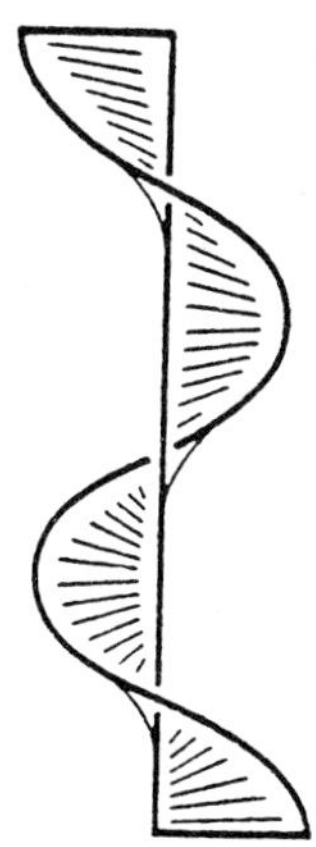

FIGURE 1.19

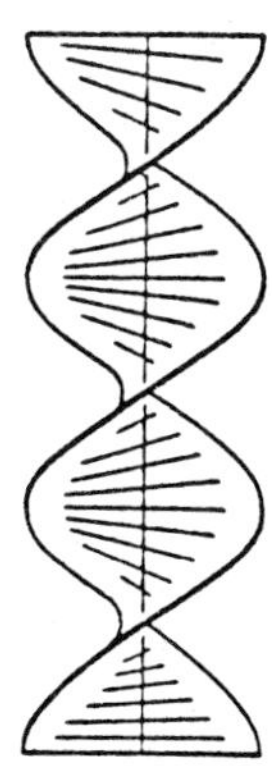

FIGURE 1.20

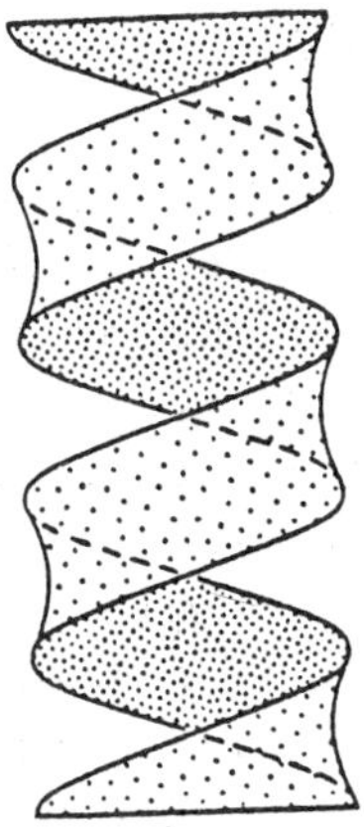

FIGURE 1.21

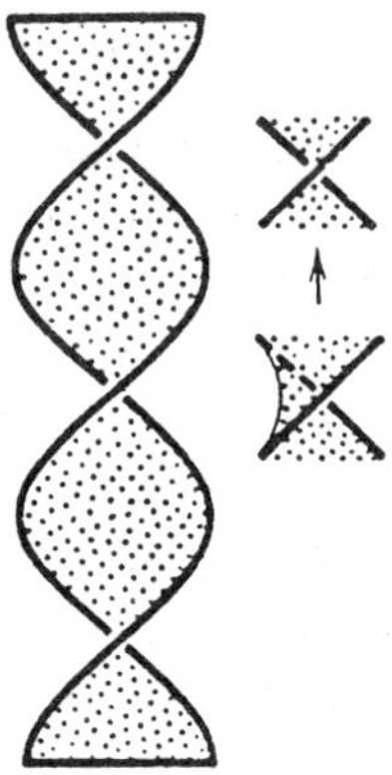

FIGURE 1.22

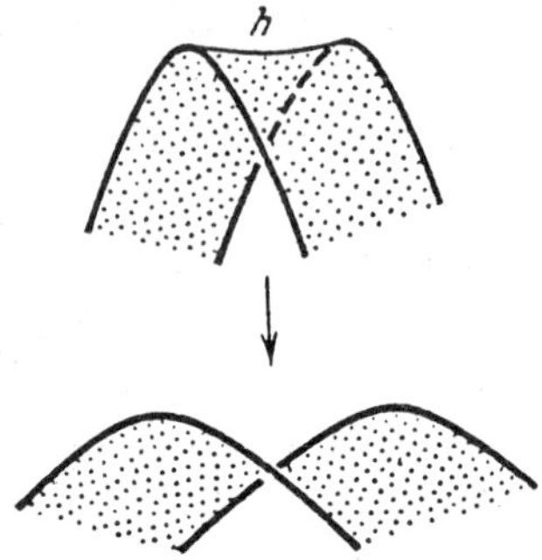

FIGURE 1.23

that is, it does not consist of line segments. Attempts to realize part of a right helicoid as a soap film that spans a closed wire contour consisting of coaxial spirals of small pitch ended unsuccessfully. Here there is an important difference between a right helicoid and the surfaces with which we have become familiar earlier. It turns out that for a small pitch it is an *unstable surface.* In real experiments we can realize only stable soap films. If we pose the problem of realizing a helicoid with a rather large pitch of the bounding spirals, there is no difficulty in constructing such a stable soap film (Figure 1.22). The proof that this film is ruled can be given in the following visual way. Clearly, the helicoid in Figure 1.22 is obtained from the minimal surface in Figure 1.21 by stretching the bounding spirals in the direction of the axis. As the pitch of the spiral increases, the soap film is deformed. Let us fix the axis of the spirals in space and the direction of our view of the film. We shall trace the deformation of the part of the film h lying in the immediate neighborhood of a crossing point of the spirals (Figure 1.23). This part of the film has in the projection on the retina of the eye the form of a curvilinear triangle, which obviously decreases as the spiral is stretched. At some instant this triangle disappears, becoming a point, and we obtain the film shown in Figure 1.22. Since the projection of the film now has the form shown in Figure 1.23, on rotating the film around the axis the spirals begin to run, for example from left to right. The crossing points of the spirals move in this direction. Since the part of the film projected (on the retina) into each crossing point is a line segment joining a point on the nearer spiral to a point on the farther spiral, the whole film consists of these line segments orthogonal to the axis, which proves that the film is ruled and coincides with a helicoid.

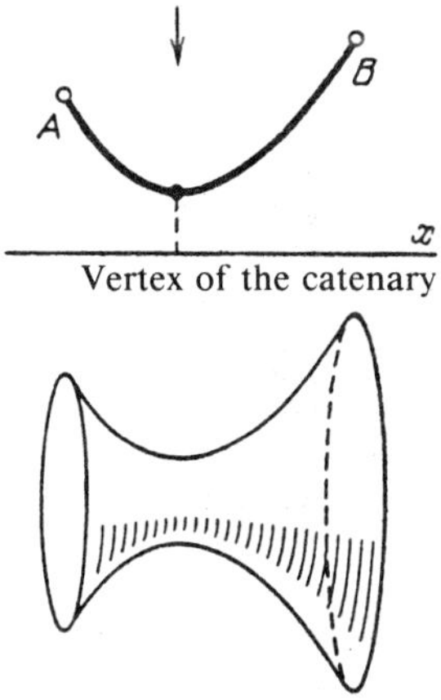

FIGURE 1.24

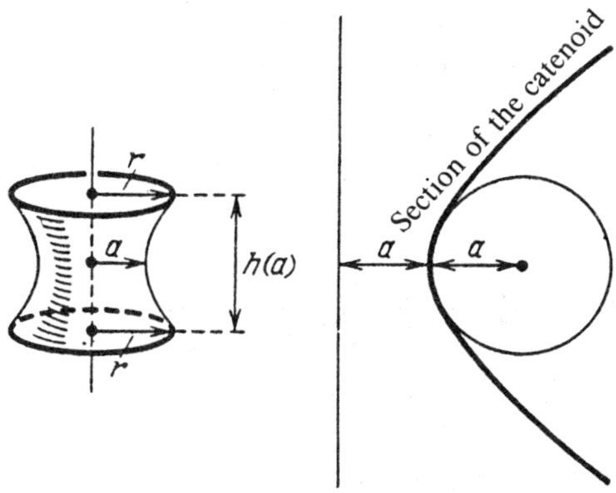

FIGURE 1.25

3.3. Physical realization of a catenoid as a soap film. Topological reconstructions of a catenoid under a deformation of the boundary contour. Consider the surface formed by rotating about the x-axis the curve given by the equation $y = a\,\mathrm{ch}(x/a)$, where a is a nonzero constant. This curve coincides in form with a sagging heavy chain fixed at two points (Figure 1.24). The force of gravity is directed downwards along the y-axis. Direct calculation shows that the mean curvature of the resulting surface of revolution is zero. We have obtained a *minimal surface,* called a *catenoid*. It can be regarded as a soap film covering two coaxial circles lying in parallel planes. If we rewrite the equation of the catenary in the form $y/a = \mathrm{ch}(x/a)$, we see that all catenaries, and consequently all catenoids, are *equivalent* in the sense that they can be taken into each other by a motion of three-dimensional space and a change of scale. A catenoid is obtained by rotating a catenary only when this curve is *at a definite distance from the axis of rotation.*

Let us consider the dependence of the structure of catenoids on the size of the boundary contour. For simplicity we consider two coaxial circles of the same radius r, and let h be the distance between the parallel planes containing them (Figure 1.25). We shall assume that the radius r is fixed. We introduce one more parameter a, the radius of the mouth of the catenoid, that is, the distance from the axis of rotation to points of the catenoid closest to it, which form a circle of radius a. Clearly, the parameters a and h are connected by some relation. We may assume that h is a smooth function of a. Let us describe the qualitative characteristics of the function $h(a)$. Since the mean curvature of a catenoid is zero, the

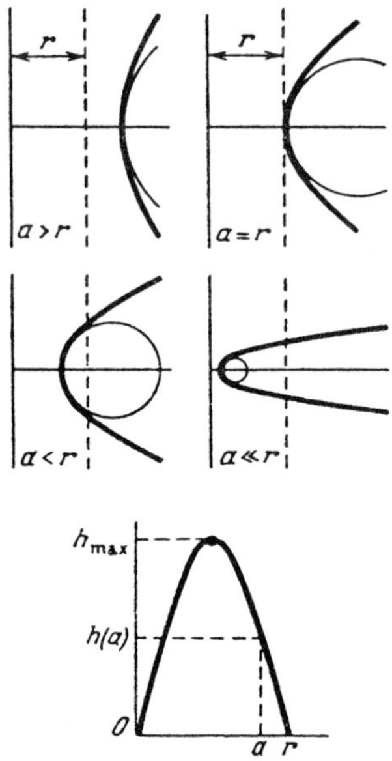

FIGURE 1.26

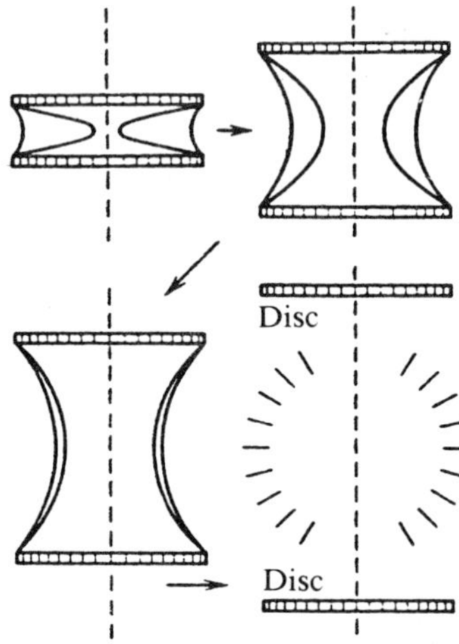

FIGURE 1.27

radius of curvature, that is, the radius of the osculating circle at the vertex of the catenary (and the point closest to the axis of rotation), is equal to the radius of the circle described by the vertex of the curve under rotation around the axis, that is, it is equal to a. Consequently, the smaller a is, the larger is the curvature of the catenary at its vertex. The parameter a varies in the interval from zero to infinity. If $a > r$, it is obvious that there is no minimal surface spanned by the two circles (Figure 1.26). In the figure a bold line shows a section of the catenoid by a vertical plane passing through the axis of symmetry, and a thin line shows the osculating circle at the vertex of the catenary. When $a = r$ it is obvious that $h = 0$, that is, the two circles touch. When $a < r$ the distance h starts to increase, the circles move apart and a minimal surface hangs on them. As a increases further, the distance h continues to increase for a time and then, as we see from Figure 1.26, having attained some maximal value $h_{\max}$, it starts to decrease. As a tends to zero, h also tends to zero, and the catenary approaches the a-axis even closer. For the resulting graph of the function $h(a)$ see Figure 1.26. When a is close to zero, the radius of curvature at the vertex of the catenary is also small, which forces this curve to approach the horizontal axis. The maximal value $h_{\max}$ is approximately equal to $4r/3$. When the distance h between the circles exceeds h_{max}, the catenoid breaks up and the minimal surface becomes two flat discs spanning the boundary circles. In this case there are no other soap films with this boundary. This transition from one solution of the equation of minimal surfaces to another solution proceeds by a jump, which essentially changes the topology of the film. It is useful to consider the evolution of the catenoid as h increases from zero to h_{max}. Figure 1.27 shows several subsequent positions of the soap film. A fact that is interesting and important for what follows is the existence, for any h such that $0 < h < h_{max}$, of two soap film catenoids: the *outer* and the *inner*. As h increases these two films start to sag in the direction of the axis of symmetry and come together. Finally, when h exceeds the critical values, the two films merge, split, and re-form into a pair of discs spanning circles. Thus, for any value of h in the interval $0 < h < h_{max}$ there are *three* soap films with a given boundary: *two catenoids and a pair of discs*. Plateau conjectured that the inner catenoid is *unstable*, in contrast to the outer catenoid and the two discs, but he could not prove it. The assumption of instability of the inner catenoid turns out to be true.

3.4. The general problem of investigating the many-valued graph that describes the values of a functional calculated on its extremals. The nature of the dependence of a minimal surface on the bounding contour. We first state a general problem discussed in [136]. Consider the space $C^\infty(M, N)$ of smooth maps $f: M \to N$, where N is a *fixed* Riemannian manifold, and M are *all possible* smooth Riemannian manifolds with nonempty boundary. On the space $C^\infty(M, N)$ we consider a functional $F(f)$ specified in the form $F(f) = \int_M L(f)d\sigma$, where L is a certain Lagrangian that depends on the map f and on its derivatives of different orders. Here $d\sigma$ is the Riemannian volume form on M. As basic examples of functionals F we shall consider the multidimensional volume functional $\mathrm{vol}_k f(M)$ and the Dirichlet functional. We then consider the *Euler-Lagrange equation* $I(f) = 0$ for the functional F. The extremals, that is, the critical (stationary) points f_0 of F, are solutions of the Euler-Lagrange equation. The extremals of the volume functional are locally minimal surfaces. To specify some extremal f_0 we need to specify the boundary conditions, that is, to specify a map $g: \partial M \to N$. To specify a two-dimensional soap film in $\mathbf{R}^3$ we need to fix in $\mathbf{R}^3$ a closed contour γ, that is, to fix a map of a circle (or system of circles) in $\mathbf{R}^3$.

Thus, the *set of all boundary conditions* is naturally identified with the space $C^\infty(\partial M, N)$ of maps $g: \partial M \to N$. The specification of a boundary condition g determines, generally speaking, *several* extremals f_0. In other words, generally speaking there are *many solutions* f_0 of the Euler-Lagrange equation for a specified and fixed boundary condition g, that is, such that $f_0|_{\partial M} = g$. With each boundary condition g we associate the set K_g of extremals f_0 corresponding to it. For each extremal $f_0 \in K_g$ we calculate the value of $F(f_0)$, that is, the value of the functional F on the given extremal. Associating with each boundary condition g the set of numbers $\{F(f_0)\}$, where $f_0 \in K_g$, we obtain, generally speaking, a *many-valued function* $g \to \{F(f_0), f_0 \in K_g\}$, defined on the space $C^\infty(\partial M, N)$. This function has branch points and other singularities. We have thus associated with each functional F its *graph*, defined on the space $C^\infty(\partial M, N)$. In other words, we have associated with each Euler-Lagrange equation $I(f) = 0$ the graph of a many-valued function $g \to F(f_0)$. Different functionals (defined on the same space of boundary conditions) are represented by different many-valued functions. Each of them is characterized by its system of singularities, and to different Euler equations (that is, to different functionals F) there correspond, generally speaking, different types of singularities. A detailed study of this correspondence would enable us to split the set of all functionals into types, and to give a classification of them in terms of the theory of singularities. It is natural to consider two functionals as belonging to the same type if they have identical systems of singularities. This problem is of interest also in connection with the fact that these types of singularities characterize the Euler-Lagrange equations of the corresponding functionals. In many cases for a family of deformations in general position the answer is given by the well-known theory of V. I. Arnol′d (see [14]). It would be interesting, for example, to determine which types of singularities correspond to the equation of minimal surfaces and to the equation of harmonic surfaces (maps).

For a concrete investigation of many-valued functions $g \to F(f_0)$ we can simplify the problem in the first stages by considering not the whole space $C^\infty(\partial M, N)$, but only the finite-dimensional submanifolds of $C^\infty(\partial M, N)$ to which the function $g \to F(f_0)$ is restricted. In our example, which we shall present below, we study a two-dimensional submanifold of $C^\infty(\partial M, N)$ for the case of the two-dimensional area functional.

3.5. The many-valued function of lengths of geodesics on Riemannian manifolds. As an illustration we consider the quite simple case where F is the length functional on the space of piecewise-smooth paths joining two fixed points on a Riemannian manifold N. Its extremals are geodesics with fixed ends. The many-valued function F is defined on the direct product $N \times N$. The sheets of its graph intersect if two geometrically distinct geodesics have the same length.

Let N be a homogeneous Riemannian space, that is, it has the form $N = G/H$, where G and H are compact Lie groups. Then it is obviously sufficient to study the function of lengths of geodesics joining any point $x \in N$ to a fixed point $e \in N$. Suppose, in addition, that N is a *symmetric space*. Then it is well known that it always contains a maximal submanifold, the invariant metric on which coincides with the flat metric of an ordinary Euclidean torus T^r of dimension r. The torus T^r is called a *maximal torus* in the symmetric space, and its dimension is called the *rank of the space* N. Suppose that $e \in T^r$ and $x \in T^r$. Let γ be an arbitrary geodesic in N joining e and x. We can show that there is always an isometry of the symmetric space N that leaves e and x fixed and takes γ into the maximal torus T^r; see Figure 1.28. Thus, the problem of studying the many-valued function of lengths of geodesics on a symmetric space actually reduces to the study of this function for the case of the torus T^r. Suppose, for simplicity, that this torus is two-dimensional (that is, we consider spaces of rank 2). For this case the properties of the function F have been studied by I. V. Shklyanko. Clearly, the main role in the study of the graph of F is played by the singular points of the function, that is, those at which at least two distinct sheets of the graph intersect.

Thus, consider the torus T^2, fix on it a point O and consider geodesics joining it to a variable point with coordinates (x, y). The point (x, y) is said to be *singular* if two distinct sheets of F intersect over it. This means that there are two distinct geodesics of the same length joining O to (x, y). It is convenient to consider the universal covering over the torus T^2, which is a two-dimensional Euclidean plane. The geodesics of the torus unroll into ordinary Euclidean lines of the plane. Let $\mathbf{Q}$ be the set of rational numbers, and I the set of irrational numbers. We state the result of Shklyanko [426].

PROPOSITION 1.3.1. *A point of the form* $(x, y) \in \mathbf{Q} \times \mathbf{Q}, \mathbf{Q} \times I, I \times \mathbf{Q}$ *is singular for the function of lengths of geodesics on the torus. The regular points form a set of full measure in the plane. Through each point* $(x, y) \in \mathbf{Q} \times \mathbf{Q}$, *that is, through a point whose coordinates are both rational, there passes an everywhere dense pencil of singular lines* (*that is, consisting of singular points*).

Thus, each rational point in the plane is the center (vertex) of a pencil of singular lines emanating from this point in an everywhere dense set of directions (Figure 1.29).

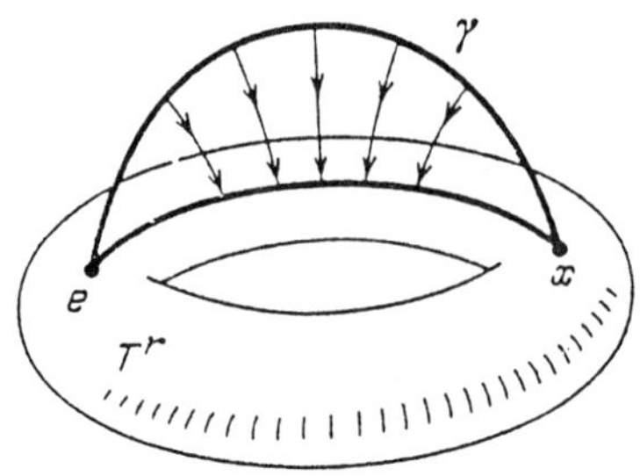

FIGURE 1.28

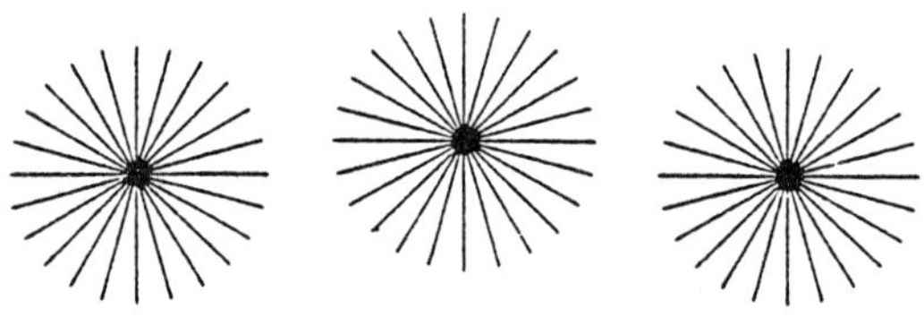

FIGURE 1.29

3.6. The many-valued function of areas of minimal surfaces that arises under deformation of the boundary contour. Distinct sheets of a graph correspond to stable and unstable minimal surfaces. The appearance of singularities of "swallow-tail" type in the theory of minimal surfaces. In this subsection we describe an interesting effect connected with the geometry of two-dimensional minimal surfaces. This work arose as a result of the analysis of a curious observation of T. Poston (see [362]). This effect has been investigated in detail by A. A. Tuzhilin and A. T. Fomenko [111], and M. J. Beeson and A. J. Tromba [164], which made it possible to complement Poston's observation.

On the same contour we can, generally speaking, stretch several distinct minimal surfaces (see above). The uniqueness of a soap film can be guaranteed, for example, for *flat* closed non-self-intersecting contours, that is, those realized as a system of curves embedded in one plane.

We can consider two variational problems: to find films of least area for a given contour and to find minimal films of least topological genus. A film having *least possible area* need not have minimal topological genus, that is, it may be homeomorphic to a sphere with a certain number of handles. We recall that a surface of zero topological genus does not have handles.

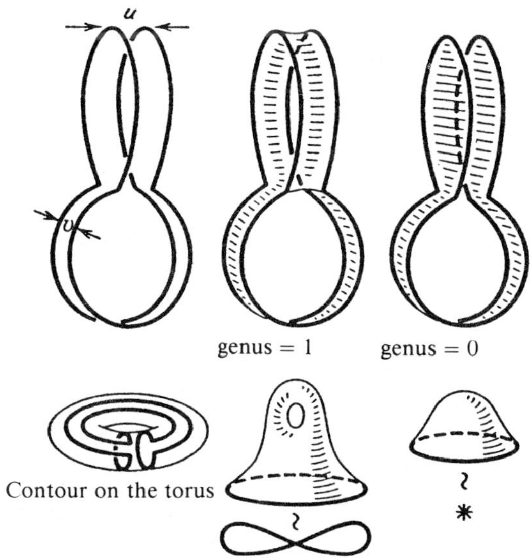

FIGURE 1.30

Conversely, a minimal film of least genus need not realize an absolute minimum of the area in the class of all films with a given boundary. In other words, *the absolute minimum of the area functional may be attained on films that are not of minimal topological genus* (Figure 1.30). If the pairs of circles shown in the figure are sufficiently close, then the area of the first film is less than that of the second. At the same time, the first film is homeomorphic to a torus with a hole, that is, it has topological genus 1, and the second film is homeomorphic to a disc and has topological genus 0 (see Figure 1.30). We shall therefore talk for now about minimal films, understanding by this their *local minimality* and not raising for now the question of their *global minimality* or the connection between absolute minima of the area functional and the topological genus of the film.

Consider the contour in Figure 1.30. We call it the *Douglas contour*, since its properties were first studied by Douglas in [201], [202]. It turns out that this contour is quite convenient for tracing nontrivial modifications of minimal surfaces and the change of their topological type. This contour is homeomorphic to a circle and it can be deformed without self-intersections into a circle embedded in the plane in the standard way. Suppose it is realized in $\mathbf{R}^3$ by means of a thin wire contour. We denote the distance between the upper parallel annuli by u (Figure 1.30), and the

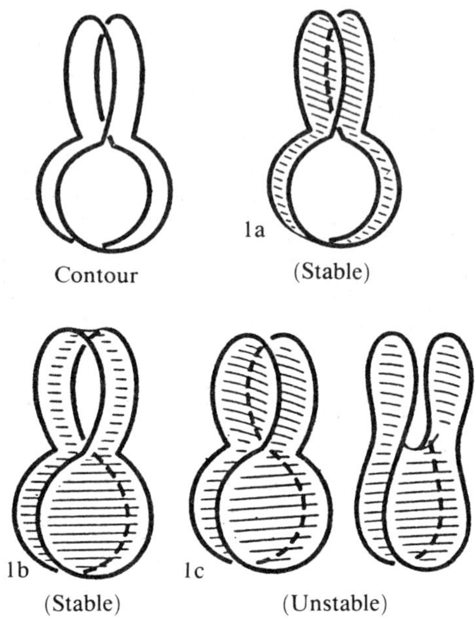

FIGURE 1.31

distance between the lower annuli by v. A contour with fixed values of u and v is called a *state of the contour* (u, v). We can identify the set of all possible states of the contour (u, v) with the square $[0, u_{\max}] \times [0, v_{\max}]$. For each fixed state of the contour there are, generally speaking, several minimal surfaces with the given boundary. Our aim is to investigate their behavior and reconstructions under a continuous deformation of the contour.

Since a formal analytic investigation of the behavior of such surfaces in the large is very difficult, A. T. Fomenko and A. A. Tuzhilin were sometimes forced to resort to *physical experiments*, realizing minimal surfaces and deformations of them as soap films.

In particular, the soap films of the form 1a, 1b, 1c (Figure 1.31) were obtained. In order to make the (unstable) structure of film 1c clearer to the reader, we have shown it in Figure 1.31 in two distinct foreshortenings. It is now clear that in reality film 1c is specified by a *saddle surface*.

Poston [362] found the following reconstruction of discs, which arises under a deformation of the contour. We take film 1a and deform the contour, thereby causing a deformation of the film stretched on it, as shown in Figure 1.32. We move the two lower circles apart. We may assume that this process approximately coincides with the stretching of a catenoid when its two rims are moved apart. For a small initial distance of the circles

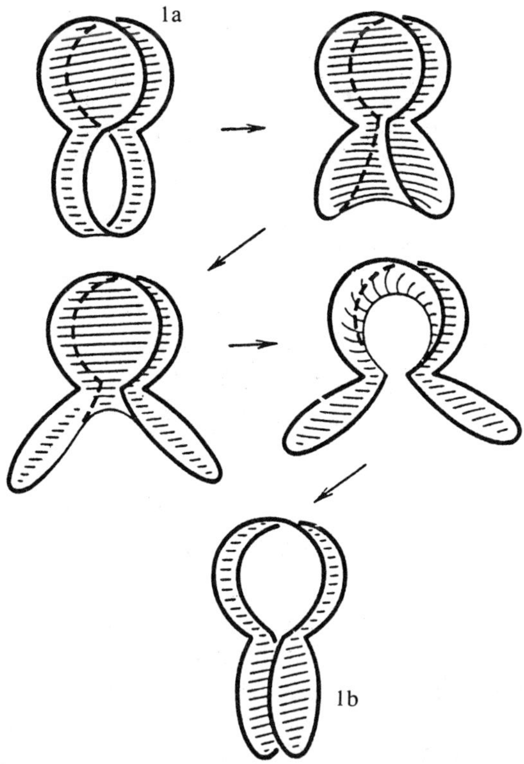

FIGURE 1.32

from each other the soap film is a band spanning each pair of circles, and it is well approximated by a catenoid. As the catenoid is stretched there occurs a critical instant when the catenoid "flops down", re-forms, and becomes the pair of discs spanning the two circles that have moved away from each other. However, in the case of the given contour these two discs are joined by a narrow band stretched on the two upper circles, which are affected by the deformation to a lesser degree (Figure 1.32).

Now, moving the two lower circles in the opposite direction and returning them to the original position, we do not change the topological type of the film and as a result we arrive at film 1b.

We have thus constructed a *continuous deformation of the contour*, as a result of which there occurs a *stepwise (jump-like) change in the film*; it changes its type: from position 1a we have gone over to position 1b.

Using the symmetry of the contour, in a similar way we can carry out the reverse stepwise transition from film 1b to film 1a. It is sufficient to carry

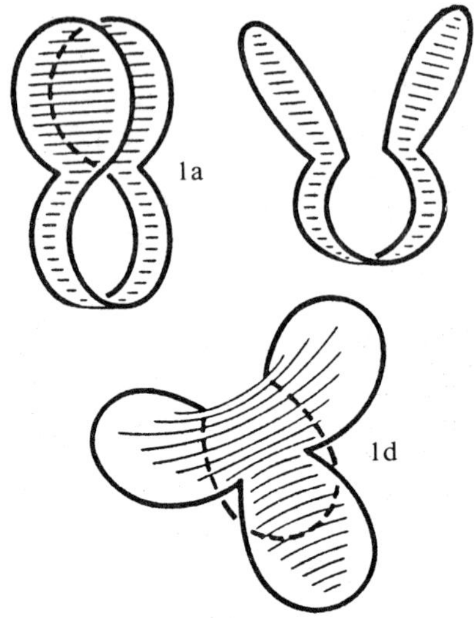

FIGURE 1.33

out the operation described above, but just with the two upper circles.

But together with the *stepwise* reconstruction of the film that we have described there is another remarkable deformation of the bounding contour that leads to a *smooth* variation of the film *with the same final result* (!). Consider film 1a and start to move the upper circles apart, as shown in Figure 1.33. The topological type of the film does not change. Then start to move the lower circles apart. We have obtained a symmetrical film 1c of saddle type: its center is a saddle point. This deformation is smooth, *without jumps*. Since the film 1c is symmetrical, applying the same operation to it we can now go over smoothly from it to film 1b. Similarly, repeating all the steps in the opposite order, we can smoothly change film 1b into film 1a without jumps and topological reconstructions.

Thus, in the space of all surfaces with homeomorphic boundary contours of the given type, we obtain *two essentially different paths* joining positions 1a and 1b. The first path was realized by deforming the contour, in the course of which the film underwent a stepwise reconstruction. The second path is realized by another deformation, in the course of which the film changes without jumps.

In the first approximation, we may assume that the deformations of contours can be described by changing two real parameters: u, the distance between the two upper circles, and v, the distance between the two lower circles. Poston [362] proposed a conventional scheme (the so-called Whitney fold or cusp) that represents the deformations.

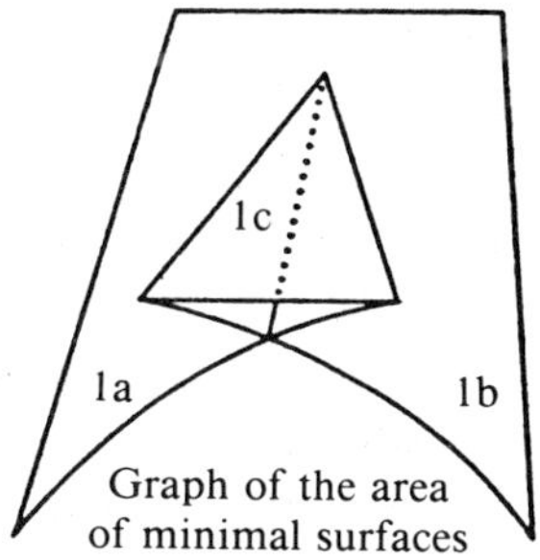

FIGURE 1.34

A. T. Fomenko and A. A. Tuzhilin decided to approach this question from a slightly different point of view, namely to investigate the properties of the *graph of the area of soap films.* It turned out that the resulting picture is described by the so-called *swallow-tail* (a known singularity).

Consider the square in the (u, v)-plane that represents all possible states of the bounding contour. For each point (u, v) we have a corresponding contour on which we can, generally speaking, stretch several minimal surfaces of type 1. Calculating their areas, we obtain a set of numbers and therefore obtain the *graph of a many-valued function.* The films are stable and unstable. Among the films listed above, the film 1c is unstable and the rest are stable. Let us consider the unstable films more carefully. We shall rely on one of the so-called principles of Plateau, according to which the part of the minimal film bounded by an arbitrary closed contour drawn on the film is also minimal with respect to this contour. For more details on Plateau's principles see the following subsections.

PROPOSITION 1.3.2 (Tuzhilin and Fomenko [111]). *The graph of the area of minimal surfaces of type* 1 *has the form shown in Figure* 1.34. *This surface (and the singularity corresponding to it) is called a "swallow-tail". It splits into the domains corresponding to stable and unstable minimal films. The unstable films* 1c *fill the upper triangle of the swallow-tail, and the wings hanging below correspond to the stable films* 1a *and* 1b. *The domains of stable and unstable films are joined along two singular edges emanating from the singular point. To the right edge there are joined the films* 1c *and* 1a, *and to the left edge there are joined the films* 1c *and* 1b. *On each singular edge the smooth sheets of the graph touch with order of tangency* 3/2 *(as in the classical "swallow-tail").*

For properties of the "swallow-tail" see [14]. We recall that a minimal surface is said to be *stable* if any deformation of it in a compact domain does not decrease its area. A. V. Pogorelov [86] gave a sufficient condition for instability of a minimal surface, from which there follows, in particular, the theorem of Do Carmo and Peng [183] that *the plane is the only complete simply-connected stable minimal surface.*

Let us turn to a presentation of the plan of the proof of Proposition 1.3.2. We first study the behavior of the catenoid when its height is increased. The boundary of the catenoid consists of two coaxial circles. Suppose they have the same radius r, and that h is the distance between them. Suppose also that a is the radius of the mouth of the catenoid.

LEMMA 1.3.1. *The function* $h(a)$ *is nonnegative,* $h(0) = 0$ *and* $h(r) = 0$, *and the derivative* $h'(a)$ *vanishes on the interval* $(0, r)$ *at only one point and this point is a maximum point of* $h(a)$.

The proof has actually been given above. Thus, for each h, where $0 < h < h_{\max}$, there are two (!) catenoids (with the same boundary): the outer and the inner.

LEMMA 1.3.2. *The outer catenoid is stable and the inner catenoid is unstable.*

The lemma is proved by direct calculation, which we omit here. Thus, of the three soap films with the same boundary two are stable (the outer catenoid and the pair of flat discs), while the third (the inner catenoid) is unstable.

LEMMA 1.3.3. *The graph of the dependence of the area of a film on the parameter* v *has the form shown in Figure* 1.35.

We now consider the set of all processes $F(u)$ of the form

$$(u, v_0, 1\mathrm{a}) \xrightarrow{v \text{ increases}} (u, v_1, 1\mathrm{b}) \xrightarrow{v \text{ decreases}} (u, v_0, 1\mathrm{b}),$$

parametrized by u. As u increases the jump appears "later", that is, the saddle at the critical instant (critical saddle) is situated closer to the center of the contour.

Let us turn to the unstable films 1c. In fact they are either films of the form 1a or films of the form 1b, where the saddle is situated closer to the center. We note that for each state of the contour the critical saddle of the film is closer to the side where there is a larger distance between the annuli. When $u = v$, we obtain a symmetrical unstable film. It is also seen that when the parameters (say v) increase, the area of the unstable film increases. On nearing the instant of the jump the distance between the saddles in the stable and unstable films decreases. At the instant of the jump the stable and unstable films merge. This process is analogous to the process of interaction of stable and unstable catenoids.

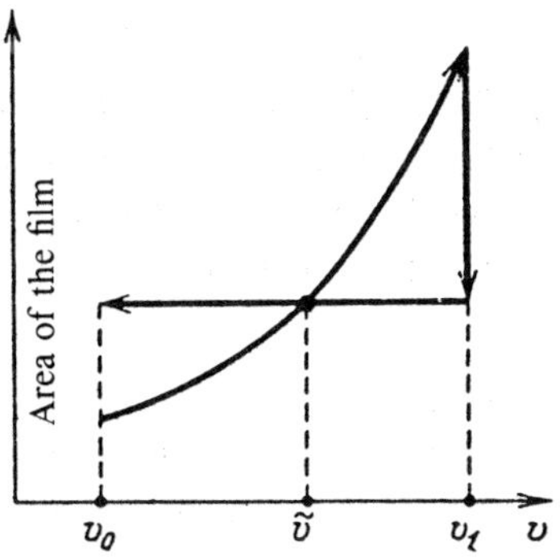

FIGURE 1.35

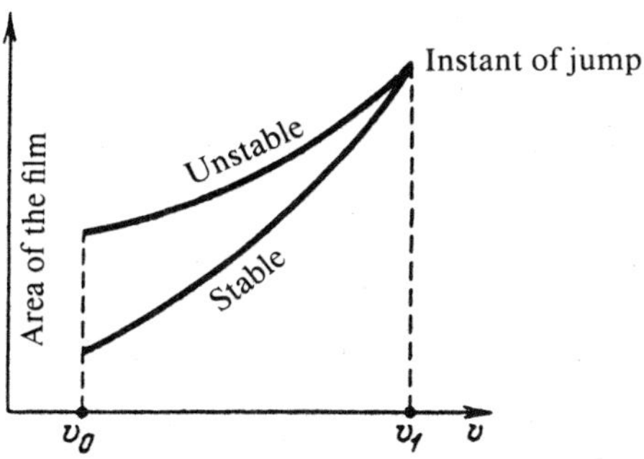

FIGURE 1.36

LEMMA 1.3.4. *With the growth of* v *the area of both the stable film* 1a *and the unstable film* 1c *increases monotonically. In particular, the graph of the dependence of the area of the films* 1a *and* 1c *on* v *has the form shown in Figure* 1.36. *The intersection of the two branches of this graph, which corresponds to the instant of fusion of the stable and unstable films, is therefore a singular point. As the parameter* u *varies, these singular points fill an edge of the two-dimensional surface along which the smooth sheets of the graph of the area join, corresponding to the stable and unstable films. There are exactly two such edges and they meet in one singular point of the graph of the area.*

PROOF. From the observations made above it follows that it is sufficient to verify this fact for the inner and outer catenoids, that is, to verify that the growth of their area as their height h increases is monotonic. Let Oz be the axis of symmetry of the catenoid. We write this equation in the form $r = a\,\mathrm{ch}(z/a)$, where $r = \sqrt{x^2+y^2}$, $|z| \le h/2$, $r(h/2) = R$, where R is the radius of the bounding circles of the catenoid. If S is the area of the catenoid, then $S = S(h)$ and $a = a(h)$. Direct calculation shows that $S = \pi a(h)(h + 2R\sqrt{R^2/a(h)^2 - 1})$. Further, $dS/dh = 2\pi a(h) > 0$. Hence it follows that with the growth of h the areas of the two catenoids (both the stable and the unstable) increase strictly monotonically up to the critical value $h_{\max}$, when the catenoids merge and then burst. This proves the lemma.

LEMMA 1.3.5. *If $s_1(h)$ and $s_2(h)$ are the areas of the outer and inner catenoids (that is $s_1(h) < s_2(h)$), then as $h \to h_{\max}$, the graphs of $s_1(h)$ and $s_2(h)$ touch each other at the singular point $h = h_{\max}$ with order of tangency $3/2$.*

PROOF. It is required to find a number α such that

$$\lim_{h\to h_{\max}} \frac{s_2(h) - s_1(h)}{(h_{\max} - h)^\alpha} = c\ (\neq 0, \infty).$$

From the proof of Lemma 1.3.4 it follows that $S(h) = \int 2\pi a(h)\,dh$. Hence

$$2\pi \int_h^{h_{\max}} \frac{a_2(t) - a_1(t)}{(\Delta h)^\alpha} dt = \frac{s_2 - s_1}{(\Delta h)^\alpha}.$$

If $a_2 = a_1 = c\,(\Delta t)^\beta + o((\Delta t)^\beta)$, where $c \neq 0$, then

$$s_2 - s_1 = \frac{2\pi c}{1+\beta}(\Delta h)^{\beta+1} + 2\pi \int_h^{h_{\max}} o((\Delta t)^\beta)\,dt.$$

Let $a_{cr} = a(h_{\max})$. We can express h as a function of a; then we can verify that $d^2h/da^2|_{a=a_{cr}} = \mu \neq 0$.

If $\Delta a_2 = a_2 - a_{cr}$ for $a \geq a_{cr}$ and $\Delta a_1 = a_1 - a_{cr}$ for $a \leq a_{cr}$ and $h(a_1) = h(a_2)$, then $\Delta h = \mu(\Delta a_1)^2 + o((\Delta a_1)^2) = \mu(\Delta a_2)^2 + o((\Delta a_2)^2)$, where $\mu \neq 0$. Hence

$$(\Delta h)^\beta/\Delta a = \mu(\Delta a_2)^{2\beta-1}(1 + o(1))^\beta/(2 + o(1)),$$

that is, we obtain the condition $\beta = 1/2$, so $\alpha = 1 + \beta = 3/2$.

From Lemmas 1.3.4 and 1.3.5 there follows Proposition 1.3.2. To obtain the unstable film 1c, we have carried out each time a certain task (we have stretched the thread), that is, we have increased the area of the film. Therefore the graph of the area $s_2(h)$ of unstable films 1c lies above the graph of the area $s_1(h)$ for the stable films 1a and 1b.

Consider the set of processes $F(u)$ for unstable films. When $u \geq u_{cr}$ (where u_{cr} is the value of the parameter before which the process $F(u)$ is continuous, that is, without jumps for stable films), the picture changes: the unstable films merge with the stable films.

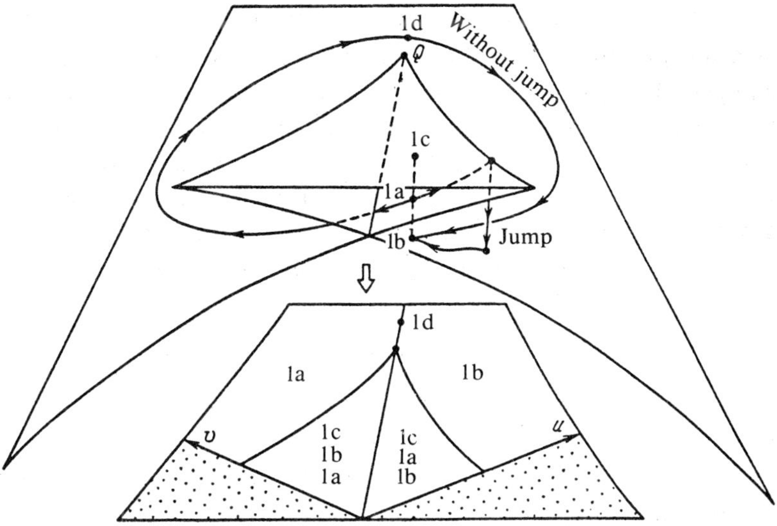

FIGURE 1.37

On Figure 1.37 we now trace the interesting possibility of transition from the state $(u_0, v_0, 1a)$ to the state $(u_0, v_0, 1b)$ both by means of a jump and in a continuous smooth way. The first path is a rise to the singular edge (the path on the right in Figure 1.37), and then a "fall" from the point on the singular edge to the point 1b. The second path is as follows: we descend smoothly from the point 1a along the sheet of the graph, go round the singularity (the point Q in Figure 1.37), and arrive at the point 1b.

Thus, the values of the area of unstable films of type 1c fill the upper triangle of the swallow-tail, and the wings going down correspond to the stable films 1a and 1b. Films of type 1d divide the regions 1a and 1b from each other and correspond to symmetrical contours (see above). Inside the "beak" there are three types of films, two of which are stable and one unstable.

Above we described the *experimental-theoretical argument* of A. A. Tuzhilin and A. T. Fomenko, applied to an arbitrary continuous two-parameter family of deformations from the class mentioned above. Then Tuzhilin noticed that for special analytic deformations of an analogue of the Douglas contour a *strict theoretical proof* of Proposition 1.3.2 can be extracted from the remarkable results of Beeson and Tromba [164] on the properties of the Enneper surface.

Consider the *Enneper surface*: $x(r, \theta) = (r\cos\theta - (r^3/3)\cos 3\theta, -r\sin\theta - (r^3/3)\sin 3\theta, r^2\cos 2\theta)$, where r, θ are polar coordinates. Let

$\gamma_{r_0} = x(r = r_0, \theta)$, *a contour on the surface.* We can show that when $r_0 < 1$ there is a unique minimal surface spanning Γ_{r_0}. When $r > 1$, the situation changes. We observe that the *Enneper contour* can be regarded as a *smoothed Douglas contour*, situated so that the parameter u corresponds to the distance between the upper ends of the Enneper contour, and v corresponds to the lower. Beeson and Tromba [164] considered the following deformation of the contour $\Gamma_1 = x(r = 1, \theta)$. Let $\Gamma_1 = x(\theta)$ and $(a, b) \in \mathbf{R}^2$ vary in a neighborhood of $0 \in \mathbf{R}^2$. We specify a deformation as follows:

$$\tilde{x}(\theta, a, b) = u(\theta) + a\cos^2\theta \cdot x_r + b\sin^2\theta \cdot x_r,$$

where x_r is the derivative with respect to r of the harmonic continuation of $x(\theta)$ to D^2 (when $r = 1$).

Since x_r is a vector in the plane to the surface, which plane is directed along the normal to the boundary, and $\cos^2\theta$ is large in a neighborhood of the vertex of the contour, and $\sin^2\theta$ is small there, and conversely, $\sin^2\theta$ is large in a neighborhood of the lower ends of the contour, and $\cos^2\theta$ is small there, *we obtain an analogue of a deformation of the Douglas contour on the Enneper surface.*

From the results of Beeson and Tromba it follows that the graph of the area of minimal surfaces spanning the *models of the Douglas contour on the Enneper surface* is a swallow-tail.

§4. Plateau's four experimental principles and their consequences for two-dimensional minimal surfaces

4.1. Minimal surfaces in three-dimensional space and Plateau's first principle. In the 30's and 40's of this century great progress was made in the study of the properties of two-dimensional minimal surfaces in $\mathbf{R}^3$. Remarkable results were obtained, in particular, by Douglas, Rado, and Courant. An important role in these investigations was played by the *Dirichlet principle*, which we state in the following form. Let G be a planar domain bounded by a Jordan curve γ, and let g be a continuous function on the closure of G, piecewise smooth in G and such that the Dirichlet integral $D[g]$ takes a finite value on this function. We recall that a *Dirichlet integral (functional)* is an expression of the form $D[\varphi] = \frac{1}{2}\iint(E + G)\,du\,dv$, where E and G are coefficients in the first quadratic form (Riemannian metric) induced on the graph of φ by the ambient Euclidean metric in $\mathbf{R}^3$. Consider the class of all functions φ continuous on the closure of G, piecewise smooth in G and taking the same values as g on the boundary of the domain. Then the problem of finding a function φ on which the Dirichlet integral $D[\varphi]$ takes the least possible value has a unique solution. This function is also the only solution of the boundary value problem for the Laplace equation $\Delta\varphi = 0$ with preassigned boundary values $\overline{g} = g|_\gamma$ on the curve γ.

In the first half of this century a change of emphasis in the approach to the solution of Plateau's problem occurred. Plateau himself had stated several principles, which we list below; here we just give the first of them.

Plateau's first principle. Suppose we are given a surface of zero mean curvature. The surface may be described by means of an equation, in the form of a radius vector or by means of some geometrical rule, like a helicoid, for example. We consider an arbitrary piecewise smooth closed non-self-intersecting contour on it, that is, we draw on the surface a closed curve without self-intersections. We assume that the contour bounds part of the surface, and that this part is *stable.* We make a wire contour, which exactly reproduces the curve we have drawn, and then slightly oxidize it in dilute nitric acid and completely immerse it in soapy water with glycerine. If we remove the contour from the liquid, then *among the soap films that can appear on this contour, there is necessarily a film that coincides with the part of the surface bounded by the curve originally drawn.*

The question of first importance is the existence of a minimal film spanning the given contour. Since the study of arbitrary contours is nontrivial, a rather special class of them was first distinguished. Consider surfaces specified in $\mathbf{R}^3$ by the graph of a single-valued function $z = f(x, y)$, that is, surfaces that admit a smooth orthogonal one-to-one projection on a two-dimensional plane or on a domain in it. Suppose that the surface M is projected into the domain G with *convex boundary* γ. If the surface is specified by the graph, the area functional has the form $A|f| = \iint_G \sqrt{1 + f_x^2 + f_y^2}\, dx\, dy$. The Euler equation for the area functional has the form $(1 + f_x^2)f_{yy} - 2f_x f_y f_{xy} + (1 + f_y^2)f_{xx} = 0$. It turns out that *for any closed piecewise smooth contour* γ' *that projects one-to-one on a plane into a convex closed curve* γ, *there is a minimal surface with the given boundary that consequently projects onto the domain bounded by the plane curve* γ.

4.2. The area functional and the Dirichlet functional. Further important progress was made by Douglas [201]–[206]. We first describe his results informally. He considered smooth maps of a standard two-dimensional disc D ($\subset \mathbf{R}^2$) into $\mathbf{R}^3$. We refer the disc to Euclidean coordinates u, v; then its image is given by the radius vector $r = r(u, v)$. If E, F, and G are the coefficients of the first form of the surface that is the image of the disc in $\mathbf{R}^3$, then for the Dirichlet and area functionals we obtain the following explicit expressions:

$$D[r] = \frac{1}{2}\iint_{D^2}(E+G)\,du\,dv\,;$$
$$A[r] = \iint_{D^2}\sqrt{EG-F^2}\,du\,dv\,.$$

The extremal critical radius vectors for these functionals are constructed as follows. For the area functional these are just those radius vectors for which the mean curvature of the corresponding surface is zero (*minimal surfaces*). The extremal radius vectors for the Dirichlet functional are just those vectors that are harmonic with respect to u, v, that is, $\Delta r = r_{uu} + r_{vv} = 0$. *If we specify conformal coordinates on a minimal surface, then its radius vector becomes harmonic. The converse is false: a surface swept out by a harmonic radius vector need not be a minimal surface.* The simplest example is the graph of the real part $a(x, y)$ (or the imaginary part $b(x, y)$) of a nonlinear complex-analytic function $f(x+iy) = a(x, y) + ib(x, y)$.

The area functional and the Dirichlet functional are connected by an inequality: $D[r] \geq A[r]$, where equality holds if and only if $E = G$, $F = 0$, that is, when the coordinates u, v are conformal.

The most important result of Douglas is *a theorem on the existence of a minimal surface in the class of surfaces of fixed topological type.*

Before stating the theorem we recall that every smooth compact closed connected two-dimensional manifold is diffeomorphic to either a sphere with a certain number of handles or a sphere with a certain number of Möbius films glued to it (Figure 1.38). A *handle* is an ordinary cylinder glued to the rims of two holes in the sphere. A *Möbius film* is an ordinary Möbius band glued along its own boundary to the boundary of a hole in the sphere. Sometimes for clarity a Möbius film is realized as a so-called *crossed cap* (see Figure 1.38). For this we need to bend the boundary of the Möbius band so that it becomes a standard plane circle. We have to "pay for" this by a complication of the configuration of the Möbius band and the appearance of self-intersections on it.

Suppose that for a two-dimensional manifold M of given topological type with boundary there is a smooth map $f\colon M \to \mathbf{R}^3$ with finite area, that is, $A[f] < \infty$, and the boundary of the surface is mapped onto a given system of closed Jordan curves in $\mathbf{R}^3$. Suppose that the greatest lower bound of the areas of such immersions of the surface in $\mathbf{R}^3$ with a fixed

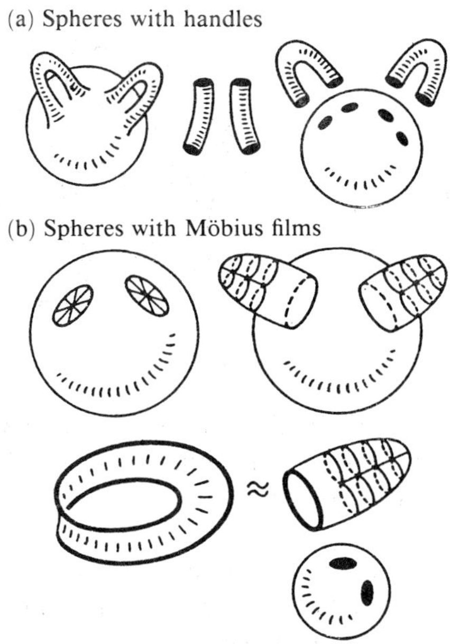

FIGURE 1.38

boundary is strictly less than the infimum of the areas of all immersions in $\mathbf{R}^3$ (with the same boundary) of all surfaces obtained from the original surface by discarding one handle or one Möbius film. Then *there is a map* $f_0: M \to \mathbf{R}^3$ *that is minimal, that is, it realizes a minimum of the area in the class of all immersions of the surface in* $\mathbf{R}^3$ *with the given boundary.*

Let us note the condition for the existence of *at least one* immersion of M in $\mathbf{R}^3$ with finite area. The fact is that *there are curves on which we cannot stretch any surface of finite area.* Minimal films may have *self-intersections* and *branch points.* Clearly, there are contours for which there is no minimal film in the class of *embeddings.* The simplest such knotted contour (a trifolium) is shown in Figure 1.39. It is not covered by any *embedded* disc in $\mathbf{R}^3$. At the same time *there is a smooth map of a disc* D^2 *in* $\mathbf{R}^3$ (with self-intersections and branches) *that realizes a minimum of the area in the class of maps of the disc with a given boundary.* This is true, incidentally, for any Jordan closed contour homeomorphic to a circle. Figure 1.39 also shows an example of a minimal film spanning a circle embedded in $\mathbf{R}^3$ and knotted in the form of a trifolium. This surface has not only self-intersections (three segments meeting at one point), but also a branch point (at the center). We should note that *this surface is not an absolutely minimal surface.*

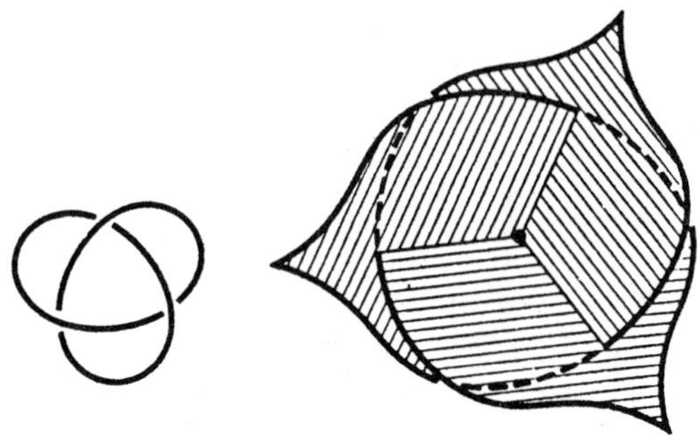

FIGURE 1.39

4.3. Singular points of minimal surfaces and Plateau's last three principles. An important property of minimal surfaces of general type is the frequent presence on them of *singular points.* The presence of singularities is the "typical situation" in the sense that *for a complicated bounding contour the soap film spanning it "almost certainly" has singularities.* The sets of singular points may be quite complicated and interesting. Quite often singular points are not isolated and fill whole intervals, which several sheets of the film approach. What is the nature of singular points? Plateau discovered experimentally three facts, which we conventionally call his second, third, and fourth principles.

Plateau's second principle. Minimal surfaces (of zero mean curvature) and systems of soap bubbles (of constant positive mean curvature) consist of fragments of smooth surfaces joined to one another along smooth arcs.

Plateau's third principle. Pieces of smooth surfaces of which films of constant mean curvature consist may be joined only in the following two ways: (a) three smooth sheets meet on one smooth curve, (b) six smooth sheets (together with four singular curves) meet at one point, the vertex.

Plateau's fourth principle. In the case when three sheets of a surface are joined along a common arc, they form angles of $120°$ with each other. In the case where four singular curves (and six sheets) meet at one vertex, these edges form angles of approximately $109°$ with each other (Figure 1.40). Each of the stable soap films shown is realized as a surface spanning the corresponding contour. It is useful to imagine one of the mechanisms for forming three sheets of a surface meeting at a common edge at angles of $120°$. Consider a contour inside a large soap bubble (Figure 1.41). Gradually expelling air from the bubble, we see that it envelops the contour and finally flops down to a surface of the required form. We say that a point of a minimal surface has *multiplicity* k if k sheets of the surface

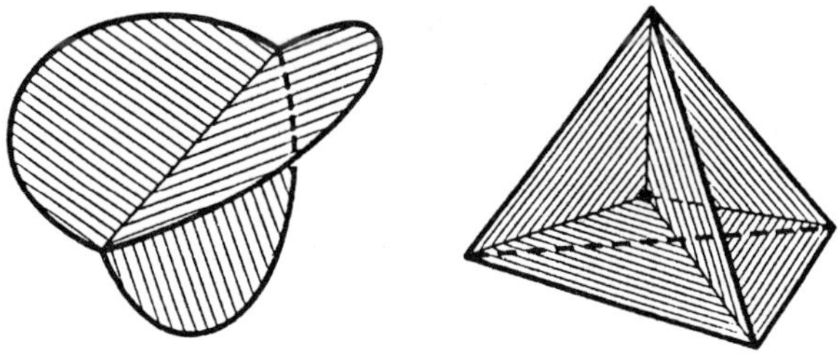

FIGURE 1.40

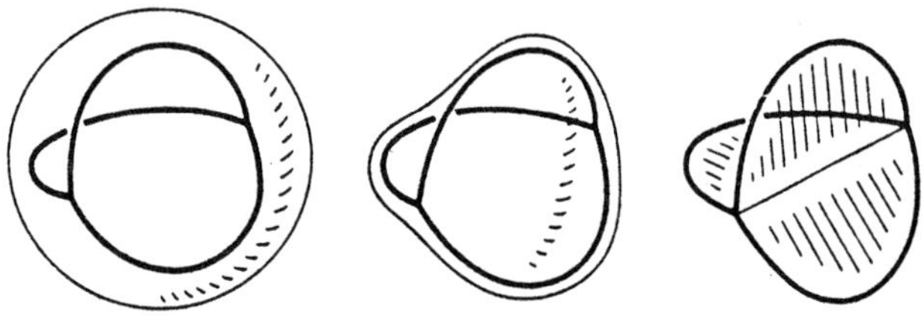

FIGURE 1.41

meet there. A regular point has multiplicity two, since two sheets meet along each edge at an angle of 180°. It is obvious from Figure 1.42 that *two sheets* of a minimal surface cannot meet at a common edge at an angle other than 180°, since otherwise there is a deformation of the film that decreases its area. There are films whose mean curvature is zero almost everywhere (at nonsingular points), and *the multiplicity of the singular points may be greater than three.* For example, two orthogonal planes intersecting along a line realize such a surface. The multiplicity of each point of self-intersection is equal to 4 here. Nevertheless, attempts to construct a *real stable* soap film for which more than three sheets meet at a singular segment have failed. It turns out that *any surface of zero mean curvature with self-intersections is unstable.* This assertion has a local character: *an arbitrarily small neighborhood of each point of self-intersection is unstable.* Figure 1.42 shows that *singular points of higher multiplicity may split in several ways into the union of threefold points.*

Above, we have considered the section of the surface by a plane orthogonal to a singular edge. The resulting deformation was uniform at all points of the edge close to the section (see Figure 1.42). However, it is easy to construct a deformation that decreases the area of the film even *when the boundary of part of the film* shown in Figure 1.42 is rigidly fixed. The fourfold edge splits into the union of two threefold bent edges, whose ends are fixed. We have thus constructed a monotonic contraction of the

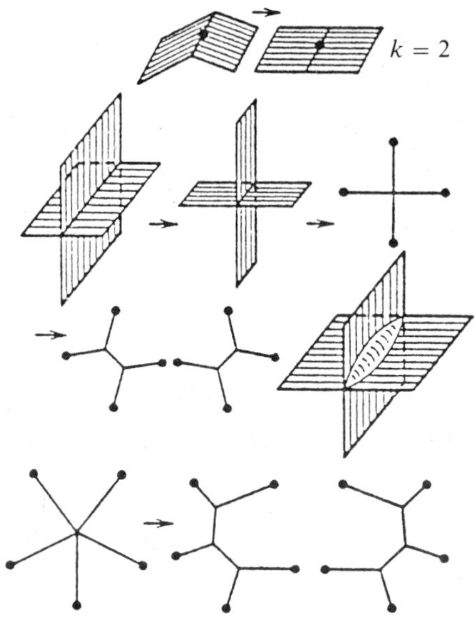

FIGURE 1.42

film, which decreases its area *for a fixed boundary*. Gluing this piece of the film to the original film of large size, we obtain the required contracting deformation.

These problems are connected with *the two-dimensional Steiner problem: to find a thread (a one-dimensional continuum) of smallest length joining a finite set of points in the plane. The thread may branch and have points of multiplicity three or more.* This problem can be solved experimentally by means of soap films. We must take two neighboring parallel planes made of thin organic glass and join them by vertical column segments situated at the points of the plane through which the required thread must pass. Dipping the construction into soapy water and removing it, we obtain a film stretched on the columns and realizing a minimum of the length of its plane section. The thread contains points of multiplicity no higher than three.

Similar mechanisms act in the cases where bubbles are formed in the soap film.

§5. Two-dimensional minimal surfaces in Euclidean space and in a Riemannian manifold

5.1. The equation of a minimal surface. Henceforth we shall make a clear distinction between the concepts of *local* minimality and *global* minimality (absolute minimality).

Let us recall the definition of *Riemannian volume* on a Riemannian manifold M^n. If d is a domain (for example, compact with piecewise smooth boundary) in M^n and $g_{ij}(x)$ is the Riemannian metric (where $x_1, \ldots, x_n$ are local coordinates on M^n), then for the volume of the domain D we take the number (integral) $\int_D \sqrt{g}\,dx_1 \wedge \ldots \wedge dx_n$, where $g = \det(g_{ij})$. Clearly, we can now define the k-dimensional volume of k-dimensional submanifolds of M^n, restricting the Riemannian metric to them.

Consider a smooth hypersurface $V^{n-1} \subset \mathbf{R}^n$ given, for example, in the form of the graph $x_n = f(x_1, \ldots, x_{n-1})$. Suppose that the domain of definition of the function f is a bounded domain D in $\mathbf{R}^{n-1}$. Consider the volume functional $\operatorname{vol}(f)$, defined on the space of all such functions f. Then it is easy to calculate that

$$\operatorname{vol}(f) = \int_D \sqrt{1 + \sum_{i=1}^{n-1} f_{x_i}^2}\, dx_1 \wedge \ldots \wedge dx_n .$$

Let us consider *the extremal surfaces* V^{n-1} for the volume functional $\operatorname{vol}(f)$, that is, *the graphs of the extremal functions* $x_n = f(x)$. *The Euler-Lagrange equation has the form*

$$\sum_{i=1}^{n-1} \frac{\partial}{\partial x_i} \left(f_{x_i} \left(1 + \sum_{j=1}^{n-1} f_{x_j}^2 \right)^{-1/2} \right) = 0 .$$

Definition 1.5.1. Surfaces that are extremal for the volume functional $\operatorname{vol}(f)$ are called *minimal* (or *locally minimal*) *surfaces*.

For a two-dimensional minimal surface given in the form of a graph $z = f(x, y)$ in $\mathbf{R}^3$ the Euler-Lagrange equation takes the form

$$(1 + f_x^2) f_{yy} - 2 f_x f_y f_{xy} + (1 + f_y^2) f_{xx} = 0 .$$

The equation of a minimal surface V^{n+1} can be written in terms of local invariants of the embedding of this surface in $\mathbf{R}^n$.

Proposition 1.5.1. *Let $V^{n-1} \subset \mathbf{R}^n$ be a smooth hypersurface. The mean curvature H is identically zero if and only if the surface V can be represented, in a neighborhood of each of its points, as the graph of an extremal function for the volume functional* (*that is, the solution of the equation of a minimal surface*).

Let us consider the case $M^2 \subset \mathbf{R}^n$. Let M^2 be given by the equations $x_i = f_i(u, v)$, where $3 \le i \le n$, $x_1 = u$, $x_2 = v$. We put $f(u, v) = (f_3(u, v), \ldots, f_n(u, v))$. Direct calculation shows that the

Euler-Lagrange equation for the functional vol_2, defined on two-dimensional surfaces in $\mathbf{R}^n$, has the form

$$(1+|f_u|^2)f_{vv} - 2\langle f_u, f_v\rangle f_{uv} + (1+|f_v|^2)f_{uu} = 0, \tag{5.1}$$

where $\langle\ ,\ \rangle$ is the Euclidean scalar product. We have obtained the equation of a two-dimensional minimal surface (in nonparametric form) in $\mathbf{R}^n$.

We give three examples of classical minimal surfaces. The *helicoid* is given (up to a multiplier) by the graph of the function $f(x, y) = \arctan(x/y)$. This is the *only* solution of (1) that is a *harmonic function*. Among all minimal surfaces *only the helicoid is a ruled surface*. This is the *only* (up to a multiplier) minimal surface of the form $z = z(\varphi)$, where φ is the polar coordinate in the plane.

The *catenoid* is specified as follows: $f(x, y) = \mathrm{Arcch}(r), r^2 = x^2 + y^2$, or $a^2(x^2+y^2) = \mathrm{ch}^2(az)$, where $a = \mathrm{const}$. The catenoid is the *only* minimal surface that is a *surface of revolution*. Here we have in mind *complete* surfaces.

The *Scherk surface* (discovered in 1834): $z = (1/a)\ln(\cos ay/\cos ax)$, where $a = \mathrm{const}$, or $e^z = \cos y/\cos x$ for $a = 1$. This is *the only minimal surface* having the form of a *surface of translation*, that is, the scalar function $f(x, y) = \ln(\cos y) - \ln(\cos x)$ is the only solution of (1) having the form $f(x, y) = g(x) + h(y)$; see Osserman [77] and Darboux [199]. Clearly, the image of a minimal surface under a rotation, a translation, or a homothety is again a minimal surface. The functions given above are not defined for all values of x and y. This is an important fact. By a theorem of S. N. Bernstein, when $n = 3$ there are no nontrivial (that is, *other than linear*) solutions of (1) that are defined for all x and y. We give an example of a minimal surface in $\mathbf{R}^{2m+2}$. Let $z = x + iy$ and let $g_1(z), \dots, g_m(z)$ be complex analytic functions. Then the surface

$$f_j(x, y) = \begin{cases} \mathrm{Re}\, g_p(z) & \text{for } j = 2p+1, \\ \mathrm{Im}\, g_p(z) & \text{for } j = 2p+2, \end{cases}$$

where $p = 1, 2, \dots, m$, is minimal.

PROPOSITION 1.5.2. *The graph of a complex analytic curve in* $\mathbf{C}^q$, *regarded as a two-dimensional surface in* $\mathbf{R}^{2q} \approx \mathbf{C}^q$, *is a minimal surface.*

5.2. Minimal surfaces and harmonic functions. The role of conformal coordinates. The Weierstrass representation. Because of shortage of space we confine ourselves to a brief survey, referring the reader for the details to the fundamental papers of Nitsche [328], Osserman [77], Courant [193], Morrey [318], and Federer [216]. Probably the first serious research in this direction is due to Lagrange [278], who began (in 1768) to consider the following problem: *to find a surface of least area stretched on a given contour.* Monge [312], [313] established in 1776 that the condition of minimality of a surface area is locally equivalent to the condition $H = 0$. We note that *a locally minimal surface may not realize an absolute minimum of the area.* The theory of two-dimensional minimal surfaces attracted the attention of nearly all the great analysts and geometers of the 19th and 20th centuries. The first general methods of integrating equation (1) were worked out by Monge in 1784 and Legendre [289] in 1787. They obtained the so-called *Monge formulas* in terms of complex characteristics of the Euler-Lagrange equation. The next step was taken by Poisson [359]–[361] in the 1830's. In 1832 Poisson announced the solution of Lagrange variational problem in the case where *the boundary of a two-dimensional surface is close to a plane curve.* In the case of a plane curve the existence of a minimal surface with a given boundary is obvious. Much attention has been paid to the construction of concrete examples of minimal surfaces. These are the *catenoid* (Euler, 1774), the *helicoid* (Euler, then Meusnier, 1776), and the *Scherk surface* (1834). In 1842 Catalan proved that the helicoid is the only ruled minimal surface. In the 1850's Bonnet made a deep investigation of minimal surfaces, gave new and simplified proofs of many facts, and discovered new properties. He proved that the catenoid is the only minimal surface of revolution, that a spherical (Gaussian) map of a minimal surface is conformal, and so on.

The transition to conformal (isothermal) coordinates on a minimal surface simplifies the investigation of its properties. It is well known that *if a regular two-dimensional surface is of class C^2, then in a neighborhood of any of its points there are always conformal coordinates. On minimal surfaces, conformal coordinates always exist in a neighborhood of any regular point.* If a minimal surface M^2 in $\mathbf{R}^n$ is given in nonparametric form, that is, in the form $x_i = f_i(u, v)$, $3 \leq i \leq n$, $u = x_1$, then the functions f_i are always real analytic functions of u and v. If M^2 is a regular surface in $\mathbf{R}^n$ given by a radius vector $r(u, v)$ of class C^2, where u and v are conformal coordinates, then $\Delta r = 2\lambda^2 H$, where H is the mean curvature, Δ is the Laplace operator, and $\lambda \geq 0$ is a scalar function such that $ds^2 = \lambda(u, v)(du^2 + dv^2)$. Hence we immediately obtain the following important assertion.

PROPOSITION 1.5.3. *Suppose that a regular surface* M^2 *in* $\mathbf{R}^n$ *is given by a radius vector* $r(u, v)$ *in conformal coordinates* u, v *on* M^2. *Then the radius vector is harmonic if and only if* $H = 0$, *that is,* M^2 *is locally minimal.*

Suppose that $M^2 \subset \mathbf{R}^n$ and that $Edu^2 + 2Fdudv + Gdv^2$ is the first quadratic form on M^2. If M^2 is given by the vector $r(u, v) = (x_1(u, v), \dots, x_n(u, v))$, then we consider the complex functions $\varphi_k(z) = \partial x_k/\partial u - i\partial x_k/\partial v$, where $z = u - iv$. Then $\sum_{k=1}^n \varphi_k^2(z) = E - G - 2iF$, $\sum_{k=1}^n |\varphi_k(z)|^2 = E + G$. Hence we obtain the following results.

1. The functions $\varphi_k(z)$ are analytic functions of z if and only if $x_k(u, v)$ are harmonic functions of u and v.
2. The coordinates u and v are conformal on M^2 if and only if

$$\sum_{k=1}^{n} \varphi_k^2(z) \equiv 0. \tag{5.2}$$

3. If u and v are conformal coordinates on M^2, then the surface M^2 is regular if and only if

$$\sum_{k=1}^{n} |\varphi_k(z)|^2 \neq 0. \tag{5.3}$$

The formulas (5.2) and (5.3) for $n = 3$ were given by Weierstrass. From Proposition 1.5.3 we immediately obtain the following important assertion.

PROPOSITION 1.5.4. *Suppose that the vector* $r(u, v)$ *defines a locally minimal surface* M^2 *in* $\mathbf{R}^n$, *and that* u *and* v *are conformal coordinates. Then the functions* $\varphi_k(z)$ *are complex analytic and satisfy the equations* (5.2) *and* (5.3). *Conversely, let* $\varphi_1(z), \dots, \varphi_n(z)$ *be complex analytic functions satisfying* (5.2) *and* (5.3) *in a simply-connected domain* $D \subset \mathbf{R}^2$. *Then there is a regular minimal surface given by the vector* $r = (x_1(u, v), \dots, x_n(u, v))$, *defined on* D, *and* $\varphi_k(z) = \partial x_k/\partial u - i\partial x_k/\partial v$.

Thus, *the study of two-dimensional minimal surfaces in* $\mathbf{R}^n$ *is equivalent to the study of complex analytic vectors satisfying conditions* (5.2) *and* (5.3).

5.3. Generalized minimal surfaces. Let us consider parametrized surfaces of class C^q, that is, pairs (M^2, r), where $r\colon M^2 \to \mathbf{R}^n$ is a map of class C^q. We recall that a topological n-dimensional manifold of class C^q is given by an atlas, that is, a collection of charts U_i and coordinate maps $\psi_i\colon D^n \to U_i$ such that the transition functions $\psi_j^{-1}\psi_i$ are homeomorphisms of class C^q. We say that a conformal structure is defined on

M if we are given an atlas, all of whose transition functions are conformal; let $z = u - iv$.

DEFINITION 1.5.2. A *generalized minimal surface* $S = (M, r)$ in $\mathbf{R}^n$ is a map $r: M^2 \to \mathbf{R}^n$ given by the vector $r(u, v) = (x_1(u, v), \dots, x_n(u, v))$ (distinct from a map to a point), where M^2 is a two-dimensional manifold with conformal structure $\{U_i; \psi_i: D^2 \to U_i\}$, and each coordinate function $x_k(u, v)$ of this vector $r: M^2 \to \mathbf{R}^n$ is harmonic on M^2 and so $\sum_{k=1}^n \varphi_k^2(z) = 0$, where $\varphi_k(z) = \partial x_k/\partial u - i\partial x_k/\partial v$.

We assume that the vector $r: M \to \mathbf{R}^n$ depends on the point $P \in M^2$, that is, $r = r(P)$, and the point P is given by the coordinates (u, v), where $P = \psi_i(u, v)$ for some i. If $r(M^2)$ is a regular minimal surface in $\mathbf{R}^n$, then on M^2 we can introduce a conformal structure. For this we need to consider an atlas consisting of local charts with conformal coordinates. Thus, *every regular minimal surface can be regarded as a generalized minimal surface.* In this sense *the theory of generalized surfaces is "wider" than the theory of regular surfaces.* If we are given a generalized minimal surface, then since the map $r(P)$ is not a map to a point, at least one coordinate function $x_k(z)$ is not a constant and the analytic function $\varphi_k(z)$ corresponding to it can have only isolated points. They are called *branch points of the minimal surface.* If we remove these points from the surface, then the remaining part of it determines a *regular surface.*

Many further theorems about the existence of minimal surfaces will be stated *for generalized surfaces*, for example, the theorem of Rado and Douglas. *If in some problem we have proved the existence of a generalized surface, it does not follow that the solution we have found is regular* (*in the usual sense*).

The *classical Plateau problem* is the problem of finding a surface of least area (either locally or globally minimal) with a given boundary. The first solutions of it were obtained in the 19th century for special polygonal contours (Plateau [356], [357], Darboux [199]). Together with the problem with a fixed boundary, from 1816 *the problem of minimal surfaces with a partially free boundary* has been considered (the *Gergonne problem*). It was required to find a minimal surface under the condition that part of its boundary is given and fixed, and the remaining part can slide along a fixed surface. The first existence theorems for such minimal surfaces were proved for the case where the fixed part of the boundary consists of line segments, and the free part of the boundary slides along a system of planes (Courant [193], [196], Nitsche [328]). The problem was also considered of finding a minimal surface of which part of the boundary is free. We consider a nonclosed contour whose ends are joined by a flexible inextensible filament N. Then on this "composite contour" there is stretched a soap film. Under the action of the forces of surface tension, the filament N itself takes up some position. The problem consists in determining the resultant minimal surface. In the modern literature this question and its generalizations have been investigated in detail in the deep papers of Hildebrandt and Nitsche [260]–[263].

In the first half of the 20th century fundamental results were obtained concerning the existence of two-dimensional minimal surfaces with a given boundary (Douglas, Rado, Courant, McShane, and others). These remarkable theorems essentially use the connection between minimal surfaces and harmonic vectors. This connection enables us to apply the theory of critical points, which was done in a cycle of papers by Morse, Tompkins, Courant, Shiffman, and others (see [321]–[325], [192]–[196]).

5.4. The theorems of Douglas and Rado. Solution of the classical Plateau problem in the class of images of two-dimensional manifolds with a fixed boundary. Further generalizations. Let us state a remarkable theorem of Rado and Douglas *which solves, in particular, the Plateau problem in the class of maps of a disc with a given boundary.* Henceforth we shall always consider only boundary contours for which there is at least one surface $f(M)$ with boundary γ and the condition $D[f] < \infty$.

THEOREM 1.5.1 (Douglas and Rado [203]). *Let γ be a closed Jordan curve (a circle) in $\mathbf{R}^n$. Then there is always a simply-connected generalized minimal two-dimensional surface with boundary γ.*

Thus, there is always a map $r: \overline{D}^2 \to \mathbf{R}^n$, where D^2 is a standard open flat disc in $\mathbf{R}^2$ with boundary $S^1 = \partial D^2$, and $\overline{D}^2$ is the closure of D^2, such that r is continuous on $\overline{D}^2$, the restriction of r to the boundary S^1 is a homeomorphism of ∂D^2 on γ, and the map r determines a minimal surface $r(D^2)$ in $\mathbf{R}^n$.

How many minimal discs can have the same boundary? Up to now there is no answer to the following question. If we are given a *simple smooth regular Jordan curve*, can there exist a nontrivial family of continuum many minimal discs bounded by this curve? P. Lèvy and Courant constructed an example of rectifiable Jordan curve that is smooth with the exception of one point and *bounds uncountably many minimal discs.*

Morgan [315] constructed an example of a *continuous family* of distinct minimal surfaces whose boundary consists of four nonintersecting circles. For these examples, see below. If a curve is *real-analytic*, then Tomi [400]

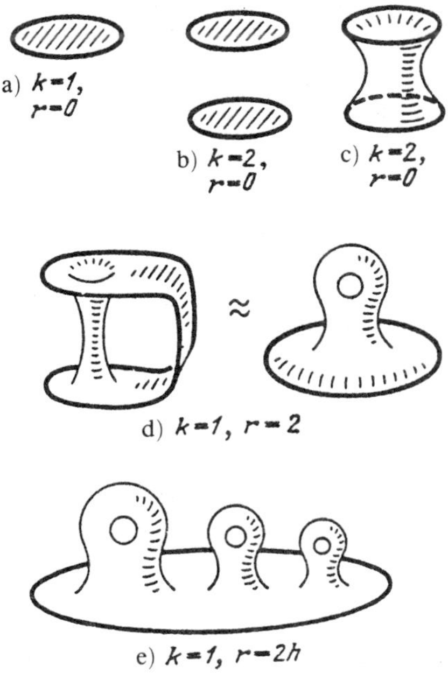

FIGURE 1.43

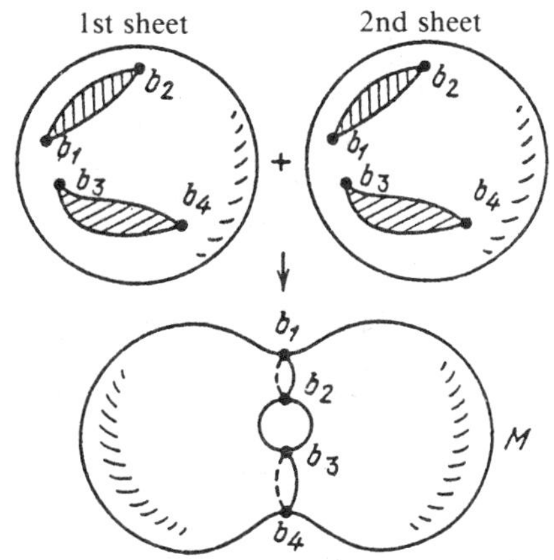

FIGURE 1.44

proved that it can bound only *finitely many* locally minimal discs. Simon and Hardt [385], [386] proved that for a $C^{4,\alpha}$-Jordan curve γ in $\mathbf{R}^3$ there are only finitely many two-dimensional rectifiable currents with boundary γ that absolutely minimize the area.

The existence of a minimal surface was also proved by Douglas [205] for *arbitrary topological type*.

We first give a precise formulation for the orientable case. Consider the class of two-dimensional orientable surfaces M_r^2 whose boundary consists of k oriented circles, $k \geq 1$. Let r be the topological characteristic of a connected surface M^2 with nonempty boundary ∂M^2, that is, r is twice the number of handles. Let us comment on the definition. We may also assume that r is the maximal number of linearly independent circles (one-dimensional cycles) on M, where a circle is taken to be zero if it splits the surface into two parts. If M is disconnected, then we define its topological characteristic as the sum of the characteristics of its connected components. In Figure 1.43 in case (a) we have $k = 1$, $r = 0$; in case (b) $k = 2$, $r = 0$; in case (c) $k = 2$, $r = 0$ (in contrast to case (b), the surface is connected here); in case (d) $k = 1$, $r = 2$; in case (e) $k = 1$, $r = 2h$, where h is the number of handles.

Similarly, calculating the number of Möbius films, we can define the

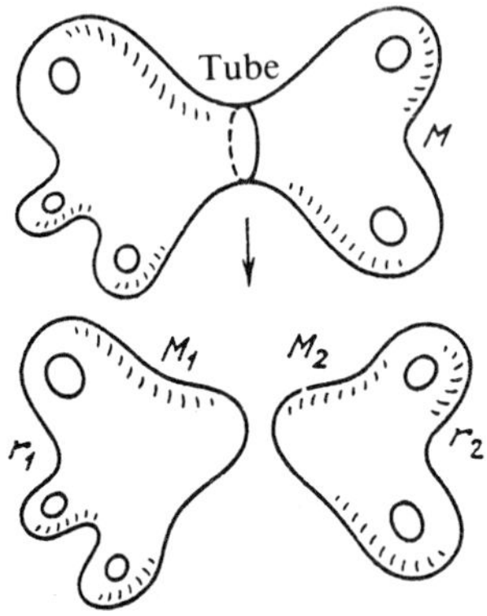

FIGURE 1.45

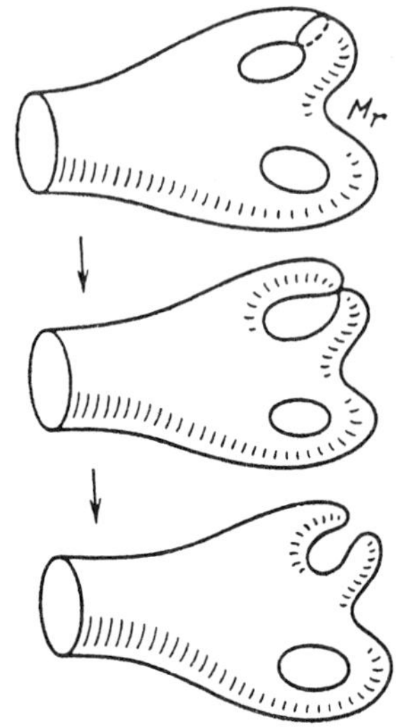

FIGURE 1.46

topological characteristic for nonorientable surfaces, see [205]. Moreover, Douglas defined an elementary reduction of the surface M_r. Each oriented two-dimensional closed surface is represented in $\mathbf{C}^2(w, z)$ as the Riemann surface of the algebraic function $w = \sqrt{P_n(z)}$, where $P_n = (z-b_1)\cdots(z-b_n)$ is a polynomial without multiple roots. The number h of handles of M_r is equal to $[\frac{1}{2}(n-1)]$ (the integer part). The singular points of the function $w = \sqrt{P_n}$ split into pairs. If $n = 2p$, then all the branch points are roots of P_n. If $n = 2p + 1$, then we need to add the point at infinity. Each pair of branch points determines one tube of the surface M_r (Figure 1.44). The number of tubes is one less than the number of handles. If two roots tend to each other and merge, then the tube corresponding to these points vanishes. The Riemann surface M turns into a new surface M'. The formal definition goes as follows. We shall say that M *has undergone an elementary reduction* if:

1. Two simple branch points merge. If this pair of points is the only pair at which two sheets of the Riemann surface are glued, then M' splits into two connected components $M' = M_1' + M_2'$, and the complete topological characteristic does not change, that is, $r = r' = r_1 + r_2$ (Figure 1.45). If M has a boundary consisting of k circles, then they are distributed between the two surfaces M_1' and M_2', that is, $k = k_1 + k_2$. If $k_2 = 0$, then M_2' is closed and it can be ignored. If M_2' is closed (that is, $k_2 = 0$), then we call the reduction $M \to M_1'$ *elementary* only when $r_2 = 2$ (that is, one handle is unfastened). If $r_2 = 0$ (a sphere is unfastened), then we shall

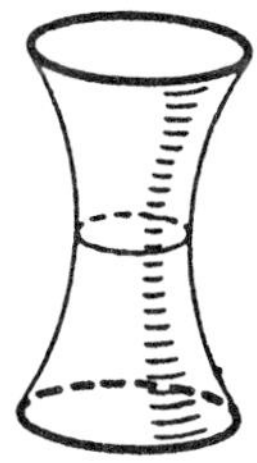

FIGURE 1.47

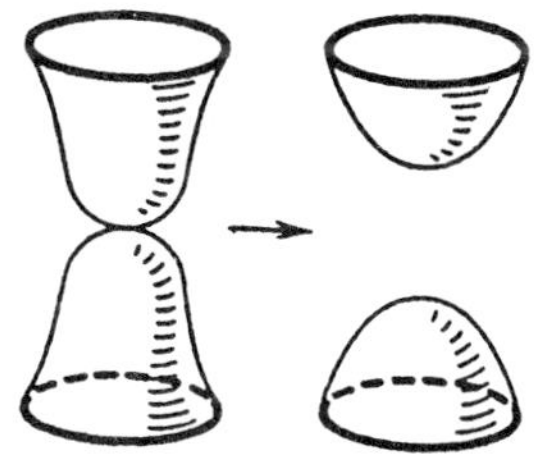

FIGURE 1.48

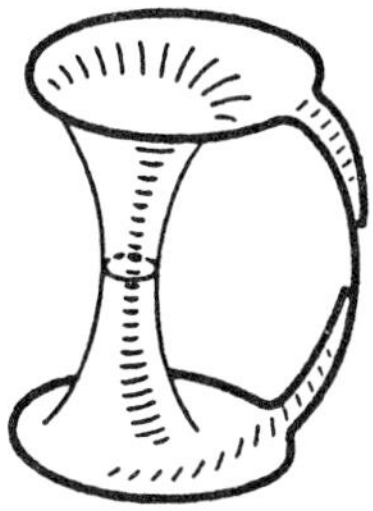

FIGURE 1.49

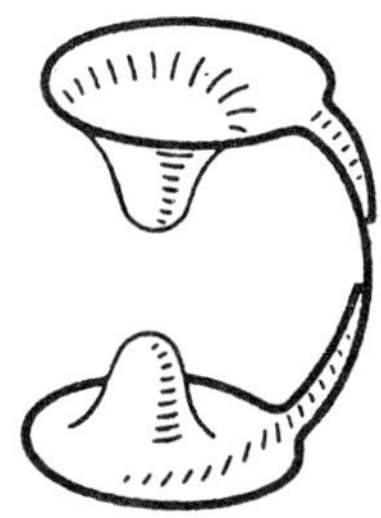

FIGURE 1.50

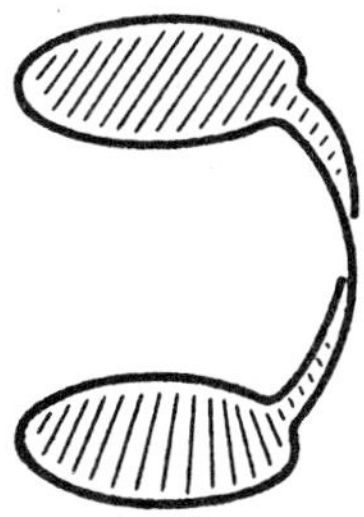

FIGURE 1.51

assume that no reduction has occurred. Ignoring M_2', we shall denote M_1' by M' and talk of an elementary reduction $M \to M'$.

2. Two simple branch points merge. If there are other branch points, determining tubes joining the same sheets as the tube corresponding to the vanishing pair of points, then the topological characteristic r is reduced by exactly 2, that is, $r' = r - 2$, where r' is the characteristic of M'.

Visually the operation of reduction looks like this. On the surface M_r (of characteristic r) we draw a circle corresponding to one of the generators of some tube. We pull the surface over this circle, that is, assemble it at a point in $\mathbf{R}^3$, and after this we separate the two sheets of the surface, that is, we cut it at the resulting singular point. Figure 1.46 shows the elementary reduction that destroys one handle. Figures 1.47 and 1.48 show the reduction of a cylinder to the disconnected union of two discs. Figures 1.49–1.51 show a torus with a hole reduced to a disc.

Let γ be a contour in $\mathbf{R}^n$ consisting of a finite number k of piecewise smooth circles. Consider a surface M_r of fixed characteristic r with boundary γ. We define the number $a(\gamma, r) = \inf \operatorname{vol}_2 M_r$. We consider the parametrized surfaces. We then consider the surfaces $M_{r'}'$ obtained from M_r by elementary reductions. We define $a(\gamma, r') = \inf \operatorname{vol}_2 M_{r'}'$. Clearly, $a(\gamma, r) \leq a(\gamma, r')$.

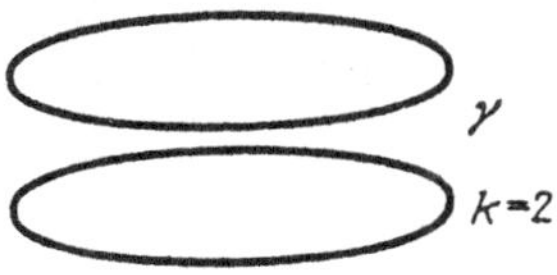

FIGURE 1.52

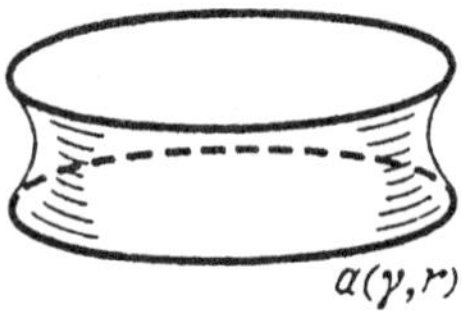

FIGURE 1.53

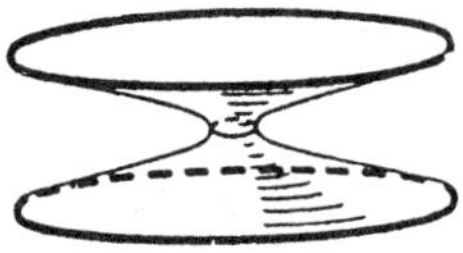

FIGURE 1.54

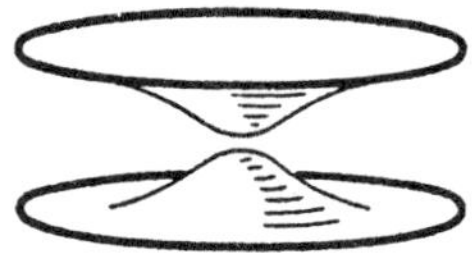

FIGURE 1.55

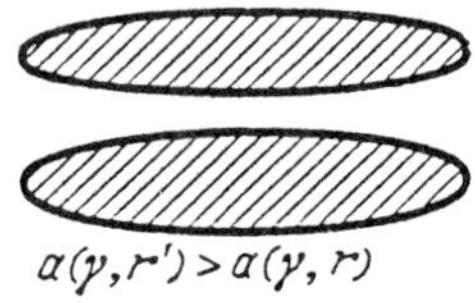

FIGURE 1.56

THEOREM 1.5.2 (Douglas [205]). *Suppose that the boundary contour* γ *is such that the number* $a(\gamma, r)$ *is finite and* $a(\gamma, r) < a(\gamma, r')$. *Then there is always a minimal surface* M_r *of topological characteristic* r *bounded by* γ. *The area of* M_r *is minimal in the class of all topologically equivalent surfaces with the same boundary. We have* $\mathrm{vol}_2 M_r = a(\gamma, r)$.

In the nonorientable case the theorem is stated in the same way except that the operation of removing a handle is replaced by the operation of removing a Möbius film. The condition $a(\gamma, r) < a(\gamma, r')$ is important. It means that the removal of handles (or tubes) from M_r strictly increases the minimum of the area of the surface. Figure 1.52 shows a contour consisting of two closely situated coaxial and parallel circles. Clearly, the minimum of the area in the class of surfaces of characteristic zero is attained by the catenoid (Figure 1.53). Its area is equal to $a(\gamma, r)$, where $r = 0$. Carrying out the reduction, we pull the catenoid over its mouth and convert it into two discs (Figures 1.54–1.56). The minimum of the area in the class of surfaces of this type is attained by the two flat discs, whose area is (in total) $a(\gamma, r')$. Clearly, $a(\gamma, r) < a(\gamma, r')$ here. The hypothesis of Theorem 1.5.2 is satisfied, and so *the minimum of the area in the class of connected surfaces of characteristic zero is attained by the catenoid*.

We now draw the bounding circles a long way apart (Figure 1.57). The minimum of the areas of connected surfaces of characteristic zero is attained by a surface consisting of two flat discs and the degenerate interval joining them. Here $a(\gamma, r) = a(\gamma, r')$. The hypothesis of Theorem 1.5.2 is not satisfied; in fact, in the class of cylinder-tubes there is no minimal regular (or generalized minimal) surface. The appearance of such "degenerate intervals", to which the part of the surface tending to minimize its

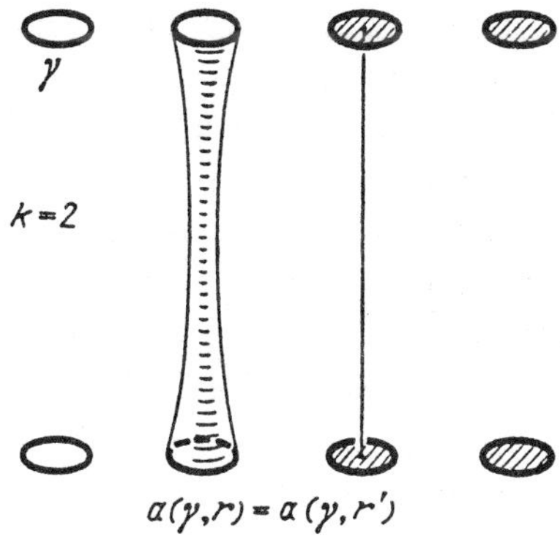

FIGURE 1.57

area is stuck, is a quite general situation. It is very difficult to wrestle with such zones of degeneracy. It is particularly difficult to do this in the *multidimensional Plateau problem*, where we run into not just one-dimensional intervals, but also stratified manifolds of low dimension. However, we can avoid this difficulty. A theorem on the existence of a multidimensional minimal surface in the class of surfaces parametrized by the spectra of manifolds with a fixed boundary was proved by Fomenko [126], [119], [129].

Douglas [205] also proved a remarkable theorem on the existence of a generalized-conformal map that minimizes the Dirichlet functional in the class of maps of surfaces of fixed topological type. Almgren and Thurston [155] constructed the following interesting example. For any positive integer g they constructed a smooth closed non-self-intersecting unknotted curve γ in $\mathbf{R}^3$ such that if M^2 is a surface smoothly embedded in $\mathbf{R}^3$, having γ as its boundary and situated inside the convex hull of the curve, then M has at least g handles (in particular, it cannot be a disc). Since a minimal surface is always contained in the convex hull of its boundary, the curve constructed by Almgren and Thurston cannot be the boundary of a minimal disc embedded in $\mathbf{R}^3$. The absolutely minimal Douglas disc has self-intersections here.

In conclusion we give a well-known theorem of Morrey. Let γ be a system $\gamma_1, \dots, \gamma_p$, where $\gamma_i \subset M^n$ are oriented closed Fréchet curves (parametrized), whose images in the smooth closed Riemannian manifold M^n we denote by $|\gamma_i|$. For example, we may assume that $\gamma_i(t)$ are piecewise smooth curves. For each $k = 0, 1, \dots$ we consider a domain G_k of class C^1 homeomorphic to a flat domain of type p, that is, a domain whose boundary is the disjoint union of p circles. We denote them by $S_{k1}, \dots, S_{kp}$, and specify on them an orientation compatible with the orientation of G_k. For each G_k there is specified a map $z_k: G_k \to M$, continuous on the closure $\overline{G}_k$. Suppose that the restrictions of z_k to each component of the boundary, that is, to S_{ki}, specify for each i the original representation of the curve $\gamma_i(t)$, that is, a map into M on $|\gamma_i|$. We define the number $a(\gamma) = \inf \lim_{k\to\infty} \inf A(G_k, z_k)$ for all sequences $\{z_k\}$, where $z_k \in C^0(\overline{G}_k)$. Here $\lim_{k\to\infty} \inf$ denotes the limit of the lower semicontinuous functional A (with respect to subsequences of $\{z_k\}$). We introduce the number $d(\gamma) = \inf\{\lim_{k\to\infty} \inf D(G_k, z_k)\}$ for all sequences $\{z_k\}$, where $z_k \in H_2^1(G_k)$. We set $d^*(\gamma) = +\infty$ in the case $p = 1$ and $d^*(\gamma) = \min \sum_{i=1}^m d(\gamma^{(j)})$, for all possible systems $\gamma^{(1)}, \dots, \gamma^{(m)}$, where each system $\gamma^{(j)}$ consists of curves γ_i for which $i \in T_j$, where $\bigcup_{j=1}^m T_j = \{1, 2, \dots, p\}$.

THEOREM 1.5.3 (Morrey [318]–[320]). *Let $\{\gamma_1, \dots, \gamma_p\} = \gamma$, where $p \geq 1$, be a set of oriented Fréchet curves for which $|\gamma_i|$ are nonintersecting Jordan curves, for example, piecewise smooth closed curves in M^n. Suppose that $d(\gamma) < d^*(\gamma)$ (the number $d(\gamma)$ is finite if $d^*(\gamma) = +\infty$). Then there is always a map $z: B \to M^n$, where B is a two-dimensional domain of type p (homeomorphic to a flat domain whose boundary is the union of p circles), $z \in C^0(\overline{B})$, and the map z is harmonic and conformal on B, and the restriction to each simple oriented closed boundary curve of B specifies the original representation of γ_i. The map z defines a surface of least area, that is, $A(B, z) = a(\gamma) = \mathrm{vol}_2\, z(B)$. Moreover, there is a map $z \in \overline{H}_2^1(\overline{B})$ such that $D(B, z) = d(\gamma)$ and $A(B, z) = a(\gamma)$, and the map is generalized-conformal and satisfies the same boundary conditions.*

We recall that there are rectifiable Jordan curves γ in $\mathbf{R}^3$ that bound an uncountable set of distinct minimal surfaces (P. Lévy [294]).

We give an example of Courant. Consider countably many Douglas contours (Figure 1.58). On each of them we can stretch two distinct minimal surfaces (Figures 1.59, 1.60). Let us denote them by 0 and 1. We take the connected sum of all the Douglas contours (Figure 1.61). As a result we obtain a connected rectifiable Jordan curve in $\mathbf{R}^3$ (homeomorphic to a circle). Clearly, we can stretch on it *very many* distinct minimal surfaces. In fact, for each small contour we can take one of the two minimal surfaces stretched on it, 0 or 1 (see Figure 1.62, for example). As a result we obtain a minimal surface uniquely determined by an infinite sequence of 0's and 1's. Clearly, to any sequence of 0's and 1's there corresponds a

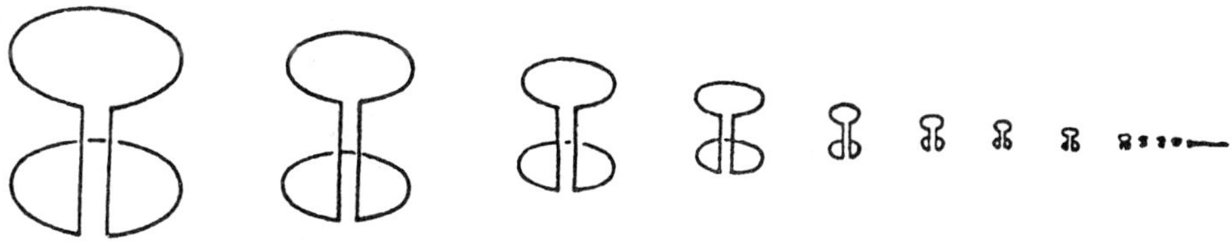

FIGURE 1.58

FIGURE 1.59

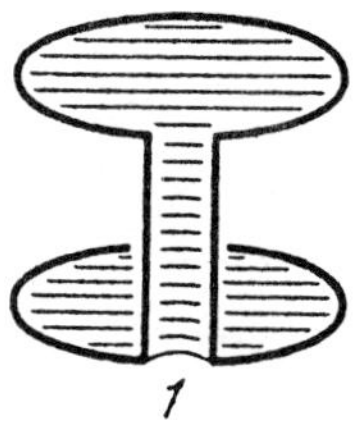

FIGURE 1.60

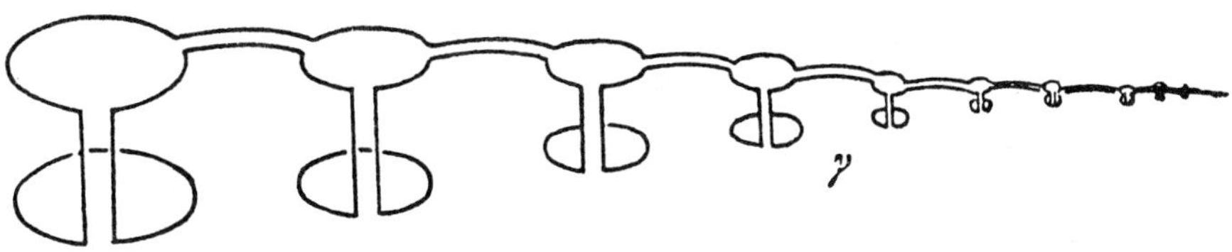

FIGURE 1.61

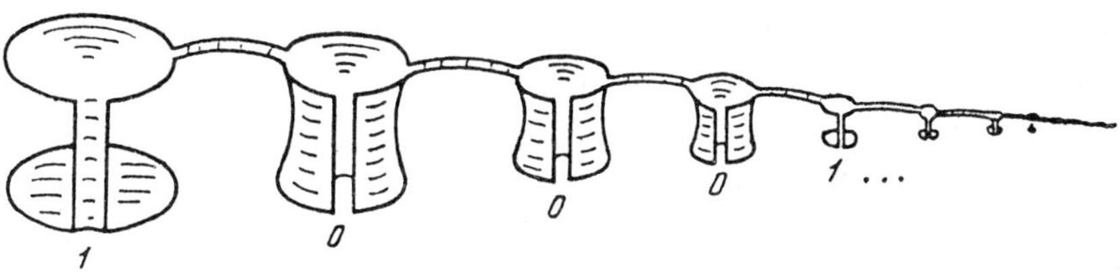

FIGURE 1.62

minimal surface. Since the set of all sequences of 0's and 1's is uncountable, we obtain *uncountably many distinct minimal surfaces with the same boundary*. True, this bounding contour has *one nonregular point*. All the surfaces we have constructed have finite area. This can be achieved by rapidly decreasing the size of the small contours from which the "large" contour is glued.

If we consider only surfaces that are *absolute minima* of the area functional, then from the results of Hardt and Simon it follows that there are

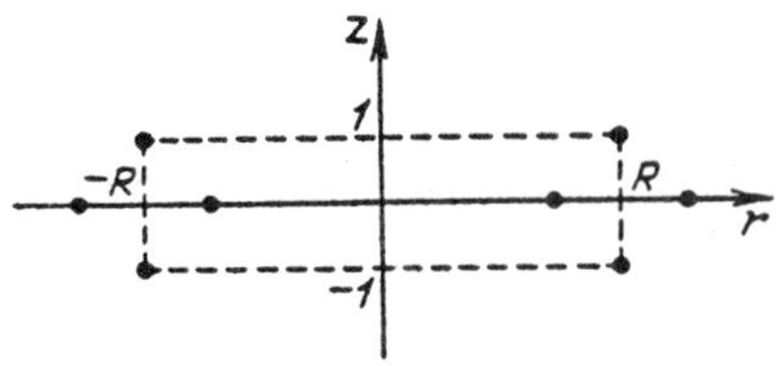

FIGURE 1.63

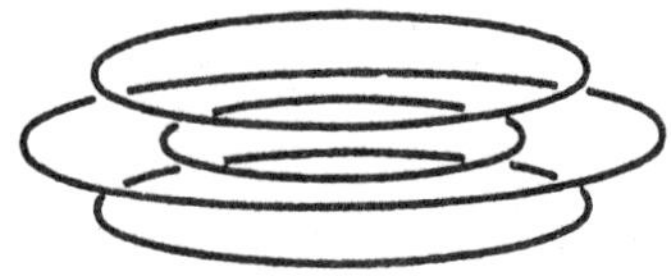

FIGURE 1.64

FIGURE 1.65

finitely many of them for sufficiently smooth curves. More precisely, *for a* $C^{4,\alpha}$*-Jordan curve* ($\alpha > 0$) *in* $\mathbf{R}^3$ *there are only finitely many surfaces spanning it that absolutely minimize the area.*

However, it is an open question whether a *smooth connected* curve in $\mathbf{R}^3$ can bound infinitely many minimal surfaces (not only absolute minima). For a *smooth disconnected* contour an example of this kind has been constructed (Morgan [315]).

Consider a contour B in $\mathbf{R}^3$ consisting of four circles. In cylindrical coordinates r, θ, z it is given by $z = 0$, $r = R + 1$, $R - 1$; $z = 1, -1$, $r = R > 20$ (Figures 1.63, 1.64).

Morgan proved that for any positive integer N there is a continuum of orientable minimal embedded manifolds with topological boundary B and Euler characteristic $-2N$. Let us describe the intuitive construction. Consider the contour B_1 in Figure 1.65. It is given by $z = 1$, $r = R$; $z = 0$, $r = R + 1$, $y \leq 0$; $z = 0$, $r = R - 1$, $y \geq 0$; $z = 0$, $R - 1 \leq |x| \leq R + 1$, $y = 0$. On it we can stretch a soap film of catenoid type (see Figure 1.65). Consider the contour B_2 centrally symmetric to this contour. On it we stretch a soap film centrally symmetric to the first. The union of the contours is $B_1 \cup B_2 = B \cup \{z = 0, y = 0, R - 1 \leq |x| \leq R + 1\}$. By the symmetry principle we can discard from the resulting contour the intervals $\{z = 0, y = 0, R - 1 \leq |x| \leq R + 1\}$ and the constructed film—the union of the two films in question—remains a minimal surface spanning the contour B. Since the contour is invariant with respect to a rotation around

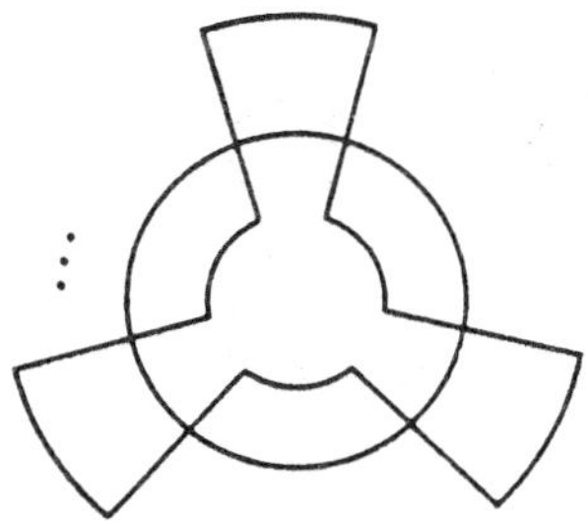

FIGURE 1.66

the z-axis, but the film is noninvariant, we have obtained infinitely many different minimal films spanning the given contour. To obtain a surface of large connectivity, we consider the map $p: \mathbf{R}^3 \to \mathbf{R}^3$, $\theta \to N\theta$, $N \in \mathbf{Z}^+$. Consider $p^{-1}(B)$. Its projection on the (x, y)-plane is shown in Figure 1.66. Similar actions lead to the construction of the required films.

There are many criteria for finiteness of the number of minimal surfaces. We give criteria for the finiteness of the number of relative minima (see Beeson [165]). Let γ be a real-analytic curve in $\mathbf{R}^3$ lying on the boundary of a convex body. Then γ bounds finitely many minimal surfaces—relative minima of the area.

A point $P \in \gamma$ is said to be *extremal* for γ if it lies on the boundary of the convex hull of γ. Let γ be a *real analytic* Jordan curve in $\mathbf{R}^3$. Suppose that each plane that contains a nonextremal point of γ intersects γ in at most eight points. Then *there are only finitely many relative minima of the area.*

No criterion for the finiteness of the number of solutions is known. Nevertheless, for almost all contours that are embedded circles and satisfy some restrictions on the total curvature there are only finitely many minimal surfaces (of class H^s under a restriction on the total curvature $\leq 2\pi(s-2)$) spanning the given contour. For surfaces of class C^∞ the result is true without restriction on the total curvature. This result was obtained by Böhme and Tromba [172], and also independently by Morgan. There are rectifiable Jordan curves γ for which the surface of least area with boundary γ must have infinite connectivity (Fleming [230]). In other words, *the absolute minimum of the area can be attained on a surface with infinitely many handles.* As the contour we can take a circle, which has finite length in $\mathbf{R}^3$. Fleming's example is in many respects analogous to Courant's example and is based on the infinite connected sum of the gradually decreasing Douglas contours. Consider the union of these contours in Figure 1.67. Then join these contours with narrow "bridges", as

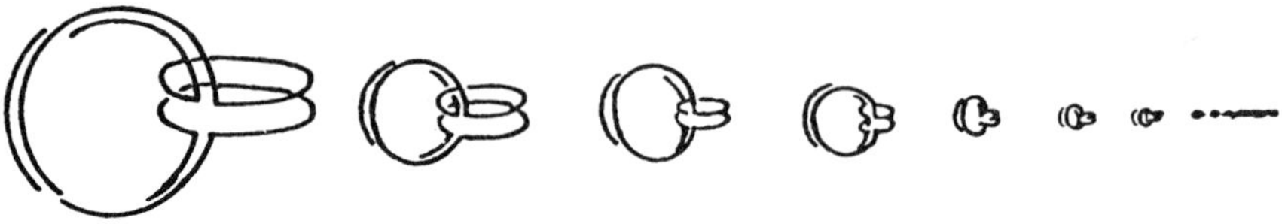

FIGURE 1.67

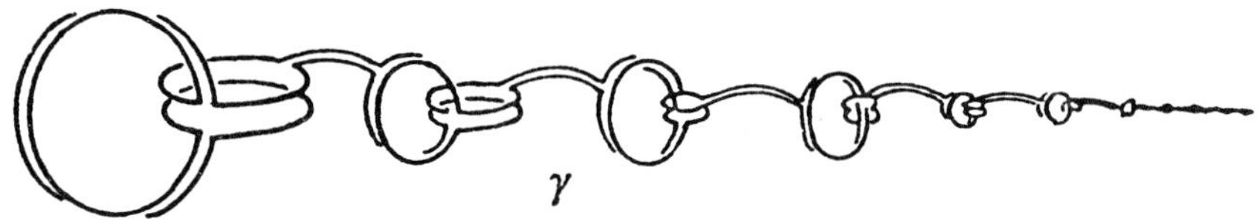

FIGURE 1.68

FIGURE 1.69

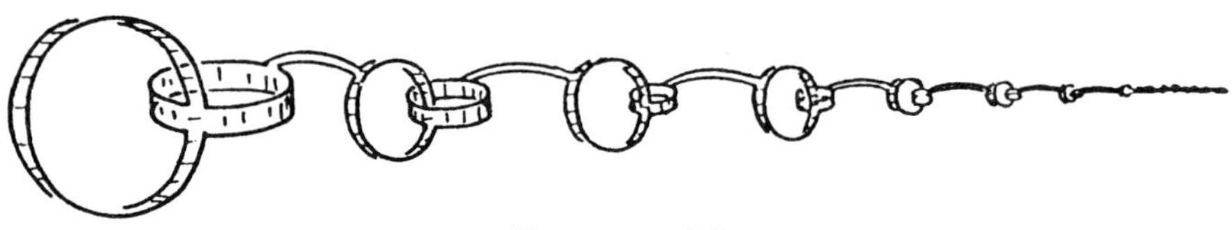

FIGURE 1.70

a result of which we obtain a Jordan curve (Figure 1.68). If we take the diameters of the annuli in the contours to be quite rapidly decreasing, then the constructed curve will have finite length.

On the Douglas contour there is stretched a minimal surface homeomorphic to a torus with a hole, or a disc with a handle (Figure 1.69). Joining these films by cross pieces (Figures 1.70–1.71), we obtain a minimal surface spanning γ and having the topological type of *a disc with countably many handles*. It is intuitively clear that if we take the annuli to be sufficiently narrow, then the surface M *realizes an absolute minimum of the area*, which was proved rigorously by Fleming.

We can sometimes prove the existence of a regular (and not generalized) minimal surface. Consider in $\mathbf{R}^n$ the graphs of the maps $f\colon B^2 \to \mathbf{R}^{n-2}$, where $B^2 \subset \mathbf{R}^2$.

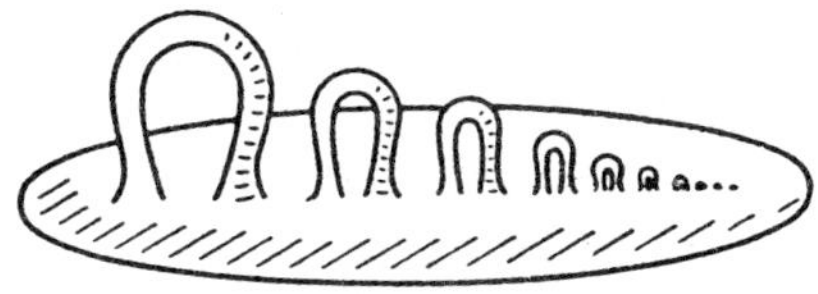

FIGURE 1.71

THEOREM 1.5.4 (see, for example, Osserman [77]). *Let* B^2 *be a convex bounded domain in the Euclidean plane of the variables* x_1, x_2 *(embedded in* $\mathbf{R}^n(x_1, \dots, x_n)$*) and* S^1 *its boundary. Let* $g_3(x_1, x_2), \dots, g_n(x_1, x_2)$ *be arbitrary continuous functions on the boundary* S^1 *(and mapping it into* $\mathbf{R}^{n-2}(x_3, \dots, x_n)$*). Then there is always a solution* $f(x_1, x_2) = (f_3(x_1, x_2), \dots, f_n(x_1, x_2))$ *of the equation of a minimal surface* (1) *in* B^2 *such that the functions* $f_k(x_1, x_2)$ *take given values* $g_k(x_1, x_2)$ *on the boundary of the domain. In other words, if a contour embedded in* $\mathbf{R}^n$ *and homeomorphic to a circle admits a one-to-one orthogonal projection onto the boundary of a convex domain in a two-dimensional plane* $\mathbf{R}^2 \subset \mathbf{R}^n$*, then there is always a regular minimal surface diffeomorphic to a disc that spans this contour and is smoothly embedded in* $\mathbf{R}^n$.

If a domain $B^2 \subset \mathbf{R}^2$ is *nonconvex*, then we have already given an example where *the equation of a minimal surface* (1) *may not have a solution in the class of graphs.*

Let $D \subset \mathbf{R}^2$ be a bounded flat domain. In order that there should exist in it a solution of the equation of minimal surfaces in $\mathbf{R}^3$ in the class of graphs (defined on D) that coincides on the boundary of the domain with an arbitrarily specified continuous function, it is necessary and sufficient that D should be convex (Finn [226]). We mention an important fact: *every minimal surface in* $\mathbf{R}^n$ *with a boundary is contained in the convex hull of its boundary values.*

5.5. Minimal surfaces in $\mathbf{R}^3$. The Gaussian map. Here we give results, true for $M^2 \subset \mathbf{R}^3$, whose generalizations to the case $\mathbf{R}^n$ are either false or need special investigation. We recall that a regular minimal surface specified in conformal coordinates determines analytic functions $\varphi_k(z)$ that satisfy the equation $\sum_{k=1}^{r} \varphi_k^2(z) = 0$. It turns out that for $n = 3$ we can describe all solutions of this equation explicitly.

PROPOSITION 1.5.5. *Let* $D \subset \mathbf{R}^2$ *be a domain in the complex* z*-plane and* $g(z)$ *an arbitrary function meromorphic in* D*. Suppose that the function* $f(z)$ *is analytic in* D*, and at the points where* $g(z)$ *has a pole of*

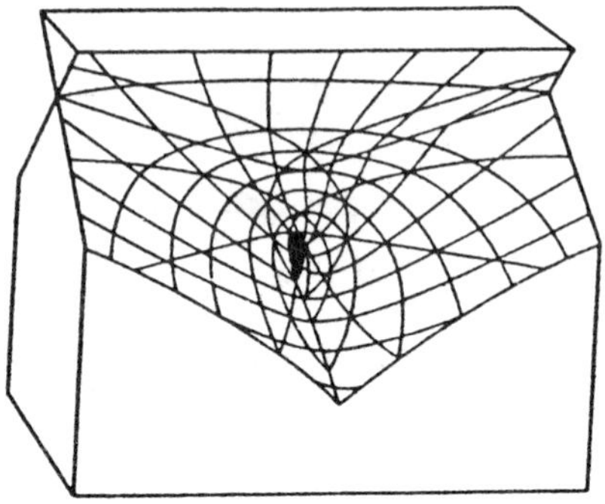

FIGURE 1.72

order m, $f(z)$ *has a zero of order at least* $2m$. *Then the functions*

$$\varphi_1 = \frac{1}{2}f(1-g^2), \qquad \varphi_2 = \frac{i}{2}f(1+g^2), \qquad \varphi_3 = fg \tag{5.4}$$

are analytic in D *and satisfy the equation* $\varphi_1^2+\varphi_2^2+\varphi_3^2=0$. *Conversely, any triple of functions analytic in* D *and satisfying the equation* $\varphi_1^2+\varphi_2^2+\varphi_3^2=0$ *can be represented in the form* (5.4) *except for the case where* $\varphi_1 \equiv i\varphi_2$ *and* $\varphi_3 = 0$.

The following important result of Enneper and Weierstrass gives a description of minimal surfaces in $\mathbf{R}^3$ and connects their properties with those of analytic functions.

THEOREM 1.5.5. *Every simply-connected minimal surface* M^2 *in* $\mathbf{R}^3$ *can be represented in the form* $(x_1(z), x_2(z), x_3(z)) = r(z)$, *where* $x_k(z) = \mathrm{Re}(\int_0^z \varphi_k(\zeta)d\zeta) + c_k$. *Here the functions* φ_k *are defined by means of the formulae* (5.4), *and the functions* f *and* g *in* (5.4) *have the properties listed in Proposition* 1.5.5. *For the domain* D (*where the function* $r(z)$ *is defined*) *we have to take the unit disc or the whole complex plane. The integral is taken along an arbitrary path joining the origin and the variable point. A minimal surface is regular if and only if* f *vanishes only at the poles of* g *and the order of each of its zeros is exactly equal to twice the order of the corresponding pole of* g.

We set $f \equiv 1$ and $g(z) = z$. Then all the given conditions are satisfied, and we obtain the *Enneper minimal surface* (Figure 1.72).

Generalized minimal surfaces M^2 in $\mathbf{R}^3$ can have self-intersections and branch points. Figure 1.39 shows a minimal surface whose boundary—a trifolium—is a knotted circle. The surface has self-intersections and a branch point. However, *it is unstable and collapses into another absolutely minimal surface that solves Plateau's problem for the given contour.* This new surface has self-intersections but no branch points. This is a reflection of a general fact. Osserman [341] stated that a minimal surface that gives (according to Rado and Douglas) a solution of the Plateau problem in the class of discs is regular at each of its interior points, that is, it has no branch points (but self-intersections are allowed). We need to distinguish "true branch points" and "false branch points". The latter are singularities of the map, and therefore under a suitable reparametrization, the image of the surface is regular. A true branch is a branch point of the image of the surface in $\mathbf{R}^3$. Osserman showed that in an arbitrarily small neighborhood of a true branch point of a minimal surface in $\mathbf{R}^3$ it can be replaced by another with the same topological type and the same boundary, but of strictly smaller area. Therefore *a globally minimal surface does not have branches.* Also, *absolutely minimal surfaces M^2 in $\mathbf{R}^3$ of arbitrary topological type do not have branch points* (although the map that specifies the surface may have singularities). Osserman's theorem was proved in full generality by Alt [158], [159] and Gulliver [247]. If a Jordan curve is *real analytic,* then Gulliver and Lesley [248] proved that *a solution of Plateau's problem does not have branch points even on the boundary.* Thus, in this case the solution of Plateau's problems is specified by an *immersed surface.* If the boundary curve is always only of class C^∞, then the existence of branch points on the boundary of the surface is an open question. For a generalization of these results to broad classes of surfaces defined by other variational problems, see the papers by Gulliver. For a deep investigation of properties of branch points of maps, see the paper by Gulliver, Osserman, and Royden [246].

Many papers have considered variational problems with "hard obstructions", that is, problems of minimizing the functional $F(u) = \int_G f(x, u, \nabla u)dx$ on a set of functions u fixed on the boundary of a domain $G \subset \mathbf{R}^n$ and satisfying in G the inequality $u \geq v$, where v is a fixed function. Deep theorems on the existence and uniqueness of the solution have been proved by Lewy and Stampacchia [296], Nitsche [335], Miranda [311], Giusti [241], and Giaquinta and Pepe [235]. O. T. Titov considered the following problem: in the class of all functions u that vanish on the boundary ∂G of a domain $G \subset \mathbf{R}^n$, $n \geq 2$, to minimize the area functional $F(u) = \int_G \sqrt{1 + |\nabla u|^2}dx$ if for a "soft obstruction" we take the condition $\int_D u dx \geq V$. Here D is an open set in G, $\overline{D} \subset G$, $V = \text{const} > 0$. Titov [109], [110] proved that for any sufficiently small V the problem has a unique solution and the extremal belongs to $C^{1,\alpha}(\overline{G})$ with Hölder exponent $\alpha \in (0, 1]$. This solution u is a classical minimal hypersurface over $G \backslash \overline{D}$ and a hypersurface of constant mean curvature over D, and they are smoothly glued onto ∂D. He gives examples to show that for large V the problem may not have a solution.

Sabitov [97] considered the problem of finding a minimal surface with semifree boundaries. He looked for a 2-connected minimal surface whose boundary must lie on given planes Π_1 and Π_2 provided that the spherical image of each boundary curve belongs to a given circle on the sphere. If Π_1 and Π_2 are parallel, he found necessary and sufficient conditions for the existence of the required surface. If Π_1 and Π_2 are not parallel, then under some geometrically simple condition he proved the absence of a solution. This problem is interesting as an example of a successful application in the theory of minimal surfaces of the methods of boundary-value problems for functions of a complex variable.

Minimal surfaces appear in very different branches of geometry that originally seem to be far from them. For example, in the theory of infinitesimal bendings, among the 12 *Darboux surfaces* that arise on bending a sphere there are two minimal surfaces, the so-called rotation diagram and displacement diagram. Conversely, with any given minimal surface X we can associate an infinitesimal bending of the sphere for which X is the rotation diagram. This fact was known to Liebmann (1920). For a new proof, see Sabitov [100]. Some new connections between minimal surfaces and the theory of infinitesimal bendings were established by Gavril′chenko [30].

Many properties of minimal surfaces in $\mathbf{R}^3$ are stated in terms of the *Gaussian map*. Considering the unit normal $n(P)$ to a surface at a point $P \in M^2$ and moving its origin by means of a parallel translation to the coordinate origin, we obtain the map $g: M^2 \to S^2$, called the *Gaussian (spherical) map*. If a map $r(z): D \to \mathbf{R}^3$ determines a regular minimal surface in conformal coordinates, then its Gaussian map $z \to r(z) \to n(z)$ determines a complex analytic transformation of D to S^2. Here S^2 is regarded as the augmented complex plane. In particular, *the Gaussian map of minimal surfaces in* $\mathbf{R}^3$ *is conformal*. If a surface is a generalized minimal surface, then the Gaussian map can nevertheless be defined by extending it continuously (and even analytically) to all the branch points.

PROPOSITION 1.5.6. *Let* M^2 *be a complete regular minimal surface in* $\mathbf{R}^3$. *Then either it is a plane or the unit normals to it form an everywhere dense set on the sphere, that is, the Gaussian map covers the sphere* S^2 *almost everywhere.*

This theorem with the additional assumption that M^2 is simply-connected was stated as a conjecture by Nirenberg and proved by Osserman. From this there follows a theorem of S. N. Bernstein. In fact, if a minimal surface $M^2 \subset \mathbf{R}^3$ is the graph of a smooth function $z = f(x, y)$ defined on the whole (x, y)-plane, then it is a complete regular minimal surface in $\mathbf{R}^3$, the unit normals to which are obviously contained in the hemisphere. From Proposition 1.5.6 it follows that it is a plane. Osserman conjectured that in fact the image of the Gaussian map of a complete minimal surface in $\mathbf{R}^3$ cannot omit more than 4 points on S^2. Recently Xavier proved that the image cannot omit more than 11 points, and Bombieri has lowered this number to 7.

In Proposition 1.5.6 it was a question of regular surfaces. If we consider *generalized surfaces* (see 5.3), then the assertion changes. There are complete generalized minimal surfaces in $\mathbf{R}^3$ not lying in a plane whose image under the Gaussian map lies in an arbitrarily small neighborhood of a point on the unit sphere; see [77].

Since $H = \lambda_1 + \lambda_2 = 0$ on a minimal surface, the principal curvatures λ_1 and λ_2 have opposite signs (or they are both zero). Therefore *the Gaussian curvature $K = \lambda_1\lambda_2$ is always nonpositive. If a minimal surface is not a plane, then the Gaussian curvature can have only isolated zeros.* Many results proved for surfaces of *negative Gaussian curvature in* $\mathbf{R}^3$ can be applied to minimal surfaces. See, for example, the papers of Efimov [50], [51], Efimov and Poznyak [52], and Rozendorn [93], [94].

5.6. Minimal surfaces in Riemannian space. The metric induced on a two-dimensional minimal surface by its embedding in $\mathbf{R}^3$ is not arbitrary. Firstly, *the asymptotic lines on a minimal surface are always orthogonal at their intersection points.* We recall that for an asymptotic direction, the normal curvature is zero. Moreover, if K is the Gaussian curvature of a minimal surface, then $K \leq 0$ and, by Ricci condition, ds^2 specifies the metric of a minimal surface $M^2 \subset \mathbf{R}^3$ if and only if the metric $\sqrt{-K}ds^2$ has zero curvature. The generalized Ricci conditions were obtained by Calabi [179], [180], and Matsumoto Makato [300]. For the metrics of minimal surfaces in $\mathbf{R}^n$ see also the papers of Osserman and Chern. Powerful generalizations were then obtained by Aminov [4]. Aminov [7], [9] considered the question of the instability of closed minimal surfaces in a Riemannian space of positive curvature. Here we have in mind the two-dimensional sectional curvature, that is, the curvature in two-dimensional directions (elements of area).

THEOREM 1.5.6 ([7], [9]). *Let M^2 be a minimal surface homeomorphic to a sphere in a complete simply-connected Riemannian space M^n whose sectional curvature K satisfies the inequalities $1/4 < K \leq 1$. Then M^2 is an unstable minimal surface.*

If the condition that M^n is complete and simply-connected is replaced by the condition that the normal bundle of M^2 is trivial or the condition $n = 4$, then Theorem 1.5.6 remains true. The theorem is exact, since in complex projective space $\mathbf{C}P^2$ with the Fubini metric, whose curvature lies in the interval $[1/4, 1]$, there are totally geodesic surfaces homeomorphic to a sphere and giving a minimum of the area among closed surfaces close to them.

THEOREM 1.5.7. *Let M^2 be a minimal surface homeomorphic to a sphere and situated in an n-dimensional orientable locally conformally flat space M^n of positive sectional curvature. Suppose that the normal bundle on M^2 is trivial. Then M^2 is an unstable surface.*

Aminov [9] introduced a local invariant of a minimal surface in a four-dimensional Riemannian space that plays a definite role in its stability. The Ricci condition on the metric of a minimal surface, described above, is expressed as a differential equation which is satisfied by the function $K : \nabla_2 \ln \sqrt[4]{-K} = K$, where ∇_2 is the Laplace-Beltrami operator. The following theorem was proved in [4].

THEOREM 1.5.8. *If K is the Gaussian curvature of the metric of a minimal surface M^2 in a four-dimensional space of constant curvature K_0, then $\nabla_2 \ln \sqrt[4]{K_0 - K} - \frac{3}{2}K \geq -\frac{1}{2}K_0$.*

In particular, when $M^2 \subset \mathbf{R}^4$, this means that the Gaussian curvature of the metric $\sqrt[3]{-K}ds^2$ is nonpositive. If this curvature is zero, then ds^2 is immersed in $\mathbf{R}^4$ as a minimal surface whose indicatrix of normal curvature is a circle. Metrics of minimal surfaces in $\mathbf{R}^4$ with a circle as indicatrix of the normal curvature were also considered by Lumiste [299]. Aminov [4] considered an analogue of the Ricci condition for a hypersurface in a Riemannian space.

THEOREM 1.5.9. *Let M^2 be a minimal hypersurface in a space of constant curvature K_0. Then for its scalar curvature R, its Riemann tensor R_{ijkl} and its Ricci tensor R_{ij} we have the inequality*

$$\nabla_2 R + R_{ijkl}R^{ijkl} + 2R_{ij}R^{ij} + 2K_0 R(n-1) \geq 0.$$

There is an interesting question concerning *the outer diameter of a minimal submanifold*. S. S. Chern posed the following question: *is a minimal submanifold in $\mathbf{R}^n$, complete in the intrinsic metric, unbounded in space?*

Aminov [10] obtained a general estimate for the radius of a ball in $\mathbf{R}^n$ that contains a submanifold with mean curvature vector bounded in magnitude and with Ricci curvature bounded below. From this estimate it follows that *a complete minimal surface in* $\mathbf{R}^n$ *with Gaussian curvature bounded below is unbounded in space.* The condition that the Gaussian curvature is bounded is essential here, because Jorge and Xavier have constructed an example of a complete minimal surface in $\mathbf{R}^4$ that is bounded in space.

THEOREM 1.5.10 (see [11]). *A complete minimal surface with integral Gaussian curvature bounded below in an n-dimensional simply-connected Riemannian space of nonpositive curvature is unbounded in space.*

We recall that the *total curvature of a surface* M^2 is the integral (over the surface) of the Gaussian curvature. *This number* (with the opposite sign) *is equal to the area of the surface that is the image of* M^2 *under the Gaussian map in* $\mathbf{R}^3$ (see [48], for example). If the Gaussian map is not one-to-one, then the total curvature of the surface is equal to the total area of the image, taking account of multiplicity.

PROPOSITION 1.5.7. *Let* M^2 *be a complete minimal surface in* $\mathbf{R}^3$. *Then its total curvature can take only values equal to* $4\pi m$, *where* m *is a nonpositive integer or* $-\infty$. *In addition, the total curvature never exceeds* $2\pi(\chi - k)$, *where* χ *is the Euler characteristic of the surface, and* k *is the number of connected components of the surface boundary. There are only two complete regular minimal surfaces in* $\mathbf{R}^3$ *whose total curvature is* -4π. *These are the catenoid and the Enneper surface. Of all complete regular minimal surfaces only for the catenoid and the Enneper surface is the Gaussian map one-to-one.*

For a proof and discussion, see [77]. See the same survey for results (up to 1966) on nonparametrized minimal surfaces, on the external Dirichlet problem, on singularities of minimal surfaces, and on properties of the generalized Gaussian map.

Masal′tsev [66] described all two-dimensional minimal surfaces in $\mathbf{R}^5$ whose Gaussian image in the Grassmann manifold $G_{5,2}$ has constant intrinsic curvature k. It was established earlier by other authors that the values of k can only be 1, 2, and $1/2$. Depending on the value of k, the surface lies either in $\mathbf{R}^3 \subset \mathbf{R}^5$ (for $k = 1$), or in $\mathbf{R}^4 \subset \mathbf{R}^5$ (for $k = 2$), or essentially in $\mathbf{R}^5$ (for $k = 1/2$), and in all cases it is possible to give a complete description of the corresponding minimal surfaces. The results essentially rely on the deep work of Hoffman, Osserman, Lawson, and Calabi.

It is very important to find conditions on the curve γ under which a minimal surface M^2 in M^n bounded by γ and solving Plateau's problem (according to Douglas, Rado, and Morrey) is an *embedded surface.* Osserman conjectured that this is true in the case where $\gamma \subset \mathbf{R}^3$ and *lies on the boundary of a convex domain.* Important progress in the solution of the problem was made by Gulliver and Spruck [245], Tomi and Tromba [401], and Almgren and Simon [156]. Almgren and Simon proved that there is always an embedded minimal surface M^2 in a uniformly convex open set $A \subset \mathbf{R}^3$, where ∂A is a C^2 surface, $\partial M^2 = \gamma \subset \partial A$, and γ is a Jordan C^3 curve (a circle) in ∂A. The surface $M^2 \subset \overline{A}$ is diffeomorphic to a disc D^2 and minimizes the area in the class of all surfaces of the form $f(D^2)$, where $f\colon D^2 \to f(D^2)$ is a diffeomorphism and $\partial(f(D^2)) = \gamma$.

A definitive solution of Osserman's problem for arbitrary three-dimensional convex manifolds was obtained by Meeks and Yau [302]. Moreover, these authors recently proved some interesting results about two-dimensional surfaces in three-dimensional manifolds, generalizing Osserman's problem (Meeks and Yau [304]). Consider an M^3 with boundary ∂M, which may be nonsmooth. We shall assume that M^3 is a compact domain in another smooth manifold $\overline{M}^3$, where for a suitable choice of curvilinear triangulation on $\overline{M}$ the boundary ∂M is a two-dimensional subcomplex in $\overline{M}$ consisting of smooth two-dimensional simplexes $\sigma_1, \dots, \sigma_l$ with the following properties:

(1) each simplex σ_i is a C^2 surface in $\overline{M}$ whose mean curvature with respect to the outward normal is nonnegative;

(2) each surface σ_i is a compact subset of some smooth surface $\overline{\sigma}_i$ in $\overline{M}$, where $\overline{\sigma}_i \cap M = \sigma_i$ and $\partial\overline{\sigma}_i \subset \partial\overline{M}$.

THEOREM 1.5.11 (Meeks and Yau [304]). *Let M be a three-dimensional compact manifold with piecewise smooth boundary properties* (1) *and* (2) (*see above*). *Let γ be a Jordan curve in the boundary ∂M and suppose that γ is contractible in M. Then there is a ramified minimal immersion of the disc D^2 in M^3 with boundary γ, smooth on the interior of D^2 and having minimal area among all such immersions. Moreover, any ramified minimal immersion of the form mentioned above is necessarily an embedding.*

In particular, we may assume that the boundary of M^3 is convex, or (more generally) is a smooth two-dimensional manifold with nonnegative mean curvature. The following assertion was also proved in [304]. Let M be a three-dimensional compact manifold with piecewise smooth boundary satisfying conditions (1) and (2) above. Let $f\colon N \to M$ be any ramified conformal minimal immersion of the compact surface N in M such that $f(\partial N) \subset \partial M$. Then either $f(N) \subset M$, or $f(N) \cap \partial M = f(\partial N)$. Moreover, the image $f(N)$ does not have branch points on the boundary. Meeks, Simon, and Yau [303] proved that if $N^2 \subset M^3$ is a closed smooth two-dimensional incompressible surface embedded in a three-dimensional irreducible compact complete (not necessarily orientable) manifold, then it is always isotopic to an embedded incompressible surface of least area. This result enabled them to prove that *a covering over any irreducible orientable three-dimensional manifold is always irreducible.* In addition, in [303] they studied the topology of compact three-dimensional manifolds with nonnegative Ricci curvature. All these manifolds have been classified except for the case when the manifold was covered by an irreducible homotopy sphere. The following theorem was proved in [302] (the *generalized Dehn lemma*). It is a *definitive solution of Osserman's problem.* Let M^3 be a three-dimensional convex manifold. Let γ be a Jordan curve (a circle) on the boundary ∂M contractible in M. Then (1) for the contour γ there is always a solution of Plateau's problem with finite area; (2) any solution of Plateau's problem is an embedded surface; (3) for any two solutions of Plateau's problem for γ, either they differ from each other only by a conformal reparametrization, or the images of the surfaces intersect only along the curve γ.

Let G be a compact Lie group, and $\mathfrak{g}$ its Lie algebra, on which G acts adjointly. From each point of the Lie algebra there "grows" an orbit of action of the group. The following problem arises: let V be an orbit of maximal volume of the adjoint action of G on a sphere of radius R in the Lie algebra; it is required to establish when the cone over V is a locally minimal surface in the class of all surfaces with boundary V. As we shall show below, this problem is closely connected with the problem of minimal cones and with Bernstein's problem. We assume that the Lie algebra is endowed with the invariant Killing metric. It follows that a Euclidean sphere of radius R in the Lie algebra goes into itself by the adjoint representation of the group. Orbits of maximal volume are locally minimal submanifolds of this sphere.

THEOREM (Balinskaya-Novikova [424]). *For each of the compact classical Lie groups $SO(n)$, $SU(n)$, $Sp(n)$ with $n \le 8$, $n \le 4$, $n \le 3$ respectively, the cones with vertex at the origin over the orbit of maximal volume are locally minimal, but unstable. For the remaining n the corresponding cones are locally stable, in particular, locally minimal.*

CHAPTER II

Information about Some Topological Facts Used in the Modern Theory of Minimal Surfaces

§1. Groups of singular and cellular homology

Results obtained in the last few years in connection with the solution of topological variational problems show the important role played in them by *homology groups*. We therefore recall briefly the information we need about these groups.

Consider the space $\mathbf{R}^{k+1}$ with Cartesian coordinates $x_1, \dots, x_{k+1}$ and a standard simplex Δ^k of dimension k defined thus: $x_1 + \cdots + x_{k+1} = 1$, $x_i \geq 0$.

DEFINITION 2.1.1. A *singular simplex* f^k of dimension k of a topological space X is a continuous map f of a standard simplex Δ^k in X. An *integral k-dimensional singular chain* c of the space X is a formal linear combination of singular simplexes f^k of X with integer coefficients, only finitely many of which are nonzero: $c = \sum_i a_i f_i^k$. The set of all k-dimensional chains obviously becomes an Abelian group (with respect to addition) $C_k(X)$, called the *group of k-dimensional chains of the space* X. The *boundary operator* $\partial_k \colon C_k(X) \to C_{k-1}(X)$, is a homomorphism defined on the generators of the group as follows: $\partial_k f^k = \sum_{i=0}^k (-1)^i f_i^{k-1}$, where f_i^{k-1} is defined as follows. Consider a standard simplex Δ^{k-1} and for each face with number i of Δ^k we fix a standard embedding α_i of the simplex Δ^{k-1} into Δ^k onto this ith face. Then we set $f_i^{k-1} = f \circ \alpha_i$.

Clearly, $\partial_k \partial_{k+1} \equiv 0$, so $\operatorname{Ker} \partial_k \supset \operatorname{Im} \partial_{k+1}$.

DEFINITION 2.1.2. The *k-dimensional singular homology group* $H_k(X)$ of a space X is the factor group $\operatorname{Ker} \partial_k / \operatorname{Im} \partial_{k+1}$. The elements of the subgroup $\operatorname{Ker} \partial_k$ are called *cycles*, and the elements of the subgroup $\operatorname{Im} \partial_{k+1}$ are called *boundaries*.

From this definition it easily follows that *singular homology groups* are topologically and homotopically invariant, that is, they do not change under a homeomorphism of the space or under a replacement of it by a homotopically equivalent space.

Let Y be a closed subspace of X. Then we can consider the *group of relative chains* $C_k(X, Y) = C_k(X)/C_k(Y)$. Clearly, the boundary operator induces the operator $\partial\colon C_k(X, Y) \to C_{k-1}(X, Y)$. Consequently, we have defined the groups $H_k(X, Y) = \operatorname{Ker}\partial/\operatorname{Im}\partial$, called the *relative homology groups of the space* X *modulo the subspace* Y.

A space X is called a *cell complex* if it is represented as the union of disjoint subsets σ^k called *cells* and having the following properties. The closure $\overline{\sigma}^k$ is the image of a closed k-dimensional disc D^k under a continuous (*characteristic*) map that is a homeomorphism on the interior of the disc. The boundary of each cell is contained in the union of finitely many cells of lower dimension. A subset of X is closed if and only if all its complete inverse images (under characteristic maps) of intersections with cells are closed. In particular, a characteristic map introduces an *orientation* in a cell. A complex is said to be *finite* if it consists of finitely many cells. The union of cells of dimension at most k is called the *k-dimensional skeleton of the complex.*

Consider the group $P_k(X)$ whose elements are identified with linear combinations of the form $\sum a_i\sigma_i^k$, where σ_i^k are k-dimensional cells of the complex, and a_i are elements of an Abelian group G. The group $P_k(X)$ is called the *group of cellular k-dimensional chains* (with coefficients in G). Let σ^k and σ^{k-1} be two cells, and X^k and X^{k-1} respectively the k-dimensional and $(k-1)$-dimensional skeletons of the complex. Let $\chi\colon D^k \to X$ be the characteristic map of the cell σ^k. Consider the map $\partial D^k = S^{k-1} \xrightarrow{\chi} X^{k-1}/X^{k-2}$, that is, we map the boundary of the cell into the quotient space X^{k-1}/X^{k-2}, which is homeomorphic to a "bouquet" of $(k-1)$-dimensional spheres. In this bouquet the cell σ^{k-1} defines a sphere S_0^{k-1}. We map the whole bouquet onto this sphere, leaving it fixed, and mapping all the remaining spheres of the bouquet to a point. Let $[\sigma^k : \sigma^{k-1}]$ denote the degree of the compositioin map of the sphere S^{k-1} into the sphere S_0^{k-1}. This number is called the *incidence number* of the cells σ^k and σ^{k-1}.

Let σ^k be an arbitrary generator of the group $P_k(X)$. We define the boundary operator ∂_k by the formula $\partial\sigma^k = \sum[\sigma^k : \sigma^{k-1}]\sigma^{k-1}$ where the sum is taken over all $(k-1)$-dimensional cells of X.

DEFINITION 2.1.3. Let $P_k(X)$ be the group of cellular chains of the complex X, and $\partial_k: P_k(X) \to P_{k-1}(X)$ the boundary homomorphisms (operators) defined above. The groups $\operatorname{Ker}\partial_k/\operatorname{Im}\partial_{k+1}$ are called the *cellular homology groups* of X.

It turns out that *the singular and cellular homology groups of a finite cell complex are always isomorphic.* Sometimes the group of coefficients of G is stated explicitly: $H_k(X, G)$.

§2. Cohomology groups

DEFINITION 2.2.1. A *singular cochain* of a space X with coefficients in a group G is a homomorphism of the group of chains $C_k(X)$ into the group G. The set of cochains naturally becomes an Abelian group $C^k(X, G)$, called the *group of cochains.*

Let $h \in C^{k-1}(X)$ be an arbitrary cochain. Then there is defined a cochain $\delta h: C_k(X) \to G$, given by the formula $(\delta h)(\alpha) = h(\partial\alpha)$, that is, $\delta h \in C^k(X)$. We shall sometimes denote the operator $\delta: C^{k-1} \to C^k$ by δ_{k-1}.

DEFINITION 2.2.2. The operator δ_{k-1} is called the *coboundary operator.* It is adjoint to the operator ∂.

Since $\partial^2 = 0$, it follows that $\delta^2 = 0$. Consequently, we obtain a sequence of groups and homomorphisms of the following form: $C^0 \xrightarrow{\delta_0} C^1 \xrightarrow{\delta_1} C^2 \to \cdots$, where $\delta_k\delta_{k-1} = 0$. This sequence is called the *cochain complex.* Proceeding on the lines of the previous section, we consider the groups $\operatorname{Ker}\delta$ and $\operatorname{Im}\delta$, and we construct the group $H^k(X, G) = \operatorname{Ker}\delta_k/\operatorname{Im}\delta_{k-1}$.

DEFINITION 2.2.3. The groups $H^k(X, G)$ are called the *cohomology groups* of the space X with coefficients in an Abelian group G. The elements of the group $\operatorname{Im}\delta_{k-1}$ are called *coboundaries,* and the elements of the group $\operatorname{Ker}\delta_k$ are called *cocycles.*

As in the homology case, it is natural to define the relative cohomology groups. Let Y be a closed subcomplex of X; then $C_k(Y) \subset C_k(X)$. Let $C^k(X, Y)$ be the group of all homomorphisms $\alpha: C_k(X) \to G$ that are equal to zero on $C_k(Y)$. Clearly, $\delta C^k(X, Y) \subset C^{k+1}(X, Y)$, and there arise the groups $H^k(X, Y; G) = \operatorname{Ker}\delta/\operatorname{Im}\delta$, which are called the *relative cohomology groups.* As in the homology case, there naturally arises a *pair exact sequence.* Consider the embeddings $i: Y \to X$ and $j: X \to (X, Y)$. Let i_* and j_* be the homomorphisms of homology groups induced by them, and i^* and j^* the homomorphisms of cohomology groups induced by them. Then the following two sequences of groups and homomorphisms are exact, that is, the image of an entry homomorphism is equal to the kernel of an exit homomorphism (in the same place in the sequence):

$$\cdots \to H_k(Y) \xrightarrow{i_*} H_k(X) \xrightarrow{j_*} H_k(X, Y) \xrightarrow{\partial} H_{k-1}(Y) \to \cdots$$

$$\cdots \leftarrow H^k(Y) \xleftarrow{i^*} H^k(X) \xleftarrow{j^*} H^k(X, Y) \xleftarrow{\partial} H^{k-1}(Y) \leftarrow \cdots$$

Singular cohomology groups are homotopically invariant. Proceeding along the lines of §1, we define the *cellular cohomology groups.* For this we introduce the *cellular cochain groups* $P^k(X)$, defined as the groups $H^k(X^k, X^{k-1})$, where X^k is the k-dimensional skeleton of the cell complex X. The operator $\partial: P_k(X) \to P_{k-1}(X)$ induces the coboundary operator $\delta: P^{k-1}(X) \to P^k(X)$. The cellular cohomology groups are defined as the groups $\operatorname{Ker}\delta/\operatorname{Im}\delta$ for the cochain complex $\{P^k(X), \delta\}$.

THEOREM 2.2.1. *For a finite cell complex X, the singular (co)homology groups and the cellular (co)homology groups are isomorphic and are finitely generated Abelian groups if the group of coefficients is so generated.*

To study topological variational problems, for example to solve the classical Plateau problem, we sometimes need to solve the following problem.

Let Y be a closed subspace of a topological space X and suppose we are given a continuous map of this subspace into some space Z. In what cases can this map be extended to a continuous map of the whole space X into Z? Clearly, such a continuous extension does not always exist. There are topological obstructions that do not allow us in certain cases to extend the map to the whole space. By a study of all the various versions of this question we understand the *theory of obstructions*—see [137]. We also need the following *Hopf theorem.*

As the group of coefficients of the homology theory we consider the Abelian group (with respect to addition) $U = \mathbf{R}^1 \pmod 1$, that is, a circle.

THEOREM 2.2.2 (Hopf). *Suppose we are given an embedding i of the sphere S^{n-1} into a finite n-dimensional cell complex K. Suppose that the homomorphism $i_*: H_{n-1}(S^{n-1}, U) \to H_{n-1}(K, U)$ induced by this embedding is a monomorphism. Then the sphere $S_0^{n-1} = i(S^{n-1})$ is a retract of K, that is, there is a continuous map $f: K \to S_0^{n-1}$ that is the identity on the sphere S_0^{n-1}.*

The hypothesis of the Hopf theorem can be restated in terms of cohomology. We have to require that the homomorphism $H^{n-1}(K, \mathbf{Z}) \to H^{n-1}(S^{n-1}, \mathbf{Z})$ is an *epimorphism.*

CHAPTER III

The Modern State of the Theory of Minimal Surfaces

§1. Minimal surfaces and homology

1.1. The second quadratic form of a submanifold of Riemannian space. Let $i: M^k \to W^n$ be a smooth embedding of a smooth manifold M into a smooth Riemannian manifold W. We assume that W is orientable, connected, and without boundary. Let TM denote the tangent bundle of M, and T_mM the tangent plane to M at the point m. Let $\langle x, y\rangle$ denote the scalar product of a pair of vectors $x, y \in T_mM$ induced by the Riemannian metric on W. Let ∇ denote the *Riemannian connection* on TM. This connection (sometimes called the *Levi-Civita connection*) is *symmetric* and with respect to it *the Riemannian metric tensor is covariantly constant* (the covariant derivative is zero).

Let $\nabla_X P$ denote the covariant derivative of the tensor field P along a given vector field X on M. If x denotes the value of the vector field X at the point m (that is, the vector $x = X(m)$ belongs to T_mM), then we denote the covariant derivative of the tensor field P along the vector x (at the point m) by $\nabla_x P$. For $M \subset W$ we define the normal bundle NM. Namely, at each point $m \in M$ we define the plane $N_m^{n-k}M$ orthogonal to T_m^kM. There arise two natural Riemannian connections, induced on TM and NM.

DEFINITION 3.1.1. Let Y be a smooth vector field on M and $x \in T_mM$ an arbitrary vector. We define $\nabla_x Y = (\overline{\nabla}_x Y)^T$, where $\overline{\nabla}$ denotes the Riemannian connection on the ambient manifold W, and $(\ )^T$ is the orthogonal projection on the plane T_mM.

It turns out that this operation ∇ is a Riemannian connection on TM; see [387].

In exactly the same way we define a Riemannian connection on the normal bundle NM. Consider an arbitrary smooth section V of the bundle NM, that is, at each point $m \in M$ we specify a normal vector

$V(m)$. We obtain a vector field K on M. Let $x \in T_mM$. By definition we set $\nabla_x V = (\overline{\nabla}_x K)^N$, where $(\ )^N$ is the orthogonal projection on the plane N_mM.

DEFINITION 3.1.2. Let $v \in N_mM$; we include the vector v in an arbitrary smooth vector field V on the manifold W in such a way that V is orthogonal to M in some neighborhood of the point $m \in M$. We define a linear map $A^v: T_mM \to T_mM$ by the formula $A^v(X) = -(\overline{\nabla}_x V)^T$. This map turns out to be symmetric, and so it defines a bilinear form A^v, which we call the *second quadratic (fundamental) form of the submanifold* M.

In fact, we have defined a whole *family* A of forms A^v, in which the vector $v \in N_mM$ plays the role of a parameter. If $x, y \in T_mM$, we can define a form $B(x, y) \in N_mM$ by the equality $\langle B(x, y), v\rangle = \langle A^v(x), y\rangle$. We include the vector y in a smooth vector field Y on W, tangent to M. We then have $B(x, y) = (\overline{\nabla}_x Y)^N$, that is, we need to differentiate Y covariantly in the direction of x and project the result (that is, a vector at the point m) onto the normal plane.

DEFINITION 3.1.3. Consider the second fundamental form represented as a form B on the tangent space T_mM with values in the normal space N_mM. Since a scalar product has been defined on T_mM, we can consider the trace of the form B, which is (at each point m) a vector belonging to N_mM. Thus, *the trace of the form B is represented by some section H of the normal bundle NM*. This section (trace) H is called the *mean curvature of the embedded submanifold $M \subset W$*. If $e_1, \dots, e_k$ is an orthonormal basis in the plane T_mM, then $H = \sum_{i=1}^k B(e_i, e_i) \in N_mM$, that is, H is a vector.

When M is a *hypersurface* in W, Definition 3.1.3 coincides with the definition given above of the *scalar* mean curvature. The fact is that here the normal plane to M is *one-dimensional* and the mean curvature vector is determined by a point on $\mathbf{R}^1$.

1.2. Multidimensional locally minimal surfaces of arbitrary codimension. The first variation of the volume functional of a submanifold.

DEFINITION 3.1.4. Suppose we are given a smooth homotopy $f_t: M \to W$, $0 \le t \le 1$, such that each map f_t is an embedding and $f_0 = f$, where f is the original embedding of M in W. Then we call f_t an *isotopic variation*, $F: M \times [0, 1] \to W$; $F = \{f_t\}$.

The variation f_t of the embedding f induces a smooth vector field E defined on $F(M \times [0, 1])$, which is the image of the standard vector field $\partial/\partial t$ on the cylinder $M \times [0, 1]$. We are interested in the restriction of this field to the submanifold M, that is, $E(m) = dF(\partial/\partial t)$, where dF is the differential of F. The vector field $E(m)$ determines two sections: $E^T(m)$ of the bundle TM and $E^N(m)$ of the bundle NM. For this it is sufficient to project $E(m)$ orthogonally on T_mM and N_mM respectively. Consider the section E^T as a vector fields on M. Since there is a k-dimensional Riemannian volume form on M, the field E^T determines an outer differential form $\Theta(E^T)$ of degree $k-1$ (by means of the operator $*$; see [48], for example).

PROPOSITION 3.1.1. *Let M be a compact submanifold of W and let $v_k(t) = \operatorname{vol} f_t(M)$ be the k-dimensional volume form of the submanifold $f_t(M)$. Then*

$$v'(0) = -\int_M \langle E^N, H\rangle + \int_{\partial M} \Theta(E^T),$$

where ∂M is the boundary of M. Here the first integral of the function $\langle E^N, H\rangle$ is taken over M with respect to the k-dimensional Riemannian volume form, and the second integral of the form $\Theta(E^T)$ is taken over the $(k-1)$-dimensional submanifold ∂M.

DEFINITION 3.1.5. A submanifold $M \subset W$ is said to be *locally minimal* if its mean curvature H is identically zero (at all points $m \in M$).

PROPOSITION 3.1.2. *A compact submanifold $M^k \subset W^n$ is locally minimal (that is, $H \equiv 0$) if and only if $v_k'(0) = 0$ for any isotopic variation of M that vanishes on ∂M.*

The term "local minimality" means that the volume of a submanifold does not change "in the first approximation" under a small perturbation. Moreover, for small (in amplitude and support) variations the volume functional increases its value "in the second approximation". If the variation has a finite magnitude, then the volume may decrease. This happens, for example, for the standard equator on the sphere, contracted to a point over the sphere.

DEFINITION 3.1.6. A submanifold $M \subset W$ is said to be *totally geodesic* if every geodesic of M (with respect to the Riemannian metric and connection induced by the ambient Riemannian metric) is also a geodesic in the metric of W.

PROPOSITION 3.1.3. *A submanifold $M \subset W$ is totally geodesic if and only if its second fundamental form is identically zero. Thus, every totally geodesic submanifold is locally minimal.*

Let $e_1, \dots, e_k$ be a basis in T_mM, and $E_1, \dots, E_k$ smooth vector fields on M that take values $e_1, \dots, e_k$ at the point m, where $\langle E_i, E_j\rangle = \delta_{ij}$ and $\nabla_{E_i} E_j(m) = 0$. Let v be a smooth section of the bundle NM.

The operator ∇^2, defined on sections v by the formula $\nabla^2 v(m) = \sum_{i=1}^k \nabla_{E_i} \nabla_{E_i} v(m)$, is called the *Laplace operator*. It is known that ∇^2 *is a negative semidefinite selfadjoint differential operator*. Let $v \in N_m M$ and let $R_{a,b}(c)$ be the *Riemannian curvature tensor on* W. We set $R(v) = \sum_{i=1}^k (R_{e_i, v}(e_i))^N$, where $(\ )^N$ is the projection on $N_m M$. We have defined a linear map R of $N_m M$ into itself. Clearly, $\langle R(v), h\rangle = \langle R(h), v\rangle$ and the definition of R does not depend on the choice of basis. The transformation $A^v : T_m M \to T_m M$ is *linear and symmetric*. Since the Euclidean scalar product $\langle\ ,\ \rangle$ is defined in the space of all such transformations (matrices), we can consider $\langle A^v, A^h\rangle$. We specify a linear transformation $\widetilde{A}: N_m M \to N_m M$ as follows: $\langle \widetilde{A}(v), h\rangle = \langle A^v, A^h\rangle$. It is easy to verify that $\widetilde{A}$ is a *symmetric positive semidefinite operator*.

The Euclidean scalar product in a space of matrices is specified in the usual way, namely $\langle X, Y\rangle = \operatorname{tr}(XY^T)$.

1.3. The second variation of the volume functional.

ASSERTION 3.1.1. *Let* M^k *be a smooth locally minimal submanifold in* W^n *and* f_t *an isotopic variation of* M, *fixed on its boundary. Suppose that* M *is compact and let* $v = E^N$ *be the section of the normal bundle* NM *defined above. If* $v_k(t) = \operatorname{vol} f_t M$, *then*

$$v_k''(0) = \int_M \langle -\nabla^2(v) + R(v) - \widetilde{A}(v), v\rangle .$$

For a minimal submanifold $M \subset W$ we define a *bilinear form* $I(v, h)$, specified on sections v and h of NM that vanish on ∂M. We set

$$I(v, h) = \int_M \langle -\nabla^2(v) + R(v) - \widetilde{A}(v), h\rangle .$$

ASSERTION 3.1.2. *Let* M *be a minimal submanifold in* W. *Then* $I(v, h)$ *is a symmetric form on the space of sections of* NM *that are equal to zero on* ∂M. *The form* I *is expressed by a diagonal matrix* I *in terms of the scalar product* $I(v, h) = \langle I(v), h\rangle$ *and has distinct real eigenvalues* λ_i *(possibly multiple) that tend to infinity, that is,* $\lambda_1 < \lambda_2 < \cdots < \lambda_i < \cdots \to \infty$. *To each eigenvalue* λ_i *there corresponds a finite-dimensional eigensubspace.*

The *null-index* (or *degree of degeneracy*) *of a minimal submanifold* M is the dimension of the eigensubspace corresponding to a zero eigenvalue. The *index of a minimal submanifold* is the sum of the dimensions of all eigensubspaces of the operator I corresponding to negative eigenvalues.

1.4. Conjugate boundaries. The index of a minimal surface. A nonzero section v of the bundle NM (along a minimal surface M) is called a *Jacobi field* if $J(v) = 0$, that is, this section satisfies the equation

$$J(v) = -\nabla^2(v) + R(v) - \widetilde{A}(v) = 0.$$

ASSERTION 3.1.3. *Let f_t be an isotopic variation of an embedding $f: M \to W$ of a minimal submanifold in W. Suppose that all the submanifolds $f_t M$ are minimal. Then the vector field E^N (the normal component of the velocity field of the variation f_t) is a Jacobi field on M.*

Since R is only the "normal part" of the complete curvature tensor (see 1.3), in the case of a one-dimensional submanifold M (that is, in the case of a smooth curve) it is easy to calculate that the expression $R - \widetilde{A}$ goes into the standard form that determines the usual equation of Jacobi fields along geodesics (one-dimensional minimal manifolds).

Jacobi fields for geodesics are connected with the kernel of the Hessian of the action functional on the space of paths. There is a corresponding analog for minimal surfaces.

ASSERTION 3.1.4. *Let M be a compact minimal submanifold with boundary ∂M, embedded in W. Then the space of Jacobi vector fields on M that vanish on its boundary is finite-dimensional and coincides with the kernel of the form I. The dimension of this space is equal to the degree of degeneracy (null-index) of M.*

According to Morse theory the index of the geodesic $\gamma(t)$ with fixed ends p and q (not conjugate along γ) is equal to the sum of the multiplicities of points on this geodesic conjugate to its origin, for example $p = \gamma(0)$. Smale [393] generalized this result to the case of strongly elliptic selfadjoint operators. We first define conjugate boundaries, which generalize the concept of conjugate points. Let $M^k \subset W$ be a compact submanifold with nonempty boundary ∂M. Let φ_t, where $t \geq 0$, be a one-parameter family of diffeomorphisms of M into itself, and φ_0 the identity transformation, and suppose that the homotopy φ_t is smooth in m and t, where $m \in M$, $t \in \mathbf{R}$. We set $M_t = \varphi_t M$. Then $\partial M_t = \varphi_t \partial M$. Suppose that with the growth of t the manifold M_t "becomes smaller", that is, $M_t \subsetneqq M_s$ for each $t > s$. Then we say that φ_t is a *contraction* of M (over itself). Suppose that a smooth measure is defined on M, for example, M is a Riemannian manifold. We say that the contraction φ_t has *ε-type* if $\operatorname{vol} M_t < \varepsilon$ for sufficiently large t (and so for all t exceeding some t_0). The manifold does not necessarily contract to a point.

Let $E \to M$ be a vector bundle over M with a scalar product, defined on each fiber, that depends smoothly on a point of the base. Let $L\colon C^\infty_{k-1}(E) \to C^\infty(E)$ be a strongly elliptic selfadjoint operator of order $2k$, where $C^\infty(E)$ denotes the space of smooth sections of the bundle $E \to M$, and $C^\infty_{k-1}(E)$ is the space of smooth sections of the bundle $E \to M$ that vanish on the boundary ∂M of the manifold M together with the first $k-1$ derivatives.

Let β_L denote the bilinear form $\beta_L(u, v) = \int_M \langle Lu, v\rangle$ corresponding to the operator L. The *Morse index* $\operatorname{Ind} L$ of the operator L is defined as the maximal dimension of a linear subspace in $C^\infty_{k-1}(E)$ on which the form β_L is negative definite. The equivalence of the definition of the index given above for minimal surfaces in the case when L is the Jacobi operator and $E = NM$, the normal bundle, and the definition just given, follows from the variational principle (see [430]).

If we are given a contraction M_t of the manifold M, then $E_t \to M_t$ denotes the restriction of the bundle $E \to M$ to M_t, and $L_t\colon C^\infty_{k-1}(E_t) \to C^\infty(E)$ denotes the corresponding restriction of the operator L. Let $\alpha(t)$ denote the null-index of the operator L_t, that is,

$$\alpha(t) = \dim\{u \in C^\infty_{k-1}(E_t) | Lu = 0\}.$$

The nonzero sections $u \in C^\infty_{k-1}(E_t)$ are called *Jacobi fields.*

For a contraction $\varphi_t\colon M \to M_t$ of a manifold M the set ∂M_t (the image of ∂M under φ_t) is called a *conjugate boundary* if there is a Jacobi vector field equal to zero on its boundary ∂M_t. The *order (multiplicity) of the conjugate boundary* ∂M_t is defined as the dimension of the linear space of such Jacobi fields.

Let us also assume that the operator L satisfies the condition of "uniqueness in the solution of the Cauchy problem", namely, if $Lu = 0$ and $u(x) = 0$ in some open subset $U \subset M$, then $u \equiv 0$ in M.

ASSERTION 3.1.5 (Smale's theorem on the Morse index, [393]). *Suppose that the operator $L\colon C^\infty_{k-1}(E) \to C^\infty(E)$ satisfies all the requirements given above. Then there is a number $\varepsilon > 0$ such that for any smooth ε-type contraction we have*

$$\operatorname{Ind} L = \sum_{a<t\leq b} \alpha(t).$$

If the contraction is not smooth, then

$$\operatorname{Ind} L \geq \sum_{a<t\leq b} \alpha(t).$$

REMARK. Essentially Smale proved that for a smooth contraction M_t of the manifold M the eigenvalues of the operator L_t increase strictly monotonically as t increases, and in sufficiently small domains all the eigenvalues are positive. Thus, the eigenvectors corresponding to negative eigenvalues of L "drift off" to the conjugate boundaries ∂M_t under a smooth contraction of M to sufficiently small size.

Simons [387] remarked that the Jacobi operator $L = -\nabla^2 + \overline{R} - \widetilde{A}$ of a compact orientable minimal surface satisfies all the requirements of Smale's theorem if for the bundle $E \to M$ we take the normal bundle $NM \to M$. We have thus obtained a generalization of the theorem on the index of a geodesic to the case of compact orientable minimal surfaces.

ASSERTION 3.1.6 (Simons [387]). *Let M be a compact orientable minimal manifold and let $\varepsilon > 0$ be chosen in accordance with Assertion* 3.1.5. *Let φ_t be a contraction of M that has ε-type. Then there are only finitely many conjugate boundaries ∂M_{t_i}. The index of a minimal manifold is equal to the sum of the orders (multiplicities) of these boundaries of all $t_i > 0$.*

Let M^2 be a catenoid in $\mathbf{R}^3$ whose axis of rotation is the z-axis. We cut it by two planes orthogonal to the z-axis, and obtain two coaxial circles. Raising or lowering the planes, we move the circles over the fixed catenoid. If the distance between them is small, then the part of the catenoid between the circles is stable. Increasing the distance, we finally obtain an unstable surface. Consequently, the conjugate boundary consists of two coaxial circles at a definite distance from each other. *Its multiplicity is equal to* 1. The complete index of a catenoid is 1.

1.5. Two-dimensional minimal surfaces and their indices. In this subsection we calculate the indices of Morse type of a large class of two-dimensional minimal surfaces, in particular, all the classical minimal surfaces in three-dimensional Euclidean space $\mathbf{R}^3$, and also all minimal surfaces of revolution (catenoids) in three-dimensional Lobachevsky space $\mathbf{H}^3$. In addition, in $\mathbf{H}^3$ we discover a one-parameter family of stable ruled minimal surfaces (helicoids).

By a minimal surface we understand an immersed connected p-dimensional submanifold M of a Riemannian manifold W of larger dimension that has zero mean curvature. A compact orientable minimal surface M can also be defined as a critical point of the functional of p-dimensional volume (area) restricted to the space of all variations of this surface that leave its boundary fixed (see above).

As we noted earlier, minimal surfaces in $\mathbf{R}^3$ are the mathematical model of the interfaces between certain physical media, for example soap films. In fact, there are "far more" smooth compact minimal surfaces in $\mathbf{R}^3$ than there are soap films without singularities obtained in physical experiments. One of the reasons for this disparity is the instability of films described by a large class of minimal surfaces: small variations of such films may decrease their areas, which leads in practice to a sudden change of their forms. A natural question arises: is a given minimal surface stable, that is, does any variation of it that is fixed on the boundary lead to an increase in its area? Moreover, if a minimal surface is unstable, then it is interesting to estimate its degree of instability, or more precisely, what is the maximal dimension of the space of (infinitesimal) variations that decrease its area? A numerical expression for the instability of a minimal surface M is its index $\operatorname{Ind} M$, defined above in the case when M is compact and orientable.

The concept of index can also be introduced for noncompact minimal surfaces. In this case we define two types of indices: smooth and piecewise-smooth. Interest in the study of indices of noncompact minimal surfaces is explained by the fact that the basic classical minimal surfaces in $\mathbf{R}^3$ are complete, and hence noncompact.

DEFINITION 3.1.7. The *smooth* (*piecewise-smooth*) *index* of a noncompact orientable minimal surface in a Riemannian manifold is the least upper bound of the indices of compact subdomains of it with smooth (piecewise-smooth) boundaries (by a compact domain we understand a domain with compact closure).

From the variational principle it follows that these two indices coincide. Consideration of the smooth index is connected with the fact that Smale's theorem (see Assertion 3.1.5), which is a basic tool for the calculation of indices, has been proved only in the case where the manifold (minimal surface) has a smooth boundary.

We note that a noncompact minimal surface is said to be stable if its index is equal to zero (compare the definition of stability in the compact case).

Many papers have been devoted to the investigation of stability. Barbosa and do Carmo have shown (see [162]) that a minimal surface in $\mathbf{R}^3$ for which the area of the image of the Gaussian map is less than 2π is stable. In [183] do Carmo and Peng proved that a complete stable minimal surface in $\mathbf{R}^3$ can only be a plane (a generalization of Bernstein's problem). By contrast with the last result, in $\mathbf{H}^3$ there are one-parameter families of complete stable minimal surfaces that are not the Lobachevsky plane. To such families there belong, for example, the one-parameter family of hyperbolic catenoids and part of the one-parameter family of spherical catenoids (see [431]), and also part of the one-parameter family of helicoids (see below).

Although the index is always finite in the case of a compact minimal surface for noncompact surfaces this is not always so. In the papers cited below it becomes clear when the index of a complete minimal surface M in a Euclidean space $\mathbf{R}^n$ is finite. Fischer-Colbrie [432] proved that for complete minimal surfaces in $\mathbf{R}^3$ finiteness of the index is equivalent to finiteness of the total curvature. We recall that the total curvature τ of a two-dimensional surface M in $\mathbf{R}^3$ is the integral over the surface of its Gaussian curvature K, that is, $\tau = \int_M K$. Bérard and Besson [433] obtained similar results for complete p-dimensional minimal surfaces in Euclidean space $\mathbf{R}^n$ in the case $p \geq 3$. They showed that the index $\operatorname{Ind} M$ of such a minimal surface M satisfies the inequality $\operatorname{Ind} M \leq F(p, n) \cdot \int_M \|B\|^p$, where $F(p, n)$ is an explicitly computable constant that depends only on the dimensions p and n, and B is the second fundamental form of the surface M. Thus, from the finiteness of the integral $\int_M |B|^p$ there follows the finiteness of $\operatorname{Ind} M$, but this theorem does not contain the converse statement. We observe that when $p = 2$ we have $|B|^2 = -2K$, where K is the Gaussian curvature. The results of Fischer-Colbrie suggest that the restriction $p \geq 3$ is not essential. In fact, Tysk [434] proved that for complete minimal surfaces in $\mathbf{R}^3$ we have $\operatorname{Ind} M \leq 7.68183 \cdot k$, where k is the degree of the Gaussian map. We observe that for complete minimal surfaces in $\mathbf{R}^3$ the total curvature τ is equal to $-4\pi k$. Therefore, the constant $F(p, n)$ in the Bérard-Besson inequality can also be calculated for the case $p = 2$, $n = 3$: $F(2, 3) \approx 0.30565$.

The calculation of the indices of noncompact minimal surfaces is a complicated problem, and until recently these indices had not been calculated. The first numerical values of the indices of minimal surfaces in Euclidean space were apparently obtained in 1985 by Fischer-Colbrie (see [432]), who showed that the indices of two of the classical minimal surfaces, namely the catenoid and the Enneper surface, are equal to one. These results were obtained independently by A. Tuzhilin in the same year as a consequence of a general theorem (Theorem 3.1.1) given below. Lopez and Ros [435], relying on the results of Fischer-Colbrie, showed that the catenoid and the Enneper surface are the only complete orientable minimal surfaces in $\mathbf{R}^3$ with index one. The last result is another characteristic of these two classical minimal surfaces: Osserman (see [436]) proved that the catenoid and the Enneper surface are the only complete minimal surfaces in $\mathbf{R}^3$ with total curvature equal to -4π. They are also the only complete minimal surfaces in $\mathbf{R}^3$ for which the Gaussian map is a diffeomorphism with an image.

In [437] an attempt was made to calculate the indices of the classical minimal surfaces in $\mathbf{R}^3$, but it was not crowned with success, because the author of [437] confused the Jacobi equation for a minimal surface, written in isothermal coordinates (where the Euclidean Laplacian is used) with the Jacobi equation "on the surface" (where the Laplacian is metrical).

We now state the main results of Tuzhilin. We begin with the case of minimal surfaces in $\mathbf{R}^3$.

1. *Two-dimensional minimal surfaces in* $\mathbf{R}^3$. It is convenient to describe two-dimensional minimal surfaces in $\mathbf{R}^3$ by the so-called Weierstrass representation.

DEFINITION 3.1.8. A *Weierstrass representation* is a triple (M, ω, g), where M is a Riemannian surface, ω is a holomorphic 1-form on M, and g is a meromorphic function on M, where it is required that the three 1-forms $\varphi^1 = \frac{1}{2}(1-g^2)\omega$, $\varphi^2 = \frac{i}{2}(1+g^2)\omega$, and $\varphi^3 = g\omega$ are holomorphic 1-forms that have zero real periods.

If x^1, x^2, x^3 are the standard Euclidean coordinates in $\mathbf{R}^3$, then the formulas $x^k(Q) = c^k + \operatorname{Re}\int_P^Q \varphi^k$, $k = 1, 2, 3$, define an orientable two-dimensional generalized minimal surface in $\mathbf{R}^3$. Any orientable two-dimensional generalized minimal surface M in $\mathbf{R}^3$ can be obtained in this way. Here c^k are real constants, and integration of the holomorphic 1-forms φ^k is carried out along arbitrary paths in M joining a fixed point $P \in M$ and a variable point $Q \in M$ (see [436]).

REMARK 1. A point $Q \in M$ is a singular point (a branch point) if and only if $\varphi^1(Q) = \varphi^2(Q) = \varphi^3(Q) = 0$. The Weierstrass representation (M, ω, g) defines an immersed minimal surface if and only if the zeros of ω and the poles of g are arranged in the same way and the order of each zero of ω is twice the order of the corresponding pole of g.

REMARK 2. The function g is the composition of the Gaussian map $n: M \to S^2$ and the stereographic projection of the sphere S^2 onto the plane $x^3 = 0$ from the North Pole $N = (0, 0, 1)$. If we regard S^2 as the Riemann surface $S^2 = \mathbf{C} \cup \{\infty\}$, then the Gaussian map $n: M \to S^2$ can be thought of as a ramified holomorphic covering.

THEOREM 3.1.1. *Let M be a two-dimensional immersed minimal surface in $\mathbf{R}^3$ given by the Weierstrass representation $(U, f(w)dw, (aw+b)^k)$ where $U \subset \mathbf{C}$ is a domain of the complex plane $\mathbf{C}$, w is the* *standard complex coordinate, $f(w)$ is a holomorphic function on U, and k is a nonzero integer. If $U = \mathbf{C}$ or $U = \mathbf{C}\setminus\{-b/a\}$, then* $\operatorname{Ind} M = 2|k|-1$. *If $U \subset \mathbf{C}$, then* $\operatorname{Ind} M \leq 2|k|-1$.

REMARK 3. The only restriction of the function $f(w)$ holomorphic in U is imposed by the condition that the surface M is immersed. If the function $g = (aw+b)^k$ has a pole in U, that is, $k < 0$ and $\{-b/a\} \in U$, then the immersion condition is equivalent to $f(w)$ being nonzero everywhere outside the pole $\{-b/a\}$, and at this point $f(w)$ has a zero of order $2|k|$. If the function $g = (aw+b)^k$ is holomorphic in U, that is, $\{-b/a\} \notin U$ or $k > 0$, then the immersion condition is equivalent to the condition $f(w) \neq 0$ in U.

The next two theorems describe a certain class of stable and maximally unstable domains U of the minimal surfaces considered in Theorem 3.1.1.

We define a closed subset C of the sphere $S^2 = \{\sum_{i=1}^{3}(x^i)^2 = 1\}$ in one of the following ways:

(a) $C = S^2 \cap \{x^3 \leq (|k|-1)/|k|\}$, $k \in \mathbf{Z} - \{0\}$; if $k = \pm 1$, then for C we take an arbitrary closed hemisphere that does not contain the North Pole when $k = 1$ or the South Pole when $k = -1$;

(b) the set C is the part of the sphere S^2 included between two parallel noncoincident planes at a distance $\tanh t_{|k|-1}$ from the center of S^2 and not containing poles. Here $t_{|k|-1}$ is the only positive root of the equation $(|k|-1)/|k| = \tanh t \tanh(((|k|-1)/|k|)t)$.

THEOREM 3.1.2. *Let an immersed minimal surface M be defined by the Weierstrass representation as in Theorem* 3.1.1. *Suppose that the domain U contains the inverse image under the Gaussian map of the subset C of S^2 defined in* (a) *or* (b). *Then* $\operatorname{Ind} M = 2|k|-1$.

We define an open subset C' of S^2 in one of the following ways:

(a′) $C' = S^2 \cap \{x^3 < 0\}$; if $m \pm 1$, then for C' we take an arbitrary open hemisphere not containing the North Pole when $m = 1$ or the South Pole when $m = -1$;

(b′) the set C' is the part of the sphere S^2 included between the two parallel noncoincident planes at a distance $\tanh t_0$ from the center of S^2 and not containing poles. Here t_0 is the only positive root of the equation $t \tanh t = 1$.

THEOREM 3.1.3. *Let an immersed minimal surface M be defined by the Weierstrass representation as in Theorem* 3.1.1. *Suppose that the image of*

U under the Gaussian map is contained in the subset C' of S^2 defined in (a′) *or* (b′). *Then the index of M is equal to zero.*

REMARK 4. For domains C' of type (a′) Theorem 3.1.3 is a special case of the results of Barbosa and do Carmo (see [162]), but for domains C' of type (b′) this is not so.

The next theorem makes substantial use of the results of Fischer-Colbrie (see [432]).

DEFINITION 3.1.9. A meromorphic function g on a Riemann surface M is said to be *good* if on M there is a holomorphic 1-form ω such that the triple (M, ω, g) is the Weierstrass representation of a complete minimal surface. If $M = \mathbf{C}$, then the most general form of good functions known to the authors is $P/h + c/Q$ or h/P, where P and Q are arbitrary polynomials, $c \in \mathbf{C}$, and h is an arbitrary holomorphic function.

THEOREM 3.1.4. *Let (M, ω, g) be the Weierstrass representation of an immersed minimal surface M, where g is a good function. Then* $\operatorname{Ind} M$ *is finite if and only if the total curvature of M is finite. In particular, if $M = \mathbf{C}$, then the finiteness of the index is equivalent to g being a rational fraction.*

Theorems 3.1.1–3.1.4 enable us to calculate the indices of all the classical minimal surfaces in $\mathbf{R}^3$.

COROLLARY.

Minimal surface	*Weierstrass representation*	*Index*
Scherk surface (incomplete)	$(\{\lvert w\rvert < 1\}, dw/(1-w^4), w)$	0
Enneper surface	$(\mathbf{C}, dw, w)$	1
Catenoid	$(\mathbf{C} \setminus \{0\}, -dw/w^2, w)$	1
Richmond surface	$(\mathbf{C} \setminus \{0\}, w^2 dw, 1/w^2)$	3
Helicoid	$(\mathbf{C}, -ie^{-w} dw, e^w)$	∞

The indices of all complete periodic minimal surfaces, in particular the classical Schwartz-Riemann surface, are equal to infinity; the index of the complete Scherk surface is also equal to infinity.

To conclude this subsection we give two results obtained by Tuzhilin, of which substantial use is made in the calculation of indices: a generalization of the theorem of Simons (Assertion 3.1.6) to the case of noncompact minimal surfaces and the "isomorphism theorem", which enables us to reduce the calculation of the index of one minimal surface in $\mathbf{R}^3$ to that of another.

First of all, by analogy with a contraction of ε-type, we define an exhaustion.

DEFINITION 3.1.10. Let M be a noncompact connected orientable Riemannian manifold. We define an *exhaustion* of M as a family M_t, $t \in \mathbf{R}_+$, of compact subdomains of M with smooth boundaries, satisfying the following conditions:

(1) $M_t \subsetneq M_s$ for $t < s$;
(2) for any compact subset K there is a t for which $K \subset M_t$;
(3) the volume of M_t tends to zero as $t \to 0$.

An exhaustion is said to be *smooth* if the boundary ∂M_t depends smoothly on t.

REMARK 5. It is easy to see that there is not a smooth exhaustion for every noncompact manifold. As an example we consider the plane $\mathbf{R}^2$ from which countably many points with integer coordinates have been discarded. In contrast to this, in the case of a compact manifold a contraction of ε-type always exists for any $\varepsilon > 0$ (see [393]).

ASSERTION 3.1.7 (generalization of the theorem of Simons). *Let M be a noncompact orientable minimal surface in a Riemannian manifold. Suppose that for M there is a smooth exhaustion M_t. Then* $\operatorname{Ind} M$ *is equal to the sum of the multiplicities of the conjugate boundaries ∂M_t over all $t \in \mathbf{R}_+$.*

DEFINITION 3.1.11. We say that the surfaces M_1 and M_2 in $\mathbf{R}^3$ are *isomorphic* if there is a diffeomorphism $F: M_1 \to M_2$ compatible with the Gaussian maps n_1 and n_2 of M_1 and M_2 respectively. This means that the following diagram is commutative:

$$\begin{array}{ccc} M_1 & \xrightarrow{F} & M_2 \\ \downarrow{\scriptstyle n_1} & & \downarrow{\scriptstyle n_2} \\ S^2 & \xrightarrow{\Phi} & S^2 \end{array}$$

where Φ is an arbitrary isometry of the sphere S^2.

ASSERTION 3.1.8 (the "isomorphism theorem"). *The indices of isomorphic minimal surfaces in $\mathbf{R}^3$ are equal.*

COROLLARY 1. *Let* (M_1, ω_1, g) *and* (M_2, ω_2, g) *be the Weierstrass representations of two minimal surfaces. Then the indices of these surfaces are equal.*

COROLLARY 2. *Let* $F: M_2 \to M_1$ *be a biholomorphic diffeomorphism of the Riemann surfaces* M_2 *and* M_1, *and* (M_1, ω_1, g) *and* $(M_2, \omega_2, g \circ F)$ *the Weierstrass representations that define the two minimal surfaces. Then the indices of these minimal surfaces are equal.*

The "isomorphism theorem" is a consequence of the fact that the Jacobi equation of the minimal surface given by the Weierstrass representation (M, ω, g) does not depend on the specific form of ω. In fact, in explicit form in the coordinates (u, v), where $w = u + iv$ is the complex coordinate on the Riemann surface M, the Jacobi equation is

$$\frac{\partial^2 T}{\partial u^2} + \frac{\partial^2 T}{\partial v^2} + \frac{8|\dot{g}|^2}{(1 + |g|^2)^2} T = 0,$$

where $\dot{g} = dg/dw$, and the function T defines the section V of the normal bundle to M: if n is the field of unit normals to M, then $V = T \cdot n$.

2. *Two-dimensional minimal surfaces in* $\mathbf{H}^3$. We consider complete minimal surfaces of revolution in the Lobachevsky space $\mathbf{H}^3$, called catenoids. In contrast to the case of $\mathbf{R}^3$, here there are three types of catenoids: spherical, parabolic and hyperbolic. In [431] catenoids were defined in a more general case, namely in n-dimensional Lobachevsky space $\mathbf{H}^n$. Let $L^{n+1} = \mathbf{R}_1^{n+1}$ be a pseudo-Riemannian space with metric ds^2 of signature $(1, n-1)$, and suppose that $\mathbf{H}^n$ is realized as a connected component of the unit pseudosphere with induced Riemannian metric $dl^2 = -ds^2|_{\mathbf{H}^n}$.

In [431] a surface of revolution was defined in $\mathbf{H}^n$. Let P be a two-dimensional plane in L^{n+1} passing through the origin. Three cases are possible: P is Lorentzian (if the metric ds^2, restricted to P, is nondegenerate and indefinite), Riemannian ($ds^2|_P$ is definite), or degenerate ($ds^2|_P$ is degenerate). Let Π be a three-dimensional subspace of L^{n+1} such that Π contains P and the intersection $\Pi \cap \mathbf{H}^n$ is not empty (and so $\Pi \cap \mathbf{H}^n$ is a Lobachevsky plane). In $\Pi \cap \mathbf{H}^n$ we consider a curve C that does not intersect P. It is well known that all motions of $\mathbf{H}^n$ can be obtained as restrictions of all possible pseudo-orthogonal transformations of L^{n+1} that take $\mathbf{H}^n$ into itself. Let G_P denote the group of motions of $\mathbf{H}^n$ consisting of those motions that leave all points of P fixed. Then the surface of revolution of the generator C around the "axis" P is defined to be the orbit of C under the action of the group G_P. A catenoid is defined to be a complete minimal surface of revolution. There are three types of catenoids: spherical (the "axis of rotation" P is Lorentzian), hyperbolic (P is Riemannian), and parabolic (P is degenerate).

Let us consider the case $n = 3$. It was shown in [431] that the hyperbolic and parabolic catenoids are stable. Mori [438] showed that spherical catenoids form a one-parameter family M_g, where $b > 1/2$, and if $b > 17/2$, then the catenoid M_g is stable. In [431] a necessary condition was found for a complete immersed minimal surface in $\mathbf{H}^3$ with finite total curvature to be stable, and it was shown that for $1/2 < b < c_0$, where $c_0 \approx 0.69$, spherical catenoids are unstable.

Below we give the results of Tuzhilin concerning the investigation of the indices of two-dimensional surfaces in $\mathbf{H}^3$. We first construct representations, more convenient for our purposes, of catenoids of all types, and also define a one-parameter family of ruled minimal surfaces—the family of helicoids.

To specify spherical and hyperbolic catenoids, and also to define helicoids, it is convenient to use a "cylindrical" coordinate system in $\mathbf{H}^3$. Suppose that in the Minkowski space $\mathbf{R}^4_1$ we have introduced the standard coordinates (t, x^1, x^2, x^3) such that the pseudometric ds^2 has the form $ds^2 = dt^2 - \sum_{i=1}^3 (dx^i)^2$. The Lobachevsky space $\mathbf{H}^3$ is defined as the upper sheet of the two-sheeted hyperboloid (pseudosphere) given by the equation $t^2 - \sum_{i=1}^3 (x^i)^2 = 1$ $(t > 0)$. The cylindrical coordinates (r, φ, z) give the following parametrization of $\mathbf{H}^3$:

$$\begin{cases} t = \cosh r \cosh z, \\ x^1 = \sinh r \cos\varphi, \\ x^2 = \sinh r \sin\varphi, \\ x^3 = \cosh r \sinh z, \qquad r \in \mathbf{R}_+, \quad \varphi \in S^1, \quad z \in \mathbf{R}. \end{cases}$$

It follows from [431] that the trajectories of the motion of points under the action of the spherical and hyperbolic group of rotations G_P (after a suitable isometric change of coordinates) are the coordinate φ-curves and coordinate z-curves respectively. To describe a parabolic catenoid it is convenient to use the model of the upper half-space for the Lobachevsky space with coordinates (x, y, z), $z > 0$, and metric $ds^2 = (dx^2 + dy^2 + dz^2)/z^2$. In this case the trajectories of points under the action of the parabolic group or rotations (after a suitable isometric change of coordinates) are the coordinate y-curves. Now it is not difficult to represent the surface of revolution of the corresponding type. If $r = r(z)$ is the equation of a spherical catenoid, and $r = r(\varphi)$ is the equation of a hyperbolic catenoid (in cylindrical coordinates), then the functions r can be determined from the following one-parameter families of ordinary differential equations:

the spherical catenoid is given by the equation

$$\dot{r}^2 = \cosh^2 r(a^2 \sinh^2 r \cosh^2 r - 1), \quad a > 0, \quad \dot{r} = dr/dz;$$

the hyperbolic catenoid is given by the equation

$$\dot{r}^2 = \sinh^2 r(a^2 \sinh^2 r \cosh^2 r - 1), \quad a > 0, \quad \dot{r} = dr/d\varphi.$$

To describe a parabolic catenoid we choose as generator the curve $(x = t, y = 0, z = z(t))$; then a parabolic catenoid is given by the equation $\dot{z}^2 = 1/z^4 - 1$.

In the cylindrical coordinate system it is also easy to define a helicoid, that is, a surface swept out by a straight line (the generator of the helicoid) under a uniform "screw" motion of it. "Half" of such a helicoid can be obtained, as in the case of $\mathbf{R}^3$, by means of the equation $\varphi = az$, $a \neq 0$. A complete helicoid can be defined as the map of the (u, v)-plane in the Lobachevsky space $\mathbf{H}^3$ given by

$$\begin{cases} t = \cosh u \cosh v \\ x^1 = \sinh u \cos av \\ x^2 = \sinh u \sin av \\ x^3 = \cosh u \sinh v, \quad u \in \mathbf{R}, \quad v \in \mathbf{R}. \end{cases}$$

The results on stability follow from the next theorem, which follows directly from the explicit form of the Jacobi equation in $\mathbf{H}^3$.

THEOREM 3.1.5. *Let M be a two-dimensional orientable minimal surface in the Lobachevsky space $\mathbf{H}^3$. Suppose that the Gaussian curvature K of M satisfies the condition $|K| \leq 1$. Then* $\operatorname{Ind} M = 0$, *that is, the surface M is stable.*

THEOREM 3.1.6. *The Gaussian curvature K of the hyperbolic and parabolic catenoids in $\mathbf{H}^3$ satisfies the condition of Theorem* 3.1.5, *that is, $|K| \leq 1$. Therefore catenoids of these types are stable. Moreover, when $|a| \leq 1$, $a \neq 0$, the Gaussian curvature of the helicoid also does not exceed one in modulus. Therefore such helicoids are also stable.*

The next theorem completes the investigation of the indices of spherical catenoids.

THEOREM 3.1.7. *The index of an unstable spherical catenoid in* $\mathbf{H}^3$ *is equal to one. A spherical catenoid* M_{b_0} *is unstable if and only if the field of variation of* M_{b_0} *in the class of catenoids* M_b *of spherical type touches* M_{b_0} *at one point at least.*

1.6. Minimal submanifolds of the standard sphere. It is well known (Synge) that if, on a Riemannian manifold all of whose curvatures in two-dimensional directions are positive, we specify a closed geodesic along which there is a normal parallel vector field (parallel in the Riemannian connection), then it can always be deformed into a closed curve of smaller length. For example, such a normal field always exists for even-dimensional manifolds. The analogous fact is also true for minimal surfaces (Simons [387]).

ASSERTION 3.1.9. *Suppose that a Riemannian manifold* W^n *has positive Ricci curvature. Then every closed minimal submanifold* M^{n-1} *of codimension* 1 *in* W^n *can always be deformed into a surface of smaller volume (area).*

There is great interest in *minimal surfaces in a sphere.* The simplest such submanifold is a totally geodesic equator S^k in S^n. Its index is $n-k$, and the degree of degeneracy is $(k+1)(n-k)$; see [387].

ASSERTION 3.1.10 (Simons [387]). *Let* M^k *be a compact closed* k-*dimensional minimal submanifold of* S^n. *Then the index of* M^k *is at least* $n-k$ *and it is equal to* $n-k$ *if and only if* $M = S^k$. *The degree of degeneracy of* M^k *is always at least* $(k+1)(n-k)$, *and equality is attained only when* $M = S^k$.

For multidimensional minimal surfaces in S^n the following assertion about properties of the Gaussian image is true. Suppose that $M^{n-1} \subset S^n$ and let $n(m)$ be the unit normal to M in S^n at a point $m \in M$. Since S^n is embedded in the standard way in $\mathbf{R}^{n+1}$, the vector $n(m)$ can be moved parallel to itself so that its origin combines with the origin in $\mathbf{R}^{n+1}$. We obtain a map $g\colon M^{n-1} \to S^n$, called the *Gaussian map.*

ASSERTION 3.1.11 (Simons [387]). *Let* M^{n-1} *be a closed minimal submanifold of codimension* 1 *in* S^n. *Then either its image under the Gaussian map* g *consists of one point (then* M^{n-1} *is a standard equator* S^{n-1}), *or the image* $g(M)$ *is not contained in any open hemisphere* S^n, *that is, it is quite intricately embedded in* S^n.

For generalization of this fact to the case of large codimension, see [387]. A standard embedding $f: S^k \to S^n$ in the form of a totally geodesic surface has "outer rigidity". Namely, no sufficiently close perturbation f' of the map f in the class of embeddings determines a minimal embedding of S^k in S^n except for the obvious cases where f' is obtained from f by orthogonal transformation (rotation of a sphere). Lawson [283] proved that any two-dimensional surface, except $\mathbf{R}P^2$, can be embedded in the standard sphere S^3 in the form of a locally minimal surface. Moreover, Lawson [285] proved that if $N_g^2 \subset S^3$ is a compact embedded minimal surface of genus g in the standard spher. S^3, then there is always a diffeomorphism $f: S^3 \to S^3$ such that $f(N_g^2)$ is a standard embedded minimal surface (defined uniquely). For minimal submanifolds of a sphere see also the papers of Lawson [283], [284], and Hsiang and Otsuki [345], [270].

1.7. Global minimality of complex submanifolds. Federer's theorem. Let $M^k \subset W^n$ be a smooth compact orientable closed submanifold. We say that a bordism-deformation of it is specified if there is specified a $(k+1)$-dimensional smooth compact orientable submanifold $Z \subset W$ with boundary $\partial Z = M \cup (-P)$, where P is a smooth compact orientable closed submanifold of W. We denote by $-P$ the manifold P with the opposite orientation, induced on P by the orientation of W (Figure 3.1). We call P a *bordism-variation of* M. In the case of a noncompact submanifold $M \subset W$ we shall say that a *bordism-deformation* of it is specified if in W there is specified a submanifold P that coincides with M outside some compact domain and there is also specified a $(k+1)$-dimensional submanifold Z with a piecewise smooth boundary $\partial Z \subset M \cup (-P)$ (Figure 3.1).

We recall that a complex manifold W is called a *Kähler manifold* if its Riemannian metric $\sum g_{ij} dz_i d\bar{z}_j$ (where $z_1, \dots, z_n$ are local coordinates) defines a closed outer differential form of degree two $\omega = \frac{i}{2} \sum g_{ij} dz_i \wedge d\bar{z}_j$.

THEOREM 3.1.8 (Federer [216]). *Let W be a Kähler manifold of complex dimension n and $M \subset W$ a complex k-dimensional submanifold of it. Consider all possible real bordism-deformations of this submanifold in W, that is, the bordism-deformations where the film Z^{2k+1} is a real $(2k+1)$-dimensional submanifold of W. Let P^{2k} be a bordism-variation of M. Then* $\mathrm{vol}_{2k} M$ *does not exceed* $\mathrm{vol}_{2k} P$ *if M and P are compact. If M and P are not compact, then we have in mind the volumes of those*

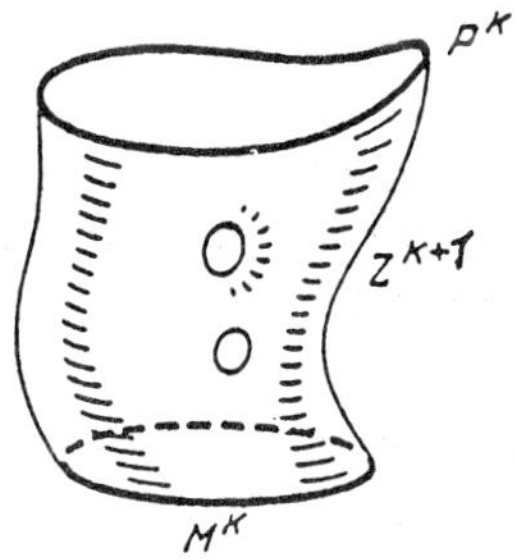

FIGURE 3.1

domains (on M and on P) where M and P are distinct (do not coincide). Moreover, if the volume of P coincides with the volume of M, then P is also complex (in W).

Thus, *in Kähler manifolds compact complex subdomains are globally minimal submanifolds.* Thus, for example, in the complex projective space $\mathbf{C}P^n$ standard submanifolds $\mathbf{C}P^k$, $k < n$, are globally minimal submanifolds. The complex space $\mathbf{C}^n$ is a Kähler manifold, and therefore any complex submanifold X of $\mathbf{C}^n$ is minimal with respect to any perturbations that are fixed outside some bounded domain in X. We recall that all complex submanifolds of $\mathbf{C}^n$ are noncompact. In particular, *the volume of a compact Kähler submanifold of a Kähler manifold never exceeds the volume of any submanifold homologous to it (with the same boundary).* Since a Kähler submanifold realizes an absolute minimum of the volume in its bordism-variation (or homology) class, its index is zero.

ASSERTION 3.1.12 (Simons [387]). *Let $M^p \subset W^n$ be a compact Kähler submanifold with a boundary in a Kähler manifold W with* $\dim \partial M = 2p - 1$. *Then its degree of degeneracy is zero (like the index). Let M be a closed (without boundary) compact Kähler submanifold. Then the index of M is zero, and the degree of degeneracy of M is equal to the dimension of the space of global holomorphic sections of the bundle NM.*

1.8. The complex Plateau problem. Let X be a real submanifold of a complex manifold W, which is the boundary of a complex submanifold V. If $\dim_{\mathbf{R}} X = 2p - 1$, then at any point $z \in X$ we have $\dim_{\mathbf{R}} T_z X \cap J(T_z X) = 2p-2$, where J is an (almost) complex structure in $T_z W$, that is, multiplication by $i = \sqrt{-1}$. The submanifold X^{2p-1} which satisfies the condition $\dim_{\mathbf{R}} T_z X \cap J(T_z X) = 2p-2$ at all points $z \in X$ is said to be *maximally complex. The condition of maximal complexity for real submanifolds of odd dimension is necessary in order that the complex*

Plateau problem should be solvable for this submanifold, that is, in order that this submanifold should be the boundary of a complex submanifold of W. Suppose that $p > 1$. *These conditions are also in a certain sense sufficient for the complex Plateau problem to be solvable.* Suppose that X is a compact orientable submanifold of a Stein manifold W. Let $[X]$ be a $(2p-1)$-current in W, specified by integrating differential forms over X. Similarly, if V is a component of an analytic subset of dimension p, then $[V]$ denotes the $2p$-current specified by integrating outer forms over the set of nonsingular points of V. The *current* $[X]$ is the boundary of $[V]$ in the sense of currents (this is written as $d[V] = [X]$) if $[X](\alpha) = [V](d\alpha)$ for all $(2p-1)$-forms α on W.

A *holomorphic p-chain* is understood to be a locally finite sum $T = \sum_i n_i[V_i]$, where $n_i \in \mathbf{Z}\setminus 0$ and V_i are irreducible analytic sets of dimension $2p$, and $\operatorname{supp} T = \bigcup V_i$.

THEOREM 3.1.9 (Harvey and Lawson [251]). *Let X be a compact orientable submanifold of real dimension $2p-1$ and class C^1 in a Stein manifold W. Suppose that X is a maximally complex submanifold, and $p > 1$.*

Then there is a unique holomorphic p-chain T in $W \setminus X$ with support $\operatorname{supp} T \subset W$ and finite mass for which $dT = [X]$ in W. Moreover, there is a compact nowhere dense subset $A \subset X$ such that for any point of $X \setminus A$ around which X is a submanifold of class C^k, $1 \le k \le \infty$, there is a neighborhood in which $(\operatorname{supp} T) \cup X$ is a C^k-regular submanifold with boundary (if $k \ge 2$, then A can be chosen to have zero $(2p-1)$-dimensional Hausdorff measure). In particular, if X is connected, then there is a unique precompact irreducible analytic p-set in $W \setminus X$ for which $d[V] = \pm[X]$ with the same regularity on the boundary as above.

1.9. Different approaches to the concepts of a surface and the boundary of a surface. Different versions of the statement of Plateau's problem. In the following subsections we attempt to give a brief description of the development of research on Plateau's problem beginning with the 1960's, when an important qualitative jump occurred in this field. Here we do not by any means pretend to a complete survey of all these new trends associated with the remarkable results of such well-known authors as Federer, Fleming, Almgren, Miranda, Reifenberg, Morrey, Bombieri, Giusti, de Giorgi, Osserman, Nitsche, Lawson, Allard, Siu, Yao, Hsiang, and many others. Fairly complete information and a bibliography is contained, for example, in [147]–[152], [217], [218], [220]–[229], [119].

In the previous section we did not specially draw the reader's attention

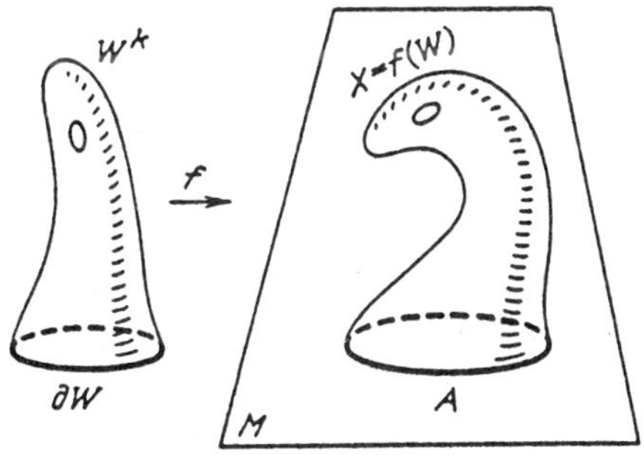

FIGURE 3.2

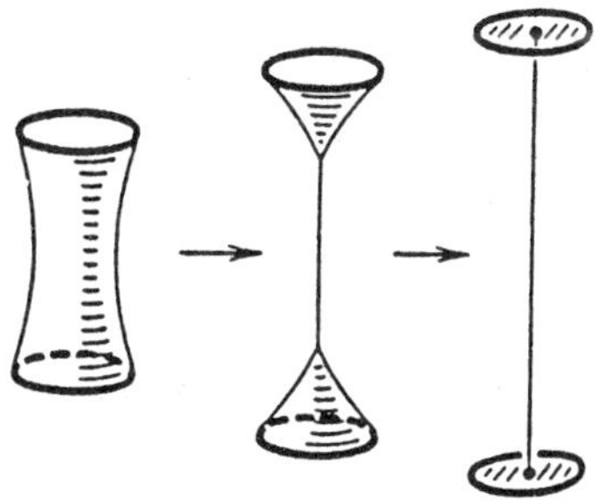

FIGURE 3.3

to a discussion of the concepts of a "film" and its "boundary", confining ourselves to either an intuitive idea or to a consideration of the class of two-dimensional surfaces with a boundary, smoothly mapped into $\mathbf{R}^n$, or to the concept of bordism-variation. *However, the concepts of a surface and its boundary are not as simple and unambiguous as might appear at first sight.* Although in the case of a *one-dimensional* contour we can confine ourselves to the ideas presented above, in the case of the *multidimensional Plateau problem*, connected with the study of the multidimensional volume functional, the very statement of these concepts requires special investigation. As a surface X^k in a manifold M^n it is natural to consider the continuous (or piecewise smooth) image of a smooth manifold W^k with boundary ∂W under a map $f\colon W \to M$ which maps ∂W homeomorphically onto a fixed closed submanifold A^{k-1} in M^n, which plays the role of a "contour" (Figure 3.2). However, it is easy to see that in an attempt to minimize the volume of such surfaces there arise, generally speaking, gluings, which lower the dimension of the surface on some parts of it (Figure 3.3). These "gluings" do not affect the k-dimensional volume functional, but it is impossible to remove them, since this can violate the topological properties of the film, for example, change its topological and even homotopical type. Thus, as we see, *we need to make precise the statement of the multidimensional Plateau problem.* The progress made in this direction, beginning in the 1960's, was due to a significant extent by the production of sufficiently flexible ideas of a surface and its boundary.

The class of smoothly embedded or immersed submanifolds is not sufficiently large to contain a minimal surface (see the example above). In this connection Reifenberg remarked: "... while it is intuitive that any set which is a surface of minimal area in some sense will be locally well-behaved, this is a result which it would be nice to prove and this cannot be done if we only consider locally well-behaved surfaces in the first place. In other words it would be nice to investigate the structure of sets of minimum area without prejudging the issue by assuming that the sets are manifold like" [365].

1.10. The cohomological boundary of a surface and the role of the group of coefficients of the homology group. The solution of Plateau's problem in the class of homology surfaces with fixed homological boundary. Since we wish to *extend the concept of a surface*, we need correspondingly to extend the concept of its area or volume. As a basic functional, defined on closed k-dimensional subsets in $\mathbf{R}^n$, we need to take the *k-dimensional spherical Hausdorff measure*, which we denote by vol_k. For its definition and properties see below and in [216]. If a subset X is a smooth submanifold or a stratified algebraic submanifold, then $\mathrm{vol}_k X$ coincides with the usual k-dimensional Riemannian volume of a submanifold. As "surfaces" we can now consider *measurable* (that is, having finite Hausdorff measure) *compact subsets of* $\mathbf{R}^n$.

It remains to formulate the concept of *boundary*. Let us fix a homology theory with coefficients in an Abelian group G. If a surface X is a finite cell complex or a smooth manifold, then as the q-dimensional homology group $H_q(X, G)$ we can take the usual cellular or simplicial homology. If X is only a measurable compact set, then as $H_q(X, G)$ we need to consider the so-called *spectral homology* (*Čech homology*). *In the case of finite cell complexes and smooth manifolds these* (*spectral*) *homology groups coincide with the cellular* (*simplicial*) *homology groups.* As applied to variational problems these questions were analyzed in a book by Fomenko [119].

Following Adams and Reifenberg, we shall say that a closed k-dimensional set $X \subset \mathbf{R}^n$ is a *G-surface spanning a closed $(k-1)$-dimensional set A for a given Abelian group of coefficients* G, if, firstly, A is contained in X, and, secondly, the homomorphism $i_* : H_{k-1}(A, G) \to H_{k-1}(X, G)$, induced by the embedding $i: A \to X$ is identically zero.

This means that all $(k-1)$-dimensional cycles contained in the boundary A must "dissolve", namely be annihilated in the surface X. In this sense A is the *homological boundary of* X.

The given definition of the boundary of a surface agrees with our intuitive idea.

1. Let M^k be a compact orientable manifold with boundary A, which is a $(k-1)$-dimensional oriented manifold. Consider an embedding $i: A \to M$. Then the homomorphism $H_{k-1}(A) \to H_{k-1}(M)$ is trivial.

FIGURE 3.4

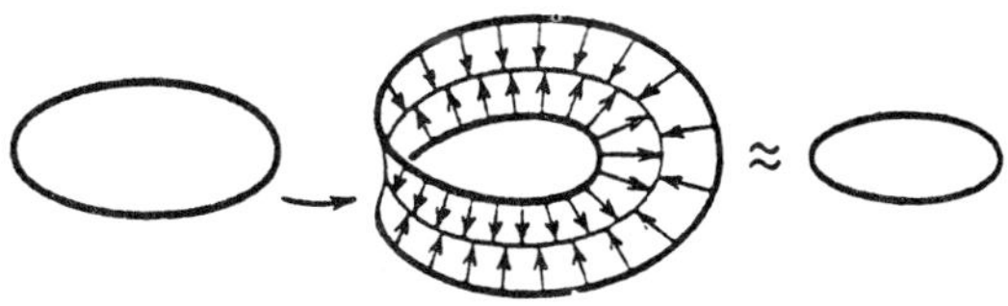

FIGURE 3.5

2. Let A be an arbitrary finite cell complex and X a cone over A (Figure 3.4). Then the induced homomorphism $H_{k-1}(A) \to H_{k-1}(X)$ is trivial (for all k).

The concept of a G-surface with homological boundary A depends essentially on the choice of the group G. The set A can be the boundary of X for one group and not be the boundary for another group.

3. If A is a circle, and X is a Möbius band (a nonorientable surface), then the homomorphism $i_*: H_1(A, \mathbf{Z}) \to H_1(X, \mathbf{Z})$ induced by the embedding has zero kernel, that is, it is a monomorphism $i_*: \mathbf{Z} \to \mathbf{Z}$, $i_*(1) = 2$, that is, i_* multiplies each cycle by 2 (Figure 3.5). Thus, the Möbius band is not a $\mathbf{Z}$-surface with a circle as boundary, although it is a smooth manifold with a circle as boundary. At the same time, the Möbius band is a $\mathbf{Z}_2$-surface with a circle as boundary.

4. Let A be a circle, and X a triple Möbius band, obtained as follows. Consider an unknotted circle in $\mathbf{R}^3$ and a "trifolium"—three segments of the same length meeting at some point of a circle orthogonal to it and making equal angles of 120° with each other. We shall move this trifolium along the circle, keeping it orthogonal to the circle, but rotating it around the circle so that after a full rotation the segments have undergone the permutation $1 \to 2 \to 3 \to 1$ (Figure 3.6). The boundary of the resulting surface is a circle going three times around the vertical axis. The surface is not a smooth manifold, since it contains a circle consisting of singular points. If the trifolium were small in diameter, then this surface would be realized as a stable soap film stretched on the bounding contour A. However, X is not a $\mathbf{Z}$-surface or a $\mathbf{Z}_2$-surface with boundary A. This follows from the fact that X is homotopically equivalent to a circle and the homomorphism $i_*: H_1(A, \mathbf{Z}) \to H_1(X, \mathbf{Z})$ takes the generator 1 of the group $\mathbf{Z}$ into the element $3 \in \mathbf{Z}$. Consequently, the triple Möbius band is a $\mathbf{Z}_3$-surface.

One of the main results proved in this direction is Reifenberg's theorem for the case of compact surfaces in $\mathbf{R}^n$, later extended by Morrey to the case of arbitrary smooth Riemannian manifolds.

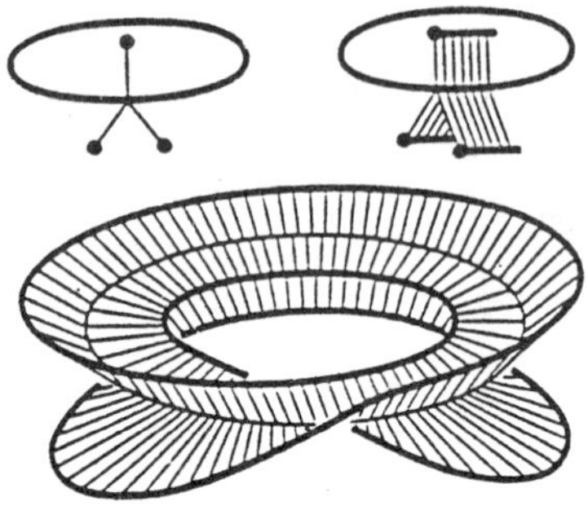

FIGURE 3.6

THEOREM 3.1.10 (Reifenberg [365]–[367], Morrey [318]). *Let M be a smooth complete Riemannian manifold and $A \subset M$ a compact measurable $(k-1)$-dimensional subset. Let G be a compact Abelian group and $H_{k-1}(A, G) \neq 0$. Suppose there is at least one measurable k-dimensional subset X that is embedded in M and is a G-surface with boundary A. Consider the class $\{X\}$ of all such surfaces and set $d = \inf \mathrm{vol}_k(X \setminus A)$, $X \in \{X\}$, that is, d is the greatest lower bound of the volumes of all G-surfaces with boundary A. Then we assert that $d > 0$ and there is a minimal surface X_0 in the class $\{X\}$ such that $d = \mathrm{vol}_k(X_0 \setminus A)$. With the exception of a set S of points of k-dimensional measure zero, this surface $X_0 \setminus A$ is an open smooth minimal submanifold in the sense of ordinary differential geometry, that is, its mean curvature is zero. If the ambient manifold M is analytic, then the submanifold $X_0 \setminus (A \cup S)$ is also analytic.*

Such surfaces, which realize an *absolute minimum of the volume*, are naturally called *globally minimal*, or *absolutely minimal, surfaces*.

1.11. Examples of minimal surfaces showing that the class of homological surfaces and homological boundaries does not cover the whole class of stable minimal surfaces that can be realized in physical experiments. The main restriction that limits the applicability of the remarkable Proposition 3.1.4 is the fact that *there are natural soap films spanning a "good" boundary A that are nevertheless not G-surfaces with boundary A for any Abelian coefficient group G.*

We now turn to an example of Adams. Consider the circle A in Figure 3.7(1), which is the boundary of a stable soap film X shown in Figure 3.7(2) and obtained if we glue a double and a triple Möbius band, joining them by a thin strip or bridge. Obviously this film is realized by a physical soap film and is a perfectly "reasonable" surface.

However, *this film is not a G-surface with a circle as boundary for any*

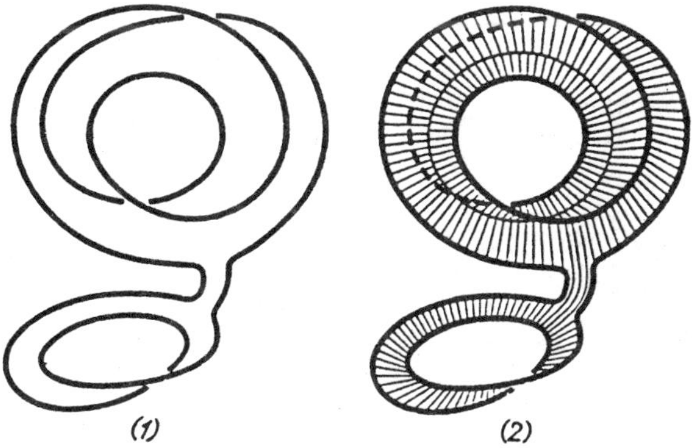

FIGURE 3.7

Abelian group G. The fact is that *the film is retracted onto its own boundary*! We recall that a subspace A of a space X is called a retract of X if there is a continuous map $f: X \to A$ that keeps all points of A fixed.

Let $i: A \to X$ be an embedding, and $f: X \to A$ a retraction. The composition homomorphism $H_1(A) \to H_1(X) \to H_1(A)$ is the identity and, in particular, i_* is a monomorphism. We prove that A is a retract of X, using *Hopf's theorem*. We compute the homomorphism $i_*: H_1(A, U) \to H_1(X, U)$; $U = \{e^{i\varphi}\} = \mathbf{R} \pmod 1$. Since X is homotopically equivalent to a bouquet of two circles, we have $H_1(X) = U \oplus U$ and $i_*(\lambda) = (2\lambda, 3\lambda)$, where $0 \leq \lambda \leq 1$. Hence it is obvious that i_* is a *monomorphism*, as required. We note that although A is a retract of X, A is not a deformation retract of X. We recall that a subspace $A \subset X$ is called a *deformation retract* if there is a continuous homotopy $\varphi_t: X \to X$ such that φ_0 is the identity map of X onto itself, and φ_1 maps X into A, while for all t the map φ_t leaves A fixed. Since the composition $\varphi_1 i: A \to X \to A$ is the identity map, and the composition $i\varphi_1: X \to A \to X$ is homotopic to the identity map, A and X are homotopically equivalent. In our case A and X are not homotopically equivalent.

It is useful to imagine visually *how a retraction looks geometrically*. This construction was carried out by T. N. Fomenko on the basis of his geometrical proof of Hopf's theorem. Consider a cell complex X. Clearly, $X = \sigma^0 \cup \sigma_1^1 \cup \sigma_2^1 \cup \sigma^2$, where σ^2 is the only two-dimensional cell. We cut the soap film along the axes b and a of the double and triple Möbius bands respectively, and along the short segment c, which is the axis of the bridge-membrane joining these two Möbius bands (Figure 3.8). The film X is represented in the form $X = \sigma^0 \cup a \cup b \cup c \cup k^2$, where k^2 is an annulus. After the cut the film becomes a domain homeomorphic to the annulus k^2, whose boundary consists of two circles, the outer,

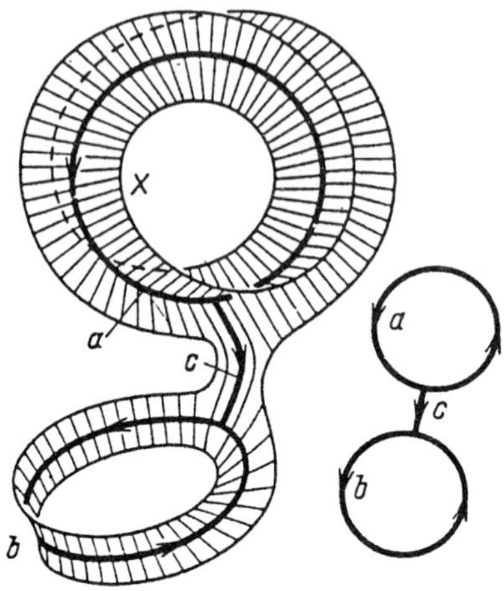

FIGURE 3.8

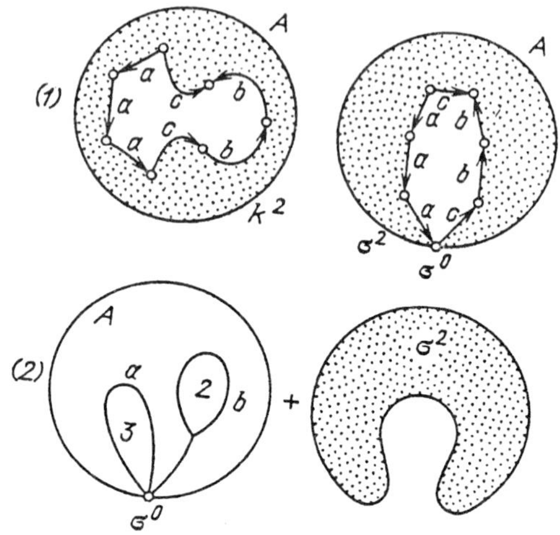

FIGURE 3.9

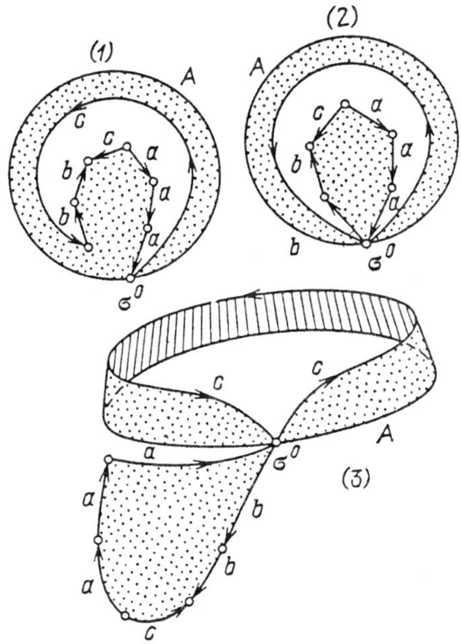

FIGURE 3.10

which is the original boundary A, and the inner, which is the union of seven arcs—three copies of a, two copies of b, and two copies of c (Figure 3.9(1)). We carry out a deformation retraction, as a result of which the inner boundary touches the outer boundary at one point of it (Figure 3.9(2)). The resulting complete boundary bounds the cell σ^2. For the details see [427].

Let us fix A and expand the side c so that it winds on A with degree one (Figure 3.10(1)). We shall not identify A and c but preserve the narrow band bounded by the circles A and c in Figure 3.10(2). The remaining part of the cell forms a disc attached to the point σ^0 and the band between A and c. We take the edge a and wind it around A: see Figures 3.10(3) and 3.11(1). The disc hanging from the point σ^0 is rotated around A. We obtain a bag with boundary a, and in the bag there is a hole with boundary $bbc^{-1}aa$. This word is written out under a counterclockwise motion on the boundary of the hole, and the degree $+1$ or -1 refers to the orientation of the corresponding edge (Figure 3.11(2)). We extend the winding of the two-dimensional cell. We take b and also wind along A (Figure 3.12(1)). We obtain the picture shown in Figure 3.12(2), and the result of this operation is the surface shown in Figure 3.13(1). The boundary of the hanging disc has the form $a^{-1}a^{-1}cb^{-1}$. We take the next copy of a and wind it along A; see Figures 3.13 and 3.14.

Thus, we successively alternate the winding of edges of type a and type b (Figure 3.14(2)). We again obtain a bag with boundary a and a hole bca. We wind the remaining copy of the edge b along A (Figure 3.15(1)).

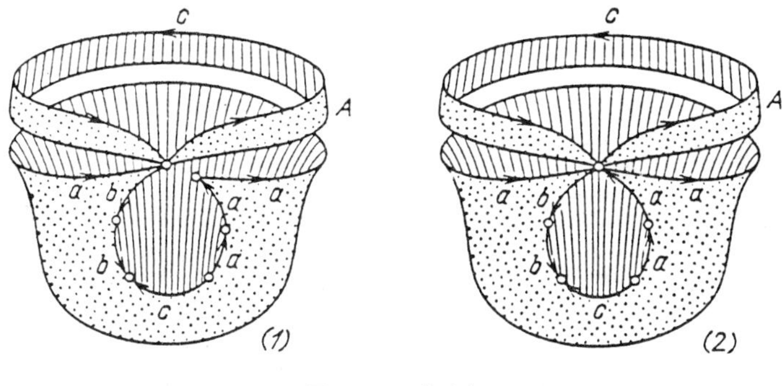

FIGURE 3.11

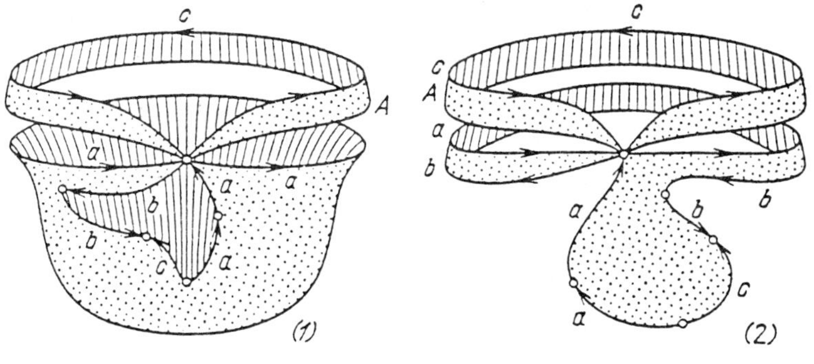

FIGURE 3.12

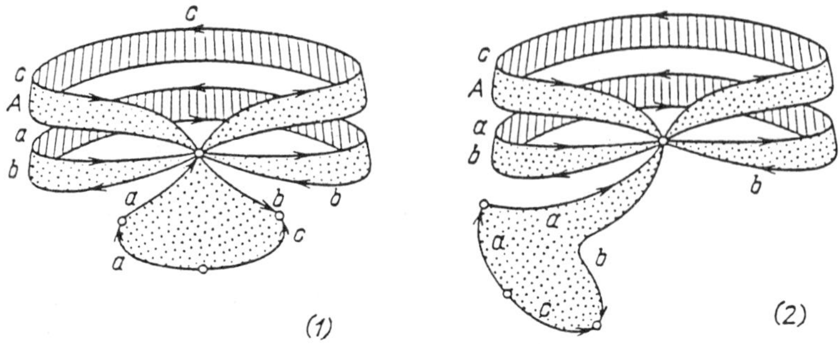

FIGURE 3.13

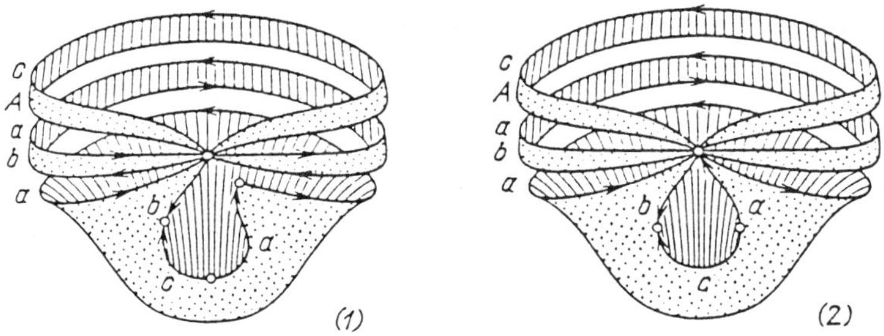

FIGURE 3.14

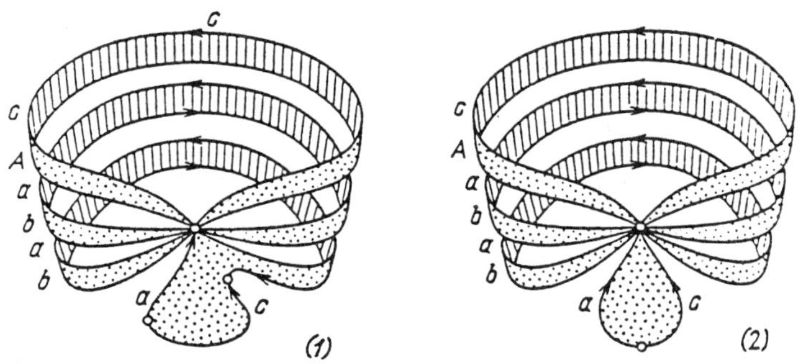

FIGURE 3.15

As a result we obtain the surface in Figure 3.15(2). We take the last copy of a and repeat the winding (Figure 3.16(1)). We obtain the surface in Figure 3.16(2). We extend the side c, blow a hole, and arrange that the bag turns into a narrow band covering A (Figure 3.17(1)). The resulting continuous deformation of the cut film X is remarkable for the fact that it now consists of four narrow bands joined at one point σ^0 and situated along the circle A. Now all the edges a are oriented in the same direction with respect to A and all the edges b are also oriented in the same way.

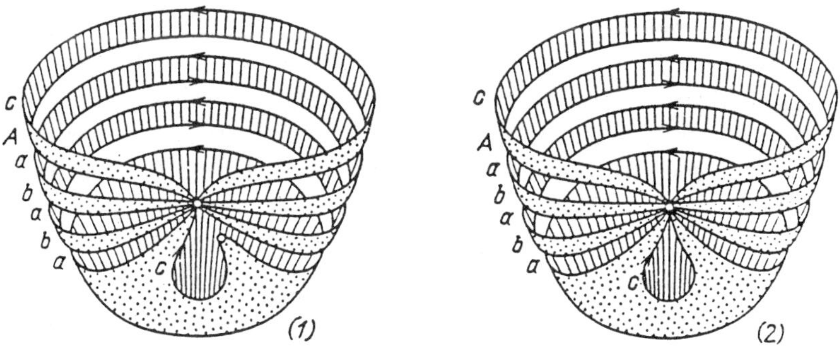

FIGURE 3.16

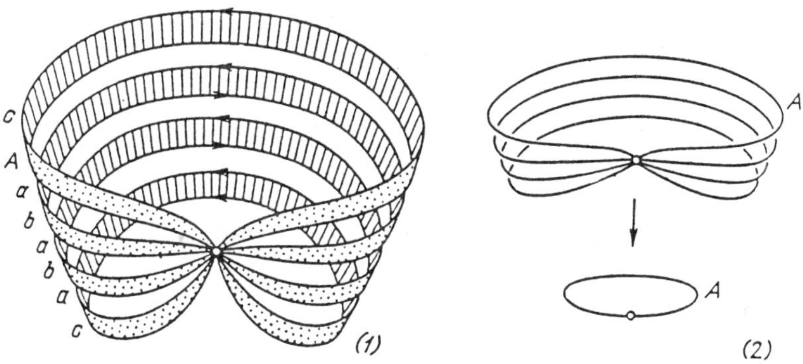

FIGURE 3.17

The two edges c are also oriented in the same way. We gather together each of the four bands on its mean curve, and obtain the four circles in Figure 3.17(2), which we map on a in the same way.

The resulting map is *continuous*. This completes the construction of the required retraction.

1.12. Description of cases where a minimal surface that covers a contour does not contain closed "soap bubbles". In real experiments we often obtain two-dimensional soap films that contain parts of surfaces of zero mean curvature and also soap bubbles of a surface of positive curvature that *bound closed volumes.*

There arises the question: in which cases does a multidimensional minimal film consist only of surfaces with boundary, and in which cases does it include closed surfaces, that is, bubbles? Simple criteria exist that a boundary A must satisfy in order that a film that realizes an *absolute minimum of the volume* (area) should not contain bubbles. These criteria were obtained by A. T. Fomenko.

The mathematical concept of a *homological cycle* is an excellent model for real soap bubbles. From the topological viewpoint of the question of whether a k-dimensional surface X_0 contains "bubbles" or not can be restated thus: does the surface X_0 contain nonzero k-dimensional cycles? In other words, a k-dimensional surface X_0 does not contain "bubbles" if and only if $H_k(X_0, G) \neq 0$.

PROPOSITION 3.1.4 (Fomenko [129], [130]). *Let M be a complete smooth Riemannian manifold and $A \subset M$ be a compact measurable $(k-1)$-dimensional subset, for example, a $(k-1)$-dimensional submanifold of M. Let G be an Abelian group that is a one-dimensional vector space over a vector field F. For example, G is the group $\mathbf{R}^1$ of real numbers, or $G = \mathbf{Z}_p$, where p is prime. Suppose there is at least one G-surface in M with boundary A. Then a globally minimal G-surface X_0 with boundary A has the property that $H_k(X_0, G) = 0$, that is, it does not contain any k-dimensional cycles, that is, "bubbles".*

The fact that the mean curvature of a multidimensional minimal surface is zero almost everywhere by no means implies that the homology group of maximal dimension is trivial.

If the ambient manifold M has a more complicated structure than Euclidean space, then it may contain nontrivial k-dimensional cycles—"bubbles"—that is, the group $H_k(M)$ may be nonzero. In applications, we often meet cases where a minimal film gluing some boundary $A \subset M$ realizes, at the same time, some k-dimensional cycles in M.

In $H_k(M, G)$, we fix a nontrivial subgroup P. We say that *a subset $X \subset M$ realizes the subgroup P*, that is, *it realizes the cycles of this subgroup*, if the subgroup P is contained in the image of the homomorphism $j_*: H_k(X, G) \to H_k(M, G)$ induced by the embedding $j: X \to M$.

Consider the class $[P]$ of all compact subsets of X (where $A \subset X \subset M$) that realize P. It turns out that in this class there is always a globally minimal surface X_0, that is, a surface such that its volume (area) $\operatorname{vol}(X_0 \setminus A)$ is equal to $\inf \operatorname{vol}(Y \setminus A)$, $Y \in [P]$. See [120], [129]–[131].

From visual considerations it is natural to expect that a minimal surface X_0 that realizes P does not contain any "superfluous" k-dimensional cycles, but only those that cover *nonzero* cycles of the group $H_k(M, G)$. We do not exclude the case where the surface X_0 realizes not only cycles of P, but also some additional cycles that do not belong to P. In other words, the homomorphism $j_*: H_k(X_0, G) \to H_k(M, G)$ must be a monomorphism. It is intuitively clear that if some cycle of the film X_0

turns out to be superfluous, that is, it becomes homologous to zero after embedding the film in a manifold, then the part of the film corresponding to this cycle (its support) can be omitted without affecting the realizability of cycles of the subgroup by the film and at the same time decreasing the volume (area) of the film. For a minimal film this procedure is impossible, so the film cannot contain superfluous "melting" cycles. The conjecture is justified.

PROPOSITION 3.1.5 (Fomenko [120], [131]). *Suppose that the assumptions of Proposition* 3.1.4 *are satisfied. If a minimal surface* X_0 *realizes a nontrivial subgroup* $P \subset H_k(M, G)$, *then the homomorphism* j_*: $H_k(X_0, G) \to H_k(M, G)$ *induced by the embedding* $j: X_0 \to M$ *is a monomorphism, that is, there are no "superfluous" cycles on the film.*

The existence of globally minimal surfaces X_0 follows from the papers of Reifenberg [365]–[367] and Morrey [318], and in the more general situation of generalized (co)homology from the papers of Fomenko [120], [129]–[131]. Although from the intuitive viewpoint Propositions 3.1.4 and 3.1.5 are natural, and it is difficult to imagine in what situations they could be false, nevertheless our intuition turns out to be insufficient here—in fact there are some G-surfaces that contain nontrivial k-dimensional bubble-cycles (an example of A. T. Fomenko [131]).

§2. Theory of currents and varifolds

2.1. Classical de Rham currents. We have become familiar with certain concepts of a surface and its boundary, starting from the ideas of the 17th and 18th centuries and ending with the algebraic definitions of a homological boundary. This does not exhaust the list of various approaches to such fundamental concepts as a surface and its boundary. The language developed by Federer, Fleming, Almgren and others has turned out to be very fruitful and has enabled us to prove many remarkable results in the calculus of variations. This is the language of currents; see [147]–[152], [216]–[228].

One of the precursors of the concept of an integral current was the process of integrating Dirac's δ-function. If $f(x)$ is a function defined on some domain, then we can consider the δ-function concentrated at some point x_0 of this domain and represent the value $f(x_0)$ at this point as the result of integrating the function $f(x)\delta(x - x_0)$ over the domain, that is, $f(x_0) = \int f(x)\delta(x - x_0)dx$. If we wish we can regard this as the process of integrating the zero-dimensional form $f(x)$ over the zero-dimensional submanifold x_0 of the domain of definition. If we vary the point x_0, we also vary the result of the integration if $f(x)$ is not constant. Attempts to widen this definition and replace the zero-dimensional submanifold by a submanifold of arbitrary dimension lead to the concept of an *integral current*. One version of this widening was proposed by de Rham. Here we give the definition of Fleming and Federer.

Consider the space $\mathbf{R}^n$, with which we associate the dual (conjugate) vector spaces $\Lambda_{n,k}$ and $\Lambda^{n,k}$ formed respectively by k-vectors and k-covectors on $\mathbf{R}^n$. The direct sums $\Lambda_{n,*} = \bigoplus_{k\geq 0}\Lambda_{n,k}$ and $\Lambda^{n,*} = \bigoplus_{k\geq 0}\Lambda^{n,k}$, endowed with the operation of outer multiplication $\wedge$, form respectively the contravariant and covariant Grassmann algebras of $\mathbf{R}^n$. The scalar product and norm corresponding to it in $\mathbf{R}^n$ naturally induce in the spaces $\Lambda_{n,*}$ and $\Lambda^{n,*}$ a scalar product and norm, which we denote by $(\ ,\ )$ and $|\ |$.

The *mass of a k-vector* ξ is the quantity $\|\xi\| = \inf_B\{\sum_{\beta\in B}|\beta|\}$, where B runs through all finite sets of simple k-vectors such that $\xi = \sum_{\beta\in B}\beta$. The *comass of a k-covector* ω is the quantity $\sup\{\omega(\xi) : \xi$ is a simple k-vector and $|\xi| < 1\}$ and is denoted by $\|\omega\|^*$.

We now consider a Riemannian manifold M and denote its tangent bundle by TM, and denote by P_kM and P^kM the standard Grassmann bundles with base M whose fibers over a point $x\in M$ are respectively the Grassmann spaces of k-vectors and k-covectors of the tangent space T_xM. Differential k-forms on M can be regraded as smooth sections of the bundle P_kM. We endow the space of all differential k-forms on M with the topology of compact convergence of all partial derivatives. The linear space of differential forms of all degrees k on M can naturally be endowed with the structure of a graded differential algebra with the operations of outer multiplication $\wedge$ and outer differentiation.

The *support of a differential k-form* φ on M is the closure of the set $\{x\in M : \varphi_x \neq 0\}$, and its *comass* is the quantity $\sup\{\|\varphi_x\|^* : x\in M\}$.

DEFINITION 3.2.1. A *k-current* (with compact support) on a Riemannian manifold M is any continuous linear functional over the space of all differential k-forms on M.

For each k-current S its *support* $\operatorname{spt} S$ is the smallest closed set K such that $S(\varphi) = 0$ for any differential k-form φ with support in $M\setminus K$, and its *mass* is the quantity $m(S) = \sup_\varphi S(\varphi)$, where φ runs through all differential k-forms with unit comass.

We endow the linear space of all k-currents on M with the *compact weak topology* specified by seminorms $S \to \sup|S(\varphi)|$, $\varphi\in A$, where A is an arbitrary finite set of differential forms on M.

The *boundary of a k-current* S is the $(k-1)$-current ∂S, defined by $\partial S(\varphi) = S(d\varphi)$ for any differential $(k-1)$-form φ. A k-current S is said to be *closed* if $\partial S = 0$, and *exact* if S is the boundary of a $(k+1)$-current. For each differential k-form φ on M we denote by $\|\varphi\|^*$ a continuous function such that $\|\varphi\|^*(x) = \|\varphi_x\|^*$ for any point $x \in M$. If the k-current S has *finite mass*, then there corresponds to it a *variational measure* $\|S\|$ defined by the formula $\|S\|(f) = \sup_\varphi S(\varphi)$ for an arbitrary nonnegative continuous function f on M, where the supremum is taken over all possible differential k-forms φ such that $\|\varphi\|^* \leq f$. A k-current of finite mass admits a natural extension (also denoted by S) to the set of all bounded $\|S\|$-measurable k-forms on M.

The following assertion is due to Federer and Fleming (see [216], [217]).

Let S be a k-current on a Riemannian manifold M, where $m(S) < \infty$. Then for almost all $z \in M$ in the sense of the measure $\|S\|$ there is a k-vector $\vec{S}_z$ of the space T_xM such that $\|\vec{S}_z\| = 1$ and the integral formula $S(\varphi) = \int \varphi(\vec{S}_z)d\|S\|(z)$ holds, where φ is an arbitrary differential k-form on M.

If for a point $z \in M$ there is a k-vector $\vec{S}_z$, then we call $\vec{S}_z$ the *tangent k-vector of the k-current S at the point z*. The assertion shows that the tangent k-vector $\vec{S}_z$ of the k-current S is uniquely defined for almost all $z \in M$ in the sense of the measure $\|S\|$.

2.2. Rectifiable currents and flat chains. Let M be a Riemannian manifold. We recall that a simplex $\Delta^k \subset M$ is called a *simplex of class C^r* if Δ^k is a submanifold of class C^r in M. Each k-dimensional simplex of class C^r in M can be identified with the k-current specified by the operation of integration over Δ^k. Let h^i denote the i-dimensional Hausdorff measure on M.

DEFINITION 3.2.2. A bounded h^i-measurable subset S of M for which $h^iS < \infty$ is called an (orientable) *i-rectifiable subset* if for each $\varepsilon > 0$ there is an (orientable) i-dimensional submanifold $N \subset M$ of class C^1 such that $h^i[(S \setminus N) \cup (N \setminus S)] < \varepsilon$.

Let S be an orientable i-rectifiable subset of M. For almost all $x \in S$ (in the sense of the measure h^i) there is an i-dimensional tangent space to S at x. We denote the i-vector associated with this space by $\vec{S}_x$. For each differential i-form ω the function $x \to \omega(\vec{S}_x)$ is h^i-measurable and determines the i-current $[S](\omega) = \int_S \omega(\vec{S}_x)\,dx$.

DEFINITION 3.2.3. A k-current is said to be *rectifiable* if it is a convergent (in the norm m) sum of k-currents of the form $[S]$, where S is an

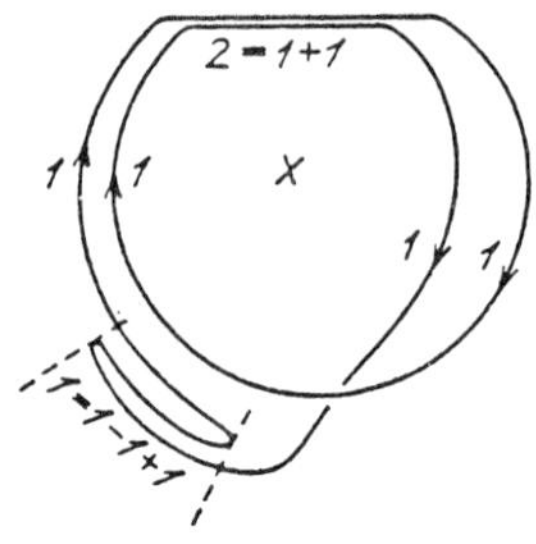

FIGURE 3.18

orientable k-rectifiable subset. In the integration we take account of the multiplicity of the sheets (Figure 3.18).

It is obvious that *a k-current is rectifiable if and only if it is the limit of a sequence of integral simplicial chains of class C^1 in the norm m.*

For the space of k-currents on a Riemannian manifold we introduce the *flat norm* F by setting $F(S) = \sup_\varphi S(\varphi)$, where φ runs through all differential forms (of degree k) such that $\|\varphi\|^* \leq 1$ and $\|d\varphi\|^* \leq 1$.

A *real* (resp., *integral*) *flat k-dimensional chain* is a k-current of the form $P + \partial Q$, where P and Q are a k-current and a $(k+1)$-current that have finite masses (resp., are rectifiable). The real and integral flat chains are the limits of convergent (in the norm F) sequences of simplicial chains of class C^∞ with real and integral coefficients respectively.

2.3. Normal and integral currents.

DEFINITION 3.2.4. A k-current S on a Riemannian manifold M is said to be *normal* if its mass and the mass of its boundary are finite, that is, $m(S) < \infty$ and $m(\partial S) < \infty$.

Normal k-currents on M form a linear space, which we denote by $N_k M$. The direct sum $N_* M = \bigoplus_{k \geq 0} N_k M$ is a *chain complex* with boundary operator ∂.

DEFINITION 3.2.5. A k-current is said to be *integral* if it is rectifiable together with its boundary.

Integral currents of dimension k on a manifold M form a linear space, which we denote by $I_k M$. The direct sum $I_* M = \bigoplus_{k \geq 0} I_k M$ is a subcomplex of $N_* M$. *Normal rectifiable currents are integral.*

A set A is called a *locally Lipschitz neighborhood retract* in a Riemannian manifold M if there is a neighborhood U of A in M and a locally Lipschitz map $f: U \to A$ such that $f(x) = x$ for each $x \in A$.

2.4. Different formulations of the theorem on the existence of minimal currents. The solution of Plateau's problem in various classes of currents.

DEFINITION 3.2.6. Suppose we are given a continuous function l: $P_kM \to \mathbf{R}$ such that $l(\lambda\xi) = \lambda l(\xi)$ for any $\lambda > 0$. The formula

$$L(S) = \int l(\vec{S}_x)d\|S\|(x)$$

defines a functional L over the space of k-currents of finite mass on M. We call it the *parametric k-dimensional integrand on M given by the Lagrangian l*. If the restriction of l to $P_kM \setminus \{\xi : \|\xi\| = 0\}$ is a function of class C^r, we say that L is an *integrand of class C^r*. The integrand L is said to be *positive* if $l(\xi) > 0$ for any ξ such that $\|\xi\| \neq 0$.

In particular, setting $l(\xi) = \|\xi\|$ for any $\xi \in P_kM$, we obtain $L(S) = m(S)$ for each k-current of finite mass.

Let x be a point of M. The tangent space T_xM is a Euclidean space. We denote by l_x the restriction of l to the fiber $(P_kM)_x$ over the point x of the bundle P_kM. We say that the integrand L is *elliptic* (*semi-elliptic*) at x if there is a number $c > 0$ (resp., $c \geq 0$) such that the inequality

$$\|P\|(l_x \circ \vec{P}) - \|Q\|(l_x \circ \vec{Q}) \geq c[m(P) - m(Q)]$$

holds for any rectifiable k-currents P and Q on T_xM such that $\partial P = \partial Q$ and $\operatorname{spt} Q$ is contained in a k-dimensional linear subspace of T_xM. The integrand L is said to be *elliptic* (*semi-elliptic*) if it is elliptic (semi-elliptic) at each point x. It is known (see Federer [216]) that *if the function $F(\xi) = l_x(\xi) - c|\xi|$ is convex for some $c > 0$ ($c \geq 0$), then the integrand L is elliptic* (*semi-elliptic*) *at x*. The following basic theorem on the existence of minimal k-currents holds.

THEOREM 3.2.1 (Federer and Fleming, see [216], [217]). *Let A and B be compact Lipschitz neighborhood retracts in M, where $B \subset A$, $X \in I_kA$, $\alpha \in H_k(I_*A/I_*B)$. Then any positive semi-elliptic integrand L of dimension k on M attains an absolute minimum on each of the following sets of k-currents: $I_kA \cap \{T : \partial(T - X) \subset B\}$, $I_kA \cap \{T : T - X \in \alpha\}$.*

Integral K-currents that provide minima for the integrand L on the sets of k-currents mentioned in Theorem 3.2.1 *naturally generalize the classical concepts of minimal surface-chains in the calculus of variations*, namely, those that minimize the functional L in the class of all surface-chains of variable topological type with a "movable" boundary. In fact, we have the assertion (see [217]) that every integral k-current T whose boundary ∂T is the true $(k-1)$-dimensional surface-chain differs from the true k-dimensional surface-chain only by the boundary of an integral

$(k+1)$-current S such that the quantity $m(S) + m(\partial S)$ is arbitrarily small and $\operatorname{spt} S$ is entirely in an arbitrarily small neighborhood of $\operatorname{spt} T$. If B is empty, then *Theorem* 3.2.1 *gives the existence of a minimum of the integrand* L *in the class of all integral* k-*currents with a given fixed boundary in the class of all homological integral* k-*currents with a given fixed boundary.*

2.5. Varifolds and minimal surfaces. Almgren ([147], [150]) developed the theory of integral varifolds and successfully applied it to the multidimensional calculus of variations.

Let X be a locally compact topological space. Consider the linear space of all continuous functions $C(X)$ on X and endow it with the topology of uniform convergence on every compactum $K \subset X$. A *Radon measure* (with compact support) on X is any continuous linear functional on the linear space of continuous functions $C(X)$. For the details see [27], for example. The *support of a measure* μ on X is the smallest closed set $K \subset X$ such that $\mu(f) = 0$ for any function f such that $f(x) = 0$, $x \in K$. For a continuous linear functional the support is actually compact. If we are given a continuous map $h\colon X \to Y$, then a measure μ with compact support on X induces a measure $h_*\mu$ with compact support on Y, namely $h_*\mu(f) = \mu(fh)$, where f is an arbitrary continuous function on Y.

Let $\Gamma_{n,k}$ denote the usual Grassmann manifold of all k-dimensional subspaces of Euclidean space $\mathbf{R}^n$. Let $\Gamma_k M$ be the standard bundle $p\colon \Gamma_k M \to M$ with base M whose fiber over the point x is the Grassmann manifold of k-dimensional subspaces of the tangent space $T_x M$.

DEFINITION 3.2.7. A k-*varifold* on a Riemannian manifold M is any Radon measure with compact support on $\Gamma_k M$.

For a k-varifold V on M we set $\|V\| = p_* V$. The support of the measure $\|V\|$ is naturally called the *support of the* k-*varifold* V and denoted by $\operatorname{spt} V$. Further, we shall call the norm of the measure $\|V\|$ the *mass of the* k-*varifold* V and denote it by $m(V)$.

Every k-rectifiable subset S of M can be associated with a k-varifold S as follows. We denote by $\vec{S}$ the map of S into $\Gamma_k M$ that takes a point x of S into the tangent space $\vec{S}_x$ to S at x. Clearly, $\vec{S}$ is an h^k-measurable map, and so there is defined an induced measure $\vec{S}_* h^k$ with compact support on $\Gamma_k M$. Thus, $\vec{S}_* h^k$ is a k-varifold on M, and we set $[S] = \vec{S}_* h^k$. In particular, if S is a compact surface, then $S = \operatorname{spt}[S]$ and $m[S]$ coincides with the k-dimensional volume of S.

Further, a k-varifold V is said to be *rectifiable* if it can be represented as a convergent (in the compact weak topology) sum of the k-varifolds

$[S_i]$, where S_i are k-rectifiable subsets of M and $\bigcup_i S_i$ is bounded. In particular, if $S = \sum_i \alpha_i \Delta_i$ is a k-dimensional integral simplicial chain of class C^1 in M, then $[S] = \sum_i \alpha_i [\Delta_i]$ is a rectifiable k-varifold and we again call it a *k-dimensional integral simplicial chain of class C^1*. In this case the *boundary of the k-varifold* $[S]$ is the $(k-1)$-varifold $\sum_i \alpha_i [\partial \Delta_i]$, which is denoted by $\partial[S]$. Obviously, $\partial(\partial[S]) = 0$. A rectifiable k-varifold V on M is said to be *integral* if there is a $(k-1)$-varifold W on M and for each $\varepsilon > 0$ there is a k-dimensional integral simplicial chain T of class C^1 such that $m(V - T) < \varepsilon$ and $m(W - \partial T) < \varepsilon$. We observe that if such a $(k-1)$-varifold W exists, then it is uniquely defined, so we naturally call W the *boundary of the integral k-varifold V*.

Let $l\colon \Gamma_k M \to \mathbf{R}$ be a continuous nonnegative function. If S is a k-rectifiable subset, we set $L([S]) = \int_S l(\vec{S}_x)\, dh^k(x)$. We call the functional L thus defined over the space of rectifiable k-varifolds the *k-dimensional integrand on M given by the Lagrangian l*.

The tangent space $T_x M$ can be identified with $\mathbf{R}^n$, and its Grassmann bundle $\Gamma_k(T_x M)$ with the direct product $\mathbf{R}^n \times \Gamma_{n,k}$. We define the Lagrangian $l_x\colon T_x M \times (\Gamma_k M)_x \to \mathbf{R}$ by setting $l_x(z, \pi) = l(\pi)$ for any $z \in T_x M$ and any $\pi \in (\Gamma_k M)_x$. We also denote by L_x the integrand on $T_x M$ given by l_x.

Let B be a compact subset of M, and let $\sigma \in H_k(M, B, \mathbf{Z})$. We say that a compactum $S \subset M$ *covers* σ if $i_* \sigma = 0$, where $i_*\colon H_k(M, B, \mathbf{Z}) \to H_k(M, B \cup S, \mathbf{Z})$ is the homomorphism induced by the pair embedding $i\colon (M, B) \to (M, B \cup S)$. We also say that a k-dimensional integrand L on M is *elliptic* if there is a continuous positive function c on M such that for any $x \in M$, any k-dimensional disc $D \subset T_x M$, and any compact k-rectifiable set $S \subset T_x M$ covering σ we have $L_x(S) - L_x(D) \geq c(x)[h^k(S) - h^k(D)]$.

THEOREM 3.2.2 (Almgren [147]). *Let L be a k-dimensional elliptic integrand on M given by a Lagrangian l of class C^3 such that $0 < \inf_\pi l(\pi) \leq \sup_\pi l(\pi) < \infty$. Let $B \subset M$ be a compact $(k-1)$-rectifiable subset and let $\sigma \in H_k(M, B, \mathbf{Z})$. Then there is a compact k-rectifiable subset $S \subset M$ such that S covers σ and $L(S) \leq L(T)$ for any compact k-rectifiable subset T of M that covers σ.*

We proceed to a description of the approach of de Giorgi [238] applied, generally speaking, *only to hypersurfaces in* $\mathbf{R}^n$. For each set $E \subset \mathbf{R}^n$ we denote by ∂E its boundary, which consists of all boundary points of E, and by $\overline{E} = E \cup \partial E$ its closure. Further, if E is a Borel set, then its *oriented boundary* $\partial_0 E$ is the set of all points $z \in \mathbf{R}^n$ such that

$0 < h^n(E \cap B(z, \varepsilon)) < h^n(B(z, \varepsilon))$ for any $\varepsilon > 0$, where B is a ball. Clearly, $\partial_0 E \subset \partial E$. In what follows we denote by φ_E the characteristic function of E, and by $B(x, r)$ and $C(x, r)$ respectively the open and closed balls of radius r with center at x.

Let E be a *Borel set* in $\mathbf{R}^n$, and suppose that $n \geq 2$. We set

$$\varphi_\lambda(x) = (\pi\lambda)^{-n/2} \int_{\mathbf{R}^n} \exp(-(x-\xi)^2/\lambda)\varphi_E(\xi)\, d\xi$$

for each $\lambda > 0$. We can verify that there is always a (finite or infinite) limit

$$P(E) = \lim_{\lambda \to 0} \int_{\mathbf{R}^n} |\operatorname{grad} \varphi_\lambda(x)|\, dx,$$

which we call the *perimeter of* E.

Let $A \subset \mathbf{R}^n$ be a Borel set, and $E \subset \mathbf{R}^n$ a Borel set of finite perimeter. We shall say that *the set* E *has an oriented boundary of minimal measure with respect to* A if $P(E) \leq P(B)$ for any Borel set $B \subset \mathbf{R}^n$ such that $B \setminus A = E \setminus A$. De Giorgi [240] and Triscari [402] showed that *if* A *is closed, then* E *has an oriented boundary of minimal measure with respect to* A *if and only if*

$$\int_A |\operatorname{grad} \varphi_E| = \inf\left\{\int_A |\operatorname{grad} \varphi_B|;\ P(B) < \infty,\ B \setminus A = E \setminus A\right\}.$$

THEOREM 3.2.3 (de Giorgi [238], Triscari [402]). *Let* $A \subset \mathbf{R}^n$ *be a closed set, and* $E \subset \mathbf{R}^n$ *a Borel set of finite perimeter. Then there is a Borel set* M *such that* $M \setminus A = E \setminus A$ *and* $P(M) \leq P(B)$ *for any Borel set* $B \subset \mathbf{R}^n$ *such that* $B \setminus A = E \setminus A$. *In particular,* M *has an oriented boundary of minimal measure with respect to* A.

2.6. Internal regularity of minimal surfaces and the structure of their singular points. As we have seen, the existence of minimal surfaces (a solution of Plateau's problem) has been proved in different formulations, using various definitions of surface and boundary. No smoothness properties are required for surfaces and their boundaries. In this connection the investigation of internal and boundary regularity of minimal surfaces is an exceptionally important problem. We first describe regularity (almost everywhere) of oriented boundaries of minimal measure.

De Giorgi [238] showed that *the minimal hypersurfaces described above are real analytic manifolds everywhere except possibly for a set of measure zero.*

THEOREM 3.2.4 (de Giorgi [238], Miranda [308]). *Let* $A \subset \mathbf{R}^n$ $(n \geq 2)$ *be an open set, and* $E \subset A$ *a Borel set of finite perimeter which has an oriented boundary of minimal measure with respect to any closed bounded* *subset G of A. Then the intersection $\partial_* E \cap A$ is an analytic $(n-1)$-dimensional submanifold of $\mathbf{R}^n$, where $\partial_* A \subset \partial_0 E$, $h^{n-1}[(\partial_0 E - \partial_* E) \cap A] = 0$.*

The *regularity almost everywhere* of surfaces of arbitrary codimension in $\mathbf{R}^n$ that minimize volume was proved by Reifenberg [365], where minimal surfaces were understood to be minimal compacta that glue cycles of a given homology subgroup of some bounding set. Reifenberg's theorem is a special case of a theorem of Fomenko on the regularity of minimal surfaces in variational classes described by generalized homology and cohomology theories.

2.7. Regularity almost everywhere of supports that minimize the elliptic integrand of k-currents and k-varifolds. Here we give a brief description of important results of Federer.

THEOREM 3.2.5 (Federer [216]). *Let L be a positive elliptic k-dimensional integrand of class C^{q+1} $(q+1 \geq 3)$ on $\mathbf{R}^n$, $A \subset \mathbf{R}^n$ an open set, and $S \in I_k A$ a current such that $L(S) \leq L(S')$ for any $S' \in I_k A$, where $\partial S' = \partial S$. Then A contains an open set U such that $h^k[(\operatorname{spt} S \setminus \operatorname{spt} \partial S) \setminus U] = 0$, and $U \cap \operatorname{spt} S$ is a k-dimensional submanifold of class C^q in $\mathbf{R}^n$.*

A similar regularity theorem for a minimal k-varifold was stated and proved earlier by Almgren [147], where he considered another approach to the concept of minimality.

2.8. Interior regularity of hypersurfaces that minimize volume and the existence of minimal cones of codimension one. The conjecture was made long ago that any minimal hypersurface in Euclidean space $\mathbf{R}^n$ is analytic at all its interior points. This conjecture was proved by Fleming [228] and Reifenberg [365]–[367] for $n = 3$, by Almgren [151] for $n = 4$, and by Simons [387] for $n = 5, 6, 7$. The conjecture was then disproved for $n \geq 8$ by Bombieri, de Giorgi, and Giusti in their remarkable paper [170]. They constructed $(n-1)$-dimensional cones of minimal volume in $\mathbf{R}^n$ for $n \geq 8$.

Let E be a Borel set of finite perimeter in $\mathbf{R}^n$. A cone $CG \subset \mathbf{R}^n$ over a set $G \subset \partial B(0, 1)$ with vertex at the origin $O \in \mathbf{R}^n$ is called the *limiting cone of the set E at the point z* if there is a sequence $r_i \to 0$ such that

$$\lim_{i \to \infty} h^n[(CG \setminus f_i(E) \cap \overline{B(0, 1)}) \cup (f_i(E) \cap \overline{B(0, 1)} \setminus CG)] = 0,$$

where $f_i(x) = (x - z)/r_i$. In this case we call $\partial_0(CG)$ the *tangent cone of the oriented boundary $\partial_0 E$ at the point z*. Similarly we define the tangent cone in the case of surfaces of finite codimension. It turns out that *if z is* *a singular point of a minimal surface, then its tangent cone is also minimal with respect to its boundary.* This determines the role of minimal cones in Plateau's problem.

THEOREM 3.2.6 (Almgren [151], Simons [387]). *Let G be a closed $(n-2)$-dimensional locally minimal submanifold of the sphere $\partial B(0, 1) \subset \mathbf{R}^n$ that is not a totally geodesic sphere. Then for $n \leq 7$ the cone CG over G with vertex at the origin $O \in \mathbf{R}^n$ is unstable with respect to its boundary G. In particular, CG does not minimize the volume among all hypersurfaces in $\mathbf{R}^n$ with boundary G.*

When $n = 3$, the assertion is trivial, since any closed locally minimal curve without singularities on a two-dimensional standard sphere is a geodesic (a great circle). When $n = 4$, Theorem 3.2.6 was proved by Almgren [151] by means of the methods of complex analysis on two-dimensional manifolds and the Gauss-Bonnet theorem. Theorem 3.2.6 was then proved in the general case (that is, $n \leq 7$) by Simons [387].

THEOREM 3.2.7 (Almgren [151], Simons [387]). *Let $A \subset \mathbf{R}^n$ $(n \leq 7)$ be an open set, and $E \subset \mathbf{R}^n$ a Borel set of finite perimeter having an oriented boundary of minimal measure with respect to A. Then $\partial_0 E \cap A$ is an $(n-1)$-dimensional analytic submanifold of $\mathbf{R}^n$ satisfying the equation of a minimal surface.*

In $\mathbf{R}^8$ there are cones that are not only locally but also globally minimal. Consider an embedding of $G = S^n \times S^n$ in the sphere S^{2n+1}, given by the formula

$$G = \left\{ x \in \mathbf{R}^{2n+2} : \sum_{i=1}^{2n+2} x_i^2 = 1, \sum_{i=1}^{n+1} x_i^2 = \sum_{i=n+2}^{2n+2} x_i^2 \right\}.$$

THEOREM 3.2.8 (Bombieri, de Giorgi, and Giusti [170]). *When $n \geq 3$ the cone over $S^n \times S^n$ is the only surface that minimizes the $(2n+1)$-dimensional volume in the class of all hypersurfaces with the same boundary.*

Broad series of new examples of minimal cones of codimension one in Euclidean spaces of high dimension were then obtained by Lawson [281], Hsiang [267], Tyrin and Fomenko (see [119]) in research on the equivariant Plateau problem.

§3. The theory of minimal cones and the equivariant Plateau problem

3.1. Every singular point of a minimal surface always determines a minimal cone. Relying on Plateau's principles we may assume that in a sufficiently small neighborhood of every singular point a *two-dimensional* minimal surface consists of several smooth pieces, which we can assume to be

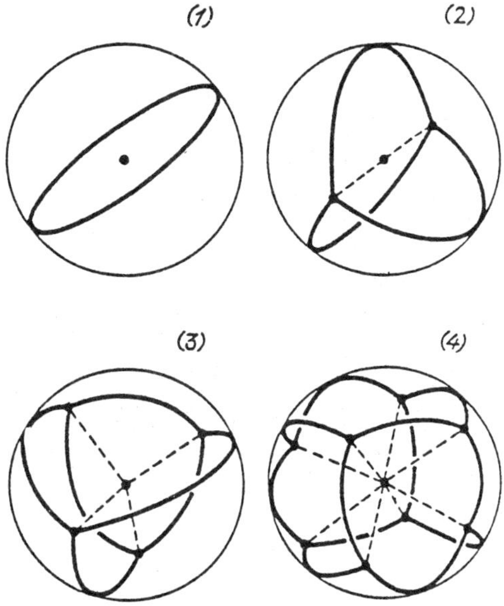

FIGURE 3.19

approximately flat. As we have seen, this property may not hold for minimal cones of dimension greater than two.

Consider a two-dimensional sphere of small radius with center at a singular point and study its intersection with a minimal surface. We may assume that the intersection consists of finitely many smooth arcs contained in the sphere and forming on it a net, since the arcs may meet.

It is easy to verify that every open interval of such an arc must be an interval of some great circle. A soap film included inside the sphere is formed by all possible radii emanating from the center of the sphere and ending on a one-dimensional net. Some radii may be singular edges of the film. There is a simple connection between the length l of the one-dimensional net and the area S of the film included in the sphere. Clearly, $S = r_0 l/2$, where r_0 is the radius of the sphere. Thus, if at some point of the sphere several smooth arcs of the net meet, then there are three of them and they form angles of 120° with one another.

What are the one-dimensional nets obtained by the intersection of a soap film with a small sphere whose center is situated at some singular point of the film? We can give a complete list of all such nets on the sphere; see [148]. There are exactly 10 such configurations (Figures 3.19, 3.20, and 3.21). To verify that there are no other nets, it is sufficient to use elementary arguments of spherical trigonometry.

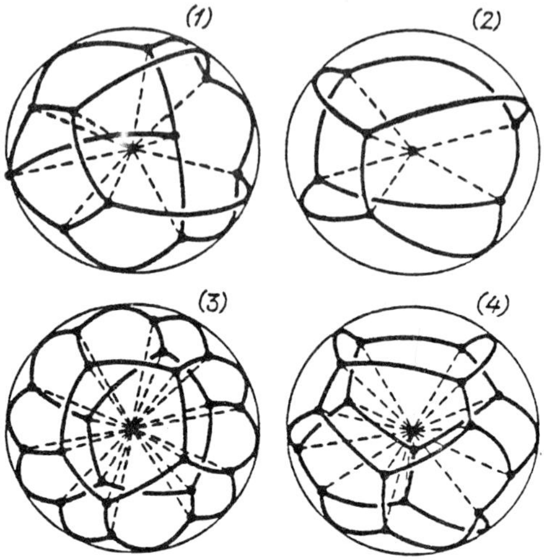

FIGURE 3.20

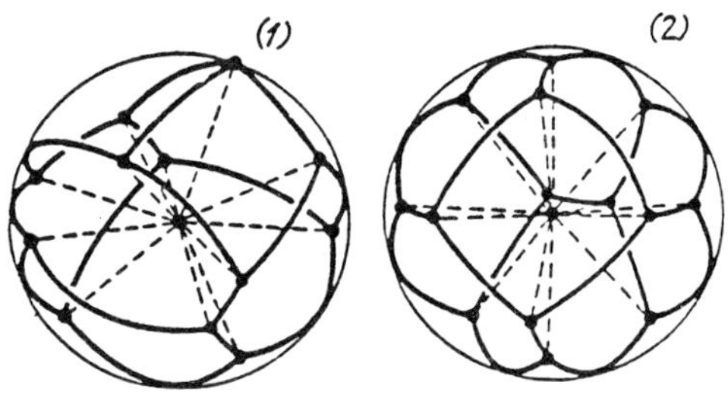

FIGURE 3.21

Are all these configurations actually realized as the intersections of some stable soap films with a small sphere centered at the singular point? It turns out that most of the configurations are not of this type. Of course, if we consider rectilinear cones over these nets with vertex at the center of the sphere, then they have zero mean curvature at all their regular points (see Figures 3.19, 3.20, 3.21). If we neglect a set of singular points of measure zero, then they are minimal surfaces. The first three surfaces, which span respectively a flat circle, three arcs meeting at equal angles at two vertices, and a regular tetrahedron, are stable. The remaining cones are unstable, that is, the vertex of the cone splits, it "blows up" into a topologically more complicated formation, but then having only singular points satisfying Plateau's principles. See, for example, Figures 3.22 and 3.23.

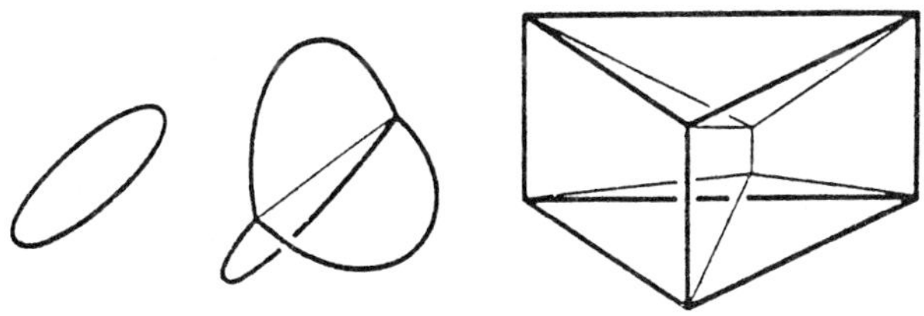

FIGURE 3.22

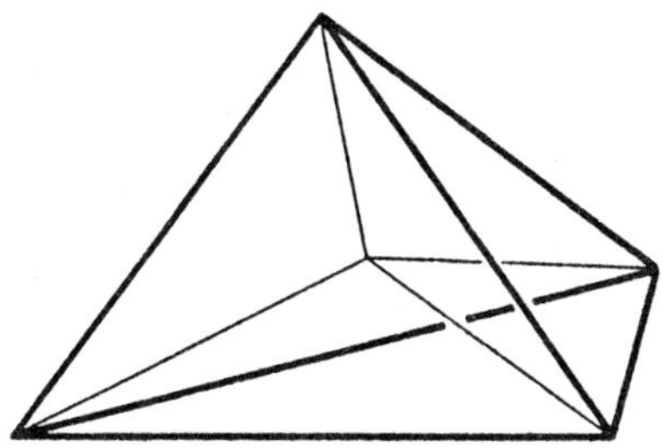

FIGURE 3.23

3.2. Multidimensional minimal cones over submanifolds of a standard sphere. Consider in $\mathbf{R}^n$ the sphere S^{n-1} of radius R with center at the origin, and let A^{n-2} be a smooth $(n-2)$-dimensional compact submanifold of the sphere. Consider the cone CA with vertex at O and base A. The cone is formed by radii.

LEMMA 3.3.1. *If the cone CA is a minimal surface in $\mathbf{R}^n$, then its boundary A is a minimal surface in the sphere S^{n-1} at all its regular points.*

The volume of the cone CA is connected with the $(n-2)$-dimensional volume of its boundary A by the relation $\operatorname{vol}_{n-1} CA = (R/(n-1)) \operatorname{vol}_{n-2} A$, where R is the radius of the sphere. Consequently, any small perturbation of the boundary that decreases its volume induces (along the radii) a similar perturbation of the whole cone that decreases its volume.

For simplicity we limit ourselves here to the case when the boundary of the cone is a smooth closed submanifold of the sphere. Question: do there exist in $\mathbf{R}^n$ cones with vertex at the origin that are minimal in the sense that any small perturbation of them increases the volume? The simplest such cone is a standard flat $(n-1)$-dimensional Euclidean disc that intersects the sphere in an equator. So let us make the question precise: do there exist nontrivial minimal cones other than a standard disc? All geodesics on a sphere are exhausted by equators. With increase of dimension, in S^{n-1} there appear locally minimal submanifolds A^{n-2} that are not equatorial hyperspheres. Thus, for example, in S^3 apart from ordinary equators there is a locally minimal two-dimensional torus embedded as follows. Suppose that $S^3 \subset \mathbf{R}^4 = \mathbf{C}^2(z, w)$; then the torus T^2 is defined as the intersection of the sphere $S^3 = \{|z|^2 + |w|^2 = 1\}$, with the surface $\{|z| = |w|\}$. However, we can verify that a minimal surface situated inside a sphere and having this torus as its boundary *is not a cone.* It turns out that in $\mathbf{R}^4$ there are no nontrivial 3-dimensional minimal cones with center at the origin. The simplest manifolds after spheres that admit simple locally minimal embeddings into a sphere are products of spheres $S^p \times S^q$, where $p+q = n-2$. Suppose that $n = 2$ and consider the circle S^1 and $A = S^0 \times S^0$ (Figure 3.24). In this case the cone with boundary A is the union of two diameters. This one-dimensional "surface" is not minimal at all its points, since the central four-fold point splits into the union of two three-fold points. The real minimal trajectory with boundary A is the union of two parallel segments—a "one-dimensional cylinder" (Figure 3.24). Now suppose that $n = 3$. For A in the sphere S^2 we take $S^1 \times S^0$. Clearly, the two-dimensional cone with boundary A is again not minimal, since there is a contracting deformation in a neighborhood of its vertex. As a result of this variation the vertex of the cone splits and becomes a circle, which is the mouth of the catenoid. Comparing the two-dimensional minimal film—the catenoid—with the one-dimensional film (that is, two parallel line segments), we notice an interesting effect. The two-dimensional film sags in the direction of the origin in contrast to the one–dimensional film. It turns out that the sag of the minimal film increases with the growth of dimension. Figures 3.25–3.27 in the first column cones with boundary $S^p \times S^q$ are shown in the conventional way, in the second column the real minimal surfaces with this boundary, and in the third column the plane sections (generators) of these minimal surfaces by two-dimensional half-planes passing through the axis of symmetry of the surface. The section has the form of a trajectory, which with the growth of dimension of the film sags more and more in the direction of the origin. Consequently, *in this monotonic process there comes a time when the minimal surface sags so much that its mouth breaks up and tightens to a point, and the film becomes a globally minimal cone.* This happens from dimension 8 onwards.

THEOREM 3.3.1 (Almgren [151], Simons [387]). *Let A^{n-2} be a smooth closed locally minimal submanifold of the sphere S^{n-1}, embedded in the standard way in $\mathbf{R}^n$. Suppose that A is not a totally geodesic sphere, that is, an equator. Then for $n \leq 7$ the cone CA, that is, the $(n-1)$-dimensional*

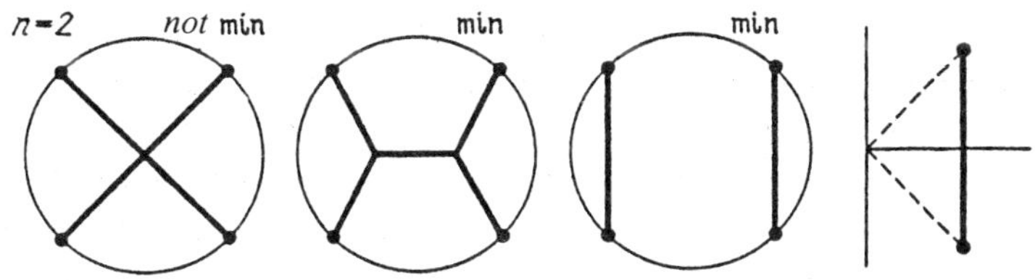

FIGURE 3.24

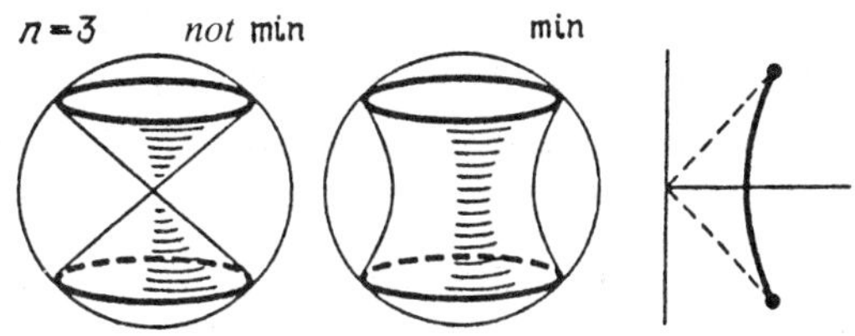

FIGURE 3.25

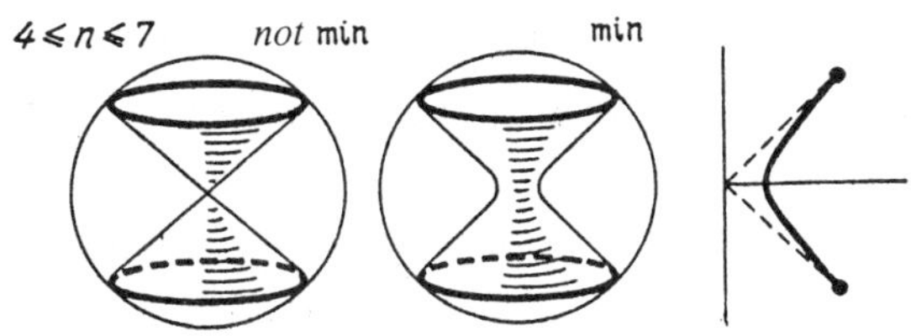

FIGURE 3.26

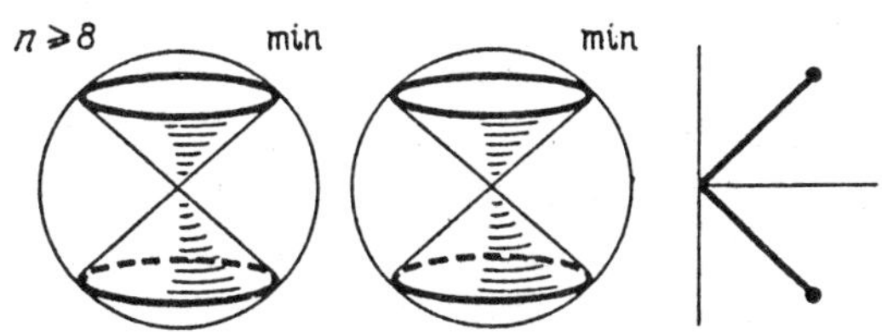

FIGURE 3.27

surface formed by all radii emanating from the origin to a point of A, is not minimal, that is, there is a contracting deformation of the cone that decreases its volume.

This assertion is true not only for Euclidean space, but also for simply-connected complete Riemannian manifolds M^n of constant sectional curvature (sphere and Lobachevsky space). Moreover, this assertion is true for any Riemannian spaces obtained from $\mathbf{R}^n$ by multiplying the Euclidean metric by a positive function $f(R)$ that depends only on the radius R (see Khor′kova [138]).

3.3. Minimal surfaces that have symmetry groups. The solution of the equivariant Plateau problem. Consider a manifold M on which there acts a group of isometries of it. Let I_0 be the connected component of the identity of this Lie group and let G be a connected compact subgroup of I_0. Let $X \subset M$ be a surface invariant under the action of G. Such surfaces are said to be *G-invariant*; they are stratified into the orbits of action of G. Consider the multidimensional Plateau problem. Suppose that in M there is fixed a "contour"—a closed submanifold A^{n-2}, invariant under the action of G. The *equivariant Plateau problem* is formulated as follows: *in which cases can we guarantee the existence of a G-invariant minimal surface X^{n-1} that spans (in the senses mentioned above) a G-invariant boundary?*

Using the methods of differential geometry, we can prove (see [267], for example) that *a G-invariant submanifold X of a manifold M is locally minimal with respect to all sufficiently small variations if and only if it is locally minimal only with respect to all sufficiently small equivariant variations (that is, invariant under the action of the same group).*

Let us consider a partition M into the union of the orbits $G(x)$ of the action of G, where $x \in M$.

We recall that every orbit $G(x)$ admits a representation as a *homogeneous space $G/H(x)$*, where $H(x)$ is the *stationary subgroup of x*, that is, the aggregate of all transformations in G that leave this point fixed. In M we can distinguish a maximal open everywhere dense smooth submanifold $\widetilde{M}$ such that for any two points x and y of it the corresponding stationary subgroups $H(x)$ and $H(y)$ are conjugate in G, that is, there is an element $g \in G$ such that $H(y) = gH(x)g^{-1}$. The orbits $G(x)$ corresponding to points x of $\widetilde{M}$ are called *principal orbits* or *orbits in general position. All these orbits are diffeomorphic and have the same dimension, which is the largest possible.* All the other orbits are called *singular orbits. The set of singular orbits has measure zero in the space of all orbits.*

Let X^q be an arbitrary G-invariant surface in M^n. We set $k = q - s$, where s is the dimension of a principal orbit, that is, k is the codimension of orbits of general position in X.

Every principal orbit $a = G(x)$ is an s-dimensional submanifold. We denote its s-dimensional volume by $v(a)$. Consider the orbit space M/G. Generally speaking, it is not a manifold and may contain singularities. However, it contains an open everywhere dense subset $\widetilde{M/G}$ that is a manifold. Let $\pi: M \to M/G$ be the projection that assigns to a point

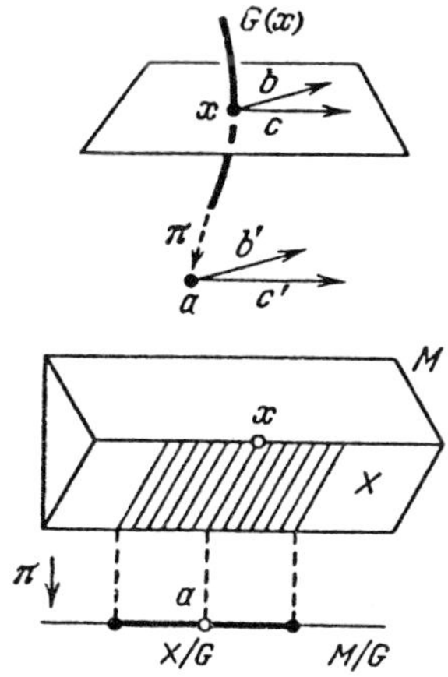

FIGURE 3.28

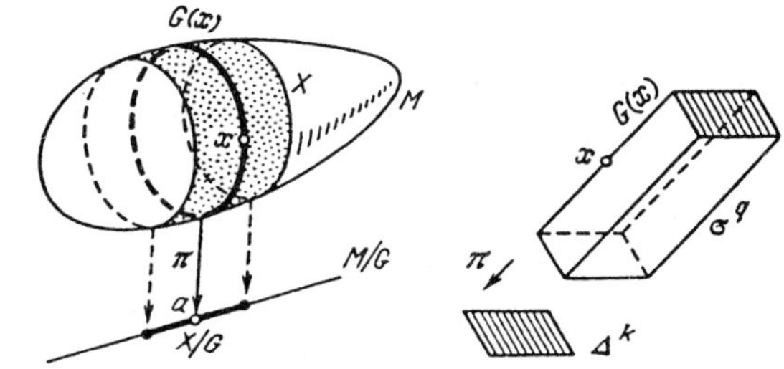

FIGURE 3.29

x its orbit $a = G(x)$. The restriction of π to the open submanifold $\widetilde{M}$ of M defines a smooth bundle with fiber $G(x) = \pi^{-1}(a)$ over the base $\widetilde{M}/G$. Let ds be an invariant metric on M and $d\tilde{s}$ its projection (which is again a Riemannian metric) on $\widetilde{M}/G$ (Figures 3.28 and 3.29). Finally, we define a metric dl on $\widetilde{M}/G$ by $dl = v(a)^{1/k}d\tilde{s}$. Since $v(a)$ is a positive smooth function on $\widetilde{M}/G$, dl is a nondegenerate positive definite metric. Let X be a G-invariant surface in M in general position with respect to the set of singular orbits. Then $\operatorname{vol} X = \operatorname{vol} \widetilde{X}$, where $\widetilde{X} = X \cap \widetilde{M}$.

LEMMA 3.3.2 (Lawson [281]). *If dl is the metric introduced above on $\widetilde{M}/G$, then the q-dimensional volume of a G-invariant surface X^q in M with respect to the metric ds is exactly equal to the k-dimensional volume of the quotient surface $\widetilde{X}/G$ in the quotient space $\widetilde{M}/G$ with respect to the metric dl, where $k = \dim X - \dim G(x)$, and $G(x)$ is a principal orbit.*

PROPOSITION 3.3.1 (Lawson [281]). *Let X^q be a G-invariant minimal submanifold of M. This submanifold is locally minimal (that is, its mean curvature is zero) in M with respect to the metric ds if and only if the quotient manifold $\widetilde{X}/G$ is locally minimal in the quotient manifold $\widetilde{M}/G$ with respect to the metric dl.*

THEOREM 3.3.1′ (see [281]). *Let G be a closed subgroup of the group of proper orthogonal rotations of $\mathbf{R}^n$, that is, $G \subset SO(n)$. Let A^{n-2} be a smooth closed submanifold of the standard sphere S^{n-1}, invariant under the action of G. Then there is a minimal surface (in the generalized sense* *of integral currents)* X_0^{n-1} *with boundary* A^{n-2}, *invariant under the action of* G. *If this G-invariant surface* X_0^{n-1} *is unique, then it is automatically also globally minimal in the class of all surfaces (currents) with the given boundary, not necessarily G-invariant.*

Thus, if the boundary admits a nontrivial symmetry group G, then to find an absolutely minimal surface with this boundary it is sometimes sufficient to find a minimal surface only in the class of G-invariant films; see [281]. It is impossible to reject the condition that the transformations in G preserve the orientation of $\mathbf{R}^n$ (see Theorem 3.3.1′), as simple examples show. Theorem 3.3.1′ was then extended by Brothers [178] to the case of Riemannian manifolds (replacing $\mathbf{R}^n$ by M^n).

3.4. Reduction of the equivariant Plateau problem to the problem of finding a shortest geodesic on a two-dimensional surface with a boundary. Let us turn to the analysis of the cone problem. Suppose that a connected boundary $A^{n-2} \subset S^{n-1}$ is a principal orbit (that is, an orbit in general position) of the action of a compact Lie group G on $\mathbf{R}^n$. Under the projection π onto $\mathbf{R}^n/G$ the boundary A goes into a point $a = \pi A$.

We now turn to *the problem of a complete classification of minimal cones of codimension one in Euclidean space that are G-invariant, where the group* G *acts in* $\mathbf{R}^n$ *with codimension two.* Most of this problem was brilliantly solved by Lawson and Hsiang in [267], [281], but some special cases were not studied, among which, as it turned out, there were some globally minimal and previously unknown cones. A solution of this problem and a *classification theorem* were obtained by A. V. Tyrin and A. T. Fomenko, B. Solomon, L. M. Simon, G. Lawlor on the basis of gauge geometry and a study of conjugate points and the qualitative behavior of the pencil of geodesics emanating from a point that is critical for the function of volumes of orbits.

THEOREM 3.3.2 (Theorem on the classification of minimal cones of codimension one). *Let* $G \subset SO(n)$ *be a connected compact subgroup such that the principal orbits* A *of its action on* $\mathbf{R}^n$ *have codimension two.*

The only globally minimal surfaces of codimension one with boundary A *are the cones over the following manifolds* $A = G/H$:

(1) $S^{r-1} \times S^{s-1}$ *in* $\mathbf{R}^{r+s}$, $r+s \geq 8$, $s, r \neq 2$;
(2) $SO(2) \times SO(k)/\mathbf{Z}_2 \times SO(k-2)b$ *in* $\mathbf{R}^{2k}$, $k \geq 8$;
(3) $SU(2) \times SU(k)/T^1 \times SU(k-2)$ *in* $\mathbf{R}^{4k}$, $k \geq 4$;
(4) $Sp(2) \times Sp(k)/(Sp(1))^2 \times Sp(k-2)$ *in* $\mathbf{R}^{8k}$, $k \geq 2$;
(5) $U(5)/SU(2) \times SU(2) \times T^1$ *in* $\mathbf{R}^{20}$;

(6) $Sp(3)/(Sp(1))^3$ *in* $\mathbf{R}^{14}$;
(7) $F_4/\mathrm{Spin}(8)$ *in* $\mathbf{R}^{26}$;
(8) $\mathrm{Spin}(10)\times U(1)/SU(4)\times T^1$ *in* $\mathbf{R}^{32}$.

Recently, by means of calibration forms on the orbit space, A. O. Ivanov has proved that *several series of symmetric cones of codimension two are minimal*. More precisely, he proved the following result.

The cones CA of codimension two in $\mathbf{R}^N$ over the following locally minimal submanifolds $A^{N-3}=G/H$ in S^{N-1} are the only globally minimal cones with boundary A in $\mathbf{R}^N$:

(1) $S^{r-1}\times S^{r-1}\times S^{r-1}$ in $\mathbf{R}^{3r}$, $r\geq 7$;
(2) $SO(r)\times SO(3)/SO(r-1)\times \mathbf{Z}_2\times\mathbf{Z}_2$ in $\mathbf{R}^r\times\mathbf{R}^5$, $r\geq 53$;
(3) $SO(r)\times SU(3)/SO(r-1)\times T^2$ in $\mathbf{R}^r\times\mathbf{R}^8$, $r\geq 39$;
(4) $SO(r)\times Sp(3)/SO(r-1)\times(Sp(1))^3$ in $\mathbf{R}^r\times\mathbf{R}^{14}$, $r\geq 74$;
(5) $SO(r)\times Sp(2)/SO(r-1)\times T^2$ in $\mathbf{R}^r\times\mathbf{R}^{10}$, $r\geq 51$;
(6) $SO(r)\times G_2/SO(r-1)\times T^2$ in $\mathbf{R}^r\times\mathbf{R}^{14}$, $r\geq 75$;
(7) $SO(r)\times F_4/SO(r-1)\times\mathrm{Spin}(8)$ in $\mathbf{R}^r\times\mathbf{R}^{26}$, $r\geq 74$;
(8) $SO(r)\times\mathrm{Spin}(10)\times U(1)/SO(r-1)\times SU(4)\times T^1$ in $\mathbf{R}^r\times\mathbf{R}^{32}$, $r\geq 136$.

In the case of a cone with a boundary invariant under the action of the subgroup $G\subset SO(N)$, the question of its minimality reduces to the problem of the minimality of a geodesic on the orbit space with metric $dl^2=f^2ds^2$, where the function f is the volume function of the orbit. Therefore the natural wish arises to learn how to obtain the answer in terms of the function f itself.

This turns out to be possible if instead of global minimality we investigate the stability of cones, assuming that the vertex of the cone is fixed. (A surface X with boundary A is said to be stable if the volume of X does not change under small variations with arbitrary support.)

More precisely, A. O. Ivanov obtained the following sufficient condition for stability of cones of arbitrary codimension in a Euclidean space $\mathbf{R}^N$ with invariant boundary.

Let G be a compact connected Lie group acting in $\mathbf{R}^N$ as a subgroup of $SO(N)$. Suppose that the orbit space of this action $\mathbf{R}^N/G\equiv W\subset\mathbf{R}^n$, where W is a subset bounded by hyperplanes passing through the origin 0.

Let $\tilde{f}\colon W\to\mathbf{R}$ be the volume function of an orbit, defined on the orbit space W. Suppose that $\tilde{f}$ is homogeneous with respect to the radius with degree of homogeneity M, that is, if $(r,\tilde{\varphi}_1,\dots,\tilde{\varphi}_{n-1})$ are coordinates in $\mathbf{R}^n \supset W$, where r is the distance from the origin, then

$$\tilde{f}(r, \tilde{\varphi}_1, \dots, \tilde{\varphi}_{n-1}) = r^M \cdot \tilde{f}(1, \tilde{\varphi}_1, \dots, \tilde{\varphi}_{n-1}) \equiv r^M \cdot f(\tilde{\varphi}_1, \dots, \tilde{\varphi}_{n-1}).$$

Suppose that $a \in S^{n-1}(1) \subset \mathbf{R}^n$, where a is a maximum point of the function f on the sphere $S^{n-1}(1) \subset \mathbf{R}^n$.

THEOREM 3.3.3 (Ivanov [439]). *Let $\tilde{f}$ be the volume function of orbits on the orbit space W, homogeneous with respect to the radius with degree of homogeneity M, and f the restriction of the volume function of orbits to the standard sphere $S^{n-1}(1)$.*

Consider the cone CA, stretched on the orbit A, corresponding to the point $a \in S^{n-1}(1) \cap W$, where a is a maximum point of f on the sphere S^{n-1}. If

$$-D_l^2(f)|_a < \frac{f(a)}{4}(M+1)^2,$$

where $l \in T_a S^{n-1}(1)$ is an arbitrary unit tangent vector, then the cone CA is a stable minimal surface with boundary A in $\mathbf{R}^N$ (the vertex of the cone is fixed).

We introduce spherical coordinates $(r, \theta, \varphi_1, \dots, \varphi_{n-2})$ in $\mathbf{R}^n$ so that $a = (1, 0, \varphi_1^0, \dots, \varphi_{n-2}^0)$, that is, θ is the angular distance from the point a on the unit sphere.

The proof of Theorem 3.3.3 is based on the construction in some cone-shaped neighborhood of the segment of the geodesic Oa corresponding to the cone CA in the orbit space of a calibration form of type dg, where

$$g(r, \theta) = f(a) \cdot \frac{r^{M+1}}{M+1} \cos^p(M+1)\theta$$

(a form of this type has been used by I. S. Balinskaya and A. V. Bolsinov).

REMARK. In the statement of Theorem 3.3.3 we have assumed the existence of a maximum point of the volume function of orbits f on the orbit space W. This requirement is natural, since to orbits of maximal volume there correspond locally minimal cones, and only for these does it make sense to verify stability.

In addition, it is not difficult to prove the following assertion.

PROPOSITION. *Suppose that the function f on the Euclidean space $\mathbf{R}^n$ has the form $\prod_{i=1}^m \alpha_i(x)$, where $\alpha_i(x)$ are linear homogeneous functions on $\mathbf{R}^n$ (the volume functions of orbits often have this form). Consider f*

on the set $W = \{x \in \mathbf{R}^n | \alpha_i(x) > 0\}$. *Then the function* f *has a unique extremum point on the set* $W \cap S^{n-1}$, *which is a maximum point.*

It turns out that if in the hypothesis of Theorem 3.3.3 we establish the stability of a segment of the geodesic Oa in some tubular neighborhood of Oa, then we shall thereby prove that Oa, and hence the cone CA, is globally minimal. Therefore in our case the requirement of stability with fixed end is in some sense an optimal weakening of the condition of global minimality. More precisely, we have the following result.

THEOREM 3.3.4 (Ivanov and Bolsinov [439]). *Suppose that* $G \subset SO(N)$ *acts on* $\mathbf{R}^N$ *and that* $W \subset \mathbf{R}^N$, *the orbit space, is a domain bounded by hyperplanes passing through the origin. Let* f *be the volume function of orbits on* W, *homogeneous with respect to the radius, let* $a \in W \cap S^{n-1}(1)$ *be a maximum point of* f *on the sphere* $S^{n-1}(1)$, *and suppose that a segment of the geodesic* Oa *is stable in some tubular neighborhood of* Oa. *Then* Oa *is a globally minimal geodesic among all those emanating from* a *on the boundary of* W.

If the maximum point of the volume function of orbits f is calculated explicitly, then it is easy to use Theorem 3.3.3. Let us give a few examples of this kind.

1. The group $G = SO(q_1) \times \cdots \times SO(q_n)$ acts on $\mathbf{R}^N$, $N = \sum_{i=1}^n q_i$, $n \geq 2$. The orbit space W has the form $W = \{(x_1, \ldots, x_n) | x_i \geq 0\}$, the volume function is $f = \prod_{i=1}^n x_i^{q_i - 1}$, and the degree of homogeneity is $M = \sum_{i=1}^n q_i - n$. The maximum point of f on the sphere $S^{n-1}(1) \cap W$ is $a = (\sqrt{(q_1 - 1)/M}, \ldots, \sqrt{(q_n - 1)/M})$. From Theorem 3.3.3 it follows that for $n = 2$ the cone over the orbit of maximal volume is stable when $q_1 + q_2 \geq 8, q_1 \geq 2$, $q_2 \geq 2$, and for $n \geq 3$ it is stable when $q_i \geq 3$ for all i.

2. Ivanov [423] constructed several series of examples of globally minimal cones of codimension two (see above). Awkward calculations led to the fact that strong restrictions on the dimension of the ambient space were introduced. It turned out that if we replace the requirement of global minimality by the requirement of stability, then Theorem 3.3.3 enables us to weaken these restrictions significantly and gives several new series of stable cones in Euclidean space of codimension two.

In the series of examples given below the orbit space is three-dimensional and has the form $W = \mathbf{R}^1(z \geq 0) \times C(\alpha)$, where $C(\alpha) = \{(x, y) \in \mathbf{R}^2 | 0 \leq \tan y/x \leq \alpha\}$. The volume function of orbits on W has the form $f = r^{k+pm} \cos^k \theta \sin^{pm} \theta \sin^m(p\varphi)$. From Theorem 3.3.3 it follows that the following cones of codimension two are stable (compare [423]):

The group $G = SO(k+1) \times SO(3)$ acts in $\mathbf{R}^{k+1} \times \mathbf{R}^5$ with stationary subgroup $H = SO(k) \times \mathbf{Z}_2 \times \mathbf{Z}_2$. In this case $\alpha = \pi/3$, $p = 3$, $m = 1$. The cone over the orbit of maximal volume is stable when $k > 3$.

The group $G = SO(k+1) \times SU(3)$ acts in $\mathbf{R}^{k+1} \times \mathbf{R}^8$ with stationary subgroup $H = SO(k) \times T^2$, $\alpha = \pi/3$, $p = 3$, $m = 2$. The cone over the orbit of maximal volume is stable when $k > 2$.

The group $G = SO(k+1) \times Sp(3)$ acts in $\mathbf{R}^{k+1} \times \mathbf{R}^{14}$, $H = SO(k) \times (Sp(1))^3$, $\alpha = \pi/3$, $p = 3$, $m = 4$. The cone over the orbit of maximal volume is stable when $k > 2$.

The group $G = SO(k+1) \times Sp(2)$ acts in $\mathbf{R}^{k+1} \times \mathbf{R}^{10}$, $H = SO(k) \times T^2$, $\alpha = \pi/4$, $p = 4$, $m = 2$. The cone over the orbit of maximal volume is stable when $k > 2$.

The group $G = SO(k+1) \times G_2$ acts in $\mathbf{R}^{k+1} \times \mathbf{R}^{14}$, $H = SO(k) \times T^2$, $\alpha = \pi/6$, $p = 6$, $m = 2$. The cone over the orbit of maximal volume is stable when $k > 4$.

The group $G = SO(k+1) \times F_4$ acts in $\mathbf{R}^{k+1} \times \mathbf{R}^{26}$, $H = SO(k) \times \mathrm{Spin}(8)$, $\alpha = \pi/3$, $p = 3$, $m = 4$. The cone over the orbit of maximal volume is stable when $k > 2$.

The group $G = SO(k+1) \times \mathrm{Spin}(10) \times U(1)$ acts in $\mathbf{R}^{k+1} \times \mathbf{R}^{32}$, $H = SO(k) \times SU(4) \times T^1$, $\alpha = \pi/3$, $p = 3$, $m = 10$. The cone over the orbit of maximal volume is stable when $k > 2$.

In the case when the extremum point a of the function f is unknown, in order to apply the theorem we need to estimate the second derivative of f in a neighborhood of the maximum point.

For instance, Ivanov obtained the following example: the group $G = SO(3) \times SO(k)$ acts in $\mathbf{R}^{3k}$ (representation $\rho_3 \otimes \rho_k$, where ρ_k is the standard representation of $SO(k)$). In this case the orbit space has the form $W = \{(x_1, x_2, x_3) | x_1 \geq x_3, x_1 \geq x_2, x_2 \geq x_3\}$, and the volume function of orbits is $f = (x_1 x_2 x_3)^{k-2}(x_1^2 - x_2^2)(x_1^2 - x_3^2)(x_2^2 - x_3^2)$. To find the maximum of f on the sphere $S^2(1)$ we need to solve an equation of the eighth degree with coefficients that depend on the parameter k. But by means of rather awkward estimates we can prove the following: there is a number n_0 such that for all k greater than n_0 the cone over the orbit of maximal volume is stable.

We now dwell on a general method of investigating cones for global minimality, suggested by Lawlor [440].

Let $B^{k-1} \subset S^{n-1} \subset \mathbf{R}^n$ be a smooth submanifold of the standard unit sphere in Euclidean space, and let CB, the cone over B, be a surface of dimension k. To prove that CB is globally minimal it is sufficient to construct a map $\Pi: \mathbf{R}^n \to CB$ such that $\Pi|_{CB} = \mathrm{id}$ and Π does not increase the volume of any k-dimensional surface (by a surface we understand an integral current). This means that for any k-dimensional surface S we have $\mathrm{vol}(S) \geq \mathrm{vol}\,\Pi(S)$.

In fact, if such a map Π has been found, then for any surface S with boundary B we obtain

$$\mathrm{vol}(S) \geq \mathrm{vol}(\Pi(S)) \geq \mathrm{vol}(C_1 B),$$

where $C_1 B$ is the intersection of the cone CB with the unit ball, since $\Pi(S) \supset C_1 B$.

To construct such a map we need the following definitions.

Let B be a smooth $(k-1)$-dimensional submanifold of S^{n-1}, and suppose that $p \in B$. A geodesic normal of length α is an arc of a great circle δ orthogonal to B at the initial point $\delta(0) = p$. If the end point $\delta(\alpha)$ is discarded, we shall speak of an open geodesic normal.

When $p \in B$ and $\alpha > 0$ we denote by $U_p(\alpha)$ the union of all open geodesic normals of length α. The normal wedge $W_p(\alpha)$ is defined as the cone over the set $U_p(\alpha)$. If the origin is discarded, we shall speak of an open normal wedge.

For small α the union of all normal wedges forms a conical neighborhood of CB in $\mathbf{R}^n$; for large α it covers the whole of $\mathbf{R}^n$. We shall be interested in the case when open normal wedges are disjoint.

Suppose that the codimension of a cone is equal to one. We first define our map Π on each wedge W_p. It will project W_p onto the ray Op along curves γ_p whose general form is shown in Figure 3.30. Each curve γ_p projects into its unique point of intersection with Op. The faces of the wedge go into the origin O. We shall also assume that Π preserves a homothety, that is, $\Pi(tx) = t\Pi(x)$, and so in each wedge W_p it is sufficient to specify just one curve γ_p. Having thus defined the map Π on each wedge W_p, we define it on the union of all wedges. The part of $\mathbf{R}^n$ that does not occur in this union is projected into the origin.

Thus, if the wedges W_p are disjoint, the projection Π is well defined, and it remains to choose the curves γ_p so that Π does not increase the volume.

In the case of codimension two and higher, instead of the curves γ_p we need to take surfaces of revolution S_p generated by such curves.

Let us sum up. Suppose we are given a family of curves $\{\gamma_p : p \in B\}$, defined in polar coordinates (r, θ) so that $r(0) = p$, $r(\theta) \to \infty$ as $\theta \to \theta_0(p)$. Let S_p be the curve γ_p or the surface passing through p generated

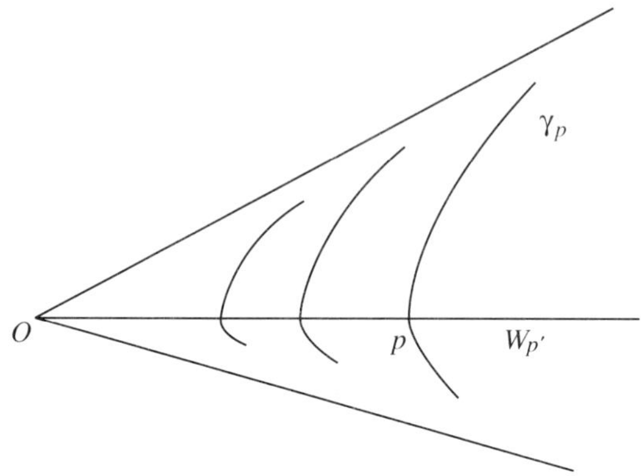

FIGURE 3.30

by the curve γ_p constructed in all two-dimensional planes spanned by the vector $\vec{Op}$ and $\vec{\nu}$, where $\vec{\nu}$ runs through all the normals to the cone CB at the point p.

Consider an open normal wedge $W_p(\theta_0(p))$ and define in it a map Π_p by $\Pi_p(S_p) = p$ and $\Pi_p(tx) = t\Pi_p(x)$ for $t \geq 0$, $x \in S_p$. Then if $W_p(\theta_0(p)) \cap W_q(\theta_0(q)) = \varnothing$ for any $p, q \in B$, we define a map $\Pi\colon \mathbf{R}^n \to CB$ by $\Pi|_{W_p} = \Pi_p$ and $\Pi \equiv 0$ outside $\bigcup_{p\in B} W_p(\theta_0(p))$.

We must now learn how to choose the curves γ_p so that the map Π that we have constructed does not increase the volume of k-dimensional surfaces.

It turns out (see [440]) that for γ_p we need to take a curve satisfying the following differential equation (if such a curve exists):

$$(1)\quad \frac{dr}{d\theta} = r\cdot\sqrt{r^{2k}(\cos\theta)^{2k-2}\cdot \inf(\det(I-\tan(\theta)h^{\nu}_{ij}))^2 - 1},$$

$$(2)\quad r(0) = 1,$$

where (r, θ) are polar coordinates in the plane, $k = \dim B$, h^{ν}_{ij} is the matrix obtained by projecting the second quadratic form h of the manifold B on the vector $\vec{\nu}$, and $\vec{\nu}$ is an arbitrary unit normal to the cone CB at the point p.

We observe that at the point with coordinates (1,0) the hypothesis of the theorem on the existence and uniqueness of the solution of our equation is violated, so there may be several solutions with this initial condition. We need to choose (if possible) a solution such that $r \to \infty$ as $\theta \to \theta_0$, and the smaller the angle θ_0 the better.

If we have succeeded in constructing for each point $p \in B$ a plane curve

γ_p satisfying equation (1) with initial conditions (2) such that $r \to \infty$ as $\theta \to \theta_0(p)$, then there arises a new function on B, namely $\theta_0 \colon p \mapsto \theta_0(p)$. We shall call $\theta_0(p)$ the vanishing angle.

THEOREM 4.3.5 (Lawlor [440]). *Let B be a smooth minimal $(k-1)$-dimensional submanifold of $\mathbf{R}^n$, not necessarily oriented, and let CB be the cone over B. Suppose that on B there is defined a function $\theta_0(p)$, the vanishing angle (see above). For each point $p \in B$ we construct an open normal wedge $W_p(\theta_0(p))$. If no two wedges intersect, then the cone CB is a globally minimal surface in $\mathbf{R}^n$ with boundary B in the class of orientable and nonorientable surfaces.*

It is clearly complicated to use the theorem in this form. In [440] a method is given that makes it possible, if we know the greatest value taken by the norm of the projection of the second fundamental form of the surface B and the dimension of the cone CB, to obtain an estimate for the function $\theta_0(p)$, the vanishing angle. If M is the smallest value of the normal radius of the cone CB, and if $2\theta_0(p) < M$, then the cone CB satisfies the condition of the theorem, and so it is globally minimal. (The normal radius of the cone CB at the point p is the largest angle α such that the wedge $W_p(\alpha)$ intersects the cone only along the ray Op.) This version of the theorem enables us to prove the minimality of certain specific series of cones (see below).

It turns out that the method described above of constructing a retraction Π is equivalent to searching for a calibration form for the cone CB that vanishes outside some (conical) neighborhood of CB. We first construct a singular calibration for the k-plane P that is nullified outside an angular neighborhood N of radius θ. (The point (x, z) lies in N if the angular distance from the plane $P = \{z = 0\}$ is less than θ, that is, $\arctan |z|/|x| < \theta$.)

We begin with the standard form ω that distinguishes the plane P in $\mathbf{R}^n$:

$$\omega = dx_1 \wedge \cdots \wedge dx_k .$$

If we multiply our form ω by the function that is equal to one on P and to zero outside N, we obtain a form which, as before, has unit comass, but it ceases to be closed. We therefore proceed in a different way.

1. We choose a form ψ such that $d\psi = \omega$:

$$\psi = \frac{1}{k}(x_1\, dx_2 \wedge dx_3 \wedge \cdots \wedge dx_k - \cdots \pm x_k\, dx_1 \wedge \cdots \wedge dx_{k-1}) .$$

We observe that $d\psi = \omega$ and $\omega = \frac{k}{r} dr \wedge \psi$, where $r^2 = \sum_{i=1}^k x_i^2$.

2. We multiply ψ by a function $g(t)$ that vanishes outside N. For the parameter t we take $t = |z|/r$, and we need $g(0) = 1$, $g'(0) = 0$, and $g(t) \equiv 0$ for $t \geq \tan\theta$, where θ is the angular radius of N.

3. We put $\varphi = d(g\psi)$. We note that the form $g\psi$ is not continuously differentiable, but it is a Lipschitz form, so (see Federer [219] and also Lawlor [440]) it can be used as a calibration form.

The form $\varphi = d(g\psi)$ is automatically closed. The requirement that the comass is one gives an ordinary differential equation of the form

$$\left(g - \frac{t}{k}g'\right)^2 + \left(\frac{g'}{k}\right)^2 = 1.$$

This differential equation can be integrated. When $k \geq 3$ it has solutions satisfying the requirements of part 2. The smallest angle θ for which $g(\tan\theta) = 0$ is approximately $\pi/2k$ as $k \to \infty$. We note that the calibration we have constructed enables us to prove the minimality of the sheaf of k-dimensional planes intersecting at the origin if the angular distance between them is sufficiently large.

To construct a calibration form for the cone we need the following definition. Let M be a k-dimensional surface in $\mathbf{R}^n$. Let $e\colon U \subset \mathbf{R}^n \to M$ be regular coordinates on M. We extend these coordinates to a neighborhood of M in $\mathbf{R}^n$ by setting $f = (e, e')$, where e' is an isometry from $(x, \mathbf{R}^{n-k})$ to the normal space to M at the point $e(x)$. If the resulting map f is of class C^∞, then f is called an *isonormal extension* of the coordinates $e = (e^1, \dots, e^k)$.

Let e be a coordinate map of a neighborhood $U \subset S^{k-1}$ into $V \subset B$, chosen (see Moser [441]) so that it preserves area. We extend the coordinates to the cone CB by setting $e^k = r$, and $e(ax) = ae(x)$ for $x \in U$. Let $f = \{e^1, \dots, e^n\}\colon CU' \to CV'$ be an isonormal extension of the coordinates e, where U' is a neighborhood of U in S^{n-1}, and V' is the corresponding neighborhood of V. For convenience we choose f so that $f(az) = af(z)$.

Let ψ be the $(k-1)$-form in $\mathbf{R}^n$ defined above. Consider the form

$$\tilde{\psi} = (f^{-1})^{\#}\psi$$

carried over by means of the coordinate map. For a form that calibrates the cone CB we take the outer derivative of the Lipschitz form $g(t)\tilde{\psi}$:

$$\varphi = d(g\psi).$$

THEOREM 4.3.6 (Lawlor [440]). *Let B be an orientable smooth submanifold of the sphere S^{n-1}, and CB the cone over B. Suppose that for each* *point* $p \in B$ *there is a function* $g_p(t)$ *such that* $g_p(0) = 1$ *and*

$$\left(g_p - \frac{t}{k} g_p'\right)^2 + \left(\frac{g_p'}{k}\right)^2 \leq \det(I - t h_{ij}^{\nu})^2$$

for all unit normals ν *to* CB *at* p. *If* $\theta(p)$ *is a function on* B *such that* $g_p(\tan\theta(p)) = 0$ *and the normal wedges of radius* $\theta(p)$ *are disjoint, then the form constructed above is a well-defined calibration form and* CB *is a globally minimal surface in* $\mathbf{R}^n$ *with boundary* B.

It turns out that the differential equations in Theorems 4.3.5 and 4.3.6 give an equivalent result. The calibration form φ is connected with the retraction Π (see above) as follows.

PROPOSITION ([440]). *The k-dimensional plane dual to the calibration form* φ *at each point where* $\varphi \neq 0$ *is the orthogonal complement to the tangent plane to the retraction surface* S_p.

REMARK. Theorems 4.3.5 and 4.3.6 are sufficient conditions for the global minimality of cones in Euclidean space. Lawlor showed that, generally speaking, they are not necessary. However, if we introduce some additional restrictions on the boundary of the cone B, then Theorems 4.3.5 and 4.3.6 turn into criteria for global minimality (for the details see [440]).

It is still not clear whether the results obtained by Ivanov (Theorem 3.3.3) follow from the results of Lawlor, although formally they do not, since, for example, for the surface $B = S^1 \times S^5$ Theorem 3.3.3 immediately gives stability, while the hypotheses of Theorems 4.3.5 and 4.3.6 are not satisfied and we need additional investigation of the differential equations, which, however, can be carried out.

Using Theorems 4.3.5 and 4.3.6, Lawlor obtained the following results:

1. THEOREM 4.3.7 (classification of cones over products of spheres). *Let* C *be a cone over a product of two or more spheres. Then*

if $\dim C > 7$, *the cone* C *is globally minimal*;

if $\dim C < 7$, *the cone* C *is not stable*;

if $\dim C = 7$, *then the cone* C *is stable. In this case the cone is globally minimal if and only if one of the spheres is not a circle.*

This theorem completes the investigations of Bombieri, de Giorgi, and Giusti [170], Lawson [281], Federer [219], Fomenko and Tyrin [136], Simons [429], Bindschadler [169], Ivanov [423], and Cheng [442].

2. *An example of a minimal cone over an unorientable manifold.* Let B be the image in the space $\mathbf{R}^{k^2}$ of the unit sphere $S^{k-1} \subset \mathbf{R}^k$ under the map

$$F: X \mapsto X \cdot X^T,$$

where X is a column vector in $\mathbf{R}^k$ and $X \cdot X^T$ is a $(k \times k)$-matrix. It is easy to show that $B \subseteq \mathbf{R}P^{k-1}$, where $\mathbf{R}P^{k-1}$ is a projective space of dimension $k-1$. If $k-1$ is even, then B is unorientable. Lawlor proved that the cone CB is a globally minimal surface in the class of unorientable surfaces with boundary B when $k \geq 4$.

3. Cheng [442] showed that cones over the following compact matrix groups, realized in the space of all matrices as submanifolds of a sphere, are special Lagrangian surfaces, and therefore globally minimal: the cone over $SU(m)$ in $\mathbf{R}^{2m^2}$ and the cone over $U(m)$ in $\mathbf{R}^{m^2}$ when $m \geq 4$. Lawlor [440] improved this result by showing that the cones over $O(m)$, $SO(m)$, $U(m)$ and $SU(m)$ are globally minimal in $\mathbf{R}^{m^2}$ or $\mathbf{R}^{2m^2}$ when $m \geq 2$.

4. If A and B are minimal submanifolds of a sphere, and the cone over $A \times B$ is globally minimal, then the dimension is not less than six [440].

If B is a minimal five-dimensional submanifold of a sphere and the norm of the second quadratic form does not exceed $\sqrt{5}$ at each point, then the cone over $S^1 \times B$ is stable, but not globally minimal [440].

There are many surfaces satisfying these conditions, for example, $\mathbf{R}P^2 \times \mathbf{R}P^3$, $\mathbf{R}P^2 \times SO(3)$, $\mathbf{R}P^5$, $S^2 \times SO(3)$.

As we have seen, in the solution of the equivariant Plateau problem it is important to take the volumes of orbits of smooth actions of Lie groups H on homogeneous spaces of the form $V = G/H$. For many cases of large codimension this problem was solved by Balinskaya (Novikova) [424].

Apart from the cones of codimension two described above, several new series of minimal cones of high codimension are known.

Consider the simple classical Lie group $\exp G$ that acts by means of the adjoint representation on its Lie algebra G, identified with Euclidean space $\mathbf{R}^M$. The Euclidean metric in G can be regarded as the Killing metric of G. Then there arises a natural variant of the spherical Bernstein problem (for more details about this problem see the next Subsection 3.5), namely: to describe minimal submanifolds of high codimension lying in the Euclidean sphere S^{M-1} (in the Lie algebra G) that are invariant with respect to the adjoint action of the group $\exp G$. Clearly, cones over such manifolds will be locally minimal in the Lie algebra. The vertex of the cone is situated at the origin. Question: how to describe globally minimal cones of this type?

Let A be a locally minimal orbit in general position (in the Lie algebra) and CA the cone over it (with vertex at the origin). I. S. Balinskaya (Novikova) has obtained the following important results.

THEOREM 3.3.8 (see [424], [425], [429], [446]). (1) *In the Lie algebra G (where G is one of the Lie algebras $so(n)$, $sp(n)$, $su(n)$) there is a unique locally minimal principal orbit A (i.e. in general position).*

(2) *This orbit has maximal volume in the class of all orbits in general position.*

(3) *For small values of the dimension n, namely for $n \leq 4$, $n \leq 3$, $n \leq 8$ for the groups $SU(n)$, $Sp(n)$, $SO(n)$ respectively, the cone CA is unstable. For the remaining n the cone CA is stable.*

(4) *For each series of simple compact Lie groups $SU(n)$, $Sp(n)$, $SO(n)$, starting with some n_0, the cone CA is globally minimal. This number n_0 does not exceed* 37, 17, 36 *respectively.*

We give a very brief plan of the proof of the theorem.

By means of the theorems of Lawson [281] described above the problem reduces to the problem on the orbit space, and takes the following form.

On the Cartan subalgebra $H^r \subset G$, to orbits of the action of $\exp G$ on the algebra G there correspond the points of intersection of these orbits with the Cartan subalgebra. Nonuniqueness is eliminated if we consider not the whole Cartan subalgebra but only the Weyl chamber, in which each orbit intersects the Cartan subalgebra only once. Since the volume function of orbits is homogeneous with respect to the radius R, the problem can be restricted to the sphere $S^{r-1}(1) \subset H^r$, more precisely to the part of the sphere that lies in the chosen Weyl chamber. To the cone CA in the Weyl chamber there corresponds the interval th_0, where h_0 is the point of intersection of the orbit A with the Cartan subalgebra.

LEMMA 1. *The interval th_0 is geodesic in the sense of the metric $ds^2 = v^2 dl^2$ if and only if h_0 is an extremum point of the volume function v, restricted to the sphere $S^{r-1} \subset H^r$.*

LEMMA 2. *The volume function of orbits has the form (Weyl formula)*

$$v(h) = c \cdot \prod_{\alpha \in \Sigma^+} \alpha^2(h),$$

where c is a constant independent of the volume, α are the roots of the algebra G, and Σ^+ is the system of positive roots.

LEMMA 3. *Suppose that on the linear Euclidean r-dimensional space L we are given a function of the form $f(x) = \prod_{i=1}^{s} \alpha_i(x)$, where $\alpha_i \in L^*$ are linear functions. The sphere $S^{r-1} \subset L$ is split by the planes $\{\alpha_i = 0\}$ into certain "polyhedra". We assert that in each such polyhedron there is a unique extremum point of the function $f|_{S^{r-1}}$.*

The proof of these three lemmas presents no great difficulty. Assertions (1) and (2) of Theorem 3.3.8 follow automatically from them.

To prove Assertions (3) and (4) of Theorem 3.3.8 we need some additional arguments of a fundamentally different character.

LEMMA 4. *At the extremum point h_0 of the volume function $v|_{S^{r-1}}$ we have the estimate $\|B\| \le 2rN$, where B is the Hessian of v, $r = \dim H$, and $N = \dim G - r$ is the dimension of an orbit in general position.*

The proof of the lemma consists in rather awkward calculations, which are carried out for each series of classical Lie algebras separately, although there is probably a general (single) proof for all simple compact Lie algebras.

In each case separately, starting from the basic equality $\operatorname{grad} v(h_0) = \lambda h_0$, where h_0 is the extremum point and λ is the Lagrange multiplier, we obtain intermediate estimates for the Hessian $B(h_0)$. The main idea is as follows.

To obtain a lower bound for the spectrum of the Hessian we need to subtract from λ a necessarily positive definite matrix so as to obtain a diagonal matrix. We look for such a necessarily positive matrix in the form MM^T. For example, in the case $SO(2n)$ the basic equality is

$$\sum_j \left(\frac{1}{x_i - x_j} + \frac{1}{x_i - x_j} \right) = \lambda x_i .$$

The Hessian is given by

$$B_{ij} = \frac{1}{(x_i - x_j)^2}, \qquad i \neq j,$$

$$B_{ii} = -\sum_{k \neq i} \frac{1}{(x_i - x_k)^2} .$$

For a matrix M we have

$$m_{ik} = \frac{2x_i}{x_i^2 - x_k^2}, \qquad m_{ii} = dx_i ,$$

where d is the degree of homogeneity of the function $v(h_0)$. For the smallest eigenvalue μ of the Hessian B we have the estimate

$$\mu \geq -d(2n-1).$$

From this there follows the assertion of the lemma: $\|B\| \leq 2rN$.

This estimate has a key significance for the proof of stability and global minimality of cones in the given problem.

LEMMA 5. *The cone CA is unstable if there is at least one eigenvalue μ of the Hessian of the volume function B for which*

$$-\mu < \tfrac{1}{2}v(h_0)(d+1)^2.$$

The proof of the lemma follows from an investigation of the Jacobi field along the geodesic th_0. Assertion (3) of Theorem 3.3.8 follows from Lemma 5 together with the estimates of Lemma 4.

To investigate global minimality we construct a calibration form ω', that is, a form such that on the geodesic $\rho(t) = th_0$ we have

(1) $|\omega'(a)| \leq |a|$ for any tangent vector a,

(2) $|\omega'(a)| = |a|$ if a is a vector tangent to the geodesic $\rho(t)$.

It is known that if a form ω' exists, then the geodesic $\rho(t)$ is a shortest curve.

We look for the calibration form ω' in the form df, where from general arguments we have

$$f = \begin{cases} 0 & \text{for } c_2\varphi > \pi/2, \\ c_1 R^{N+1} \cos^p c_2\varphi & \text{for } c_2\varphi \leq \pi/2. \end{cases}$$

Here R is the distance of the point h from the origin, and $\varphi = \varphi(h)$ is the angle between the vectors h and h_0; c_1, c_2, p are certain constants, chosen so that the support of f is entirely contained in the Weyl chamber. If we set $pc_2 = N+1$, we have

$$|\operatorname{grad} f| = c_1(N+1)R^N \cos^{p-1} c_2\varphi.$$

Since $|\operatorname{grad} f(th_0)| = v(th_0)$, it follows that $c_1 = v(h_0)/(N+1)$.

For simplicity we can assume that $p = 2$, although this choice is not optimal, and all the estimates carried out below can be improved.

From the form of $|\operatorname{grad} f|$ it follows that the conditions on the form ω' at the point h_0 are satisfied automatically.

Since $|\operatorname{grad} f|$ and v coincide at the point h_0, and their derivatives vanish, we can estimate not the functions themselves, but their logarithmic derivatives, which is much simpler. We need to verify when the following are satisfied:

$$\left|\frac{d^2}{dt^2}\ln|\operatorname{grad} f|\right| \le \left|\frac{d^2}{dt^2}\ln v\right|,$$

$$\left|\frac{d^2}{dt^2}\ln|\operatorname{grad} f|\right| \le \frac{(N+1)^2}{4},$$

$$\frac{d^2}{dt^2}\ln v(\gamma(t)) = -N - \sum_{\alpha\in\Sigma^+}\frac{\alpha^2(\dot\gamma)}{\alpha^2(\gamma)}.$$

If we define the bilinear symmetric form

$$\widetilde{B}(a, b) = -\sum_{\alpha}\frac{\alpha(a)\cdot\alpha(b)}{\alpha^2(\gamma)},$$

then it is easy to see that $\widetilde{B}$ is the restriction to the sphere of the analogous form $B(a, b)$, which is defined on the Weyl chamber and is the Hessian of the volume function $v(h)$ in Cartesian coordinates. Thus we need to prove that $N+\|\widetilde{B}\| \le (N+1)^2/4$. We now have to verify that the supports of f and ω' lie in the Weyl chamber.

The next lemma connects the norm of the bilinear function $\widetilde{B}$ with the distance from h to the nearest wall of the Weyl chamber.

LEMMA 6. *Let* $\varepsilon = \min_{\alpha\in\Sigma^+}\alpha(h)$. *Then*

$$c_1\varepsilon^{-2} \le \|B\| \le c_2\varepsilon^{-2},$$

where $\|B\| = \max_{|a|=1}\sum\alpha^2(a)/\alpha^2(h)$, *and* c_1, c_2 *are constants that do not depend on the type of Lie algebra or its dimension.*

PROOF. We have $B(a, a) = -\sum_\alpha \alpha^2(a)/\alpha^2(h)$, $|a| = 1$. Grouping the terms according to the height of the roots and estimating each sum separately, we obtain

$$|B(a, a)| \le \frac{4|a|^2}{\varepsilon^2} + \frac{4|a|^2}{(2\varepsilon)^2} + \cdots \le \frac{4|a|^2}{\varepsilon^2}\sum\frac{1}{n^2} \le \frac{c}{\varepsilon^2}.$$

The estimate in the other direction is almost obvious. We just need to take for a the vector $h_\alpha/|h_\alpha|$, which completes the proof.

LEMMA 7. *For large values of* r *and* N *the support of the form lies entirely in the Weyl chamber* W, *and for any point* $h \in S^{r-1}$ *that occurs in the support we have* $1/\alpha^2(h) \le 2rN$.

The lemma is proved by means of Lemma 4 and the triangle inequality for points lying in the support.

Now the inequality $N + \|\widetilde{B}\| \le (N+1)^2/4$ for large r and N follows automatically for points lying in the support of ω'.

We can estimate the dimension starting from which the cones certainly become globally minimal by proceeding from the inequality $N + 2c_2 rN \le (N+1)^2/4$; this is condition (1) on the form ω'. We have for

$$\begin{array}{ll} SU(n), & N = r(r+1); \\ SO(n), & N = 2r(r+1); \\ Sp(n) \text{ and } SO(2n+1), & N = 2r(r+2). \end{array}$$

Let us replace the inequality by the rougher inequality $N/r \ge 37$. Then for

$$\begin{array}{ll} SU(n), & r \ge 36; \\ SO(n), & r \ge 18; \\ SO(2n+1) \text{ and } Sp(n), & r \ge 17. \end{array}$$

This completes the proof of Theorem 3.3.8. Obviously the estimates in the theorem are not optimal and can be improved.

3.5. In the Euclidean space of large dimension there are minimal hypersurfaces that are the graphs of nonlinear functions. Let us consider in $\mathbf{R}^n$ the graph of a smooth function $x_n = f(x_1, \dots, x_{n-1})$, defined *on a whole hyperplane* $\mathbf{R}^{n-1}$. The graph is a smooth submanifold of $\mathbf{R}^n$. Suppose that it is *locally minimal*. The question arises, *when is f linear, and so its graph is a hyperplane*? S. N. Bernstein gave a positive answer to this question for $n = 3$. He proved that *if $z = f(x, y)$ is the graph of a smooth function defined on the whole plane $\mathbf{R}^2$, and if the Gaussian curvature K of this graph is nonpositive and there is a point at which K is strictly negative, then* $\sup|f| = +\infty$, $(x, y) \in \mathbf{R}^2$. The solution of the problem described above follows easily from this. Then a similar question (the *Bernstein problem*) was posed in large dimensions. Of course, the arguments given above do not work in dimensions higher than three, so for further progress in the question of the existence of nonlinear locally minimal hypersurfaces in $\mathbf{R}^n$ we need to develop new mathematical technique. As a result, in a well-known paper of Bombieri, de Giorgi, and Giusti [170] (see also the survey in [387]) it was discovered that a definitive answer depends on the dimension n.

The following remarkable result gives an exhaustive answer to the question posed above. The lower bounds on the dimension in the theorem were obtained by Almgren and Simons (see [387]), and the upper bounds were obtained by Bombieri, de Giorgi, and Giusti [170].

THEOREM 3.3.9 (see [170]). *Let $x_n = f(x_1, \dots, x_{n-1})$ be a smooth function defined on a whole hyperplane $\mathbf{R}^{n-1}$ in $\mathbf{R}^n$. Suppose that its graph is locally minimal, that is, it has zero mean curvature. Then for $n \leq 8$ the function f is linear. If $n \geq 9$, then there are nonlinear functions whose graphs are not only locally but even globally minimal surfaces.*

In the case of large codimensions Bernstein's conjecture can be generalized in several directions. In particular, there has been quite a lot of talk about properties of so-called tubular minimal surfaces of large codimension, investigated in detail in a series of interesting papers by Miklyukov [68]–[70].

In 1969 Chern [188], [189] stated the following so-called *spherical Bernstein problem.* Suppose that a sphere S^{n-1} is embedded as a minimal hypersurface in the standard sphere S^n. Does it follow from this that S^{n-1} must be an equator? When $n = 2$, the answer is obviously yes. When $n = 3$, the answer is also yes; this was known earlier from the work of Almgren [151] and Calabi [179]. Finally, in 1982 there appeared an important paper of Hsiang [270], in which he produced infinitely many distinct examples of minimal embeddings of S^{n-1} in S^n that are not equators for $n = 4, 5, 6$. The construction was carried out on the basis of the methods of the equivariant Plateau problem (Hsiang and Lawson [267]). Minimal spheres $S^{n-1} \subset S^n$ turned out to be G-invariant (for suitable actions of groups G on S^n with two-dimensional orbit space); see also [271]. Then Hsiang also constructed embeddings of spheres for $n = 7, 8, 10, 12, 14$. It is likely that such minimal embeddings of spheres exist for all $n \geq 4$. The following questions were posed in [271]. (1) Let M^n be a simply-connected compact symmetric space. Is there a minimal embedding of S^{n-1} in M^n? (2) Suppose that there is a minimal embedding of S^{n-1} in M^n. Is it true that any other minimal embedding of S^{n-1} in M^n is necessarily congruent to the original embedding? However, in view of a negative answer to the spherical Bernstein problem (for $M^n = S^n$) the next question is of particular interest. (3) Let M^n be a compact simply-connected symmetric space with $n \geq 4$. Are there infinitely many distinct (noncongruent) minimal embeddings of S^{n-1} in M^n? The answer is yes for $M^n = S^n$, $n = 4, 5, 6, 7, 8, 10, 12, 14$, for $M^{2n} = S^n \times S^n$ with $n = 2$ or 3 (the radii of the spheres are assumed to be equal), and for all $\mathbf{C}P^n$ with $n \geq 2$. It was proved in [271] that there are infinitely many distinct (noncongruent) minimal embeddings of S^{2n-1} in $\mathbf{C}P^n$ when $n \geq 2$.

Kesel′man [57] proved that if a surface in $\mathbf{R}^3$, explicitly defined over a whole plane, has nonpositive curvature and its Gaussian (spherical) map is quasiconformal with $Q < \sqrt[3]{4/3}$, then this surface is a plane (Bernstein's theorem for $Q = 1$, that is, for a conformal Gaussian map).

3.6. The Jacobi equation on minimal homogeneous surfaces in homogeneous Riemannian spaces. Classification of stably minimal simple subgroups of classical Lie groups. We briefly present some recent results of Le Khong Van.

Let $M = G/K$ be a homogeneous space with G-invariant Riemannian metric, and N a minimal homogeneous submanifold H/L of M, where $L = K \cap H$. Let $m^\perp$ denote the normal fiber over the point $x_0 = \{eL\}/L$, $m^\perp \subset T_{x_0} G/K \subset lG$. Obviously, the normal bundle $n(H/L)$, on which H acts on the left, is H-equivalent to the bundle $H_{\otimes \mathrm{Ad}} m^\perp$, modulo the action of $\mathrm{Ad}\, L$. With each section $\psi \in \Gamma(n/H)$ we associate the $m^\perp$-valued function $\overline{\psi} \in C^\infty(H, m^\perp)$ on H such that

$$\overline{\psi}(h) = h^{-1}\psi(h/L). \tag{3.1}$$

Clearly, $\overline{\psi}$ satisfies the condition

$$\overline{\psi}(hl) = \mathrm{Ad}(l^{-1})\overline{\psi}(h). \tag{3.2}$$

Let $C^\infty(H, m^\perp)_L$ denote the subspace of $C^\infty(H, m^\perp)$ distinguished by the equation (3.2). From the above it follows that the correspondence between the space of normal sections $\Gamma(n(H/L))$ and $C^\infty(H, m^\perp)_L$ given by (3.1) is one-to-one. Therefore, any operator (in particular, the Laplace operator and the Jacobi operator) $I\colon \Gamma(n(H/L)) \to \Gamma(n(H/L))$ can be lifted to an operator $\overline{I}\colon C^\infty(H, m^\perp)_L \to C^\infty(H, m^\perp)_L$.

Before stating the main theorem we introduce some new notation. Consider the orthogonal decompositions

$$lG = lk + m, \quad lH = lL + p, \quad lL = lK \cap lH, \quad m = p_m + m^\perp,$$

where p_m is the orthogonal projection of the tangent space $p = T_{x_0}(H/L)$ on the tangent space $m = T_{x_0}(G/K)$.

We set $m' = p + m^\perp$. It is clear that the projection π_m is a 1-1 map from m' onto m.

We also define a linear operator $\theta\colon m \to \mathrm{End}(m)$ as follows:

$$\langle \theta(v)X, Y\rangle = \langle [v, X]_m, Y\rangle + \langle [Y, v]_m, X\rangle + \langle [Y, X]_m, v\rangle \tag{3.3}$$

where $\langle\ ,\ \rangle$ denotes the K-invariant metric on m which is the restriction of the G-invariant Riemannian metric to the tangent space $T_{x_0}(G/K) = m$. We set $e_0 = (1/2)\sum_i \theta(e_i)e_i$, where e_i, $i = 1, \dots, s$, runs through the orthonormal basis in p_m.

PROPOSITION. (a) *The second fundamental form at a point* $x_0 \in H/L$ *is expressed as* $A(W)(X, Y) = (1/2)\langle -\theta(X_m)W, Y_m\rangle$.

(b) *The Ricci transformation in the normal bundle* $m^{\perp}$ (*where* x_m (*resp.,* $x_{m^{\perp}}$) *is the orthogonal projection of* x *on* m (*resp.,* $m^{\perp}$)) *is expressed as*

$$\overline{R}(e_i, W)e_i = (([\theta(e_i), \theta(W)] - \theta[e_i, W])e_i)_{m^{\perp}}.$$

THEOREM 3.3.10. *The lift of the Jacobi operator* I *to the space* $C^{\infty}(H, m^{\perp})_L$ *has the form*

$$\overline{I}(\overline{\psi}) = -C^{\mathrm{diff}}_{H,L}(\overline{\psi}) - \sum_{1}^{s} E_i(\theta(e_i)\overline{\psi}) + E_0(\overline{\psi}) + \left(\tfrac{1}{2}\theta(e_0) - \tfrac{1}{4}\sum_{1}^{s}\theta^2(e_i) + \overline{R} - \widetilde{A}\right)(\overline{\psi}).$$

Here $\{E_i\}$ *are left-invariant fields on* H, *regarded as first-order differential operators, and* $\{e_i = E_i(x_0)_m,\ i = 1, \dots, s\}$ *form an orthonormal basis in* p_m; $C^{\mathrm{diff}}_{H,L} = \sum_1^s E_iE_i$, $\overline{R}$ *and* $\widetilde{A}$ *are the curvature operator and the operator of the second fundamental form respectively.*

Clearly, the operator $\overline{I}$ can be extended by linearity to the space $C^{\infty}(H, m^{\perp}_{\otimes\mathbf{C}})_L$. By the Peter-Weyl theorem and the Frobenius duality principle, the latter is isomorphic to the direct sum $\bigoplus_{\lambda\in D(H)} \tau(V_\lambda \otimes \mathrm{Hom}_L(V_\lambda, m^{\perp}_{\otimes\mathbf{C}}))$, where $\tau(v_\lambda \otimes T)(h) = T(\lambda(h^{-1})v_\lambda)$. Here $D(H)$ is the set of all complex irreducible representations λ of the group H, and Hom_L is the set of L-invariant linear operators.

THEOREM 3.3.11. *The subspace* $\tau(V_\lambda \otimes \mathrm{Hom}_L(V_\lambda, m^{\perp}_{\otimes\mathbf{C}}))$ *is an invariant subspace of the Jacobi operator* I, *which acts on it as* $\overline{I}\tau(v_\lambda \otimes T) = \tau(v_\lambda \otimes \overline{I}_*T)$, *where* $\overline{I}_*$ *is a linear operator on the space* $\mathrm{Hom}_L(V_\lambda, m^{\perp}_{\otimes\mathbf{C}})$.

If we consider the special case where H/L is a totally geodesic submanifold of the space G/K, endowed with the Killing metric, then we have $p_m = p$, $\theta(X) = (\mathrm{ad}\, X)_m$, $\widetilde{A} = 0 = e_0$, so Theorems 3.3.10 and 3.3.11 can be restated as follows.

COROLLARY. *In the case under consideration, the lift of the Jacobi operator* I *has the form*

$$\overline{I}(\overline{\psi}) = -C^{\mathrm{diff}}_{H,L}(\overline{\psi}) - \sum_{1}^{s}(E_i[e_i, \overline{\psi}]_m - [e_i, [e_i, \overline{\psi}]_{lk}]).$$

The induced operator $\overline{I}_*$ *on* $\operatorname{Hom}_L(V_\lambda, m^\perp_{\otimes \mathbf{C}})$ *has the form*

$$(\overline{I}_* T)v = -\sum_1^s (T(\lambda^2(e_i)v) + [e_i, T(\lambda(e_i)v)]_m + [e_i, [e_i, Tv]_{lk}]_m).$$

REMARK. The idea of using the space $C^\infty(H, m^\perp)_L$ to calculate distinct invariant operators goes back to Smith [394]. In that paper and also in the recent paper of Ohnita [463] only cases of a totally geodesic embedding of H/L in a symmetric space G/K with canonical metric are considered.

Let us consider the special case where M is a compact Lie group G, endowed with the Killing metric, and N is a compact subgroup H of G. As a symmetric Riemannian space G (resp., H) can be represented as $G \times G/\operatorname{Diag}(G \times G)$ (resp., $H \times H/\operatorname{Diag}(H \times H)$). Clearly, H is a totally geodesic submanifold of G, and in this case the corollary looks like this.

PROPOSITION. *The Jacobi operator* I *on* $C^\infty(H \times H, m^\perp)_{\operatorname{Diag}(H\times H)}$ *has the form*

$$\overline{I}(\overline{\psi}) = -C^{\text{diff}}_{H\times H}(\overline{\psi}) + C^{\text{alg}}_{H\times H}(\overline{\psi}),$$

where $H \times H$ *acts on the algebra* $lG + lG$ *by means of the adjoint representation* $\operatorname{Ad}(\rho \otimes \rho)$, *where* ρ *is an embedding* $H \to G$, *and* $C^{\text{diff}}_{H\times H}$ (*resp.,* $C^{\text{alg}}_{H\times H}$) *is the Casimir differential* (*algebraic*) *operator on* $H \times H$. (*See also* §2 *of Chapter* XI.)

Let m denote the orthogonal complement to the subalgebra $\rho(lH)$ in the algebra lG. The subspace m splits into irreducible components of the representation $\operatorname{Ad}_G \rho$, $m = \oplus_i m_i$. It is well known that the Casimir operator of the representation $\operatorname{Ad}\rho$ of the group H on m_i is diagonalized, that is, we have

$$C^{\text{alg}}_H|_{m_i} = a_i \cdot \operatorname{Id}. \tag{3.4}$$

We recall that the embedding of the tangent subspace $lM \to l(G \times G)$ is antidiagonal, that is, $lM = \{(x, -x)\}$. Clearly, the normal fiber over a point $e \in \rho(H \times H/H)$ is the subspace $\overline{m} = \{(x, -x), x \in m\} \subset lM$.

We set $\overline{m}_i = \overline{m} \cap (m_i \oplus m_i)$. The next lemma follows immediately from the proposition.

LEMMA. *The space* $C^\infty(H \times H, \overline{m})_H$ *is a direct sum of subspaces each of which is an invariant subspace of the Jacobi operator* $\overline{I}$.

Next, on the subspace $C^\infty(H \times H, m_i)_H$ the Jacobi operator has the form $\overline{I}(\overline{\psi}) = -C^{\text{diff}}_{H\times H}(\overline{\psi}) + a_i\overline{\psi}$, where a_i is an eigenvalue of the Casimir operator with respect to the irreducible representation $\mathrm{Ad}_G \rho$ of the group H on the subspace m_i (see (3.4)).

By means of the proposition and the lemma, and using facts from the theory of representations of compact Lie groups (see §2 of Chapter XI) we obtain the following classification theorem (Le Khong Van).

THEOREM 3.3.12. *Let G be a classical Lie group, and $\rho: H \to G$ an embedding of the simple compact group H. The subgroup $\rho(H)$ is a stably minimal submanifold of G if and only if*

(a) $G = SU_{m+1}$, $H = SU_{n+1}$, *and ρ is a canonical embedding, or $H = Sp_n$ and ρ is a canonical embedding.*

(b) $G = SO_m$, $H = SU_n$ *or* Sp_n, *and ρ is a composition of canonical embeddings ρ_1, ρ_2, where $\rho_1: SU_n \to SO_{2n}$ (or $\rho_1: Sp_n \to SU_{2n} \to SO_{4n}$), $\rho_2: SO_{2n} \to SO_m$ (or $\rho_2: SO_{4n} \to SO_m$ for $H = Sp_n$).*

Or: $H = SO_n$ and ρ is a canonical embedding. When $n = 7, 8$, or 16, we have additional semispinor embeddings.

Or: $H = G_2, F_4, E_8$, and ρ is a representation of smallest dimension.

Or: $H = E_6, E_7, F_4$, and ρ is a composition of the adjoint representation Ad_H and the canonical embedding $SO(lH) \to SO_m$.

(c) $G = Sp_m$, $H = Sp_n$, *and ρ is a canonical embedding.*

CHAPTER IV

The Multidimensional Plateau Problem in the Spectral Class of All Manifolds with a Fixed Boundary

§1. The solution of the multidimensional Plateau problem in the class of spectra of maps of smooth manifolds with a fixed boundary. An analog of the theorems of Douglas and Rado in the case of arbitrary Riemannian manifolds. The solution of Plateau's problem in an arbitrary class of spectra of closed bordant manifolds

1.1. The classical statement of the multidimensional Plateau problem is equivalent to the minimization of the volume of a submanifold in the class of manifolds bordant to it. In this chapter, we briefly present the statement and complete solution, due to A. T. Fomenko, of the multidimensional Plateau problem *in the class of spectra of all manifolds with a fixed boundary* (spectral version of the Plateau problem).

Consider a Riemannian manifold M and distinguish in it a smooth compact closed (that is, without boundary) $(k-1)$-dimensional submanifold A, a "contour". Suppose there is at least one k-dimensional smooth manifold W with boundary A. Consider all possible pairs of the form (W, f), where W is a compact manifold with boundary ∂W homeomorphic to A, and $f: W \to M$ is a continuous (or piecewise smooth) map that is the identity on the boundary $\partial W \to A$.

PROBLEM A. *Among all pairs of the form* (W, f) *described above,* is it possible to find a pair (W_0, f_0) such that the map f_0 or the film $X_0 = f_0(W_0)$, that is, the image of W_0 in M, has reasonable minimality properties? In particular, it is required that $\mathrm{vol}_k X_0 \le \mathrm{vol}_k X$, where $X = f(W)$ is any other film of the given class.

This is the problem of finding an absolute minimum of the volume functional over all films that glue the given contour in the classical sense (Figure 4.1).

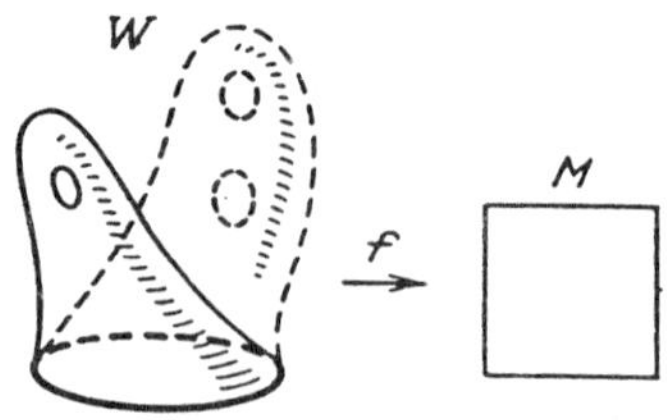

FIGURE 4.1

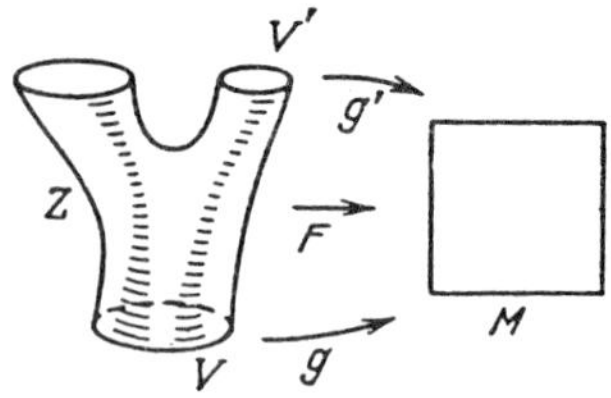

FIGURE 4.2

By "reasonable" minimality properties of the image $X_0 = f_0(W_0)$ in M, in addition to the inequality $\mathrm{vol}_k X_0 \leq \mathrm{vol}_k X$ we can, for example, understand the following property: there is a nowhere dense subset Z of singular points of the film X_0 such that every point $P \in X_0 \backslash Z$ has a neighborhood U in M for which $(X_0 \backslash Z) \cap U$ consists of smooth manifolds V_α of dimension not exceeding k (the maximal possible dimension of the image $X_0 = f_0(W_0)$), and every manifold V_α is a minimal submanifold in the sense of classical differential geometry.

Let us make precise the problem of finding an absolute minimum in the class of all bordism-variations of a given *closed* submanifold $V^k \subset M^n$.

PROBLEM B. Let (V, g) be a pair, where V is a compact closed k-dimensional manifold, $g: V \to M$ is a continuous (or piecewise smooth) map of it into M^n, and $X = g(V)$ is the image of V in M. We say that a pair (V', g') is a bordism-variation of the pair (V, g) if there is a compact manifold Z with boundary $\partial Z = V \cup (\pm V')$ and a continuous map $F: Z \to M$ such that $F|_V = g$, $F|_{V'} = g'$ (Figure 4.2). Among all pairs (V, g) of the given form, can we find a pair (V_0, g_0) such that the image $X_0 = g_0(V_0)$ has reasonable minimality properties, in particular, that $\mathrm{vol}_k X_0 \leq \mathrm{vol}_k X$, where $X = g(V)$ is any film of the given class?

This problem, like Problem A, is the problem of finding an *absolute minimum in the class of all bordism-variations of the given pair* (V, g).

PROBLEM A$'$. Among all pairs of the form (W, f), where W is a fixed manifold with boundary A, and $f: W \to M$ are all possible continuous (or piecewise smooth) maps that are *homotopic to some fixed map* f' and are the identity on the boundary A (that is, coinciding with a fixed homeomorphism of A onto itself), can we find a pair (W, f_0) such that the map f_0 or the film $X_0 = f_0(W)$, the image of W in M, has minimality properties? In particular, we require that $\mathrm{vol}_k X_0 \leq \mathrm{vol}_k X$, where $X = f(W)$ is any film of the given homotopy class.

This is the problem of finding a minimum in each homotopy class. In this sense we also talk of a *relative minimum*, in contrast to the *absolute minimum* in Problem A, which must be found among all homotopy classes.

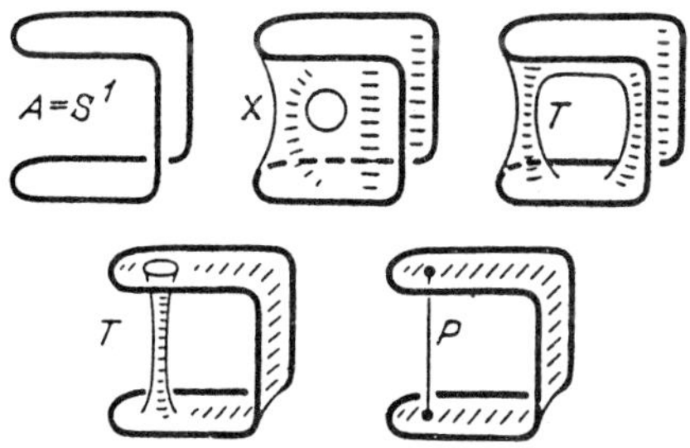

FIGURE 4.3

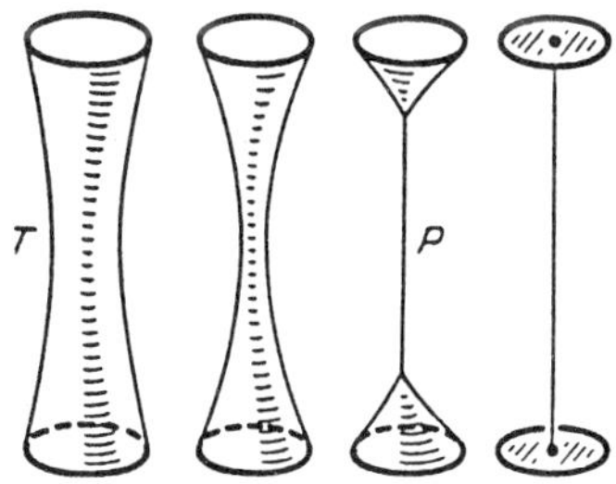

FIGURE 4.4

PROBLEM B′. Among all maps $g: V^k \to M^n$ (where V is a *fixed* manifold) *homotopic to some original map* $f: V \to M$, can we find a map g_0 that has a minimality property? In particular, so that $\mathrm{vol}_k\, g_0(V) \leq \mathrm{vol}_k\, g(V)$, where g is any representative of the given class of maps.

Within the framework of Problems A′ and B′, we consider the problem of finding a minimum in the *homotopy class of given map* (in case A′ with fixed boundary). Within the framework of Problems A and B, we consider the variational problem *in the wider class of admissible variations*, since it is permitted to change the topological type of the film.

Let us fix the terminology: problems of type A or A′ are called *problems of gluing the contour*, and Problems B and B′ are called *problems of realization*, for example of a nontrivial homotopy class.

The minimal surfaces that we discover in these problems are said to be *globally minimal*, in contrast to *locally minimal*.

We now describe the difficulties that arise when minimizing the volume functional vol_k for $k > 2$, and also the appearance of unremovable strata of small dimensions. Consider the contour $A = S^1$ shown in Figure 4.3 and the film $X_t = f_t(W)$, which tends to take up in $\mathbf{R}^3$ a position corresponding to the least two-dimensional area. Clearly, at some instant there occurs a gluing of the film, where instead of a thin tube T, there appears a segment P joining the upper and lower bases of the film. In the two-dimensional case, it is easy to get rid of this by continuously mapping P into a two-dimensional disc gluing the contour A. Another way is to discard P, preserving only the "two-dimensional part" of the minimal film. Similar difficulties arise in minimizing the film in Figure 4.4.

In the case $k > 2$, where for the contour there appears a $(k-1)$-dimensional submanifold A of M, a situation analogous to that described complicates sharply the problem of minimization. When the k-dimensional volume vol_k of the deformed film X_t tends to a minimum,

FIGURE 4.5

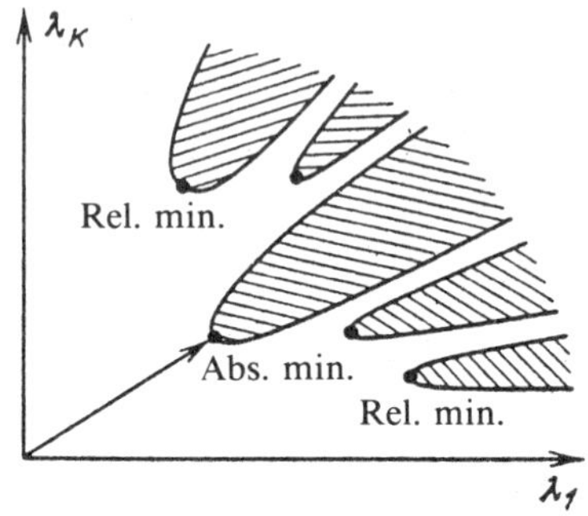

FIGURE 4.6

the film $X_t = f_t(W)$ tends to take up the corresponding "minimal position" and in it there may arise gluings, that is, a map $f_1 \colon W \to M$ homotopic to the original map f may lower the dimension on some subsets open in W (Figure 4.5). This leads to the appearance in the image $X_1 = f_1(W)$ of pieces P of dimensions s, where $s \le k-1$. In contrast to the two-dimensional case, such pieces cannot, generally speaking, either be removed or continuously mapped into a "massive" k-dimensional part $X^{(k)}$ of the film X_1. We recall that we wish to solve the minimum problem in the class of films $X = f(W)$, that is, those that admit a continuous parametrization by means of manifolds W, and so in any version of removing "low-dimensional pieces" $P \subset X$, we must guarantee that the film $\widetilde{X}$ obtained after such a reconstruction must as before admit a continuous parametrization $\widetilde{X} = \tilde{f}(\widetilde{W})$, possibly by means of another manifold $\widetilde{W}$. The removal of low-dimensional pieces P, that is, transition from X to the film $X^{(k)} = \widetilde{X} = \overline{X \backslash P}$ (the bar denotes closure), that is, to a k-dimensional part of X, does not satisfy this requirement. It is easy to construct examples where the "remainder" $\widetilde{X} = X^{(k)}$ does not admit any continuous parametrization $\tilde{f} \colon \widetilde{W} \to X^{(k)}$, where $\tilde{f} \colon \partial \widetilde{W} \to A$ is a homeomorphism. In general, it is also impossible to map low-dimensional pieces continuously into $X^{(k)}$.

In view of the fact that serious topological difficulties arise in handling pieces of films of lower dimension, to simplify the problem we could temporarily ignore low-dimensional pieces P, limiting ourselves for now to a consideration of only the k-dimensional volume functional, from the viewpoint of which all these low-dimensional pieces of the film are unimportant, since they have zero k-dimensional volume. In other words, we can first solve a restricted problem: to minimize only the *leading k-dimensional volume* of a film X concentrated on the set $X^{(k)}$. However, it turns out that even in such a simplified formulation, *the solution of the problem nevertheless requires the study of the behavior of low-dimensional pieces of the film* (see below).

Consequently, the multidimensional Plateau problem forces us to introduce:

(a) *stratified surfaces* $X = X^{(k)} \cup X^{(k-1)} \cup \cdots$, where each subset (stratum) $X^{(s)}$ is an s-dimensional surface in M of dimension s at each point and $X^{(s)}$ is contained in $X \setminus \bigcup_{i=s+1}^{k} X^{(i)}$, where $X^{(s)} \cap (\bigcup_{i=s+1}^{k} X^{(i)}) = \varnothing$;

(b) the *stratified volume of the surface* $SV(X) = (\mathrm{vol}_k X^{(k)}, \mathrm{vol}_{k-1} X^{(k-1)}, \dots)$ represented by a *vector*, each component of which is equal to the volume of the corresponding stratum.

To find a *minimal* surface X means to find a stratified surface *whose stratified volume is least in the sense of lexicographical ordering.* In other words, by a minimization of the stratified volume vector we understand the following. We first need to minimize its first coordinate, that is, the leading k-dimensional volume. Then, having fixed this minimal value of the first coordinate, we need to minimize the second coordinate. Then, having fixed the minimal values of the first two coordinates, we minimize the third, and so on. Consider, for example, Problem A of finding a minimal surface with a fixed boundary. Associating with each stratified surface $X = \bigcup_i X^{(i)}$ its *stratified volume vector* $SV(X) = (\mathrm{vol}_k X^{(k)}, \dots)$, we can represent $SV(X)$ by a vector in $\mathbf{R}^k$. Varying the surfaces X, we obtain a set of vectors $(\lambda_k, \lambda_{k-1}, \dots)$ that sweep a certain subset of $\mathbf{R}^k$ (Figure 4.6). The specification of a lexicographical ordering in the set of all vectors enables us to define the *least* (in this sense) *vector.* The problem consists in finding a surface X *whose stratified volume is exactly equal to this minimal vector.*

1.2. The simplest properties of bordism groups. For a successful realization of this programme we need a language in which we can precisely pose both the question of gluing a contour and the problem of realization. This language must differ from the language that uses the usual homology theory (see above).

Let V^{k-1} be a closed oriented manifold. Let $-V$ denote the oriented manifold that differs from V only in the orientation.

DEFINITION 4.1.1. A closed oriented manifold V^{k-1} is said to be *bordant to zero,* $V \sim 0$, if there is a compact oriented manifold W^k whose boundary ∂W is diffeomorphic to V and preserves the orientation of V.

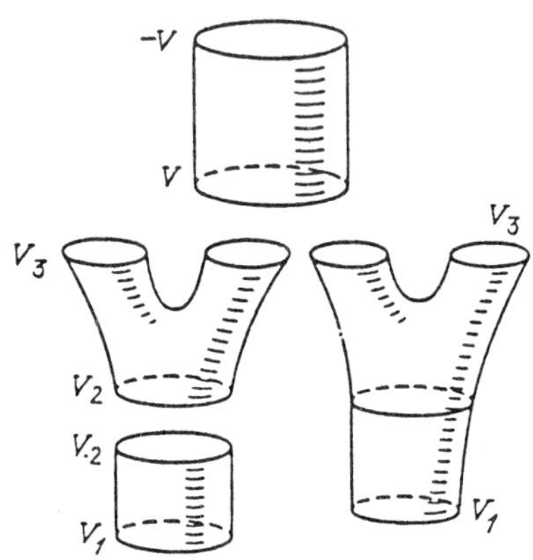

FIGURE 4.7

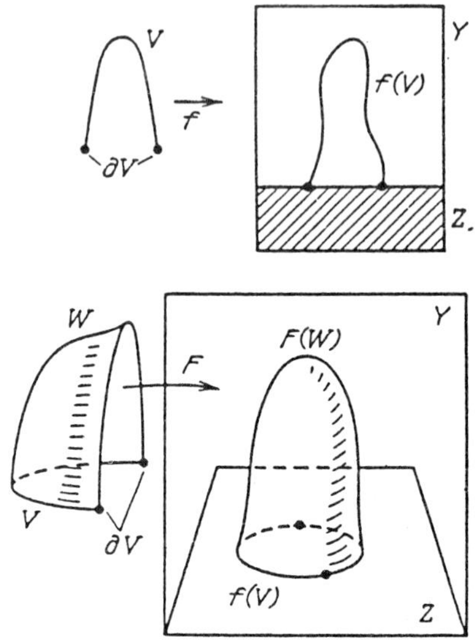

FIGURE 4.8

Two closed oriented manifolds V_1 and V_2 are said to be *bordant*, $V_1 \sim V_2$, if their disjoint union $V_1 \cup (-V_2)$ is bordant to zero.

The bordancy relation is an *equivalence relation* in the class of closed oriented $(k-1)$-dimensional manifolds. The set Ω_{k-1} of equivalence classes is an *Abelian group* in which addition is induced by the disjoint union of manifolds (Figure 4.7).

The direct sum of groups $\bigoplus \Omega_{k-1}$ is usually denoted by Ω_*. It turns out that together with the *theory of oriented bordisms* Ω_* a major role in variational problems is played by the *theory of nonoriented bordisms* $N_* = \bigoplus N_{k-1}$. To construct it we use *all* closed manifolds (both orientable and nonorientable). This gives the possibility of constructing the group N_{k-1} (the analog of the group Ω_{k-1}).

Suppose that a space Y and a subspace Z of it are specified.

DEFINITION 4.1.2. The *oriented singular manifold* of the pair (Y, Z) is the pair (V, f), where V^{k-1} is a compact oriented manifold with boundary ∂V, and f is a continuous map $(V, \partial V) \to (Y, Z)$, that is, (Figure 4.8) $f(V) \subset Y$, $f(\partial V) \subset Z$. If $Z = \varnothing$, we set $\partial V = \varnothing$. The oriented singular manifold (V, f) of the pair (Y, Z) is said to be *bordant to zero* if there is a compact oriented manifold W^k and a continuous map $F: W \to Y$ such that: (a) V is a regular submanifold of the boundary of W, that is, $V \subset \partial W$ and the orientation of V coincides with the orientation induced on it by the orientation of W, and (b) $F|_V = f$, $F(\partial W - V) \subset Z$.

For two singular oriented manifolds (V_1, f_1) and (V_2, f_2), the *disjoint union* is defined as the pair $(V_1 \cup V_2, f_1 \cup f_2)$, where $V_1 \cap V_2 = \varnothing$.

DEFINITION 4.1.3. Two oriented singular manifolds (V_1, f_1) and (V_2, f_2) of the pair (Y, Z) are said to be *bordant* if and only if the disjoint union $(V_1 \cup V_2, f_1 \cup f_2)$ is bordant to zero.

The class of bordisms of a singular oriented manifold (V, f) is usually denoted by the symbol $[V, f]$ and called a *singular bordism*. The set of all such classes is denoted by $\Omega_{k-1}(Y, Z)$. The group $\Omega_i(Y, Z)$ is called the i-dimensional oriented bordism group of the pair (Y, Z). The direct sum $\bigoplus_i \Omega_i(Y, Z)$ is denoted by $\Omega_*(Y, Z)$. If we consider pairs (V, f), where V are compact closed manifolds but *not necessarily orientable* (just as the films W may be nonorientable), then the construction we have described leads us to the group $N_*(Y, Z)$ of *nonoriented bordisms*.

Let us consider *the problem of gluing in Plateau's problem*. Let A be a compact oriented submanifold of M, and $i: A \to X$ an embedding, $A \subset X$, where X is some compactum in M. Then the classical Plateau problem A admits an equivalent reformulation.

PROBLEM A. Among all compacta X containing A and having the property that the singular bordism (A, i) is equivalent to zero in X, can we find a compactum X_0 with minimality properties?

Since the identity map $e: A \to A$ determines an element $\sigma \in \Omega_{k-1}(A)$ (and an element $\sigma \in N_{k-1}(A)$ in the case of nonoriented bordisms), the class of film-compacta X introduced above is characterized by the fact that $i_*(\sigma) = 0$, where $i_*: \Omega_{k-1}(A) \to \Omega_{k-1}(X)$ (respectively, $i_*: N_{k-1}(A) \to N_{k-1}(X)$) is the homomorphism induced by i.

PROBLEM B. Among all singular manifolds (V, g), $g: V \to M$, bordant (equivalent) to a given singular manifold (V', g'), where $g': V' \to M$, can we find a singular manifold (V_0, g_0) such that the film $X_0 = g_0(V_0)$ has minimality properties? In other words, among all representatives (V, g) of a given bordism class $\alpha \in \Omega_k(M)$ (or $N_k(M)$) can we find a (V_0, g_0) such that the film $X_0 = g_0(V_0)$ is minimal in M?

EXAMPLE 1. Consider a minimal two-dimensional film X_0 (Figure 4.9). This film is homeomorphic to a smooth two-dimensional manifold with boundary A (a circle), that realizes an absolute minimum of two-dimensional area in the sense of Problem A. This film annihilates the singular manifold $(A, e) \in \Omega_1(A)$ in the sense of bordisms. We can detect films of this oriented type by means of the groups $\Omega_*(X, A)$.

EXAMPLE 2. The minimal film X_0 shown in Figure 4.10 is a Möbius band and is most naturally detected by means of the groups of nonoriented bordisms.

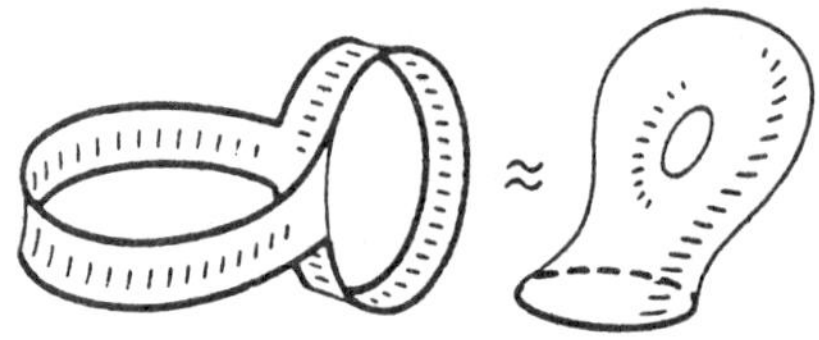

FIGURE 4.9

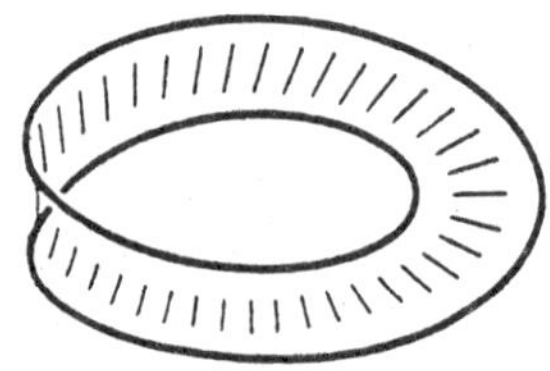

FIGURE 4.10

EXAMPLE 3. There is a class of minimal films X_0 for which it is most natural to use the so-called *bordisms modulo* p. Let us consider a triple Möbius band (see Figure 3.6) with boundary $A = S^1$ (a circle). Clearly, this film realizes an absolute minimum of two-dimensional area for a given boundary A. However, to describe such films the language of oriented or nonoriented bordisms is insufficient. It is therefore useful to consider new groups, denoted by Ω_*^p and called *groups of singular bordisms modulo* p (see [119]).

Thus, the *classical Plateau problem* requires the use of the bordism groups $\Omega_*(A)$, $N_*(A)$, $\Omega_*^p(A)$ and homomorphisms of them into the bordism groups $\Omega_*(X)$, $N_*(X)$, $\Omega_*^p(X)$ respectively that arise under the embedding $A \to X$.

The introduction of bordism groups in the statement of multidimensional variational problems of the type of Plateau's problem turns out to be successful, since the bordism groups Ω_*, N_*, and Ω_*^p satisfy the *Eilenberg-Steenrod axioms.* Consequently, they form *extraordinary* (*generalized*) *homology theories.* For definiteness we consider the groups Ω_*. The arguments are completely analogous in the other cases. If $\varphi: (X, A) \to (X_1, A_1)$ is a continuous map, then there is defined the induced homomorphism $\varphi_*: \Omega_k(X, A) \to \Omega_k(X_1, A_1)$, where $\varphi_*[V^k, f] = [V^k, \varphi f]$. In addition there is defined the boundary homomorphism $\partial: \Omega_k(X, A) \to \Omega_{k-1}(A)$, where $\partial[V^k, f] = [\partial V^k, f]$.

PROPOSITION 4.1.1. *The triple* $\Omega_*, \varphi_*, \partial$ *satisfies the first six Eilenberg-Steenrod axioms but does not satisfy the seventh—the point axiom. For a one-point space* x *the group* $\Omega_*(x)$ *is isomorphic to the oriented bordism group.*

1) If $i: (X, A) \to (X, A)$ is the identity map, then i_* is the identity homomorphism.

2) If we are given two maps $\varphi: (X, A) \to (X_1, A_1)$ and $\psi: (X_1, A_1) \to (X_2, A_2)$, then we have $(\psi\varphi)_* = \psi_*\varphi_*$.

3) For any map $\varphi: (X, A) \to (X_1, A_1)$ we have $\partial\varphi_* = \varphi_*\partial$.

4) If a map φ_0 is homotopic to a map φ_1, then $\varphi_{0*} = \varphi_{1*}$.

5) For any pair (X, A) the sequence of groups and homomorphisms

$$\cdots \to \Omega_k(A) \xrightarrow{i_*} \Omega_k(X) \xrightarrow{j_*} \Omega_k(X, A) \xrightarrow{\partial} \Omega_{k-1}(A) \to \cdots$$

is exact.

6) (The excision axiom) The embedding $i: (X \backslash U, A \backslash U) \to (X, A)$ for any U such that $U \subset \operatorname{Int} A$ induces the group isomorphism i_*: $\Omega_k(X \backslash U, A \backslash U) \xrightarrow{\approx} \Omega_k(X, A)$.

The last 7th axiom (the point axiom) is not satisfied for bordism groups, since $\Omega_k(x) \not\equiv 0$ for $k > 0$, where x is a point. *This strongly distinguishes bordism theory from the ordinary homology theory H_*. We recall that the ordinary homology of a point is zero in all dimensions other than zero.*

For an accurate construction of the theory of minimization we need to be able to compute bordism groups not only for manifolds, cell complexes, but also for *surfaces with singularities.* The fact is that minimal films contain, as we have seen, *rather complicated singular points*, which fill subsets of measure zero. This extension of the concept of a bordism group to the wider class of compact subsets is carried out by means of a *spectral process*, as in the case of constructing ordinary spectral homology. As a result we obtain spectral bordism groups, which can be computed for any compact sets. We recall that an element of the spectral bordism group of a space X is a consistent sequence (spectrum) of continuous maps $f_i: W_i \to X_i$, where W_i are manifolds, and X_i are simplicial complexes, the so-called nerves of coverings of X, and $X = \lim X_i$. This means that with the growth of i the complexes X_i approximate the space X better and better. The classical origin and analog of this construction is spectral homology (Čech homology). For the details see [212], [119]. If a space X is a finite cell complex, then the spectral bordism groups coincide with the usual bordism groups. Everywhere below, when we consider compact spaces and talk about their singular bordism groups, we shall always have in mind spectral singular bordism groups, without mentioning it again each time (see Fomenko [126], [119]). For simplicity we consider nonoriented bordisms.

1.3. The solution of Plateau's problem in the class of maps of a spectrum of manifolds with a fixed boundary. We are now ready for a statement of the theorem that solves Plateau's problem in terms of a spectrum of manifolds with a fixed boundary, that is, in a form sufficiently close to the classical two-dimensional Plateau problem. In $\mathbf{R}^n$, we fix a "contour" A^{k-1}, which we take for simplicity as a closed smooth manifold. Let $A(\sigma)$

be the class of all measurable compact subsets X of $\mathbf{R}^n$ that contain A and that annihilate the bordism $\sigma = [A] = (A, e)$, where $e: A \to A$ is the identity map. In other words, the singular manifold (A, i), where $i: A \to X$ is an embedding, is bordant to zero in X, that is, $i_*\sigma = 0$, where i_* is the homomorphism that maps $(k-1)$-dimensional bordisms of the boundary A into $(k-1)$-dimensional spectral bordisms of X. We denote by vol_k either the Riemannian volume of X or its k-dimensional Hausdorff measure, when the film contains sufficiently complicated singularities.

THEOREM 4.1.1 (Fomenko [128]). *Let A^{k-1} be an arbitrary smooth $(k-1)$-dimensional closed submanifold of Euclidean space $\mathbf{R}^n$. Let us suppose that the variational problem "makes sense" in the class of surfaces with a fixed boundary, i.e., the infimum of stratified volume "is finite". For the strong definition see* [128]. *Then in the class $A(\sigma)$ of all surfaces with boundary A (that is, that annihilate the bordism $[A]$—see the description above), there is always a globally minimal surface $X_0 = A \cup S^k \cup S^{k-1} \cup \cdots$, stratified by strata S^i, where the stratum S^i has dimension i (if it is not empty), and $S^i \subset X \backslash (A \cup \bigcup_{\alpha=i+1}^{k} S^\alpha)$. This surface solves the spectral Plateau problem, since it has least leading volume vol_k. Moreover, this surface X_0 has the property that its stratified volume $SV(X_0) = (\mathrm{vol}_k S^k, \mathrm{vol}_{k-1} S^{k-1}, \ldots)$ is least in lexicographical ordering among the stratified volumes of other surfaces X of the same class $A(\sigma)$. In other words, the minimal film X_0^k spanning the given "contour" A^{k-1} is minimal in all its dimensions simultaneously. In addition, each of the strata S^i of the surface X_0 is, except possibly for a set of i-dimensional measure zero, an analytic minimal submanifold of $\mathbf{R}^n$.*

If we trace the process of minimization over the dimensions, then we can present the theorem in the following way.

1) Let B_k be the class of all compacta X such that $X \in A(\sigma)$ and $\mathrm{vol}_k X = d_k = \inf_Y \mathrm{vol}_k Y$, $Y \in A(\sigma)$. Then the class B_k is not empty, $d_k > 0$, and each compactum $X \in B_k$ contains a uniquely defined k-dimensional subset S^k (that is, it has dimension k at each of its points), $S^k \subset X \backslash A$, such that $A \cup S^k$ is a compactum in $\mathbf{R}^n$, and the set S^k contains a "singular" subset Z_k, where $\mathrm{vol}_k Z_k = 0$ and $S^k \backslash Z_k$ is a smooth minimal submanifold of $\mathbf{R}^n$. In fact, $S^k \backslash Z_k$ is even an analytic submanifold; it is open and everywhere dense in S^k, and $\mathrm{vol}_k S^k = \mathrm{vol}_k X = d_k > 0$.

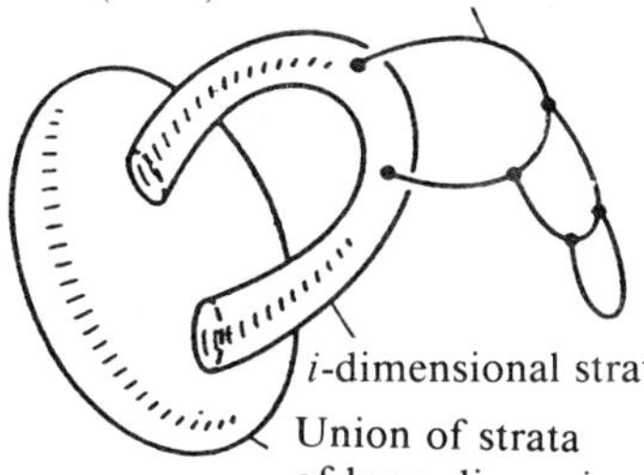

FIGURE 4.11

2) Further, if B_{k-1} is the class of all compacta X such that $X \in A(\sigma)$, $X \in B_k$, and $\mathrm{vol}_k X \backslash S^k = d_{k-1} = \inf_Y \mathrm{vol}_{k-1} Y \backslash S^k$, $Y \in B_k$, then the class B_{k-1} is not empty, and when $d_{k-1} > 0$ each compactum $X \in B_{k-1}$ contains a uniquely defined $(k-1)$-dimensional subset S^{k-1} (of dimension $k-1$ at each of its points), $S^{k-1} \subset X \backslash (A \cup S^k)$, and the subset $A \cup S^k \cup S^{k-1}$ is a compactum in $\mathbf{R}^n$; S^{k-1} contains a "singular" subset Z_{k-1}, where $\mathrm{vol}_{k-1} Z_{k-1} = 0$ and $S^{k-1} \backslash Z_{k-1}$ is a smooth minimal $(k-1)$-dimensional submanifold of $\mathbf{R}^n$ that is everywhere dense in S^{k-1} and $\mathrm{vol}_{k-1} S^{k-1} = \mathrm{vol}_{k-1} X \backslash (A \cup S^k) = d_{k-1} > 0$. If $d_{k-1} = 0$, we set $S^{k-1} = \varnothing$.

3), ..., 4), ..., and so on, downwards in dimension.

A similar theorem was proved by A. T. Fomenko in the case of bordant closed (that is, not having a boundary) manifolds that realize a nontrivial class of bordisms α in some complete Riemannian manifold M. *In the class of surfaces $B(\alpha)$ that realize (in the sense of nonoriented spectral bordisms) a nontrivial element α of the bordism group in M there is always a globally minimal stratified surface whose leading volume is minimal and whose stratified volume is smallest (with respect to lexicographical ordering) among the stratified volumes of other surfaces of the same class.* Every nonempty i-dimensional stratum of this surface has the following properties. Except possibly for a set of i-dimensional measure zero, a stratum is analytic (if the ambient manifold M is analytic) minimal (in the sense of classical differential geometry) submanifold of M. We recall once more that the closure of each separate stratum is not contained in the closure of the union of all strata of large dimension (Figure 4.11). *It is impossible to remove any of the nonempty strata, since this immediately destroys the topological properties of the minimal surface.* In particular, the "remainder" may not admit any continuous parametrization by means of a smooth manifold. For the details see Fomenko [129] and the book [119].

Plateau's problem has actually been solved for *a substantially wider class of variational problems* (A. T. Fomenko). It turns out that we can assert the existence of minimal surfaces in those cases when the boundary conditions are specified in terms of *cohomology* groups. This immediately widens our possibilities in the search for *concrete examples* of minimal surfaces of nontrivial topological type.

§2. Some versions of Plateau's problem require for their statement the concepts of generalized homology and cohomology

2.1. Definition of generalized homology and cohomology. Suppose that each pair of compacta (X, A) where $A \subset X$, and each integer q we associate an Abelian group $h_q(X, A)$, and with each continuous map $f: (X, A) \to (Y, B)$ we associate a homomorphism of Abelian groups $f_*: h_q(X, A) \to h_q(Y, B)$. Suppose also that for each pair (X, A) there is defined a boundary operator (homomorphism) $\partial: h_q(X, A) \to h_{q-1}(A)$. We require that these groups and homomorphisms *should satisfy the six Eilenberg-Steenrod axioms* (see above).

DEFINITION 4.2.1. We shall say that a collection of groups h_* and homomorphisms f_*, ∂ satisfying axioms 1–6 determine a *generalized* (*extraordinary*) *homology theory.*

Let $f: X \to x$ be a map of the space X to a point x.

DEFINITION 4.2.2. The *reduced generalized homology group* $\tilde{h}_*(X)$ with respect to a point x is the subgroup $\operatorname{Ker} f_*$ of the group $h_*(X)$. Here $f_*: h_*(X) \to h_*(x)$.

For the *cohomology* case these definitions are stated in *dual form.*

DEFINITION 4.2.3. We shall say that a collection of groups $h^*(X, A)$ and homomorphisms f^*, δ satisfying axioms 1–6 determine a *generalized* (*extraordinary*) *cohomology theory.*

Let $i: x \to X$ be an embedding of the point x in the space X.

DEFINITION 4.2.4. The *reduced generalized cohomology group* $\tilde{h}^*(X)$ with respect to a point x is the subgroup $\operatorname{Ker} i^* \subset h^*(X)$, where $i^*: h^*(X) \to h^*(x)$.

Important examples of generalized homology are *bordism groups and stable homotopy groups of topological spaces.* Examples of generalized cohomology are *groups of stable vector bundles.*

We need to distinguish explicitly one more property of generalized homology and cohomology groups that must be satisfied in order that these groups should actually be useful for variational problems. This is the *continuity property.* Let X_j be a sequence of compacta embedded in one another, that is, $X_j \supset X_{j+1}$ for $j = 1, 2, \ldots$, and let $X = \bigcap_j X_j$ be their intersection. Consider the embeddings $i_\alpha: X_{\alpha+1} \to X_\alpha$. Then the collection of compacta X_α and embeddings i_α determines the so-called *inverse spectrum* $\{X_\alpha, i_\alpha\}$. The limit of this inverse spectrum is the compactum X. Consider the sequence of groups $h_q(X_\alpha)$, $\alpha = 1, 2, 3, \dots$, of homomorphisms $i_{\alpha*}: h_q(X_{\alpha+1}) \to h_q(X_\alpha)$. There arises the inverse spectrum of homology groups $\{h_q(X_\alpha), i_{\alpha*}\}$ and the limit of this inverse spectrum, that is, a group denoted by $\varprojlim \{h_q(X_\alpha)\}$. We say that *the homology theory is continuous on the set of compacta* if for each q there is an isomorphism $\varprojlim \{h_q(X_\alpha)\} = h_q(X)$.

In a similar way (replacing the inverse limit by the direct limit) we formulate the *continuity condition* for *cohomology theories.*

The supply of continuous generalized homology and cohomology theories is very large. Such theories can be constructed by means of a *spectral process,* starting from generalized homology and cohomology theories defined originally on the class of cell complexes.

2.2. The general concept of boundary (or coboundary) of a surface that covers all previous definitions. A. T. Fomenko introduced the concept of the (co)boundary of a pair of spaces (X, A) in the general case of an arbitrary generalized (co)homology theory.

DEFINITION 4.2.5. Let (X, A) be a pair of compacta and suppose that $x \in A$. The *coboundary* $\nabla^k(X, A)$ *of the pair* (X, A) *in dimension* k *with respect to the point* x is the set of all elements $\alpha \in \tilde{h}^{k-1}(A)$ such that $\alpha \notin \operatorname{Im} i^*$, where $i: A \to X$ is an embedding, and the homomorphism $i^*: \tilde{h}^{k-1}(X) \to \tilde{h}^{k-1}(A)$ is induced by this embedding. In general, the number k is not connected with the topological dimension of X. We then set $\nabla^*(X, A) = \bigcup_k \nabla^k(X, A)$.

The concept of the coboundary of a pair of spaces corresponds to the intuitive idea of the geometrical boundary of a film. For example, if $X = CA$ (a cone over A), then obviously $\nabla^*(CA, A) = \bigcup_k \tilde{h}^{k-1}(A)\backslash 0$ that is, the cone CA completely glues A. Another example: let X be a k-dimensional manifold with boundary A, where A is a closed $(k-1)$-dimensional manifold. Let $h^* = H^*$ be the ordinary cohomology theory. Then $\nabla^k(X, A) = H^{k-1}(A)\backslash 0$, which again agrees with intuition.

Suppose we are given a generalized homology theory h_*.

DEFINITION 4.2.6. Let (X, A) be an arbitrary pair and suppose that $x \in A$. The *boundary* $\Delta_k(X, A)$ *of the pair* (X, A) *in dimension* k *with respect to the point* x is the subgroup $\operatorname{Ker} i_* \cap \tilde{h}_{k-1}(A)$, where $i: A \to X$ is an embedding, and the homomorphism $i_*: \tilde{h}_{k-1}(A) \to \tilde{h}_{k-1}(X)$ is induced by this embedding. We then set $\Delta_*(X, A) = \bigcup_k \Delta_k(X, A)$.

2.3. Classes of variations of surfaces. Let $(X, A) \subset M$ be a compact pair and let $i: A \to X$ and $j: X \to M$ be embeddings. We first consider the homology case. Let $L = \{L_p\}$ be a fixed collection of subgroups $L_p \subset \tilde{h}_p(A)$, where $p \in \mathbf{Z}$, and $L' = \{L'_q\}$ a fixed collection of subgroups $L'_q \subset \tilde{h}_q(M)$.

We denote by $\boldsymbol{O} = h_*(A, L, L')$ the class of all compacta X with $A \subset X \subset M$ such that 1) $L \subset \Delta_*(X, A)$, 2) $L' \subset \operatorname{Im} j_*$.

We now consider the cohomology case. Let $L = \{L_p\}$ be a fixed collection of subsets $L_p \subset \tilde{h}^p(A)\backslash 0$, and $L' = \{L'_q\}$ a fixed collection of subsets $L'_q \subset \tilde{h}^p(M)\backslash 0$.

DEFINITION 4.2.7. We denote by $\boldsymbol{O} = h^*(A, L, L')$ the class of all compacta X with $A \subset X \subset M$ such that 1) $L \subset \nabla^*(X, A)$, 2) $L' \subset \tilde{h}^*(M)\backslash \operatorname{Ker} j^*$.

The class $\boldsymbol{O}$ consists of all surfaces $X \subset M$ that glue (co)homological "holes" L in the boundary A and at the same time realize (co)homological "holes" L' in the manifold M.

Let us consider the two limiting cases: the class $h(x, 0, L')$ and the class $h(A, L, 0)$. In the first case the class $\boldsymbol{O}$ consists only of surfaces X that realize "holes" of the manifold M without any additional boundary condition, since $A = x$. We say that such surfaces are *realizing* (Plateau's problem B). In the second case the class $\boldsymbol{O}$ consists of surfaces X that glue "holes" in the boundary A. The manifold M plays the role of an embedding space in this case. We say that such surfaces are *gluing surfaces* (Plateau's problem A).

2.4. The solution of Plateau's problem in each class of surfaces of fixed topological type, defined in terms of generalized spectral homology or cohomology. The statement of variational problems and solution of the corresponding Plateau problem presented below are due to A. T. Fomenko. Each spectral theory h^*, h_* distinguishes in M a class of surfaces of definite topological type. It turns out that in each such class there is an absolutely minimal surface, that is, Plateau's problem is solved positively.

For each surface $X \in \boldsymbol{O}$ we construct a stratification of it, $X = A \cup S^k \cup S^{k-1} \cdots$, where S^k is a maximal subset of $X\backslash A$ that has dimension k at each point, S^{k-1} is a maximal subset of $X\backslash A\backslash S^k$ that has dimension $k-1$ at each point, and so on. The subsets S^i are called *strata*. If the stratum S^i is not empty, then it does not intersect the closure of all strata of higher dimension. In other words, each nonempty stratum is situated "outside" the remaining strata (see Figure 4.11). If the strata are measurable, there is defined the *stratified volume* $(\mathrm{vol}_k S^k, \mathrm{vol}_{k-1} S^{k-1}, \ldots) = SV(X)$, represented by a vector with k coordinates. Varying the surface X in the class of admissible variations, that is, in the class $\boldsymbol{O}$, we vary the vector of stratified volume of the surface. The problem consists in finding a surface with the least (in the sense of lexicographical ordering) stratified volume. The least vector $SV = (d_k, d_{k-1}, \ldots)$ can be understood in the following sense. We first try to minimize the first coordinate of the vector $SV(X)$, that is, find in the class $\boldsymbol{O}$ a surface X_k for which $\mathrm{vol}_k S^k = \mathrm{vol}_k(X_k \backslash A) = d_k = \inf_Y \mathrm{vol}(Y \backslash A)$, $Y \in \boldsymbol{O}$. If such surfaces X_k exist, we proceed to minimize the second coordinate of the stratified volume vector. In the class of surfaces X_k with minimal first coordinate (that is, such that $\mathrm{vol}_k(X_k \backslash A) = d_k$) we look for a surface X_{k-1} such that $\mathrm{vol}_{k-1}(X_{k-1} \backslash A \backslash S^k) = \inf_Y \mathrm{vol}_{k-1}(Y \backslash A \backslash S^k) = d_{k-1}$, where $\mathrm{vol}_k(Y \backslash A) = d_k$. And so on, that is, each time we minimize the next coordinate of the stratified volume vector, provided that all the previous coordinates are already minimized and fixed. If this process is well defined (and this will be proved), then it is completed by some surface whose stratified volume is globally minimal in the class of all stratified surfaces of the given class.

The method developed by A. T. Fomenko for minimizing the volume can be applied not only to spectral bordism classes (Plateau's problem) but also to any class of the form $\boldsymbol{O}$. We give an example of a "*cohomological variational problem*" defined by the K-functor. Suppose that on a manifold M there is defined a *stably nontrivial vector bundle* ξ. Consider the class of all surfaces $X \subset M$ such that the restriction of ξ to X is *stably nontrivial as before*, that is, X is the *support of* ξ (Figure 4.12). Question: among all such surfaces, can we find one that is globally minimal (from the viewpoint of the usual volume and stratified volume)? It turns out that the answer is yes. Suppose that $\pi_1(M^n) = \pi_2(M^n) = 0$.

THEOREM 4.2.1 (Fomenko [128]). *Let M^n be a compact closed Riemannian manifold of class C^r, where $r \geq 4$, let $A \subset M$ be a fixed compactum, and $x \in A$ a fixed point. Let h be a reduced continuous relatively invariant generalized spectral (co)homology theory. Consider an arbitrary nonempty class of surfaces $\boldsymbol{O} = h(A, L, L')$, that is, a class consisting of surfaces of fixed topological type defined by the theory h and the "parameters" L and L'. Suppose that the infimum of stratified volume is finite (in the class $h(A, L, L')$, i.e. the variational problem "makes sense"). Then in each*

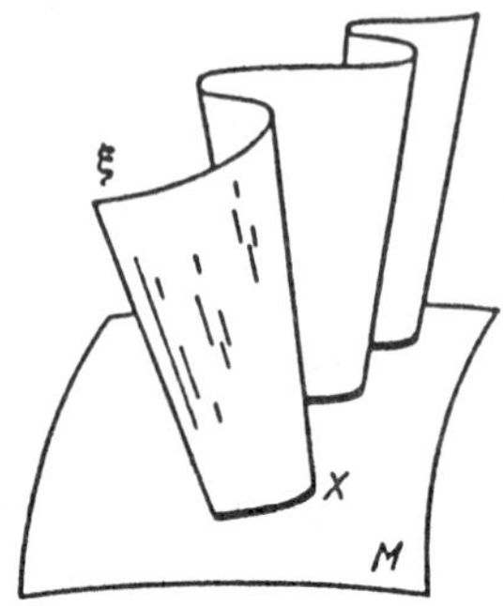

FIGURE 4.12

such class Plateau's problem can be solved positively, that is, there is always a globally minimal surface whose stratified volume (including the ordinary leading volume) is the smallest in the sense of lexicographical ordering.

We can now solve the problem of minimizing only one volume vol_k, the *leading one in dimension*, since we are not interested in pieces of the surface of lower dimensions. This problem also is completely solved by Theorem 4.2.1 (for doubly-connected manifolds).

COROLLARY 4.2.1. *In each class of surfaces* $\boldsymbol{O}$ *under the assumptions of Theorem* 4.2.1 *there is always a globally minimal surface* X_0 *whose leading volume and stratified volume* $SV = (d_k, d_{k-1}, \dots)$ *are smallest (in all dimensions). This surface* X_0 *has a uniquely defined stratification, that is,* $X_0 = A \cup S^k \cup S^{k-1} \cdots$, *where each subset* S^i *is (except for a set of measure zero consisting of singular points) a smooth (for smooth* M*) minimal submanifold of* M, *that is, the mean curvature is zero. We have* $d_i = \mathrm{vol}_i S^i$.

EXAMPLE 1. If for the generalized homology theory we take *spectral bordism groups*, we obtain a solution of Plateau's problem (on the absolute minimum) in the class of surfaces that are images of spectra of manifolds with a fixed boundary A in M. For the basic theories we can take oriented bordisms $\bigoplus\Omega$, nonoriented bordisms, and bordisms modulo p. For the details see Fomenko [119] and [126]. Thus, let h_* be one of the bordism theories listed above. The manifold A determines an element $\sigma = [A, e] \in h_{k-1}(A)$, where $e\colon A \to A$ is the identity map. Let L be the subgroup of $\tilde{h}_{k-1}(A)$ generated by the element σ.

COROLLARY 4.2.2. *Suppose that the class of surfaces* $h_*(A, L, 0)$ *is not empty. Then in it there is always a globally minimal surface* X_0 *that annihilates (glues) the element* σ. *This surface (possibly with singularities)* *is a solution of Plateau's problem in the class of all surfaces* X *that glue* A *and admit a continuous parametrization by means of the spectrum of manifolds with boundary. If we consider the second limiting case, that is, the class of surfaces* $h_*(x, 0, L')$, *we obtain the existence of a globally minimal surface* X_0 *in the class of all surfaces that realize a given bordism* $\{\sigma'\} = L' \subset \tilde{h}_k(M)$. *Also, in the case of Problem* A *the inequality* $d_k > 0$ *is always satisfied.*

In the problem of gluing a "contour" for a minimal film X_0 in the general case there is an infinite sequence of smooth manifolds $\{W_\alpha\}$ with boundary A that glue the boundary A in the film X_0. If X_0 is a cell complex, then all the manifolds W_α are homeomorphic to the same manifold W_0, which parametrizes the film X_0. In particular, X_0 is a continuous image of W_0, that is,

$$X_0 = f_0(W_0).$$

EXAMPLE 2. If for the homology we take the usual homology theory, then from Theorem 3.5.1 there follow the results of Riefenberg and Morrey (see [365] and [318]).

EXAMPLE 3. If for the generalized homology theory we take the stable homotopy groups π_i^s, *we obtain a theorem on the existence of a globally minimal surface in each spectral homotopy class* $\sigma \in \pi_i^s(M)$. *In particular, there is a global minimum in the stable spectral homotopy class of discs that glue a fixed sphere in the manifold* M.

For a detailed proof of Theorem 4.2.1 see Fomenko [129], [119], [130], and [131].

§3. In certain cases the Dirichlet problem for the equation of a minimal surface of large codimension does not have a solution

Let B be an open subset of $\mathbf{R}^n$ and let $F: B \to \mathbf{R}^{n+k}$ be a smooth immersion.

We shall say that an immersion F of a domain B in $\mathbf{R}^{n+k}$ is *non-parametric* if F has the form $F(x) = (x, f(x))$, where f is a map (vector-valued function) $f: B \to \mathbf{R}^k$. The surface $F(B)$ is the *graph in* $\mathbf{R}^{n+k}$ *of the map* f.

We now state the Dirichlet problem for vector-valued functions of the type we have described (see [280]). Suppose we are given a smooth map φ of the boundary ∂B of a domain B in $\mathbf{R}^k$. It is required to find a map $f: B \to \mathbf{R}^k$, continuous on the closure of B and locally Lipschitz in

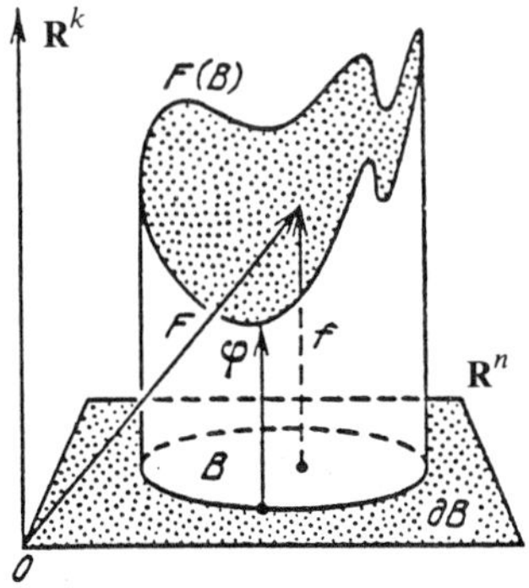

FIGURE 4.13

B and also satisfying the minimal surface equation. Finally, it is required that the volume of the surface $F(B)$, that is, the graph of the map f, should be finite (Figure 4.13).

PROPOSITION 4.3.1. *When $k = 1$ the Dirichlet problem in a convex domain $B \subset \mathbf{R}^n$ is always solvable for any continuous initial (boundary) data, that is, for any continuous function $\varphi: \partial B \to \mathbf{R}^1$. The solution $f: B \to \mathbf{R}^1$ has the following properties:* 1) *the solution is unique,* 2) *the solution is real-analytic,* 3) *the n-dimensional graph of the map $f: B \to \mathbf{R}^1$ is a globally minimal surface in $\mathbf{R}^{n+1}$, that is, its volume is least in the class of surfaces with a given boundary $(x, \varphi(x))$, where $x \in \partial B$. Surfaces are understood here in the sense of integral currents.*

For the case of *large codimensions* we need to distinguish especially *two-dimensional convex domains* $B \subset \mathbf{R}^2$ for which, as it turns out, the Dirichlet problem is also always solvable. Namely, it is known that when $n = 2$ and for any $k \geq 1$ and for any convex domain $B \subset \mathbf{R}^2$ there is always a Dirichlet solution (continuous on the closure of the domain) for any continuous boundary data and for any continuous map $\varphi: \partial B \to \mathbf{R}^k$. However, the two-dimensional solution may turn out to be nonunique, and unstable, that is, not globally minimal—see [280].

Let us consider for B a two-dimensional disc D^2 of unit radius on the Euclidean plane. As Lawson and Osserman showed, there is a real-analytic map $\varphi: \partial D^2 \to \mathbf{R}^2$ of the boundary circle ∂D^2 in $\mathbf{R}^2$ (here $n = 2$ and $k = 1$) such that there are at least three distinct solutions of the Dirichlet problem with the same boundary condition φ. One of these solutions is represented by an unstable minimal two-dimensional surface in four-dimensional Euclidean space (see [280]).

Let us consider minimal surfaces of large codimension, that is, $k \geq 2$. For the convex domain B in $\mathbf{R}^n$ we take an open disc D^n of unit radius. It turns out that when $n \geq 4$ the Dirichlet problem *may have no solution.*

PROPOSITION 4.3.2 (Lawson and Osserman [280]). *Suppose that* $n > k \geq 2$, $S^{n-1} = \partial D^n$, $S^{k-1} \subset \mathbf{R}^k$, *where* S^{n-1} *and* S^{k-1} *are standard spheres of unit radius. Let* $\varphi: S^{n-1} \to S^{k-1}$ *be any map of class* C^2 *that determines a nontrivial (nonzero) element of the homotopy group* $\pi_{n-1}(S^{k-1})$, *that is, it is not homotopic to zero. Then there is a number* R_φ *such that for every* $R \geq R_\varphi$ *there is no solution of the Dirichlet problem in the ball* D^n *for the equation of a minimal surface of codimension* k *with boundary function* $\varphi_R = R\varphi$.

From the geometrical viewpoint the following happens. Lawson and Osserman consider the graph of a map that realizes a nonzero element of the homotopy group, and then subject the graph to a simple transformation that stretches the domain with a homothety coefficient R. The center of the homothety is the origin in $\mathbf{R}^k$. Together with the stretching of the graph (and the boundary data) we can consider the inverse process of contracting the boundary function. This process is interesting in that the Dirichlet problem is always solvable for a convex domain (and for any codimension) for sufficiently small boundary data. If φ is small on ∂B, then there is a minimal surface close to B and coinciding with the graph of φ on ∂B. Hence it follows that by multiplying the boundary function by a decreasing parameter r we finally find the required minimal surface. In other words, there is a number $r_\varphi > 0$ such that for all $r < r_\varphi$ the Dirichlet problem is solvable in the domain B with boundary data $\varphi_r = r\varphi$.

As an example we consider the *Hopf map.* Consider a sphere S^3, embedded in complex space $\mathbf{C}^2(z_1, z_2)$ as the set of points (z_1, z_2) such that $|z_1|^2 + |z_2|^2 = 1$. We identify the sphere S^2 with the complex projective line $\mathbf{C}P^1$. We construct a map $p: S^3 \to S^2$ defined as follows: $p(z_1, z_2) = z_1/z_2$. Then p determines a bundle with base S^2 and fiber S^1.

It is well known that the map p is not homotopic to zero, that is, it determines a nonzero element of the group $\pi_3(S^2) = \mathbf{Z}$. Applying Proposition 3.6.2, we find that for sufficiently large R the Dirichlet problem on a four-dimensional ball with boundary values $R \cdot \varphi$ does not have a solution in the form of the minimal graph of any map $f: D^4 \to \mathbf{R}^3$. Direct calculation shows (see [280]) that for R_φ we can take 4.2 (for unit radii of the spheres). Consequently, for all $R > 4.2$ the Dirichlet problem with boundary function $R\varphi$ does not have a solution.

The lack of a solution to the Dirichlet problem does not prevent the appearance of a four-dimensional minimal surface that has as boundary contour a three-dimensional sphere embedded in $\mathbf{R}^7$ as the graph of a map $\varphi: S^3 \to S^2 \subset \mathbf{R}^3$. It follows from general existence theorems that for any continuous map φ there is always a globally minimal surface spanning this "contour". From Proposition 4.3.2 it follows that this surface cannot be represented as the graph of any map $f: D^4 \to \mathbf{R}^3$.

§4. Some new methods for an effective construction of globally minimal surfaces in Riemannian manifolds

4.1. A universal lower bound for the volumes of topologically nontrivial minimal surfaces. Let X^k be a globally minimal surface of nontrivial topological type in a manifold M, for example, X realizes some cycle of the group $H_k(M)$ or some bordism class of the group $\Omega_k(M)$. In studying properties of such surfaces there often arises the important question of a lower bound for the volume of the surface. *Exact bounds of this kind are particularly important for discovering specific globally minimal surfaces in specific Riemannian spaces.* The fact is that for the proof of global minimality of some specific surface it is sufficient to verify that its volume is exactly equal to the lower bound of the volumes of surfaces of the same topological type. For this we need to be able to calculate this lower bound effectively in explicit form. The nontriviality of the problem is connected with the fact that we need to investigate large variations, and not infinitesimal ones, as is usually done in the proof of local minimality.

It turns out that there is an effective method of calculating the smallest volumes of closed surfaces of arbitrary nontrivial topological type in terms of the metric of the ambient manifold and its curvature tensor. This method was developed by Fomenko [133], [134], [119].

Let M^n be a smooth compact orientable connected Riemannian manifold with boundary ∂M. Let $f(x)$ be a smooth function on M such that it has a unique minimum point x_0, it varies from 0 to 1 on M, it attains a maximum only on the boundary ∂M and is equal to one at all points of the boundary. Suppose that there are no critical points of f on ∂M and that f is a Morse function on the open manifold $M \backslash (\partial M \cup x_0)$ and has only finitely many critical points. Among these points there are no minima or maxima. Consider the level surface $\{f = r\}$, that is, the set of points x where $f(x) = r$. We have $\{f = 0\} = x_0$, $\{f = 1\} = \partial M$. Consider the vector field $v = \operatorname{grad} f$ on M. Let X be an arbitrary k-dimensional measurable compact subset of M. Consider the "boundary" of this compactum, that is, the set $\partial X = X \cap \partial M$ (Figure 4.14).

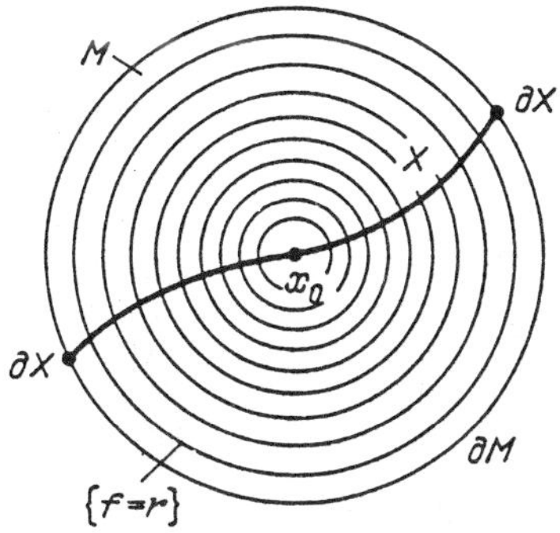

FIGURE 4.14

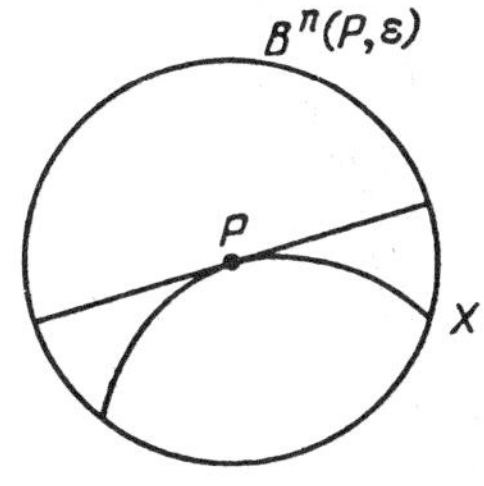

FIGURE 4.15

With each surface X we associate a *function of spherical density* $\Psi_k(P, X)$ defined as follows: $\lim_{\varepsilon\to+0}\gamma_k^{-1}(\varepsilon)\cdot\mathrm{vol}_k\, X\cap B^n(P, \varepsilon) = \Psi_k(P, X)$, where $P \in M$, $B^n(P, \varepsilon)$ is the n-dimensional ball in M with center at P and radius ε, and $\gamma_k(\varepsilon)$ is the k-dimensional volume of the standard Euclidean ball of radius ε and dimension k.

If $P \notin X$, then $\Psi_k(P, X) = 0$. If $P \in X$, then $\Psi_k(P, X) \geq 0$ (Figure 4.15). Clearly, the function $\Psi_k(P, X)$ measures the deviation of the surface X at the point P from a k-dimensional flat disc. If the surface X in a neighborhood of a point $P \in X$ is a smooth submanifold of M, then obviously $\Psi_k(P, X) = 1$, since $B^n(P, \varepsilon) \cap X$ for small ε deviates little from a k-dimensional smooth disc of radius ε.

Let $\Psi(x)$ be the spherical density function of the subset X of M. If X has the form of a sharp thin hair in a neighborhood of x (Figure 4.16), then the k-dimensional volume of the intersection of X with the ball $B_\varepsilon(x)$ is small, and so $\Psi(x) < 1$. If, on the other hand, a point $x \in X$ is a singular point, then there are cases where $\Psi(x) > 1$.

Consider a variational problem in the class of surfaces that glue a fixed contour A, where A is a $(k-1)$-dimensional measurable compact subset of ∂M such that $H_{k-1}(A, G) \neq 0$. We introduce the class of surfaces for which we solve the problem of the existence of an absolute minimum. Consider the class A^* of all compacta X with boundary $A \subset \partial M$ such that: (1) $\dim X = k$ and the compactum X is measurable; (2) X does not contain thin hairs, that is, for any point $x \in X\backslash\partial X$ we have $\Psi(x) \geq 1$; (3) the boundary of X contains A, that is, $A \subset \partial X$; (4) the embedding $i: A \to X$ annihilates the whole $(k-1)$-dimensional homology group of A, that is, the homomorphism $i_*: H_{k-1}(A) \to H_{k-1}(X)$ is trivial.

If d is the lower bound of the volume of all surfaces of the class A^*, then there is a globally minimal surface $X_0 \in A^*$ such that $\mathrm{vol}_k X_0 = d$.

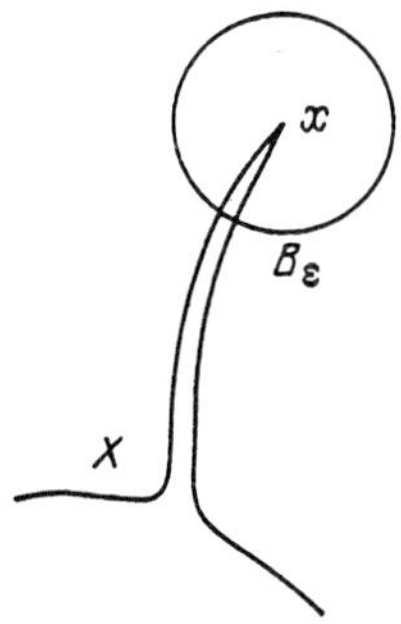

FIGURE 4.16

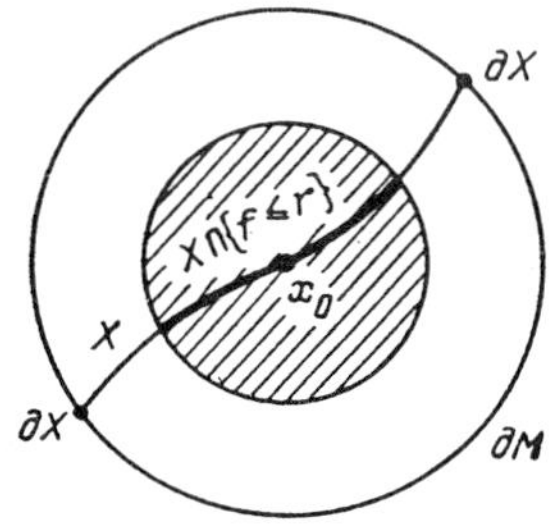

FIGURE 4.17

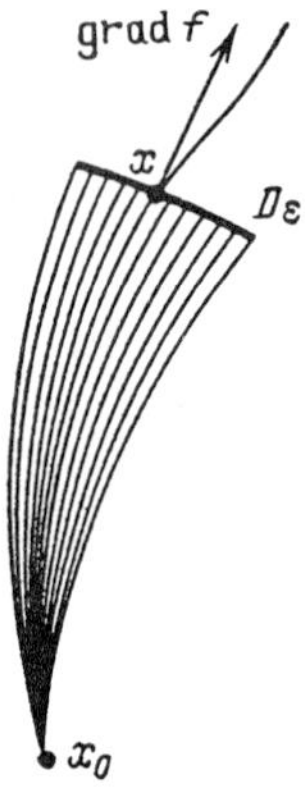

FIGURE 4.18

The results that follow are also true for cohomology (including generalized cohomology); see [133], [124], and [119]. For the surface X_0 we also have $\Psi(x) \geq 1$ for any $x \in X_0 \backslash \partial X_0$. We assume that X_0 passes through a point x_0. We construct an important function, which we call the *surface volume function*: $\psi(r) = \operatorname{vol}_k X_0 \cap \{f \leq r\}$. Let $\{f \leq r\}$ denote the n-dimensional subset of M consisting of all points x where $f(x) \leq r$ (Figure 4.17). The volume function is defined on the interval $[0, 1]$ and does not decrease as r goes from 0 to 1. We can now state the problem described at the beginning of this section as follows. *To give the greatest lower bound on the surface volume function ψ in terms of the metric of the manifold M independently of the topological type of the minimal surface X_0*. "Greatest" is understood in the sense that the required bound must become an equality for sufficiently rich series of specific minimal surfaces. For these cases A. T. Fomenko has also obtained exact formulas for their volume. Clearly, $\psi(1) = \operatorname{vol} X$.

Thus, knowing a lower bound for the function $\psi(r)$, we can find a lower bound for the number $\psi(1)$, that is, the *volume of the minimal surface*.

4.2. The coefficient of deformation of a vector field. By virtue of the properties of the function f, almost all integral trajectories of the field $\operatorname{grad} f$ that start at the point x_0 end on the boundary ∂M. We shall assume that a point $x \in M$ is a point in general position if through it there passes an integral trajectory of the field that starts at x_0 and ends on ∂M. Let us consider in M at the point x a small $(k-1)$-dimensional ball D_ε of radius ε orthogonal to the vector $\operatorname{grad} f(x)$ (Figure 4.18). Consider the set of all integral trajectories that start at x_0 and end at points of this ball. We denote the set of points swept by these trajectories by CD_ε. From the geometrical viewpoint this k-dimensional set is a "cone" with vertex at x_0 and base the ball D_ε. If x is a point in general position, then the cone is a smooth k-dimensional submanifold with boundary in M. If x is arbitrary, then the cone is, generally speaking, a *stratified surface.* Let us calculate the limit

$$\lambda(x) = \lim_{\varepsilon\to 0}(\operatorname{vol}_k CD_\varepsilon / \operatorname{vol}_{k-1} D_\varepsilon)$$

of the ratio of the volume of the cone to the volume of its base when the base contracts to a point. Let $\eta_k(x) = \sup \lambda(x)$, where the supremum is taken over all discs D_ε orthogonal to the vector $\operatorname{grad} f(x)$. The function η_k is smooth on M. We call it the *k-dimensional coefficient of deformation of the vector field* $\operatorname{grad} f$. It is closely connected with the Riemann curvature tensor of the manifold.

In many cases the coefficient η_k is easily calculated. Let us consider an example. Let M be a ball of radius R in $\mathbf{R}^n$, let $f(x) = |x|$ be the distance from the point $0 = x_0$ to the point x, and let $v = \operatorname{grad} f$. Then $\eta_k(x) = r/k$, where $r = |x|$.

4.3. Minimal surfaces of nontrivial topological type and least volume. The coefficient of deformation of a field, introduced by A. T. Fomenko, plays an important role in the calculation of least volumes of globally minimal surfaces, and *in many cases it enables us to calculate these volumes explicitly.*

THEOREM 4.4.1 (see Fomenko [133], [134], [119]). *Let f be a smooth function on a manifold M such that $0 \le f \le 1$, $f(x_0) = 0$, $f(\partial M) = 1$; f is a Morse function on $M \backslash (x_0 \cup \partial M)$ with only saddle critical points whose indices do not exceed $k-2$, where k is an integer and $k < n$. Let X_0^k be a globally minimal surface of the topological class described above and passing through x_0. Then for the volume function of this surface we have $\psi(r) \ge q(r)\cdot l$, where the constant $l = \lim_{a\to 0}[\psi(a)/q(a)]$ does not depend on the parameter r and is determined only by the structure of the minimal surface X_0 about the point x_0, and the function $q(r)$ has the form*

$$q(r) = \exp \int \left[\max_{x\in\{f=r\}} (\eta_k(x)\cdot|\operatorname{grad} f(x)|)\right]^{-1} dr.$$

In particular, $\psi(1) = \operatorname{vol} X_0 \ge l\cdot q(1)$. Thus, the behavior of the volume function $\psi(r)$ on the surface X_0 is determined by its behavior at the initial instant $r = 0$, that is, in a neighborhood of x_0, and by the geometry of the ambient manifold. This estimate is exact in the sense that there are *rich series of specific minimal surfaces for which the inequality becomes an exact equality. In these cases we obtain an exact expression for the volume function and for the volume of the minimal surface.*

Of course, the function $q(r)$, specified by an indefinite integral, is defined up to a constant multiplier, but in the formula for l there occurs the quantity $1/q(a)$, which removes the indefiniteness.

Theorem 4.4.1 follows from a more general theorem of Fomenko on the behavior of the normalized volume function on surfaces. Suppose that the triple (M, f, X) satisfies the conditions of Theorem 4.4.1; then it turns out that the piecewise continuous function ψ/q does not decrease with respect to r as r goes from 0 to 1. Hence we find that

$$\frac{\psi(1)}{q(1)} = \frac{\operatorname{vol} X_0}{q(1)} \geq \lim_{a\to 0} \frac{\psi(a)}{q(a)} = l, \quad \text{that is,} \quad \operatorname{vol} X_0 \geq l \cdot q(1).$$

A similar assertion also holds for closed minimal surfaces that realize nontrivial cycles or cocycles in the ambient manifold M.

Let us consider the consequences of Theorem 4.4.1, for example the problem of a lower bound on the volume of a minimal surface X^k passing through the center of a Euclidean ball D^n in $\mathbf{R}^n$. Among this class of surfaces we can naturally distinguish flat k-dimensional discs—sections of D^n by k-dimensional planes passing through its center. It is well known that in many cases (for example, when the surface X^k is complex-analytic in $\mathbf{R}^{2n} = \mathbf{C}^n$) the volume of its part inside the ball is bounded below by the volume of the flat k-dimensional disc. It turns out that a more general assertion is true, which holds for arbitrary (co)homological globally minimal surfaces of any codimension and for an arbitrary generalized (co)homology theory.

PROPOSITION 4.4.1. *Let X^k be a globally minimal surface in $\mathbf{R}^n$ passing through the center of a ball D^n of radius r and having its boundary on the boundary of the ball (in the sense of the class A^*; see above). Then we have the inequalities*

$$\operatorname{vol}_k X \cap D^n \geq \Psi(O) r^k \gamma_k \geq r^k \gamma_k = \operatorname{vol}_k D^k,$$

where $\Psi(O)$ is the density of the surface X at the point O, γ_k is the volume of a unit Euclidean ball of dimension k, and $r^k\gamma_k$ is the volume of a k-dimensional Euclidean ball of radius r. In addition, $\Psi(O) \geq 1$, and equality holds if and only if the point O is regular on the surface X^k. Thus, the volume of any globally minimal surface of any codimension passing through the center of the Euclidean ball and having its boundary on *the boundary of the ball is not less than the volume of a standard central plane section of the ball, that is, than the volume of a flat k-dimensional disc.*

Let M^n be a closed connected compact Riemannian manifold and x_0 a point of it. It is known that M can be represented as a cell complex containing only one n-dimensional cell homeomorphic to a disc D^n whose boundary is glued to a subcomplex C, where $\dim C \leq n-1$. Then the set $M\backslash(C \cup x_0)$ is homeomorphic to an open disc D^n with a punctured point x_0. Let us consider on M a function f, smooth on $M\backslash(C \cup x_0)$, having exactly one minimum on M at x_0 (we assume that $f(x_0) = 0$), taking its maximal value, equal to one, on C, and having no critical points on the open manifold $M\backslash(C \cup x_0)$. The set of all such functions is denoted by $F(x_0)$. Considering on $D^n\backslash x_0$ the field $\operatorname{grad} f$, where $f \in F(x_0)$, we can, by following the construction we have presented, define the function

$$q(r) = \exp \int_0^r \frac{dr}{\max_{x\in\{f=r\}}(\eta_k \cdot |\operatorname{grad} f|)},$$

which depends on the point x_0 and the choice of the function f.

DEFINITION 4.4.1. We define the *function of k-dimensional F-deficiency of a manifold M at a point* x_0 by

$$\Omega(x_0, f) = \gamma_k q(1) \lim_{a\to 0} \frac{a^k}{q(a)},$$

where $q(1) = \lim_{r\to 1} q(r)$. At each point $x_0 \in M$ we consider the number $\Omega_k(x_0) = \sup_{f\in F(x_0)} \Omega(x_0, f)$. Finally, we define the *k-dimensional F-deficiency of M* as the number

$$\Omega_k = \inf_{x_0\in M} \Omega_k(x_0).$$

This number depends on the dimension k and on M. Clearly, we always have $\Omega_k > 0$. Consider the (co)homology $H_k^{(k)}(M)$ with coefficients in a group G. Let L be an arbitrary nontrivial subgroup of the group $H_k^{(k)}(M)$. Consider a globally minimal surface X^k in M that realizes this subgroup. This means that under the embedding $i\colon X \to M$ the surface X realizes all the (co)cycles of the subgroup L. In the case of homology groups the condition $L \subset \operatorname{Im} i_*$ must be satisfied, where $i_*\colon H_k(X) \to H_k(M)$, and in the case of cohomology groups the condition $L \subset H^k(M)\backslash \operatorname{Ker} i^*$, where $i^*\colon H^k(M) \to H^k(X)$. Suppose that $x_0 \in X$ and let $\Psi(x_0)$ be the value of the density function of the surface X at x_0.

THEOREM 4.4.2 (Fomenko [133]). *Let $X^k \subset M$ be a closed globally minimal surface in M that realizes a nontrivial subgroup L. Then we always have $\operatorname{vol}_k X \geq \Psi(x_0) \cdot \Omega_k \geq \Omega_k > 0$. Hence, the number Ω_k turns out to be a universal constant, which gives a lower bound to the k-dimensional volume of any closed minimal surface that realizes a nontrivial cycle or cocycle in M. There are rich series of examples in which the bound is attained on specific minimal surfaces of nontrivial topological type, that is, the inequality becomes an exact equality. In particular, if the exact equality $\operatorname{vol}_k X = \Omega_k$ holds for a minimal surface X, then the surface is a smooth minimal compact closed submanifold of M, that is, it has no singularities.*

Let $x \in M^n$ be a *fixed point* and let $\exp: T_xM \to M$ be a standard map defined by the pencil of geodesics emanating from the point x. In other words, for a point $\xi \in T_xM$ its image in M under the map exp is the point $\gamma(t)$, where t is equal to the length of the vector O_ξ in T_xM joining O to ξ, and $\dot{\gamma}(0) = O_\xi/t$. It is well known that for small ε the map exp establishes a diffeomorphism between the disc $D^n(O, \varepsilon)$ embedded in T_xM with center at O and radius ε and its image in M, which we denote by $Q^n(x, \varepsilon)$. Consider all those values of t for which exp establishes a diffeomorphism between $D(O, t)$ and $Q(x, t)$, and let $R(x) = \sup(t)$, that is, when $t > R(x)$ the map exp ceases to be a diffeomorphism. In particular, this means that if $\overline{D}(O, R(x))$ is a closed disc in T_xM, then the map exp is no longer a diffeomorphism on it and glues together certain points on the boundary of the disc $\overline{D}(O, R(x))$. It is also clear that for all $t < R(x)$ the closure $\overline{Q}(x, t)$ of an open disc in M is homeomorphic to a closed disc in M. Thus, the number $R(x)$ is the maximal radius of an open geodesic disc $Q(x, t)$ with center at x that can be inscribed in M. The radii of this disc $Q(x, R(x))$ will be assumed to be geodesics emanating from x that are images of rays emanating from O in the tangent plane T_xM. For each point $y \in Q(x, R(x))$ there is exactly one radius joining it to x. The diffeomorphism of $D(O, R(x))$ on $Q(x, R(x))$ described above is called a *geodesic diffeomorphism*. Clearly, the disc $\overline{Q}(x, R(x))$ need not coincide with M. At the same time, for some symmetric spaces M (for example, for symmetric spaces of rank 1) this disc completely exhausts the whole manifold.

Consider a geodesic diffeomorphism and the disc $Q(x, R(x)) = Q$ corresponding to it. This disc consists of the pencil of radii (geodesics) emanating from x. On these geodesics we introduce a natural parameter r, which varies from 0 to $R(x)$. On $Q \backslash x$ we consider a smooth function $f(y) = r$, where $y = \gamma(r)$, that is, the value of this function at the point y is equal to the distance from this point to the central point x. This distance is measured along the unique radius (geodesic) joining x to y. Obviously, this function does not have critical points on $Q\backslash x$ and $0 \leq f \leq R(x)$. As the vector field v we take the field $\operatorname{grad} f$. Clearly, $|v| \equiv 1$ on $Q\backslash x$. We now consider the triple $Q\backslash x, f, v$. We can then calculate the deformation coefficient $\eta_k(v, y)$.

DEFINITION 4.4.2. The *function of* k*-dimensional geodesic deficiency* $\Omega_k^0(x)$ is the function

$$\Omega_k^0(x) = \gamma_k q_x(R(x)) \lim_{a\to 0} \frac{a^k}{q_x(a)},$$

where

$$q_x(r) = \exp \int_0^r \frac{dr}{\max_{x\in\{f=r\}} \eta_k(v, x)}.$$

We recall that $|\operatorname{grad} f| \equiv 1$. We set $\eta_k(x, r) = \max_{x\in\{f=r\}} \eta_k(v, x)$. We define the k*-dimensional geodesic deficiency of the manifold* M as the number

$$\Omega_k^0 = \inf_{x\in M} \Omega_k^0(x).$$

Since $|\operatorname{grad} f| \equiv 1$, we obtain $\lim_{a\to 0} q_x(a) = \lim_{a\to 0} a^k$. Consequently, we have proved the following assertion: $\Omega_k^0(x) = \gamma_k q_x(R(x))$.

We point out *the connection between the deficiency of a manifold and the least volumes of surfaces that realize nontrivial cycles or cocycles in the manifold.*

Fomenko [134] proved the following theorem. Let $X^k \subset M$ be a globally minimally closed (that is, without boundary) surface that realizes, under the embedding in M, a subgroup (respectively, subset) $L' \neq 0$ in the homology (respectively, cohomology) of M, that is, $X \in O(L')$. Then the following inequality always holds:

$$\operatorname{vol}_k X \geq \left(\sup_{x\in X} \Psi_k(x, X)\right) \cdot \Omega_k \geq \Omega_k \geq \Omega_k^0 > 0.$$

Hence the number Ω_k^0 *turns out to be a universal constant, which gives a lower bound to the* k*-dimensional volume of any closed (without boundary) minimal surface that realizes nonzero cycles or cocycles in* M. In the general case this bound is the best possible, that is, there are rich series of examples in which this bound is attained on specific minimal surfaces (see [134]).

Until recently there were comparatively few specific examples of globally minimal surfaces. This is connected with the surmounting of serious difficulties standing in the way of a proof of global minimality for various specific topological surfaces. We demonstrate the effectiveness of the method described above for constructing minimal surfaces. Let M be a compact symmetric space of rank 1.

1. Let $M^{2n} = \mathbf{C}P^n$ be a complex projective space, $n > 1$, and let $\mathbf{C}P^k = X^{2k}$, $1 \leq k \leq n-1$, be complex projective subspaces, embedded in the standard way, each of which realizes a generator of the $2k$-dimensional cohomology group of $\mathbf{C}P^n$.

2. Let $M^{4n} = \mathbf{Q}P^n$ be a quaternion projective space, $n \geq 1$, and let $\mathbf{Q}P^k = X^{4k}$, $1 \leq k \leq n-1$, be quaternion projective subspaces, embedded in the standard way, each of which realizes a generator of the $4k$-dimensional cohomology group of $\mathbf{Q}P^n$.

3. Let $M^n = \mathbf{R}P^n$ be a real projective space, $n \geq 2$, and let $\mathbf{R}P^k = X^k$, $1 \leq k \leq n-1$, be real projective subspaces, embedded in the standard way, each of which realizes a generator of the k-dimensional cohomology group of $\mathbf{R}P^n$.

4. Let $M^{16} = F_4/\operatorname{Spin}(9)$ be a symmetric space containing a sphere $S^8 = X^8$, embedded in the standard way (for a description of the embedding, see below), that realizes a generator of the 8-dimensional cohomology group of M^{16}.

THEOREM 4.4.3 (Fomenko [128], [129]). *Each of the submanifolds X^p listed above is not only a totally geodesic submanifold of M, but also a globally minimal surface. The exact equality $\operatorname{vol}_p X^p = \Omega_p^0(M)$ holds for them, that is, all these surfaces have the least possible volume among all minimal surfaces in this dimension (that realize nonzero cycles or cocycles). Moreover, any globally minimal surface $X^p \subset M$ that realizes some nontrivial cocycle in dimension p coincides (up to an isometry in M) with the submanifold X^p mentioned above, which realizes a generator of the cohomology group in this dimension. In particular, in each dimension p in the manifold M there can be only one globally minimal surface X^p. The list of manifolds M (see above) exhausts all the so-called symmetric spaces of rank 1.*

THEOREM 4.4.4 (Fomenko [128], [129]). 1. *Let $M = G^{\mathbf{C}}_{p+q,q}$, $1 \leq q \leq p$, be a complex Grassmann manifold (that is, the manifold of q-dimensional planes in $(p+q)$-dimensional space) and $\mathbf{C}P^s$, $1 \leq s \leq p$, a collection of totally geodesic submanifolds, embedded in the standard way, each of which is diffeomorphic to a complex projective space and realizes a* *nonzero element of the cohomology ring* $H^*(M, \mathbf{Z})$. *Then all these submanifolds are globally minimal surfaces. In addition, their volume is the least possible, that is,* $\operatorname{vol} \mathbf{C}P^s = \Omega^0_{2s}$. *Moreover, the values of the dimension* $2, 4, 6, \dots, 2p$ *are the only ones in which there is a minimal surface* $X^k \subset M$ *such that* $\operatorname{vol} X = \Omega^0_k(M)$.

2. *Let* $M = G^{\mathbf{Q}}_{p+q,q}$, $1 \le q \le p$, *be a quaternion Grassmann manifold, and let* $\mathbf{Q}P^s$, $1 \le s \le p$, *be a collection of totally geodesic submanifolds, embedded in the standard way, each of which is diffeomorphic to a quaternion projective space and realizes a nonzero element of the cohomology ring* $H^*(M, \mathbf{Z})$. *Then all these submanifolds are globally minimal surfaces and, in addition, their volume is the least possible, that is,* $\operatorname{vol} \mathbf{Q}P^s = \Omega^0_{4s}$. *Moreover, these values* $k = 4, 8, 12, \dots, 4p$ *are the only dimensions in which there is a minimal closed surface* $X^k \subset M$ *such that* $\operatorname{vol} X^k = \Omega^0_k$.

We now consider the case of general symmetric spaces. We recall that a compact manifold M is called a *symmetric space* if it can be represented as a homogeneous space G/H, where G and H are compact Lie groups, $H \subset G$, and on G there is defined an involutory automorphism σ, $\sigma^2 = 1$, such that $H = \{g \in G | \sigma(g) = g\}$. For simplicity we shall consider only simply-connected and irreducible symmetric spaces, that is, those that cannot be represented as a direct product of other symmetric spaces.

THEOREM 4.4.5 (Fomenko [128], [129], [131]). *Let* M *be a compact simply-connected symmetric space and* $X \subset M$ *a* k*-dimensional minimal closed* (*that is, without boundary*) *surface that realizes nonzero cycles or cocycles, that is,* $X \in O(L')$, $L' \neq 0$. *Suppose that the surface has least possible volume, that is,* $\operatorname{vol} X = \Omega^0_k$, *where* Ω^0_k *is the geodesic deficiency of* M. *Then the minimal surface* X *is a simply-connected compact totally geodesic submanifold of* M. *In particular,* X *is a symmetric space of rank* 1, *that is,* X *is diffeomorphic to one of the following manifolds:* 1) S^k, 2) $\mathbf{C}P^{k/2}$, 3) $\mathbf{Q}P^{k/4}$, 4) $F_4/\operatorname{Spin}(9)$.

4.4. On the minimal volume of surfaces passing through the center of a symmetric convex domain in Euclidean space. As we have seen above, *for the multidimensional volume functional there is always a globally minimal surface on which the absolute minimum of the volume is attained in the class of spectral bordisms.* Above we considered compact surfaces, closed or with boundary. However, the question of absolute minima also arises in the noncompact case. Minimality of a noncompact surface is understood here in the sense that any perturbation of it with compact support does not decrease its volume. An important example of noncompact minimal surfaces in $\mathbf{R}^n$ is that of Euclidean planes. Fomenko [133], [135] formulated the following conjecture, which generalizes a large number of earlier observations, according to which *linear* minimal surfaces realize in a certain sense the absolute minimum of the volume functional.

Let B^n be a *convex* domain in $\mathbf{R}^n$ with piecewise smooth boundary that is *centrally symmetric* with respect to a point $O \in B^n$. Let X^k be a globally minimal surface without boundary that goes off to infinity in $\mathbf{R}^n$ and passes through O. Then *the volume of its intersection with* B^n *is always at least equal to the volume of the smallest plane section of this domain by planes of dimension* k *passing through the center of the domain.* It is assumed that X^k does not have thin hairs of zero volume. This obvious stipulation can be written as follows: $\Psi_k(x) \geq 1$ at each point $x \in X$. Here Ψ_k is the spherical density function. For the precise meaning of the condition that X is globally minimal and does not have a boundary, see §4.1.

If the domain is a ball, then for the case of an arbitrary globally minimal surface *of any codimension* passing through the center of the ball and any homology or cohomology theory this conjecture was proved by A. T. Fomenko (see above). Some special cases of this assertion were proved earlier, for example a theorem of Lelong (see [134], [119]). If the domain is a standard four-dimensional cube in $\mathbf{R}^4 = \mathbf{C}^2$, and X^2 is a complex-analytic surface of codimension one (that is, a complex curve in $\mathbf{C}^2$) passing through the center of the cube, then the conjecture follows from a well-known result of Katsnel′son and Ronkin [56]; see also Ronkin [95]. Recently Le Hong Van [415] proved the conjecture for quite a wide class of domains in $\mathbf{R}^n$. We give her result, which is more simply formulated for special domains.

THEOREM 4.4.6 (Le Hong Van). *Let* B^n *be a rectilinear parallelepiped or an ellipsoid with center at a point* $O \in \mathbf{R}^n$. *Then the area of the intersection of this domain with any two-dimensional simply-connected minimal surface passing through the center of* B^n *is always at least equal to the area of a minimal two-dimensional central plane section of this domain.*

The condition that the surface is simply-connected is an additional restriction here, which is probably not difficult to remove. We emphasize that in Fomenko's conjecture it is a question of globally minimal surfaces without boundary that go off to infinity. For example, not every two-dimensional minimal surface without boundary that goes off to infinity satisfies the requirement of global minimality. Thus, for example, a catenoid can be regarded as a locally minimal unbounded surface, but "at infinity" this surface is not globally minimal. The fact is that for a sufficiently close position of coaxial circles that are the boundary of a catenoid there are two locally minimal surfaces with the same boundary. Each of them is a catenoid, but one is stable and the other is unstable. Clearly, the outer catenoid has smaller area than the inner catenoid.

4.5. Calibration forms and minimal surfaces. The method presented below was developed by Harvey and Lawson [253]. Let M be a Riemannian manifold. Suppose that on it there is defined an exterior differential p-form φ that satisfies the following two conditions:

(1) φ is closed (that is, $d\varphi = 0$);

(2) $\varphi|_\xi \leq \mathrm{vol}_\xi$ for any oriented tangent p-plane ξ on M.

Let $X^p \in M$ be an oriented submanifold of M of dimension p such that

$$\varphi|_X = \mathrm{vol}_X , \qquad (*)$$

that is, the restriction of φ to X coincides with the volume form on X. Then it turns out that X^p is a globally minimal surface in its homology class. In other words, $\mathrm{vol}_p X \leq \mathrm{vol}_p X'$ for any surface X' homologous to X and such that $\partial X = \partial X'$, that is, $[X - X'] = 0$ in $H_p(M, \mathbf{R})$, where $[Y]$ denotes the homology class of the surface Y. In fact, $\mathrm{vol}_p X = \int_X \varphi = \int_{X'} \varphi \leq \mathrm{vol}\, X'$.

Thus, having constructed on M a closed form normalized by condition (2), we can try to find globally minimal surfaces corresponding to it.

A form φ satisfying conditions (1) and (2) is called a *calibration form on* M, and M is called a *calibrated manifold*. An oriented surface X for which the condition $(*)$ is satisfied is called a *φ-surface*.

Let us give a simple example. Consider the plane $\mathbf{R}^2(x, y)$, where (x, y) are Cartesian coordinates (the metric is Euclidean), and a 1-form dx on $\mathbf{R}^2$. It is easy to verify that dx is a calibration form in $\mathbf{R}^2$. The corresponding dx-manifolds are horizontal lines oriented like the x-axis. Hence it follows that the curve of least length joining the points $A(x_1, y_0)$ and $B(x_2, y_0)$ is the horizontal line joining them.

We now turn to statements in the language of the theory of currents. The Grassmannian $G(p, T_xM)$ is embedded in the natural way in the vector space $\Lambda^p T_xM$ of p-vectors:

$$G(p, T_xM) = \{\xi \in \Lambda^p T_xM : \xi \text{ is a simple unit } p\text{-vector}\};$$

$G(p, TM)$ is the corresponding Grassmann bundle. Let φ be an exterior

p-form on M. Then at each point $x \in M$ there is defined the *comass* φ_x:

$$\|\varphi\|_x^* = \sup\{\langle \varphi_x, \xi_x \rangle : \xi_x \text{ is a simple unit } p\text{-vector}\}.$$

If $A \subset X$, we can define the *comass* φ *on* A:

$$\|\varphi\|_A^* = \sup \|\varphi\|_x^*, \quad x \in A.$$

To each rectifiable current T there corresponds the measure $\|T\|$, and for almost all $x \in M$ in the sense of the measure $\|T\|$ there is defined the "tangent plane" $\vec{T}(x) \in G(p, T_xM)$. We have $T(\psi) = \int_M \langle \vec{T}, \psi \rangle \, d\|T\|$, $m(T) = \|T\|(M)$.

Consider φ, a smooth p-form on M. Then $\|\varphi\|_M^* = 1$. We define a subset $\boldsymbol{G}(\varphi) \subset G(p, TM)$ as follows:

$$\boldsymbol{G}(\varphi) = \bigcup_{x \in M} \boldsymbol{G}_x(\varphi),$$

$$\boldsymbol{G}_x(\varphi) = \{\xi_x \in G(p, T_xM) : \langle \varphi_x, \xi_x \rangle = 1\}.$$

A *positive* φ*-current* is defined as a rectifiable current T such that $\vec{T}(x) \in \boldsymbol{G}(\varphi)$ $\|T\|$-almost everywhere on M.

Directly from the definitions there follows an assertion of Harvey and Lawson [253].

Let φ *be a smooth p-form on* M *such that* $\|\varphi\|_M^* = 1$. *Let* T *be an arbitrary p-current with compact support. Then* $T(\varphi) \le m(T)$, *and equality is attained if and only if* T *is a positive* φ*-current.*

In particular, if S is a compact oriented p-dimensional submanifold (possibly with boundary) in M, then $\int_S \varphi \le \operatorname{vol} S$, and equality is attained if and only if S is a φ-submanifold. A smooth closed p-form φ of comass 1 on a Riemannian manifold M is called a *calibration form*; the pair (M, φ) is called a *calibrated manifold*.

THEOREM 4.4.7 (Harvey and Lawson [253]). *Let* (M, φ) *be a calibrated manifold. Let* T *be a positive* φ*-current with compact support. Let* T' *be any other current with compact support that is homologous to* T *(that is,* $dT = dT'$*). Then* $m(T) \le m(T')$, *and equality is attained if and only if* T' *is a positive* φ*-current.*

The following special cases are of particular interest. Let (M, φ) be a calibrated manifold.

Any closed φ*-current* T *with compact support realizes a current of least mass in its de Rham cohomology class.*

Any other rectifiable current of minimal mass in this cohomology class is a positive φ*-current.*

We note that a positive closed φ-current T minimizes the mass in the class of all real currents homologous to T (in $H_p(M, \mathbf{R})$, and not only in the class of rectifiable currents).

COROLLARY 4.4.1. *Let T be a positive φ-current with compact support, $\partial T = A$. If $H_p(M, \mathbf{R}) = 0$, then T is a solution of the Plateau problem for A, that is, T is the current of least mass among all currents with compact support with boundary A.*

In particular, if φ is a calibration in $\mathbf{R}^N$ with the usual metric, then any φ-current T is a solution of the Plateau problem for $A = \partial T$.

Let M be of class C^k, $2 \le k \le \omega$. Every φ-submanifold S without boundary of class C^2 is a submanifold of class C^k.

Let M be a complex manifold. On it we fix a Riemannian metric $\langle\ ,\ \rangle$ such that the map $J: TM \to TM$ corresponding to multiplication by $\sqrt{-1}$ is orthogonal at each point $x \in M$. Then there arises a Kähler 2-form ω: $\omega(V, W) = \langle JV, W\rangle$. If $d\omega \equiv 0$, then M is called a *Kähler manifold*. We set $\Omega_p = \omega^p/p!$. Then $\|\Omega_p\|_M^* = 1$, $d\Omega_p \equiv 0$, and $\mathbf{G}_x(\Omega_p)$ is the Grassmannian of positive complex p-planes. We obtain a well-known result of Federer: *complex submanifolds of a Kähler manifold are globally minimal surfaces.* Let us consider *special Lagrangian manifolds.* Consider $\mathbf{C}^n(z_1, \dots, z_n)$ with the standard metric. We denote $z = x+iy$, and let $J: \mathbf{C}^n \to \mathbf{C}^n$ be multiplication by $\sqrt{-1}$. An oriented n-plane ξ in $\mathbf{C}^n$ is said to be *Lagrangian* ($\xi \in \mathrm{Lag}$) if for any $u \in \xi$ we have $Ju \perp \xi$. Obviously, a real n-plane $\xi_0 = \mathbf{R}^n$ $(y = 0)$ is Lagrangian. Moreover, any other Lagrangian plane ξ is obtained from ξ_0 by a unitary transformation: $\xi = A\xi_0$, $A \in U(n)$. An oriented n-plane ξ in $\mathbf{C}^n$ is said to be *specially Lagrangian* ($\xi \in S_0\,\mathrm{Lag}$) if: (1) ξ is Lagrangian, (2) $\xi = A\xi_0$, $A \in U(n)$, and $\det A = 1$ (that is, $A \in SU(n)$). There obviously arises the bundle $\mathrm{Lag} \approx U(n)/SO(n) \xrightarrow{\det} S^1$. We observe that $(\det_{\mathbf{C}})^{-1}(1) = S_0\,\mathrm{Lag}$. On $\mathbf{C}^n$ we define the exterior n-form $\alpha_0 = \mathrm{Re}(dz_1 \wedge \dots \wedge dz_n)$.

REMARK. We can define $S_\theta\,\mathrm{Lag} = (\det_{\mathbf{C}})^{-1}(e^{i\theta})$, $\theta \in [0, 2\pi)$; (when $\theta = 0$, we obtain $S_0\,\mathrm{Lag}$). We introduce the special Lagrangian form $\alpha_\theta = e^{i\theta} \cdot \alpha_0$. We interpret $\xi \in S\,\mathrm{Lag}$ as $\xi \in S_\theta\,\mathrm{Lag}$ for some θ; let α denote the corresponding exterior form ($\theta \in [0, 2\pi)$ is arbitrary, but fixed). It turns out that the form α is closed and has comass 1. The set $G(\alpha)$ coincides with the set $S_\theta\,\mathrm{Lag}$ of specially Lagrangian planes.

A submanifold M^n of $\mathbf{C}^n$ is said to be *specially Lagrangian* if $T_xM \in S_\theta\,\mathrm{Lag}$ at any point $x \in M$.

THEOREM 4.4.8 (Harvey and Lawson [253]). 1. *Specially Lagrangian manifolds are globally minimal surfaces.*

2. *Let* $M^n \subset \mathbf{R}^{2n}$ *be a connected Lagrangian submanifold. Then* M *is minimal* (*the mean curvature* $H \equiv 0$) *if and only if* M *is specially Lagrangian.*

Hence we obtain the following corollary: *if a Lagrangian manifold is locally minimal, then it is also globally minimal*. Harvey and Lawson [253] found many nontrivial examples of specially Lagrangian manifolds.

Consider the torus $T^{n-1} = \{\operatorname{diag}(e^{i\theta_1}, \dots, e^{i\theta_n}), \theta_1 + \cdots + \theta_n = 0\} \subset SU(n)$. Let C be a cone over the orbits of the standard action of T^{n-1} in $\mathbf{C}^n$ as a subgroup of $SU(n)$. Then C *is a specially Lagrangian submanifold of* $\mathbf{C}^n$.

Hence we obtain examples of low-dimensional globally minimal cones of *large codimension* in $\mathbf{R}^{2n} = \mathbf{C}^n$.

Consider an arbitrary submanifold $V \subset \mathbf{R}^N$. It is well known that its normal bundle $N(V)$ is canonically embedded in the cotangent bundle $T^*\mathbf{R}^N$ as a Lagrangian submanifold with respect to the natural symplectic structure $dp \wedge dq$ on T^*M.

When is $N(V)$ *specially Lagrangian*? We denote the second quadratic form of V by A. A submanifold $V \subset \mathbf{R}^N$ is said to be *simple* if $\operatorname{Spec} A^\nu = (\lambda_1, \dots, \lambda_p) = (a, -a, b, -b, \dots, c, -c, 0, \dots, 0)$ at each point $x \in V$ for any $\nu \in N_x V$. Let V be a submanifold of $\mathbf{R}^N$, and let $N(V) \subset T^*\mathbf{R}^N$ be a canonical Lagrangian embedding. Then, as Harvey and Lawson showed, $N(V)$ *is specially Lagrangian if and only if* V *is a simple submanifold*. We note the following.

1. Let $M^2 \subset \mathbf{R}^N$ be a locally minimal surface. Then M^2 is a simple submanifold of $\mathbf{R}^N$.

2. If $M^p \subset \mathbf{R}^N$ is simple, then the cone CM over M in $\mathbf{R}^N$ is also a simple submanifold of $\mathbf{R}^N$.

The results of Harvey and Lawson for minimal Lagrangian surfaces in $\mathbf{R}^{2n}$ have recently been generalized by Le Hong Van in the case of an arbitrary Hermitian manifold by means of the method of relative calibrations that she has developed [454], [457].

DEFINITION 4.4.3. Let φ^n be a differential n-form on a Riemannian manifold M, and $G_\varphi(M)$ a closed subset of the bundle of unit n-vectors $G_n(M)$. Then the pair $(\varphi^n, G_\varphi(M))$ is called a *regular relative calibration* if the following conditions are satisfied:

(A) the value of φ^n on the set $G_\varphi(M)$ is identically equal to one;

(B) the set $G_\varphi(M)$ is a critical set in $G_n(M)$ that realizes a locally maximal value of φ;

(C) $G_\varphi(M)$ has the structure of a fiber space over M.

The pair $(\varphi^n, G_\varphi(M))$ is called a *relative calibration* if it is obtained by gluing regular relative calibrations $(\varphi_i^n, G_{\varphi_i}(M))$ by means of functions of partition of unity on M: $\sum h_i = 1$, $h_i \geq 0$, $\varphi^n = \sum h_i \varphi_i^n$, $G_\varphi(M) = \bigcap_i G_{\varphi_i}(M)$.

REMARK. If we require that the maximal value of φ_i^n on $G_n(M)$ is equal to 1, and φ is closed, then we obtain the classical calibration [253], which we shall call henceforth the absolute calibration.

DEFINITION 4.4.4. A submanifold $N^n \subset M^m$ is called a *φ-submanifold* if it is an integral submanifold of the distribution G_φ of the (relative) calibration φ. A (relative) calibration φ is said to be *effective* if there is at least one closed φ-submanifold.

We show that for any submanifold $N \subset M$ there is always a relative calibration φ for which it is a φ-submanifold. Moreover, the functional $S\varphi$, which acts on surfaces N by integration of φ along N, approximates the volume functional to the first order on φ-submanifolds. If a φ-submanifold N is locally minimal, then $S\varphi$ approximates the volume functional on N to the second order. Namely, we have the following theorems (Le Hong Van).

THEOREM 4.4.9. *Let φ be a relative calibration on a Riemannian manifold M^m, and let $N^n \subset M^m$ be a φ-submanifold. Then for any normal vector field X on N we have*

$$\left.\frac{d}{dt}\right|_{t=0} \operatorname{vol}(N_t) = \left.\frac{d}{dt}\right|_{t=0} \left(\int_{N_t} \varphi\right),$$

where N_t is the image of N under the deformation induced by the vector field X.

THEOREM 4.4.10. *Let φ be a relative calibration on a Riemannian manifold M, and let N be a minimal φ-submanifold. Then for any normal vector field X with compact support on N we have*

$$\left.\frac{d^2}{dt^2}\right|_{t=0} \operatorname{vol}(N_t) \geq \left.\frac{d^2}{dt^2}\right|_{t=0} \left(\int_{N_t} \varphi\right),$$

where N_t is the image of N under the deformation induced by the vector field X.

COROLLARY. *If φ is closed and N is a minimal φ-submanifold, then it is a stably minimal submanifold.*

We call an n-dimensional surface $L \subset M^{2n}$ Φ-Lagrangian if the restriction of a Kähler 2-form to it is equal to zero. Φ-Lagrangian surfaces in Kähler manifolds are simply Lagrangian. An n-form Ω^n on a Hermitian manifold is said to be specially Lagrangian if its restriction to any tangent plane $T_xM^{2n} \cong \mathbf{R}^{2n}$ is a specially Lagrangian form. For each Φ-Lagrangian surface we construct a family of specially Lagrangian calibrations. Applying Theorem 4.3.9, we obtain a criterion for local minimality of Φ-Lagrangian submanifolds in terms of a certain differential 1-form defined on the Grassmann bundle $G_L(M^{2n})$ associated with the given manifold M^{2n}. In other words, $G_L(M^{2n})$ is a bundle of Lagrangian planes in the tangent spaces T_xM^{2n}. This form is defined as follows. Let g be the determinant of the Riemannian metric; then $\omega = \frac{1}{2}Jd(\ln g) - d\theta$, where J is the operator of the complex structure on M^{2n}, and the function θ with period 2π (that is, a map into a circle) is defined as follows: $\theta(l_x) = \sqrt{-1}\ln(\sqrt{g}\,dz_1 \wedge \cdots \wedge dz_n, l_x)$, where l_x is a Lagrangian plane in the tangent space, and $(\sqrt{g}\,dz_1 \wedge \cdots \wedge dz_n, l_x)$ is the value of the form $\sqrt{g}\,dz_1 \wedge \cdots \wedge dz_n$ on the unit multivector that defines the Lagrangian plane l_x. Let L be a Lagrangian submanifold, and $p\colon L \to G_L(M^{2n})$ the natural map that associates a Lagrangian plane T_xL with each point $x \in L$. Then the 1-form ω defines a 1-form $p^*(\omega)$ on L. The following assertion holds: a Lagrangian submanifold L of M^{2n} is locally minimal (that is, its mean curvature is zero) if and only if $p^*(\omega) = 0$. For the case $M^{2n} = \mathbf{R}^{2n} = \mathbf{C}^n$ the form ω is closed and the cohomology class of the form $(1/2\pi)\omega$ coincides with the Maslov-Arnol′d characteristic class (after the natural embedding $H^1(G_L(\mathbf{R}^{2n}), \mathbf{Z}) \to H^1(G_L(\mathbf{R}^{2n}), \mathbf{R})$).

Since $dJd(\ln g) = 2\pi\Omega$ is the first Chern form on the Hermitian manifold M, it follows immediately that the restriction of Ω to a minimal Φ-Langragian submanifold is equal to zero.

Let us give some examples of locally minimal surfaces in Kähler manifolds by means of the stated criterion.

1. Let M^{2n} be the complex projective space $\mathbf{C}P^n$ with the Fubini-Study Kähler metric. Then the real Lagrangian projective space $\mathbf{R}P^n = \{(z_0, z_1, \ldots, z_n) : z_i/z_0 \in \mathbf{R}\}$ is locally minimal. If a Lagrangian submanifold in some chart $\{z_i = 1\}$ is a cone and $p^*\,d\theta = 0$, then L is a locally minimal submanifold. An example of such cones is the submanifold

$$\left\{z_0 = 1, z_j = ke^{i\theta_j} : k \in \mathbf{R}, \sum \theta_j = 0\right\}.$$

2. If M^{2n} is the complex Grassmann manifold $G_{2,k}(\mathbf{C})$, then the real Grassmann submanifold $G_{2,k}(\mathbf{R})$ is a locally minimal Lagrangian submanifold.

A criterion for local minimality of Lagrangian surfaces in Kähler manifolds was also obtained by Bryant [461] by means of the canonical Cartan equations on the bundle of unitary bases $U(M^{2n})$ over M^{2n}. He also proved the local existence of minimal Lagrangian surfaces in Einstein Kähler manifolds.

It is well known that a Lagrangian submanifold of symplectic space $\mathbf{R}^{2n}$ has a natural topological invariant, the Maslov index, and more generally the Maslov-Arnol′d characteristic classes α_i. A. T. Fomenko made the conjecture that in many cases the minimality of a Lagrangian submanifold implies the triviality of the Maslov-Arnol′d characteristic classes. For the case $\mathbf{R}^{2n} = \mathbf{C}^n$ this conjecture turned out to be true, namely, Le Hong Van and Fomenko [453] proved the triviality of the characteristic classes α_i (mod 2) and the triviality of α_i for the case of the group of coefficients in $\mathbf{Z}$ for minimal Lagrangian manifolds in $\mathbf{R}^{2n}$ (also see above). The results of Fuks [414] were used here. The truth of Fomenko's conjecture was proved by Le Hong Van [454], using a construction of Trofimov [458].

Let L be a Φ-Lagrangian surface in M^{2n}, $x_0 \in L$, and $\mathrm{Lag}^+(x_0) = \Lambda^+(x_0)$ denote the Grassmannian of oriented Lagrangian planes in $T_x M^{2n}$. Obviously, the holonomy group H acts on $\Lambda^+(x_0)$. Let $H_0 \subset H$ be the subgroup generated by parallel displacements along loops on L. Then we can map L into the reduced Grassmannian Λ^+/H_0 that associates with each point $x \in L$ the image $\bar{u}(\overrightarrow{T_x L})$ by means of parallel translation along the path $u(x, x_0) \subset L$. This map induces the corresponding map $H^*(\Lambda^+(x_0)/H_0) \to H^*(L)$ in cohomology. Thus for each element $\alpha \in H^*(\Lambda^+(x_0)/H_0)$ there is defined the characteristic class $\bar{a} \in H^*(L)$ of the Φ-Langrangian submanifold $L \subset M^{2n}$.

THEOREM 4.4.11. *Let L^n be a minimal Φ-Lagrangian submanifold of the Hermitian manifold M^{2n}. Then*

(i) *the subgroup H_0 is contained in the group SU_n;*

(ii) *the composition* $\det j\colon L \to \Lambda^+/H_0 \to S^1$ *takes L into some point on S^1 (here* det *associates with each point $x \in \Lambda^+/H_0$, where $x = g(\overrightarrow{T_{x_0} L})$, the determinant of the matrix $g \in u_n$).*

COROLARY 4.4.3. *Let $[\alpha]$ be the generating element of the cohomology group $H^1(S^1, \mathbf{Z})$. Then the induced cohomology class $j^* \det^*[\alpha]$ is annihilated on L.*

REMARK. In the case $M^{2n} = \mathbf{C}^n$ the class $j^* \det^*[\alpha]$ is the Maslov index of the Lagrangian surface L. Thus we again obtain a theorem on the triviality of the Maslov index in this case [453]. Moreover the other Maslov-Arnol′d characteristic classes with coefficients in $\mathbf{Z}$ or $\mathbf{Z}_2$ of the minimal surface $L^n \subset M^{2n}$ are also equal to zero.

As we mentioned above, Harvey and Lawson in the fundamental paper [253] proved that any locally minimal Lagrangian submanifold of $\mathbf{R}^{2n} = \mathbf{C}^n$ with the standard Kähler metric is specially Lagrangian and therefore globally minimal. It turns out that global minimality and stability of locally minimal submanifolds of Hermitian manifolds M^{2n} depend largely on the first Chern form on M^{2n} (the Chern form on $\mathbf{C}^n$ is equal to zero). This dependence was discovered by Le Hong Van by means of the method of relative calibrations (see Definition 4.4.3). We have the following theorem (Le Hong Van).

THEOREM 4.4.12. *Let $L \subset M^{2n}$ be a minimal Φ-Lagrangian submanifold of the Hermitian manifold M^{2n} and X a normal vector field with compact support on L. Then*

$$XX\left(\int_L \mathrm{vol}\right) \geq -2\pi \int_L \Omega(X, JX),$$

where Ω is the first Chern form on M^{2n}.

COROLLARY 4.4.4. *A minimal Φ-Lagrangian submanifold of a Hermitian manifold M is stable if the first Chern form Ω on M is nonpositive (that is, $\Omega(X, TX) \leq 0$ for any vector $X \in T_* M$).*

We note that in the case of a Hermitian manifold with positive first Chern form there is an example of unstably minimal Lagrangian submanifolds (for example, $\mathbf{R}P^n \to \mathbf{C}P^n$).

As we mentioned above, the form ω on $G_L(M^{2n})$ is closed if and only if the first Chern form Ω on M^{2n} is equal to zero. This is a necessary and sufficient condition that there is a local (and global) calibration of type SL on M^{2n}. A minimal Φ-Lagrangian submanifold of a complex manifold with zero first Chern form is orientable.

In conclusion let us dwell on one more application of calibration forms, which is connected with the so-called equivariant Plateau problem. Let G be a compact connected Lie group that acts orthogonally on $\mathbf{R}^n$. Let P denote the set of principal orbits. Let p be the dimension of the principal orbits; let $a\colon P/G \to \mathbf{R}$ be the volume functions of orbits, and Ω_0 a unit p-form along the orbits.

THEOREM 4.4.13 (Harvey and Lawson [253]). *Let* ω_0 *be any closed form of degree* $N - p - 1$ *on* P *such that* $\|\omega_0\|^* \leq a$, *where* a *is the volume function of orbits. Then the smooth form* $\Omega = \frac{1}{a}\Omega_0 \wedge \omega_0$ *defines a calibration form in* $\mathbf{R}^n$.

By this method Lawson [281] investigated the minimality of equivariant cones of codimension 1 in $\mathbf{R}^n$. Recently, A. O. Ivanov proved by a similar method the minimality of several new series of symmetric cones of codimension 2.

CHAPTER V

Multidimensional Minimal Surfaces and Harmonic Maps

§1. The multidimensional Dirichlet functional and harmonic maps The problem of minimizing the Dirichlet functional on the homotopy class of a given map

1.1. Harmonic surfaces in a manifold of negative curvature and connections with holomorphic maps.

DEFINITION 5.1.1. The *Dirichlet functional* (*energy functional*) is a map that associates with each sufficiently smooth map $f: M \to N$ of a compact orientable Riemannian manifold (M, h) to a Riemannian manifold (N, g) the number $D(f) = \frac{1}{2}\int_M \|df\|^2 * 1_M$. Here h and g are metric tensors on M and N respectively, and the map f is, as usual, assumed to be of class C^q; df is the differential of f, and $df \in \mathrm{Hom}(TM, f^*TN)$. The norm of df is taken with respect to the natural Riemannian structure in the bundle $\mathrm{Hom}(TM, f^*TN)$. We can verify that $\|df\|^2 = h^{ij} f_i^\alpha f_j^\beta g_{\alpha\beta}$, where $f_i^\alpha = \partial f^\alpha / \partial x^i$, and $*1_M$ is the volume form on M.

If M is not compact, then the energy $D(f)$ is defined only for those maps f for which the integral in the definition of $D(f)$ exists.

DEFINITION 5.1.2. A map f is said to be *harmonic* if it is an extremal (critical point) of the functional D.

If $f \in C^2$ and f is harmonic, then $f \in C^\infty$ (or is real-analytic if M and N are real-analytic). A fundamental question of the theory of harmonic maps is the question of the existence of a harmonic representative in a homotopy class and a description of these representatives.

In a fundamental survey article Eells and Lemaire [210] summarize the developments in the theory of harmonic maps up to 1979. Therefore here we shall dwell, for the most part briefly, on the most fundamental results of this period and then give an account of some later achievements. We first consider the most widely studied case reflected in the publications of Al′ber [2], Eells and Sampson [209], and Hartman [250]. In the case under consideration any given smooth map can be deformed continuously into a harmonic map. If $\dim M = 1$, then this result is true without any restrictions on the curvature of N (a theorem of Hilbert).

THEOREM 5.1.1 (see [2], [209], [250]). *Let M and N be compact Riemannian manifolds, and let M be orientable. If the sectional curvature of N is nonpositive, then in each component of the functional manifold $C^2(M, N)$ the harmonic maps form a path-connected subset with the same minimal value of the Dirichlet functional for this component. If, in addition, the curvature of N is negative, then any connected component of the functional manifold of maps $C^2(M, N)$ has one of the following properties:* (1) *it has the homotopy type of a point and contains a unique harmonic map;* (2) *it has the homotopy type of a circle, and all the harmonic maps map M with the same value of the energy functional* (*Dirichlet functional*) *into the same closed geodesic of N;* (3) *it has the homotopy type of N, and every harmonic map of this component is a constant, that is, a map to a point.*

Another important direction of development in the theory of harmonic maps is connected with *Kähler manifolds*. It was observed long ago that *a holomorphic map of Kähler manifolds is harmonic.*

THEOREM 5.1.2 (Lichnerowicz [298]). *If M and N are Kähler manifolds and M is compact, then a holomorphic or antiholomorphic map $M \to N$ realizes an absolute minimum of the energy functional* (*Dirichlet functional*) *in its homotopy class.*

COROLLARY 5.1.1 (Lichnerowicz [298]). *Let M and N be Kähler manifolds, with M compact. Then a holomorphic map $M \to N$ cannot be homotopic to an antiholomorphic map $M \to N$ except for the case when they are both constant* (*maps to a point*).

Important results on the existence of harmonic maps of two-dimensional spheres were obtained in 1977 by Sacks and Uhlenbeck [368]. They proved the existence of harmonic maps that also minimize the volume, in two cases. If N is a compact Riemannian manifold for which $\pi_2(N) = 0$, then in any homotopy class of maps of a closed orientable two-dimensional surface M into N there is a harmonic map. This was also proved by Lemaire [292] and Schoen and Yau [376]. If $\pi_2(N) \neq 0$, then there is a generating set for $\pi_2(N)$ consisting of conformal branched minimal immersions of spheres S^2 that minimize the energy and area (volume) in their homotopy class.

A more general result on the holomorphic property of harmonic maps of Riemann surfaces was obtained by Eells and Wood [208]. If N is a complex manifold, let R denote the set of integral homology classes $[\gamma]$ in $H_2(N)$ that can be represented by holomorphic maps $\gamma\colon S^2 = \mathbf{C}P^1 \to N$. The theorem of Eells and Wood asserts the following. Let M be a closed Riemann *surface* and N a simply-connected Kähler manifold. If $f\colon M \to N$ minimizes the energy in its homotopy class and if $[f] \in R$, then f is holomorphic. But if $[f] \in -R$, then f is antiholomorphic. Hence we have the following assertion. If M is a closed Riemann surface and N is a simply-connected Kähler manifold for which $H_2(N) = \mathbf{Z}$ is generated by a holomorphic map $\mathbf{C}P^1 \to N$, then any map $f\colon M \to N$ that minimizes the energy is either holomorphic or antiholomorphic.

1.2. An important class of harmonic surfaces—totally geodesic surfaces in symmetric spaces. In the previous sections we have studied mainly globally minimal surfaces. We have discovered that *closed surfaces of least possible volume often turn out to be totally geodesic submanifolds.* In the class of harmonic maps there is naturally distinguished the class of maps that determine submanifolds *whose second fundamental form is identically zero.* These are, so to speak, "*strongly harmonic*" *maps* (submanifolds). At the same time, this is the definition of a *totally geodesic surface.* Thus, in studying these surfaces we are studying "strongly harmonic" extremals of the Dirichlet functional.

Totally geodesic submanifolds are always locally minimal, that is, they are extremals of the volume functional (possibly not absolute or local minima, but "saddles"). The interesting papers of Helgason [257] and Wolf [409], [410], for example, are devoted to the study of totally geodesic submanifolds.

Of independent interest is the problem of describing totally geodesic (that is, "strongly harmonic") submanifolds of *symmetric spaces,* i.e. in one of the most important classes of Riemannian manifolds. Of special interest are *submanifolds that realize nontrivial cycles or cocycles in a symmetric space.*

This problem was completely solved by A. T. Fomenko in a cycle of papers [121], [124], and [125], where a complete classification was also given of those cases where nontrivial elements of the homotopy groups $\pi_i(M)\otimes\mathbf{Q}$ (where M is a symmetric space, and $\mathbf{Q}$ is the field of rational numbers) are realized by totally geodesic spheres.

If the ambient manifold is a *Lie group*, then totally geodesic submanifolds can be described very simply. Let $[X, Y]$ denote the commutator in a Lie algebra.

Let G be a Lie algebra over the field of real numbers, and B a subspace of G. It is called a *Lie triple system* if for any three elements X, Y, Z of B the element $[[X, Y], Z]$ also belongs to B.

We denote by exp the canonical map of the Lie algebra G onto the group $\mathbf{G}$. Let V be a totally geodesic submanifold of G. We may assume that it passes through the identity e of the group.

PROPOSITION 5.1.1 (see [258], for example). *Let* $\mathbf{G}$ *be a connected compact Lie group, and* V *a totally geodesic submanifold of the group. Then the subspace* $B = T_eV$ *is a Lie triple system and* $\exp B = V$. *Conversely, if a plane* $B \subset G$ *is a triple system, then the submanifold* $V = \exp B$ *is a totally geodesic submanifold of the Lie group* $\mathbf{G}$.

Clearly, *any Lie subgroup of positive dimension is a totally geodesic submanifold of the Lie group.*

DEFINITION 5.1.3. A connected Riemannian manifold V is called a *symmetric space* if for each point $p \in V$ there is an involutory isometry $V \to V$, not the identity, for which p is an isolated fixed point.

A connected compact Lie group with an invariant Riemannian metric is a symmetric space. Let $I(V)$ be the set of all isometries of a symmetric space. Let $\mathbf{G} = I_0(V)$ denote the connected component of the identity in the Lie group $I(V)$. Then a compact simply-connected symmetric space can be represented in the form $V = \mathbf{G}/\mathbf{H}$, where $\mathbf{H}$ is a stationary subgroup, $\mathbf{H} = \{g \in \mathbf{G} \colon \sigma(g) = g\}$, where σ is an involution. We denote by $\theta = d\sigma$ the corresponding involutory automorphism of the Lie algebra G. Then the subalgebra $H = T_e\mathbf{H}$ is the subalgebra of fixed points of θ. Also, $G = B + H$, where $B = \{X \in G \colon \theta(X) = -X\}$.

A symmetric space V is said to be *irreducible* if the adjoint representation ad_H on the plane B (generated by the relation $[H, B] \subset B$) is irreducible.

PROPOSITION 5.1.2 (see [258], for example). *A compact simply-connected Riemannian symmetric space* V *splits into the product* $V_1 \times \cdots \times V_s$ *of irreducible compact symmetric spaces.*

There is a complete classification of compact irreducible symmetric spaces, obtained by E. Cartan (see [258]).

1.3. When does a totally geodesic submanifold realize a nonzero cycle in the ambient space? Consider a connected compact Lie group $\mathbf{G}$, and let V be a compact simply-connected totally geodesic submanifold of $\mathbf{G}$. The problem that will be solved in this subsection is stated as follows. When does V realize a nonzero homology cycle in $H_*(\mathbf{G}, \mathbf{R})$, that is, when is the element $i_*[V]$ nonzero? Here $[V]$ denotes the fundamental homology class of V. We first consider the special case where the totally geodesic submanifold V is a *subgroup*. Let $i: V \to \mathbf{G}$ be an embedding.

A well-known construction, which enables us to give the answer to the question of whether a subgroup $\mathbf{H}$ *of* $\mathbf{G}$ *is not homologous to zero,* is presented in a series of remarkable papers of Borel [175], H. Cartan [184], Sammelson [370], and Bott [176].

It turns out that *a Lie group that contains a totally geodesic submanifold automatically contains the isometry group of this manifold.* Let $\mathbf{G}$ be a connected compact Lie group, and $V \subset \mathbf{G}$ a compact simply-connected totally geodesic submanifold. When is the cycle $i_*[V]$ nonzero in the homology group $H_*(\mathbf{G})$? *The case when V is a subgroup has been analyzed above.* Since any totally geodesic submanifold V of $\mathbf{G}$ is a symmetric space, one of the ways of solving the problem could be the following. We must consider all ways of embedding compact symmetric spaces in Lie groups and determine which of these embeddings realize nontrivial cycles or cocycles.

THEOREM 5.1.3 (Fomenko [119], [124], [125]). *Let V be a compact simply-connected totally geodesic submanifold of a compact Lie group* $\mathbf{G}$. *Let $\{\mathbf{G}_\alpha\}$ be the set of all subgroups of* $\mathbf{G}$ *containing V. Then V splits into the direct product $V = K \times V' = V_1 \times \cdots \times V_m \times V_{m+1} \times \cdots \times V_s$, where K is a compact subgroup of* $\mathbf{G}$, *and each submanifold V_i (where $m+1 \leq i \leq s$) is a totally geodesic submanifold of* $\mathbf{G}$ *and is not a subgroup of* $\mathbf{G}$. *In addition, this decomposition has the following property. If $A(V) = \bigcap_\alpha \mathbf{G}_\alpha$, then $A(V)$ is compact and semisimple. The universal covering $\widetilde{A}(V)$ is isomorphic to the direct product of groups $\widetilde{A}(V) = K \times \widetilde{I}_0(V_{m+1}) \times \cdots \times \widetilde{I}_0(V_s)$, where $\widetilde{I}_0(V_i)$ are the isometry groups of the corresponding symmetric spaces. Here $K = V_1 \times \cdots \times V_m$, where V_α are subgroups.*

From Theorem 5.1.3 we obtain an important corollary.

COROLLARY 5.1.2 (see [119], [124], [125]). *Let V be a compact simply-connected totally geodesic submanifold of a compact Lie group $\mathbf{G} = A(V)$ and $V = V_1 \times \cdots \times V_m \times V_{m+1} \times \cdots \times V_s$ a decomposition of V into irreducible component-factors of* $\mathbf{G}$ *(see Theorem 5.1.3). Then V realizes a nontrivial cycle in the homology group $H_*(\mathbf{G}, \mathbf{R})$ if and only if all submanifolds V_i, $1 \leq i \leq s$, realize nontrivial cycles in $H_*(\mathbf{G}, \mathbf{R})$. We assume that no manifold V_i is a point.*

Corollary 5.1.2 reduces the original problem to the following. We are given an arbitrary connected compact Lie group $\boldsymbol{G}$ and a compact simply-connected totally geodesic irreducible submanifold. It is required to determine when this submanifold realizes a nontrivial real cycle. We showed above that if V is a subgroup, then the question reduces to an analysis of an algorithm constructed by Borel [175] and H. Cartan [184]. We can therefore concentrate our attention entirely on the study of an embedding in $\boldsymbol{G}$ of the submanifold V', where $V' = V_{m+1} \times \cdots \times V_m$. The embedding $i: V \to \boldsymbol{G}$ has the form of a composition: $i = i_2 i_1$, where $i_1: V \to A(V)$ and $i_2: A(V) \to \boldsymbol{G}$. Since $i_* = i_{2*} i_{1*}$, in the first place we investigate the homomorphism i_{1*}. This problem will be completely solved.

The submanifold V realizes a nontrivial homology cycle in $A(V)$ if and only if a nontrivial cycle in the homology realizes a submanifold $\widetilde{V}$ diffeomorphic to V in the universal covering $\widetilde{A}(V)$. It is therefore sufficient to study embeddings of V in the simply-connected group $\widetilde{A}(V) = \widetilde{I}_0(V)$. *This embedding can be described canonically*, using the existence of an involutory automorphism θ defining the symmetric space V. An embedding of an irreducible symmetric space V in the group $\widetilde{A}(V)$ generates in the algebra G an involutory automorphism θ. Since $\widetilde{A}(V)$ is simply-connected, the automorphism θ can be extended to an involutory automorphism σ of the whole group. The totally geodesic submanifold $V \subset \widetilde{A}(V)$ consists of those elements for which $\sigma(g) = g^{-1}$.

The converse is also true: any irreducible symmetric space V admits an embedding in the group $\widetilde{I}_0(V)$ in the form of a totally geodesic submanifold, which is called the Cartan model of the symmetric space and was first discovered by E. Cartan.

PROPOSITION 5.1.3 (see [258], for example). *Let σ be an involutory automorphism of a compact connected Lie group* **G**. *We denote the set of all its fixed points by* **H** *and suppose that an invariant metric is specified on* **G**. *Then the map $g\boldsymbol{H} \to g\sigma g^{-1}$ is a diffeomorphism of the manifold* **G/H** *on a closed totally geodesic submanifold V of* **G**, *which is a symmetric space.*

If $\boldsymbol{G}$ is simply-connected, then $\boldsymbol{G}/\boldsymbol{H}$, where $\boldsymbol{H}$ is the set of fixed points of σ, is also simply-connected. Since any embedding $i: V \to \widetilde{A}(V)$ determines one and only one automorphism θ (up to conjugacy), it follows from Proposition 5.1.3 that this embedding is a Cartan embedding in the sense that V admits a representation in the form $\{g\sigma g^{-1}\}$, $g \in \widetilde{A}(V)$.

1.4. Description of totally geodesic and topologically nontrivial cycles. The main goal of this subsection, *the description of (co)homologically nontrivial totally geodesic submanifolds of compact Lie groups* (and, more generally, of symmetric spaces), reduces to the solution of the following problem. When does an irreducible compact symmetric space V, embedded in its own isometry group $\widetilde{I}_0(V)$ as a Cartan model, realize a nontrivial (co)cycle in the homology group $H_*(\widetilde{I}_0(V))$? We recall that compact irreducible symmetric spaces of type I have the form $\boldsymbol{G}/\boldsymbol{H}$, where $\boldsymbol{G}$ is a simple Lie group. Symmetric spaces of type II are homeomorphic to the group $\boldsymbol{G}$ and admit a representation $\boldsymbol{G}/\boldsymbol{H}$, where $\boldsymbol{G} = \boldsymbol{H} \times \boldsymbol{H}$ and the embedding of $\boldsymbol{H}$ in $\boldsymbol{G}$ has the form $h \mapsto (h, h)$, $h \in \boldsymbol{H}$.

THEOREM 5.1.4 (Fomenko [124], [125], [134]). *Let $i\colon V \to \widetilde{I}_0(V) \approx \widetilde{A}(V)$ be an embedding of a Riemannian compact irreducible simply-connected symmetric space V as a totally geodesic submanifold of a simply-connected maximal connected isometry group $\widetilde{I}_0(V)$. Then this embedding is necessarily a Cartan embedding, that is, it is obtained from the standard Cartan embedding (see above) by an automorphism of the Lie group $\widetilde{I}_0(V)$. Below we list all cases where this embedding is (co)homologically nontrivial, that is, where the submanifold realizes nonzero cycles or cocycles.*

1. *Of the symmetric spaces of type* I *only the following spaces V actually realize nontrivial (co)cycles in the (co)homology group of the space $\widetilde{I}_0(V)$:*

(a) $SU(2m+1)/SO(2m+1)$, $m \geq 1$, *in* $SU(2m+1)$;

(b) $SU(2m)/Sp(m)$, $m \geq 2$, *in* $SU(2m)$;

(c) S^{2l-1} = (sphere) = $SO(2l)/SO(2l-1)$, $l \geq 4$, *in the group* Spin$(2l)$;

(d) E_6/F_4 *in the group* E_6.

All the remaining spaces of type I *realize trivial cocycles in the group $H^*(\widetilde{I}_0(V), \mathbf{R})$, that is, $i_*[V] = 0$.*

2. *Any symmetric space V of type* II *always realizes a nontrivial (co)cycle in the cohomology group $H^*(\widetilde{I}_0(V), \mathbf{R})$.*

THEOREM 5.1.5 (see [124], [125], [134]). *Let V be an arbitrary compact simply-connected totally geodesic submanifold of a compact Lie group $\boldsymbol{G}$, and let $V = K \times V_{m+1} \times \cdots \times V_s = K \times V'$ be a decomposition of V into a direct product. Under the embedding $i\colon V \to \widetilde{A}(V)$ the manifold V realizes a nontrivial cocycle in the cohomology group $H^*(\widetilde{A}(V), \mathbf{R})$ if and only if the subgroup $K \times \boldsymbol{H}$, where $\boldsymbol{H}$ is a stationary subgroup of V' (we recall that $\boldsymbol{H} \subset \widetilde{A}(V') \subset \widetilde{A}(V)$), is not homologous to zero in the group $\widetilde{A}(V)$ for real coefficients.*

We now give a classification of cocycles realized by totally geodesic spheres in compact Lie groups. Here A. T. Fomenko solved a problem which is in a certain sense inverse to the one described above. Up to now we have started from a given totally geodesic submanifold V of a compact Lie group $\mathbf{G}$ and determined when it realizes a nontrivial cycle or cocycle in the group $H_*(\mathbf{G})$. The solution of this question was obtained above. We now fix a manifold V_0, and let $\mathbf{G}$ be a Lie group. We pose the question: *which nontrivial cycles $x \in H_*(\mathbf{G})$ can be realized in $\mathbf{G}$ by totally geodesic submanifolds diffeomorphic to V_0?*

As a representative manifold V_0 we choose a sphere. A sphere is the simplest of the symmetric spaces. In addition, realization of cycles by spheres is closely connected with the realization of nontrivial elements of homotopy groups π_i. In the case where the ambient manifold is not a group, but a symmetric space of type I, we shall solve the problem of realizing nontrivial elements of the homotopy groups $\pi_i(M) \otimes \mathbf{R}$ by totally geodesic spheres. Since we can regard a totally geodesic sphere as an extremal for the multidimensional Dirichlet functional, we solve the problem of finding harmonic maps of a sphere. As A. T. Fomenko discovered, the problem of realizing nontrivial elements of homotopy groups by totally geodesic spheres is closely connected with the problem of finding the maximal number of linearly independent smooth vector fields on spheres, which was definitively solved by J. F. Adams. Let S^{n-1} be a sphere. We represent n in the form $n = (2q+1)2^{4\nu+\beta}$, where $\beta = 0, 1, 2, 3$. Then the maximal number $s(n)$ of independent vector fields on the sphere S^{n-1} is given by $s(n) = 8\nu + 2^\beta - 1$.

We denote the integer part of a number N by $[N]$.

THEOREM 5.1.6 (Fomenko [124], [125], [134]). *Let $\mathbf{G}$ be a compact simple Lie group, and let $k = k(n) = [1 + \log_2 n]$. Then the only (up to multiplication by real numbers and modulo the kernel of a homomorphism $H^*(\mathbf{G}) \to H^*(S)$ induced by an embedding of the sphere S in $\mathbf{G}$) elements $x \in H^*(\mathbf{G}, \mathbf{R})$ that can be realized in $\mathbf{G}$ by totally geodesic spheres are the following.*

1. *In the cohomology group $H^*(SU(n), \mathbf{R})$, $n \geq 2$, these are the elements $x_3, x_5, x_7, \dots, x_{2k-1}$.*
2. *In the cohomology group $H^*(SO(n), \mathbf{R})$, $n \geq 8$, these are the elements*

a) $x_3, x_7, x_{11}, \dots, x_{2k-1}$ *if* $k \equiv 0 \pmod 4$,
b) $x_3, x_7, x_{11}, \dots, x_{2k-3}$ *if* $k \equiv 1 \pmod 4$,
c) $x_3, x_7, x_{11}, \dots, x_{2k-5}$ *if* $k \equiv 2 \pmod 4$,
d) $x_3, x_7, x_{11}, \dots, x_{2k-3}$ *if* $k \equiv 3 \pmod 4$.

3. *In the cohomology group* $H^*(Sp(n), \mathbf{R})$, $n \geq 1$, *these are the elements*

a) $x_3, x_7, x_{11}, \dots, x_{2k-1}$ *if* $k \equiv 0 \pmod 4$,
b) $x_3, x_7, x_{11}, \dots, x_{2k+1}$ *if* $k \equiv 1 \pmod 4$,
c) $x_3, x_7, x_{11}, \dots, x_{2k-1}$ *if* $k \equiv 2 \pmod 4$,
d) $x_3, x_7, x_{11}, \dots, x_{2k-3}$ *if* $k \equiv 3 \pmod 4$.

4. *In the cohomology group* $H^*(G_2, \mathbf{R})$ *this is the element* x_3.
5. *In the cohomology group* $H^*(F_4, \mathbf{R})$ *this is the element* x_3.
6. *In the cohomology group* $H^*(E_6, \mathbf{R})$ *these are the elements* x_3, x_9.
7. *In the cohomology group* $H^*(E_7, \mathbf{R})$ *these are the elements* x_3, x_{11}.
8. *In the cohomology group* $H^*(E_8, \mathbf{R})$ *this is the element* x_3.

1.5. Description of totally geodesic spheres, not homotopic to zero, in symmetric spaces. The results of the previous subsections describe totally geodesic spheres in compact irreducible symmetric spaces of type II. Since the study of totally geodesic spheres in a compact simply-connected symmetric space reduces to a consideration of irreducible spaces of types I and II (see above), to determine the general picture it remains for us to study irreducible spaces of type I, which we shall now do. All these spaces are well known (see [258], for example). Combining certain series of these spaces, we obtain the following list (not containing singular series, that is, spaces whose groups of motions are singular Lie groups):

$$SU(p+q)/S(U(p)\times U(q)), \quad SO(p+q)/S(O(p)\times O(q)),$$
$$Sp(2p+2q)/Sp(2p)\times Sp(2q), \quad SU(n)/SO(n),$$
$$SU(2n)/Sp(2n), \quad SO(2n)/U(n), \quad Sp(2n)/U(n).$$

Let $k = k(n) = [1 + \log_2 n]$. We define an important numerical function:

$$f_i(n) = f_i(2^{k-1}) = \begin{cases} 2k-i-1 & \text{if } k \equiv 0 \pmod 4, \\ 2k-i-3 & \text{if } k \equiv 1 \pmod 4, \\ 2k-i-5 & \text{if } k \equiv 2 \pmod 4, \\ 2k-i-3 & \text{if } k \equiv 3 \pmod 4. \end{cases}$$

THEOREM 5.1.7 (Fomenko [125], [134]). (A) *Let* V *be a compact irreducible symmetric space of type* I, *whose group of motions is not a singular Lie group. Then the only nonzero elements* x (*up to a scalar multiplier*) *of the groups* $\pi_i^r = \pi_i(V)\otimes\mathbf{R}$ *that can be realized by totally geodesic spheres are the following* (*here* $k = k(n)$):

(1) *if* $V = SO(2n)/U(n)$, $k \geq 5$, *then the elements* $x \in \pi^r_{4\alpha+2}$, *where* $2 \leq 4\alpha + 2 \leq f_1(2n)$, *are realized*;

(2) *if* $V = SU(2n)/Sp(2n)$, $k \geq 5$, *then the elements* $x \in \pi^r_{4\alpha+1}$, *where* $5 \leq 4\alpha + 1 \leq f_2(4n)$, *are realized*;

(3) *if* $V = Sp(2n)/Sp(n) \times Sp(n)$, $n = 2s$, $k \geq 8$, *then the elements* $x \in \pi'^{r}_{4\alpha}$, *where* $4 \leq 4\alpha \leq f_3(4n)$, *are realized*;

(4) *if* $V = SU(2n)/S(U(n) \times U(n))$, $k \geq 3$, *then the elements* $x \in \pi'^{r}_{2\alpha}$, *where* $1 \leq \alpha \leq k$, *are realized*;

(5) *if* $V = Sp(2n)/U(n)$, $k \geq 5$, *then the elements* $x \in \pi'^{r}_{4\alpha+2}$, *where* $2 \leq 4\alpha + 2 \leq f_5(8n)$, *are realized*;

(6) *if* $V = SU(n)/SO(n)$, $k \geq 5$, *then the elements* $x \in \pi'^{r}_{4\alpha+1}$, *where* $5 \leq 4\alpha + 1 \leq f_6(8n)$, *are realized*;

(7) *if* $V = SO(2n)/S(O(n) \times O(n))$, $k \geq 6$, *then the elements* $x \in \pi'^{r}_{4\alpha}$, *where* $4 \leq 4\alpha \leq f_7(16n)$, *are realized.*

(B) *Let* V *be a compact irreducible symmetric space of type* I *whose group of motions is a singular Lie group. Then the only elements* $x \in H^*(V, \mathbf{R})$ *that can be realized by totally geodesic spheres are the following:*

(1) *the element* x_2 *in* $H^*(\mathrm{Ad}\, E_6/T^1 \cdot \mathrm{Spin}(10))$;

(2) *the element* x_9 *in* $H^*(E_6/F_4)$;

(3) *the element* x_2 *in* $H^*(\mathrm{Ad}\, E_7/T^1 \cdot E_6)$.

We now turn to *the connections between the maximal number of linearly independent vector fields on spheres and the number of elements of homotopy groups of symmetric spaces realized by totally geodesic spheres.* It is easy to represent the group $SU(2p)$ as a smooth submanifold of the standard sphere S^{8p^2-1} of radius $\sqrt{2p}$. Similarly we embed the group $SO(2p)$ in the sphere S^{4p^2-1} of radius $\sqrt{2p}$. For this it is sufficient to realize these groups as matrix groups and to consider an embedding of the matrices in the linear space of all matrices of the corresponding size, which must be identified with Euclidean space. The Killing metric on groups goes into the Euclidean metric on the space of all matrices.

We define *central plane sections of the groups* $SU(2p)$ and $SO(2p)$ as intersections $\Pi_s \cap \boldsymbol{G}$, where Π_s is a subspace of dimension s passing through the origin. It turns out (see [125], [134]) that the maximal possible dimension of a sphere S^{s-1} that is a central plane section of the group $\boldsymbol{G}$ cannot exceed the number $s(4p)$ for the group $SU(2p)$ and the number $s(2p)$ for the group $SO(2p)$. Here we have denoted by $s(h)$ the maximal number of linearly independent vector fields on the sphere S^{h-1}. Moreover, it turns out that these numbers are attained in many cases. A similar estimate, $s - 1 \leq s(4p)$, holds for the symplectic group $Sp(2p)$. For the details see Fomenko [119], [134]. We denote by r the largest dimension of totally geodesic spheres S^s, that is, that are plane sections in the group $SO(2^{k+1})$ and realize generators of the cohomology group $H^*(SO(2^{k+1}))$.

Then if $s(n)$ is, as before, the maximal number of linearly independent vector fields, we have the inequalities

(1) $s(2^{k+1}) = 2k+1 > r = 2k-1$, $k \equiv 0 \pmod 4$,

(2) $s(2^{k+1}) = 2k+1 = r = 2k+1$, $k \equiv 1 \pmod 4$,

(3) $s(2^{k+1}) = 2k+3 = r = 2k+3$, $k \equiv 2 \pmod 4$,

(4) $s(2^{k+1}) = 2k+2 > r = 2k+1$, $k \equiv 3 \pmod 4$,

that is, $f_i(n) \approx s(n) - i \approx r$ when $n = 2^{k+1}$.

Thus, the numerical function $f_i(n)$ constructed by A. T. Fomenko is connected in a very direct way with the maximal number of linearly independent vector fields on spheres, and this function turns out to be universal.

§2. Connections between the topology of manifolds and properties of harmonic maps

2.1. The problem of minimizing the Dirichlet functional. From the definition of the Dirichlet functional D it is obvious that locally constant maps realize an absolute minimum (equal to zero) of the functional in the whole space of maps. There naturally arises the question of the *attainability of global minima* $\inf D$ (greatest lower bounds of values of the Dirichlet functional) on separate connected components of the space $C^\infty(M, N)$, identified with the homotopy classes $[\varphi]$ of maps $\varphi: M \to N$ of smooth Riemannian manifolds. At present no definitive answer to this question has been obtained. Without pretending that our account is complete, we give a number of interesting results about the connection between the problem of the existence of globally minimal (in their homotopy class) harmonic maps $f: M \to N$ and the topology of the manifolds M and N.

Let us consider the following problem (the problem of a minimal realization of a homotopy class): $D(f) \to \inf$, $f \in [\varphi] \in C^\infty(M, N)$, where $\varphi \in C^\infty$ and the two Riemannian manifolds M and N are closed (that is, compact and without boundary). The solution f_0 of this problem (if it exists)—a *globally minimal harmonic map*—is defined by the conditions $D(f_0) = \inf D(f)$; $f, f_0 \in [\varphi]$.

A classical result in the calculus of variations, which goes back to the work of Hadamard, E. Cartan, Hilbert, Birkhoff, and Synge, is an assertion about the complete solvability of the *one-dimensional problem* of minimal realization; $\dim M = 1$. In this case each free homotopy class from the set $[S^1, N]$ is realized by a smooth closed geodesic $\gamma: S^1 \to N$ with minimal action $D(\gamma)$. In one homotopy class there may be several such geodesics, and a description of them is a complicated and very interesting problem ("the periodic problem of the calculus of variations").

Next, the problem of minimization has been well studied for those spaces $C^{\infty}(M, N)$ that correspond to manifolds N *of nonpositive two-dimensional curvature* (see Eells and Sampson [209], Al′ber [1], [2], and Hartman [250]). The use of modern methods of the calculus of variations has made it possible in this situation not only to prove the existence of a solution of the problem of minimal realization in every homotopy class, but also to give possible variants of the structure of the set of homotopic globally minimal harmonic maps. It is important to note that nonconstant (for connected M and N) globally minimal harmonic maps $f: M \to N$ arise in this case only if the manifold is not simply-connected, since all closed manifolds of nonpositive curvature have trivial homotopy groups $\pi_i(N)$, $i > 1$.

Significant progress has been attained in the solution of the *two-dimensional minimization problem*: $\dim M = 2$. We say that maps $\varphi, \psi \in C^{\infty}(M, N)$ are π_1-equivalent if there are inner automorphisms A and B of the fundamental groups $\pi_1(M)$ and $\pi_1(N)$ respectively such that the following diagram

$$\begin{array}{ccc} \pi_1(M) & \overset{\varphi_*}{\to} & \pi_1(N) \\ {\scriptstyle A}\downarrow & & {\scriptstyle B}\downarrow \\ \pi_1(M) & \overset{\psi_*}{\to} & \pi_1(N) \end{array}$$

is commutative. The set of maps that are π_1-equivalent to a fixed map φ form an open-and-closed subset $\pi_1[\varphi]$ of the space $C^{\infty}(M, N)$ that includes the homotopy class $[\varphi]$ but consists, generally speaking, of several connected components. In recent papers of Sacks and Uhlenbeck [368], [369] and Lemaire [292] it was proved that for any $\varphi \in C^{\infty}(M, N)$, $\dim M = 2$, there is a map $\varphi_0 \in \pi_1[\varphi]$ that minimizes the values of the Dirichlet functional in the class $\pi_1[\varphi]$. It is well known that under an additional assumption about the triviality of the homotopy group $\pi_2(N)$ the classes $\pi_1[\varphi]$ and $[\varphi]$ coincide. In this situation, from the stated assertion there follows a theorem on the existence of a globally minimal harmonic map in each connected component of $C^{\infty}(M, N)$, $\dim M = 2$. Examples of closed manifolds with trivial second homotopy group are Lie groups. If $M = S^2$ and $\pi_2(N)$ is nontrivial, then, as Sacks and Uhlenbeck showed in the papers cited, there is at least one homotopically nontrivial harmonic map $f: S^2 \to N$ that minimizes the values of the Dirichlet functional in its homotopy class.

Important examples of solutions of the minimal realization problem are (anti)holomorphic maps of closed Kähler manifolds (Lichnerowicz [298]). The simplest example is that of (anti)holomorphic maps of the two-dimensional sphere S^2 into itself that yield globally minimal harmonic maps of S^2 into S^2 of any degree.

Quite a different (and rather unexpected) pattern is revealed when we consider the minimal realization problem in spaces $C^\infty(S^n, S^n)$ with $n \geq 3$. As Eells and Sampson [209] showed, in this situation we can construct a map of the sphere $\varphi\colon S^n \to S^n$ of any degree and with arbitrarily small Dirichlet integral $D(\varphi)$, that is, $\inf D = 0$ in any homotopy class of maps of S^n into itself when $n \geq 3$. Since only locally constant maps have a zero Dirichlet integral, the minimal realization problem is unsolvable in those nontrivial (that is, not containing locally constant maps) homotopy classes on which $\inf D = 0$. If $\inf D = 0$ on any connected component of the space $C^\infty(M, N)$ and the manifolds M and N are connected, then globally minimal harmonic maps are necessarily constant and exist only in one (trivial) homotopy class.

As further research has shown, the key point in the rise of spaces $C^\infty(M, N)$ with zero $\inf D$ *on all connected components* is the equality $\inf D = 0$, which is satisfied in the homotopy class of the *identity map* of M or N. We introduce the notation

$$D(\alpha) = \inf_\alpha D, \quad \alpha \in [M, N];$$
$$D(M) = D([\mathrm{id}_M]) = \inf_g D(g), \quad g \in [\mathrm{id}_M],$$

where id_M is the identity map of M.

THEOREM 5.2.1. *Let M be a compact smooth Riemannian manifold. Then the following three assertions are equivalent*:

(a) $D(M) = 0$;

(b) $D(\alpha) = 0$ *for all* $\alpha \in [M, N]$, *where N is an arbitrary smooth Riemannian manifold*;

(c) $D(\beta) = 0$ *for all* $\beta \in [N', M]$, *where N' is an arbitrary compact smooth Riemannian manifold.*

Theorem 5.2.1 was completely proved by Pluzhnikov [81]–[83]. A weaker version, in which assertion (c) extends only to those homotopy classes $\beta \in [N', M]$ that contain Riemannian submersions, was proved by Eells and Lemaire [211]. We give a brief outline of the proof of Theorem 5.2.1. Obviously, (a) is a consequence of each of the assertions (b) and (c). The proof of the reverse implications (a) $\Rightarrow$ (b) and (a) $\Rightarrow$ (c) is based on a consideration of the compositions $f\varphi_i$ and $\varphi_i f'$ respectively, where $f \in \alpha$ and $f' \in \beta$ are arbitrary (but fixed) maps, and $\varphi_i \in [\mathrm{id}_M]$, $i = 1, 2, \dots,$ is a sequence of maps such that $D(\varphi_i) \to 0$ as $i \to \infty$, which exists by virtue of (a). Assertion (b) follows from the simple integral estimate $D(f \circ \varphi_i) \leq (\max_M \|df\|^2) \cdot D(\varphi_i)$. If f' can be assumed to be a submersion, then to prove that (a) $\Rightarrow$ (c) it is sufficient to use Fubini's theorem. In the general case, some additional constructions are necessary. Let us assume (for simplicity) that $\partial M = \varnothing$, and choose a number k sufficiently large so that the vector bundle $(f')^*TM$ on N' induced by the map f' can be assumed to be a subbundle of the trivial bundle $N' \times \mathbf{R}^k \to N'$. We denote by π the (fiberwise) projection of this trivial bundle on $(f')^*TM$, and by B^k a closed unit disc in $\mathbf{R}^k$. We define a submersion $F \colon N' \times B^k \to M$ by the formula $F(x, a) = \exp^M_{f'(x)}(\pi a)$. From Fubini's theorem it follows that

$$\int_{B^k} D(\varphi_i \circ f'_a)\, dv \leq c_{f'} \cdot D(\varphi_i),$$

where $f'_a = F|_{N' \times \{a\}} \colon N' \to M$, and the positive constant $c_{f'}$ does not depend on i. Consequently, there is a vector $a_0 \in B^k$ such that $D(\varphi_i \circ f'_{a_0}) \to 0$ as $i \to \infty$. This proves (c), since f'_{a_0} is homotopic to $f' = f'_0 = F|_{N' \times \{0\}}$.

In view of Theorem 5.2.1 the problem posed by Koiso [336], of describing the class of closed Riemannian manifolds M such that $D(M) = 0$, has special significance. It is not difficult to verify that the equality $D(M) = 0$ does not depend on the choice of Riemannian metric on M. In [211] Min-Oo announced the following result: $D(G) = 0$ for compact simply-connected Lie groups. A full solution of this problem was then given in topological terms, which reveals deep connections between the qualitative picture of the behavior of the Dirichlet functional and the topological properties of manifolds.

THEOREM 5.2.2 (Pluzhnikov [81]–[83]). *Let M be a closed smooth Riemannian manifold. Then $D(M) = 0$ if and only if each connected component of M is doubly-connected, that is, its first and second homotopy groups are trivial.*

This result was then generalized by White in his important papers [147], [148].

LEMMA 5.2.1 (A. I. Pluzhnikov). *Let M be a connected closed doubly-connected smooth Riemannian manifold of dimension ≥ 5 or one of the standard spheres S^3, S^4. Then there is a diffeotopy φ_t $(0 \leq t < 1)$ of M such that $\varphi_0 = \mathrm{id}_M$ and $D(\varphi_t) \to 0$ as $t \to 1$.*

The proof of the lemma is carried out by explicitly constructing the diffeotopy φ_t. Here we cannot present this technically complicated construction, so we shall confine ourselves to a very general description. On a manifold M that satisfies the hypothesis of the lemma there is an exact Morse function μ with one minimum and one maximum that does not have critical points of index 1 or 2 (see Smale [105], [106]). Hence it follows that M can be obtained from a ball by successively attaching Smale handles with indices not equal to $n-1$ or $n-2$ $(n = \dim M)$. The diffeotopy φ_t models the deformation of M along trajectories of the field $\operatorname{grad}(-\mu)$ and is constructed by induction on the handles as a composition of "local" diffeotopies, each of which is concentrated on one of the handles. The resulting family φ_t is the identity on some stratified submanifold of codimension ≥ 3 in M (the union of left separatrix discs of the function $-\mu$) and "pulls" into a small neighborhood of the minimum point of μ the complement of this submanifold, which is diffeomorphic to an open disc. A model example of this construction is the one-parameter group of conformal diffeomorphisms of the sphere, which "pulls" it from one pole to another along the gradient lines of the height function.

We outline the proof of Theorem 5.2.2. Without loss of generality we may assume that M is connected. The direct assertion of the theorem follows from Lemma 5.2.1. In the case $\dim M < 5$ the diffeotopy φ_t is constructed for the manifold $M \times S^5$ and arguments are used that are analogous to those in the proof of the implication (a) $\Rightarrow$ (c) of Theorem 5.2.1. Conversely, if $D(M) = 0$, then this implication shows that $D(\beta) = 0$ for any homotopy class $\beta \in [S^n, M]$. Using the results stated above about minimal realization of the classes $\beta \in [S^n, M]$ for $n = 1$ or 2, we find that M is doubly-connected.

COROLLARY 5.2.1. *If at least one of the homotopy groups* $\pi_1(M)$ *and* $\pi_2(M)$ *of a connected closed smooth manifold* M *is nontrivial, then* $D(M) > 0$ *for any Riemannian metric on* M.

COROLLARY 5.2.2. *Let* M *be a closed connected doubly-connected smooth manifold. Then in any Riemannian metric on* M *the assertions* (b) *and* (c) *of Theorem* 5.2.1 *are satisfied.*

COROLLARY 5.2.3. *Let* M *and* N *be arbitrary connected compact smooth Riemannian manifolds and* $f \in C^\infty(M, N)$ *a globally minimal* (*in its homotopy class*) *harmonic map. Then* $f = \text{const}$ (*a map to a point*) *if at least one of the manifolds* M *and* N *is doubly-connected and does not have a boundary.*

Thus, the class of closed Riemannian manifolds M having the following property is very wide: for any compact Riemannian manifold N

the problem of minimizing the Dirichlet functional is automatically unsolvable in any nontrivial homotopy class of maps both from M to N and from N to M. This class contains, for example, closed manifolds endowed with arbitrary Riemannian metrics and homotopically equivalent to spheres S^n $(n \geq 3)$, to compact simply-connected Lie groups, to homogeneous Grassmann and Stiefel manifolds over the skew-field of quaternions, and their direct products and disjoint sums. In the class of connected closed Riemannian manifolds, the topological property that at least one of the first two homotopy groups of M and N is nontrivial is a necessary condition for the existence of nonconstant globally minimal (in their homotopy class) harmonic maps $f \in C^\infty(M, N)$.

In these results of A. I. Pluzhnikov, the unsolvability of the problem of minimal realization in nontrivial homotopy classes follows from the fact that the value of $\inf D$ is zero in these classes. However, using these results we can exhibit situations in which all four possible combinations of the value of $\inf D$ and the attainability of $\inf D$ are realized on distinct connected components of the space $C^\infty(M, N)$: (1) $\inf D = 0$ and it is attained; (2) $\inf D = 0$ and it is not attained; (3) $\inf D > 0$ and it is attained; (4) $\inf D > 0$ and it is not attained. Consider, for example, the homotopy class $\alpha = [f \times \varphi]$ of a map of the product of spheres $S^1 \times S^3$ into itself, generated by the maps $f\colon S^1 \to S^1$ and $\varphi\colon S^3 \to S^3$. It is intuitively obvious that the minimal value of $D(f \times \varphi)$ must correspond to the minimal values of $D(f)$ and $D(\varphi)$. Using the fact that $D([f]) > 0$ if $[f]$ is nontrivial and it is always attained, but $D([\varphi]) = 0$ and is not attained if $[\varphi]$ is nontrivial (see Corollary 5.2.2), we can get all the cases (1)–(4). For example, the interesting case (4) corresponds to the nontriviality of both classes $[f]$ and $[\varphi]$. These observations have obtained an exact expression in the next two theorems and the corollaries that follow from them, proved by Pluzhnikov [83].

THEOREM 5.2.3. *Consider connected closed Riemannian manifolds M, N, Q and denote by π_N and π_Q the canonical projections of the Riemannian direct product $N \times Q$ on N and Q respectively. Let Q be doubly-connected and $F \in C^\infty(M, N \times Q)$. Then* (a) $D([F]) = D([\pi_N \circ F])$; (b) *the minimization problem is solvable in $[F]$ if and only if it is solvable in $[\pi_N \circ F]$ and the class $[\pi_Q \circ F]$ is trivial.*

THEOREM 5.2.4. *Consider connected closed Riemannian manifolds M, N, P, Q and a map $f \times \varphi\colon M \times P \to N \times Q$, where $f \in C^\infty(M, N)$ and $\varphi \in C^\infty(P, Q)$. Suppose that at least one of the manifolds P and Q is doubly-connected. Then* (a) $D([f \times \varphi]) = \operatorname{vol}(P) \cdot D([f])$ (*if* $\dim P = 0$, *we put* $\mathrm{vol}(P) = 1$), (b) *the minimization problem is solvable in* $[f \times \varphi]$ *if and only if it is solvable in* $[f]$ *and the class* $[\varphi]$ *is trivial.*

COROLLARY 5.2.4. *Let* M *and* P *be connected closed Riemannian manifolds,* $\dim P > 0$, P *doubly-connected, and suppose that at least one of the groups* $\pi_1(M)$ *and* $\pi_2(M)$ *is nontrivial. Then* $\inf D$ *is positive in* $[\mathrm{id}_{M \times P}]$, *but it is not attained on a map of this homotopy class.*

Comparison of Theorems 5.2.3 and 5.2.4 with the theorems given at the beginning of this subsection on the existence of globally minimal harmonic maps gives two more corollaries.

COROLLARY 5.2.5. *Suppose that under the hypothesis of Theorem* 5.2.3 *the two-dimensional curvature of* N *is nonpositive. Then the solvability of the minimization problem in* $[F]$ *is equivalent to the triviality of* $[\pi_Q \circ F]$, *and the equality* $D([F]) = 0$ *is equivalent to the triviality of* $[\pi_N \circ F]$.

COROLLARY 5.2.6. *Suppose that under the hypothesis of Theorem* 5.2.4 *either* $\dim M = 1$, *or* $\dim M = 2$ *and* $\pi_2(N) = 0$, *or the curvature of* N *is nonpositive. Then the solvability of the minimization problem in* $[f \times \varphi]$ *is equivalent to the triviality of* $[\varphi]$, *and the equality* $D([f \times \varphi]) = 0$ *is equivalent to the triviality of* $[f]$.

2.2. Minimization of functionals of Dirichlet type. The result of A. I. Pluzhnikov on topological obstructions to the realizability of minima of the Dirichlet functional in homotopy classes (see 2.1) can be generalized to more general functionals and higher homotopy groups. For this we introduce on the space of maps of one Riemannian manifold into another the set of functionals

$$D_p(f) = \int_M \|df\|^p \, dv, \qquad 1 \le p < \infty;$$

when $p = 2$ we obtain the usual Dirichlet functional. We note that if $p = \dim M$, then the functional D_p is conformally invariant. In exactly the same way as for the usual Dirichlet functional, $D_p(f) \ge 0$ and $D_p(f) = 0$ if and only if $f = \mathrm{const}$.

The following two theorems of Tyrin [114] generalize Theorems 5.2.1 and 5.2.2 of Pluzhnikov. For similar theorems and further important generalizations see White [417], [418].

THEOREM 5.2.5. *Suppose that* $\inf D_p$ *is equal to zero in the homotopy class of the identity map of* M *onto itself. Then for any* N: (a) *in any homotopy class of maps of* N *into* M *and* (b) *in any homotopy class of maps of* M *into* N, *the greatest lower bound* $\inf D_p$ *is also equal to zero.*

Consequently, in this situation in a nontrivial homotopy class the minimum is never attained.

THEOREM 5.2.6. *For a compact Riemannian manifold M the condition* $\inf D_p = 0$ *in the homotopy class of the identity map of M onto itself is equivalent to the condition* $\pi_1(M) = \cdots = \pi_{[p]}(M) = 0$ *(here $[p]$ is the largest integer not exceeding p)*.

For the special case $N = S^n$ we can prove the following local version of the assertion about the unattainability of minima of D_p.

THEOREM 5.2.7. *A nonconstant map $f \in C^2(M, S^n)$ of a connected manifold M cannot be a local minimum of D_p when $p < n$.*

For the functional $D = D_2$ this was proved in [112]. A similar theorem for functionals D_p and maps of $C^2(S^n, M)$ is mentioned in [407].

2.3. A priori instability of harmonic maps. We recall that a harmonic map $f \in C^\infty(M, N)$ of Riemannian manifolds is said to be *stable* if the second variation form of the Dirichlet functional, calculated at the "point" f, is positive semidefinite. For example, the globally minimal harmonic maps considered in 2.1 are stable. Stability is a necessary but not sufficient condition for local minima of the Dirichlet functional. In 2.1 we gave topological conditions on a closed (compact and without boundary) Riemannian manifold M that ensure that global minima of the Dirichlet functional are not attainable on all connected components of the spaces $C^\infty(M, N)$ and $C^\infty(N, M)$, expect those containing locally constant maps, where the compact Riemannian manifold N is arbitrary. In this subsection we give a whole series of homogeneous Riemannian manifolds M such that all nonconstant harmonic maps in the spaces $C^\infty(M, N)$ and $C^\infty(N, M)$ (if they exist) are a priori unstable for all connected compact Riemannian manifolds N. In these spaces, the Dirichlet functional does not have nonconstant local minima and all its critical points are "saddle" points, except constants. It is curious that as in the case of global minima, a definite role here is played by the identity map id_M (more precisely, the instability of the identity map), which is obviously always harmonic, like any isometry of a Riemannian manifold.

The degree of instability of a harmonic map $f: M \to N$ is measured by its *index* $\mathrm{ind}(f)$, which is equal to the dimension of the maximal subspace on which the second variation form of the Dirichlet functional is negative definite. If M is closed, then $\mathrm{ind}(f)$ is finite. *The inequality* $\mathrm{ind}(f) > 0$ *is equivalent to the instability of f.*

THEOREM 5.2.8 (Tyrin [112], [113]). *Let $M = G/H$ be a compact homogeneous irreducible Riemannian manifold with G-invariant metric such that its identity map is unstable. Then all nonconstant harmonic maps $f \in C^\infty(N, M)$ of an arbitrary connected compact Riemannian manifold N are unstable.*

THEOREM 5.2.9 (Pluzhnikov [81], [83], [84]). *Let $M = G/H$ be a connected compact homogeneous irreducible Riemannian manifold with G-invariant metric such that its identity map is unstable. Then all nonconstant harmonic maps $f \in C^\infty(M, N)$ into an arbitrary Riemannian manifold N are unstable.*

At the basis of the proof of both theorems lies the same idea: averaging the second variation quadratic form over the action of a compact isometry group G of M on a suitable space of vector fields along a harmonic map f. The local variations f_t of the map f that decrease the value of the Dirichlet functional for small t have the form $\varphi_t \circ f$ in Tyrin's theorem, and $f \circ \varphi_t$ in Pluzhnikov's theorem, where φ_t is a one-parameter group of diffeomorphisms of M such that $D(\varphi_t) < D(\varphi_0)$ for small $t \neq 0$. The technical methods used in the proofs and the explicit (and convenient for calculation) formulae for the second variation of the Dirichlet functional are, of course, quite distinct.

The union of Theorems 5.2.8 and 5.2.9 is the following result, which is analogous to Theorem 5.2.1 in its structure.

THEOREM 5.2.10. *Let $M = G/H$ be a connected compact homogeneous irreducible Riemannian manifold with G-invariant metric of positive dimension. Then the following three assertions are equivalent:*

(a) $\operatorname{ind}(\mathrm{id}_M) > 0$;

(b) $\operatorname{ind}(f) > 0$ *for any nonconstant harmonic map $f: M \to N$ into an arbitrary smooth Riemannian manifold N;*

(c) $\operatorname{ind}(f') > 0$ *for any nonconstant harmonic map $f': N' \to M$ of an arbitrary connected compact smooth Riemannian manifold N'.*

The question of the instability of identity maps of compact symmetric spaces has been studied by Smith [394] and Tyrin [112], [113]. Smith obtained a list of connected simply-connected compact irreducible nonsingular symmetric spaces of type I with unstable identity map, and Tyrin obtained a similar list of connected simply-connected compact simple nonsingular Lie groups. These results together with Theorem 5.2.10 of Tyrin and Pluzhnikov lead to the following assertion.

THEOREM 5.2.11. *In the class of connected compact irreducible nonsingular symmetric spaces any of the assertions* (a) – (c) *of Theorem* 5.2.10 *is satisfied for the following manifolds and only for them* (*the metric is standard; n and m are positive integers*): (1) S^n, $n \geq 3$; (2) $Sp(n+m)/Sp(n) \times Sp(m)$; (3) $SU(2n)/Sp(n)$, $n \geq 2$; (4) $SU(n)$, $n \geq 2$; (5) $Sp(n)$.

For spheres S^n $(n \geq 3)$ assertion (b) of Theorem 5.2.10 was proved independently by Xin [411], and (c) by Leung [293]. Pluzhnikov [78], [80], [84] obtained the estimate $\operatorname{ind}(f) \geq n+1$ for nonconstant harmonic maps $f\colon S^n \to N$ $(n \geq 3)$, which is best possible in the general case by virtue of the equality $\operatorname{ind}(\mathrm{id}_s) = n+1$, proved by Smith [394] for identity maps id_s of spheres S^n $(n \geq 3)$.

We note that all the manifolds listed in Theorem 5.2.11 are doubly-connected and so assertions (a)–(c) of Theorem 5.2.1 hold for them (see Corollary 5.2.2).

2.4. Regularity of harmonic maps. Here we consider *the question of regularity of harmonic maps* that minimize the Dirichlet functional. The theory presented briefly below is in a sense parallel to the corresponding well-known results from the theory of the volume functional; the technical details of the two theories are distinct, as a rule.

Let us outline the situation more precisely. We shall assume that the manifold N is embedded isometrically in a Euclidean space of sufficiently large dimension $\mathbf{R}^k$. Then we can extend the definition of the Dirichlet functional to a wider class of maps $f \in L_1^2(M, N)$. By definition $L_1^2(M, N) = \{f \in L_1^2(M, \mathbf{R}^k)\colon f(x) \in N \text{ for almost all } x \in M\}$.

Let us pose the question of regularity (in the sense of smoothness) of maps of $L_1^2(M, N)$ that minimize D on compact subsets M^m. In general, maps of this set are not smooth or even continuous when $m \geq 2$. It is known, however, that when $m \leq 2$, maps that minimize D are always smooth. This follows from the classical results of Morrey [318]; the fact that smoothness holds when $m = 1$ is a quite elementary result. Therefore, at present the main interest lies in discovering those N for which smoothness holds in higher dimensions.

Recently Schoen and Uhlenbeck [371], [372] proved the following very general criterion for regularity. Suppose we know that any homogeneous map $F\colon \mathbf{R}^q \to N$, defined by a smooth harmonic map $f\colon S^{q-1} \to N$ by the formula $F(x) = f(x/|x|)$, that minimizes D in L_1^2 (in particular, is stable) on compact subsets of $\mathbf{R}^q$ is constant when $3 \leq q \leq l$. Then (see [371], [372]) for any map $M \to N$ that minimizes D on compact subsets

M^m the Hausdorff dimension H^{m-l-1} of the set of singular points is finite. In particular, when $m \leq l$ such maps are always regular (smooth).

Which manifolds N satisfy this criterion? First of all, as Tyrin remarked, the fact that the criterion of Schoen and Uhlenbeck is satisfied implies strong restrictions on the topology of N.

THEOREM 5.2.12 (A. V. Tyrin). *Suppose that N is such that $\pi_1(N) = 0$ and the criterion of Schoen and Uhlenbeck holds for homogeneous maps $F: \mathbf{R}^q \to N$ when $3 \leq q \leq l$. Then $\pi_2(N) = \cdots = \pi_{l-1}(N) = 0$.*

We also state (this is almost obvious) that a map from an arbitrary M to the Cartesian product of Riemannian manifolds $N_1 \times \cdots \times N_r \subset \mathbf{R}^{k_1} \times \cdots \times \mathbf{R}^{k_r}$ is harmonic (minimizing) if and only if each "coordinate" map $M \to N_i \subset \mathbf{R}^{k_i}$ $(i = 1, \ldots, r)$ is harmonic.

This criterion is satisfied, as might be expected from the results of Al′ber [2], and Eells and Sampson [209], for any l if N is a manifold of nonpositive curvature (see [371]). Schoen and Uhlenbeck [373] considered the case of regularity of minimizing maps into Euclidean spheres. Below we present the results of A. V. Tyrin, who extended the theorems of [373] to a wider class of both manifolds and metrics.

Suppose first that $N^n \subset \mathbf{R}^{n+1}$ is a complete orientable hypersurface, that μ denotes the pointwise maximum of the principal curvatures on N, and that k (k') is the pointwise maximum (minimum) of the two-dimensional sectional curvatures on N.

THEOREM 5.2.13. *If for some $l \geq 3$ we have*

$$\frac{l-2}{l-1}k \leq \frac{k'(n-1) - \mu^2}{n}, \tag{2.1}$$

then maps in $L_1^2(M, N)$ that minimize the Dirichlet functional on compact subsets of an arbitrary Riemannian manifold M^m are smooth when $3 \leq m \leq \min\{l, 5\}$.

Now let $N^n \subset S^p$ be an isometric embedding of a compact orientable manifold into a sphere with the standard metric, whose second fundamental form is A. Let $\|A\|$ be its "length", and let k be as above.

THEOREM 5.2.14. *If for some $l \geq 3$*

$$\frac{l-2}{l-1}k \leq \frac{n-2}{n} - \frac{\max\{1, \sqrt{n-2}/2\} + 1}{n} \cdot \|A\|^2, \tag{2.2}$$

then maps in $L_1^2(M^m, N)$ that minimize D are smooth when $3 \leq m \leq \min\{1, 5\}$.

REMARK. For the case $N = S^n$ with standard metrics Theorems 5.2.13 and 5.2.14 generalize the result of [373].

The results mentioned above suggest the following question: is the regularity of minimizing maps preserved under a small deformation of the metric on N (clearly, in the theorems we are considering variations of metrics of a very special kind)? Generally speaking, this does not follow, since the closeness of two metrics on the image does not ensure closeness of the harmonic maps in the image. Nevertheless, it turns out that in many cases the answer to this question is yes. We state the following result in this direction.

THEOREM 5.2.15 (A. V. Tyrin). *In the following cases for a Riemannian manifold* (N, g) *there is a* C^2*-neighborhood of the metric* g *such that for any metric* h *of this neighborhood any map* $M^m \to (N, h)$ *that minimizes* D *on compact subsets is smooth*:

(a) $N^n \subset \mathbf{R}^{n+1}$, *the inequality* (2.1) *is strict, and* $m \leq \min\{l, 5\}$;

(b) $N^n \subset S^p$, *the inequality* (2.2) *is strict, and* $m \leq \min\{l, 5\}$.

Writing out the conditions on the dimension of M^m and S^n explicitly, we find that conclusion (a) of Theorem 5.2.15 holds for $N = S^n$ with metric close to the standard metric when $m \leq d(n)$, where $d(n) = [\min(n+1)/2, 5]$ for $n \geq 3$, $d(2) = 2$. Here $[r]$ denotes the largest integer not exceeding r.

REMARK 1. The results of Schoen and Uhlenbeck [371], [372] also show that for the manifolds (N, g) of Theorem 5.2.15 the Dirichlet problem is always solvable for harmonic maps $M^m \to (N, g)$ with an arbitrary smooth "boundary condition" $\partial M \to (N, g)$, that is, L_1^2-extendable to the whole of M.

REMARK 2. Let us describe an example of Schoen and Uhlenbeck (see [373]) of a map in $L_1^2(\mathbf{R}^7, S^7)$ that is not smooth but nevertheless minimizes the Dirichlet functional. We define a map $F\colon \mathbf{R}^7\backslash 0 \to S^6$ by $F(x) = x/|x|$. We now embed S^6 in S^7 as an equator, and we obtain a map $\mathbf{R}^7\backslash 0 \to S^7$. It can be proved (see [373]) that this map minimizes the Dirichlet functional on compact subsets of $\mathbf{R}^7$. This example is an analog of the well-known Simons cones [387] in the theory of the volume functional.

§3. Some unsolved problems

For carefully chosen lists of unsolved problems see, for example, the papers of Chern [188], [189] in the years 1969–70 and the fundamental lecture of Yau in [380], which was published in 1982 and contains several dozen unsolved problems. In particular, among them are the following important problems.

1. Is it true that every compact smooth manifold can be embedded as a minimal submanifold in a sphere S^n (for some n)?

2. Prove that any three-dimensional manifold contains infinitely many distinct embedded minimal two-dimensional surfaces.

3. Prove that every smooth regular Jordan curve in $\mathbf{R}^3$ can bound only finitely many stable minimal surfaces.

4. Prove that all elements of the homotopy groups $\pi_i(S^n)$ can be represented by harmonic maps.

We *supplement* the list of problems in [188], [189], and [380] by several problems that are interesting in the authors' opinion and are open to research by modern methods.

5. Prove that in any homotopy class of a continuous map $f: W^k \to M$ of a compact smooth manifold W^k into a compact closed Riemannian manifold M^n there is always a *smooth* minimal map f', that is, such that the leading k-dimensional (or complete stratified) volume of the image $f'(W^k)$ is globally minimal in M^n. A similar question can be posed for maps of manifolds W^k with a fixed *smooth* boundary in $\mathbf{R}^n$. A proof of this assertion would enable us to apply to the study of minimal surfaces the far advanced theory of singularities of *smooth* maps.

6. Any Lie subgroup H of a Lie group G is totally geodesic, and therefore a *locally minimal* submanifold. It is required to give a classification (in some reasonable sense) of those subgroups H that are *globally minimal* in G. Clearly, the fact that a subgroup is globally minimal can be stated in terms of the weights of the representation $H \to G \subset \operatorname{Aut} \mathbf{R}^N$, for example for compact Lie groups. This question requires preparatory explanation—which of the subgroups H are *topologically nontrivial* in the ambient group G, for example, do not contract to a point or realize nonzero (co)cycles in (co)homology groups G. For some results see the cycle of papers of Fomenko [119], [121], [124], [125], and [134]; see also a criterion for global minimality of Đào Trọng Thi [42].

7. The next problem is closely connected with the previous one. It is known that every simply-connected symmetric space $V = G/H$ can be embedded as a totally geodesic submanifold in its simply-connected isometry group G (the so-called Cartan model of a symmetric space); for the details see [55], [118]. Question: which Cartan models are globally minimal in G? A description of topologically nontrivial embeddings $G/H \to G$ is given in [119], [121], [124], and [134]. Đào Trọng Thi has proved that a

Cartan model is globally minimal if and only if the stationary subgroup H is globally minimal in G.

8. Let $B^n \subset \mathbf{R}^n$ be a convex centrally symmetric domain in $\mathbf{R}^n$ and X^k a complete globally minimal surface passing through the center of B^n and going off to infinity. Then it is probable that $\mathrm{vol}_k X \cap B$ is never less than the volume of a central minimal plane section $\min \mathrm{vol}_k \mathbf{R}^k \cap B$ of B by planes $\mathbf{R}^k$ passing through its center; see [119], [95], [56], and [415].

9. Let $C^\infty(M^k, N^n)$ be the space of smooth maps f of a manifold M^k with boundary ∂M into a Riemannian manifold N and $C^\infty(\partial M, N)$ the space of maps g of ∂M.

Consider the volume of minimal surfaces $f_0(M^k)$ hanging on each boundary "contour" $g(\partial M) \subset N$. Changing the "contour", we obtain a multivalued function (of volumes of minimal surfaces) on the space $C^\infty(\partial M, N)$. Question: what is the structure of its singularities—branch points, points where sheets stick together, and so on? Thus, for deformations of the Douglas contour $S^1 \subset \mathbf{R}^3$ we obtain "swallow-tails" (see [136], [111]).

10. The one-dimensional Plateau problem on a closed Riemannian manifold M^n. Consider on M^n a *one-dimensional continuum* whose length is minimal in the class of curves obtained from it by a continuous deformation. This continuum may contain singular points—branch points. For example, for M^2 all these branches are threefold, that is, at each such singular point only three arcs meet at equal angles (see Plateau's principles). Problem: to describe all such stable minimal continua. For $M^2 = \mathbf{R}^2$ on condition that the end points of the continuum are fixed, this is Steiner's problem. For $M^2 = S^2$ the solution is known; see [148].

11. It is known that there are nonlinear minimal surfaces $M^2 \subset \mathbf{R}^{2n}$, $2n \geq 4$, where M^2 is uniquely projected onto $\mathbf{R}^2 \subset \mathbf{R}^{2n}$ (Osserman). Question: to find a complete k-dimensional minimal surface $M^k \subset \mathbf{R}^n$ for any $k < n-1$ that is uniquely projected onto $\mathbf{R}^k \subset \mathbf{R}^n$ and does not lie in a hyperplane $\mathbf{R}^{n-1}$.

12. To describe bendings and infinitesimal bendings of complete minimal surfaces. What are the conditions that such surfaces should be unbendable and rigid?

13. To give a complete classification of effective and invariant calibrations on homogeneous spaces. In connection with this problem we give the following commentary, in which we describe recent results of Le Hong Van. Let φ be a calibration on M. A current S^k is called a *φ-current* if for almost all $x \in M$ in the sense of the measure $\|S^k\|$ we have

$\langle \varphi , \vec{S}_x \rangle = \|\vec{S}_x\| = 1$, where $\| \ \|$ is the norm of the mass of k-vectors. A calibration φ is said to be *effective* (see [416]) on M if there is at least one φ-current. A calibration φ' is said to be *weaker* than a calibration φ if any φ'-current is a φ-current.

THEOREM 5.3.1 (Le Hong Van). *For any calibration φ on a compact Riemannian homogeneous space G/H there is a G-invariant calibration φ_G such that φ is weaker than φ_G. If φ is effective, then φ_G is also effective, and the comass of φ_G is minimal among all G-invariant forms cohomologous to it.*

Let $H_*(G/H, \mathbf{R})$ be the real homology of G/H, let V be the tangent space to G/H at the point $\{eH\}$, and let $\Lambda^*_H(V)$ be the space of all H-invariant forms on V. Then there is an isomorphism $H_k(G/H, \mathbf{R}) = H_k(\Lambda^*_H(V), \partial_H)$, where ∂_H is the operator adjoint to the invariant operator of outer differentiation. The next theorem generalizes a criterion of global minimality of normal currents in symmetric spaces, discovered by Đào Trọng Thi (see Ch. 10).

THEOREM 5.3.2 (Le Hong Van). *A closed current S^k on a compact homogeneous Riemannian manifold is globally minimal if and only if the following equality holds: $M(S) = M(\pi_G S) = \mathrm{vol}(G/H)\min\{\|\overrightarrow{\pi_H S_e} + \partial\theta\|\}$, $\theta \in \Lambda^{k+1}_H(V)$.*

Here $\pi_G S$ denotes the current $\int_G S\, d\mu_g$; $\overrightarrow{\pi_H S_e}$ is the covector $(M(\pi_G S)/\mathrm{vol}(G/H))\overrightarrow{\pi_G S_e}$, that represents the homology class of the current $[\pi_G S] = [S]$ under the isomorphism of homology stated above; M is the mass of the current.

THEOREM 5.3.3. *A normal current S^k minimizes the mass functional in its real homology class $[S^k] \in H_k(G/H, \mathbf{R})$ of the compact homogeneous space G/H if and only if there is a G-invariant absolute calibration φ such that S^k is a φ-current.*

We now give the definition of the type of globally minimal surfaces in terms of absolute calibrations. Two calibrations φ and φ' are said to be *equivalent* if for any point $x \in M$ we have $G_x(\varphi) = G_x(\varphi')$. With each cohomology class $\alpha \in H_k(G/H, \mathbf{R})$ we associate a characteristic set $F(\alpha)$ of classes of equivalent absolute calibrations φ^k such that φ^k calibrates globally minimal currents in the class α.

PROPOSITION. *For each cohomology class $\alpha \in H_k(G/H, \mathbf{R})$ there is a unique class of equivalent absolute calibrations $\{\varphi_\alpha\}$ with the following* *properties*: (i) *the class* $\{\varphi_\alpha\}$ *belongs to the set* $F(\alpha)$; (ii) *for any class* $\{\varphi'\} \in F(\alpha)$ *we have* $G(\varphi_\alpha) \subset G(\varphi')$. *In other words, the calibration* φ_α *is weaker than* φ'.

The class $\{\varphi_\alpha\}$ in the Proposition is called the type of globally minimal surfaces of the cohomology class α.

If $\dim H_k(G/H, \mathbf{R}) = 1$, it is not difficult to find the type of globally minimal currents of dimension k on G/H. In fact, we take an arbitrary invariant closed k-form ψ_G^k not cohomologous to zero, and consider the set P of invariant forms $\psi_G^k + \partial\theta_G^{k-1}$ cohomologous to it. Let P_0 be the subset of P consisting of forms of smaller comass. Then from Theorems 5.3.1 and 5.3.3 it follows that the set of classes $\widetilde{P}_0 = \{\varphi/\|\varphi\|^*, \varphi \in P_0\}$ is a characteristic set for any class $\alpha \in H_k(G/H, \mathbf{R})$. We choose r representatives φ_i, $i = 1, \dots, r$, of the class $\{\varphi_i\} \in P_0$ such that: (a) $\varphi_1, \dots, \varphi_r$ are linearly independent, (b) for any representative $\varphi \in \{\varphi\} \in P_0$, φ is a linear combination of the forms $\varphi_1, \dots, \varphi_r$. Then $\{\varphi_\alpha\}$, where $\varphi_\alpha = (1/r)\sum_{i=1}^r \varphi_i$, is the type of the class α.

CHAPTER VI

Multidimensional Variational Problems and Multivarifolds. The Solution of Plateau's Problem in a Homotopy Class of a Map of a Multivarifold

§1. Classical formulations

We now proceed to the solution of Plateau's problem in a homotopy class, obtained by Đào Trọng Thi. Classical calculus of variations is concerned with one-dimensional variational problems, the simplest of which is the problem of minimizing an integral of the form

$$J(x) = \int_a^b l(x(t), \dot{x}(t))\, dt$$

in a set C of curves $x = x(t)$, $a \le t \le b$, in an open subset $U \subset \mathbf{R}^n$, joining two given points (the problem of an absolute minimum), or in some class of homotopic curves from a given set C (the problem of a homotopic minimum), where $l\colon U \times \mathbf{R}^n \to \mathbf{R}$ is a Lagrangian, that is, a function of two vector-valued variables x and $\dot{x}$, and l is positively homogeneous with respect to $\dot{x}$. If the first variation of J is zero, then we have the Euler vector equation

$$l_x - \frac{d}{dt} l_{\dot{x}} = 0. \tag{1.1}$$

Consider a pair (x, p) for which $l(x, p) = 0$ and suppose that the matrix $(l^2)_{\dot{x}\dot{x}}$, consisting of the second-order partial derivatives of l^2 with respect to $\dot{x}$ and calculated at (x, p), is nonsingular. Then from the equation $y = ll_{\dot{x}} = (l^2/2)_{\dot{x}}$, we can locally express $\dot{x}$ in terms of x and y, so that the expression $\sqrt{2\dot{x}y - l^2}$ is a function of x and y, defined close to $(x, l_{\dot{x}}(x, p))$; we denote it by $h(x, y)$.

We can verify that in this neighborhood, h is positively homogeneous and $h(x, y) = 1$ if $y = l_{\dot{x}}(x, \dot{x})$ and $l(x, \dot{x}) = 1$.

Now by homogeneity we extend h to all pairs (x, y) for which $(x, \lambda y)$ is in a given neighborhood of $(x, l_{\dot{x}}(x, p))$ for some $\lambda \geq 0$. The resulting function is again denoted by $h(x, y)$ and called the *Hamiltonian* of our problem, and x and y are called *canonical variables*. With respect to canonical variables the Euler equation has the following canonical form:

$$\dot{x} = hh_y, \qquad \dot{y} = -hh_x \tag{1.2}$$

provided that the curve in question is geodesically parametrized, that is, $l(x, \dot{x}) = \text{const}$ along it.

Famous papers of Hilbert, Weierstrass, Tonelli, and L. C. Young have been devoted to the question of the existence of a global solution. Here we distinguish two important ideas. Firstly, existence theorems can automatically be obtained from the Weierstrass-Tonelli principle that any semicontinuous function attains its minimum on a compact subset. Secondly, the deep statement of Hilbert that any problem in the calculus of variations has a solution so long as the word "solution" is given a suitable meaning led L. C. Young to the discovery of the concept of a generalized curve. The introduction of a weak topology and the natural completion with respect to it of the space of curves by generalized curves ensure, on the one hand, that all minimizable functionals are semicontinuous, and on the other hand, that minimizing sequences converge, after which the existence of a "generalized solution" becomes an obvious fact.

If we consider problems with mobile ends, then we must adjoin to Euler's equation (1.1) a transversality condition, and if we consider a problem with additional restriction, then in (1.1), the Lagrangian l must be replaced by the generalized Lagrangian $\tilde{l}$ with additional variables in the form of Lagrange multipliers.

§2. Multidimensional variational problems

2.1. Classical formulations of multidimensional variational problems and classical multidimensional Plateau problems. Let G be a fixed compact closed $(k-1)$-dimensional submanifold of a Riemannian manifold M, and W a k-dimensional manifold such that $\partial W = G$ (we observe that there is not always such a manifold W whose boundary is the given manifold G). We also consider piecewise smooth or continuous maps $f: W \to M$ such that $f|_G$ is the identity, and a functional $J(f)$ on the space of maps f; we can require that the map f and the functional $J(f)$ satisfy additional "reasonable" properties.

PROBLEM A (the problem of finding minimal surfaces of variable topological type with a given boundary). Find a piecewise smooth (or continuous) map f_0 that minimizes a given functional $J(f)$ in the class of all pairs (W, f), where W runs through all possible k-dimensional compact manifolds with boundary G, and f runs through all possible piecewise smooth (or continuous) maps of W into M that leave the boundary $\partial W = G$ fixed.

PROBLEM B (the problem of finding minimal surfaces of fixed topological type with a given boundary). Let W be a fixed k-dimensional compact manifold with boundary $\partial W = G$. Find a piecewise smooth (or continuous) map f_0 that minimizes a given functional $J(f)$ in the class of all piecewise smooth (or continuous) maps $f\colon W \to M$ that are the identity on $\partial W = G$, or in a given homotopy class of such maps.

Additional properties that we require of the maps f and the functional $J(f)$ are of various kinds in accordance with the different approaches to the original concepts of a solution, a boundary condition and minimality.

Together with Problems A and B "with fixed boundary" it is natural to consider the corresponding Problems A$'$ and B$'$ "with moving boundary", for which the maps f are not the identity on $G = \partial W$, but map it into a given subset of M. Moreover, with Problem B we can associate the following problem.

PROBLEM B$''$ (the realizing problem). Let W be a closed compact k-dimensional manifold, and suppose we are given a homotopy class of piecewise smooth (or continuous) maps $f\colon W \to M$. It is required to find a map f_0 that minimizes a given functional $J(f)$ in this class.

Of course, in Problem B$''$ the interesting case is when the homotopy class is nontrivial.

If in the problems mentioned above, we choose for $J(f)$ the multidimensional volume functional $\mathrm{vol}_k(f)$, then we obtain the classical formulations of the multidimensional Plateau problem.

2.2. Partial degeneracies of minimal maps. It is well known that in the process of approaching the minimal position, a map f may undergo partial degeneracies (that is, abruptly lower its dimension in certain open subsets of W), which involve gluings in the image film $X = f(W)$, as a result of which in $f(W)$ there may arise pieces of dimension $s < k$. This effect was studied by Fomenko [119].

The rise of zones of nonmaximal dimensions in the image-film of a minimal map f destroys the conformal property of this map (of W onto the image $f(W)$) which, as we have seen in the two-dimensional case, is necessary for the values of the volume functional $\mathrm{vol}_k(f)$ and the Dirichlet functional $D(f)$ to coincide on the map f. If we admit in the solution of the problem only maps with image-film that is homogeneous in dimension, then it is well known that some such maps may not guarantee convergence of minimizing sequences of maps (see [119]).

2.3. Introduction of stratified surfaces and the classical formulation of Plateau's problem A in the language of bordism theory. As we already know, in the process of minimizing $J(f)$ in the image-film $f(W)$ of the minimal map, there may arise zones of small dimension, which cannot always be mapped into the part of it of maximal dimension while preserving the parametrizing properties of this film. If to avoid small-dimensional pieces we remove them and consider only the part of maximal dimension, then this may lead not only to a change in the topological type of the film but also to the destruction of its parametrizing properties (see [119]).

Thus, in order that Plateau's problem should have a solution, we must (in accordance with Hilbert's statement) choose a suitable concept of solution. For this it is natural to take the concept, first introduced by Fomenko [119], [232], [129], and [134], of a *stratified surface* (that is, a surface that is the union of strata of different dimensions).

Let us first recall the definition of *i-dimensional Hausdorff measure* h^i. Consider a connected Riemannian manifold M. Let $A \subset M$ be a subset and $i \geq 1$ an integer. If $A = \varnothing$, we set $h^i A = 0$. Suppose that $A \neq \varnothing$. For $\varepsilon > 0$ we set $h^i_\varepsilon A = \inf \sum_\alpha \alpha_i \varepsilon^i_\alpha$, where α_i is the Lebesgue measure of the unit ball in $\mathbf{R}^i$, and the infimum is taken over all possible covers of A by a no more than countable family of open balls $B(x_\alpha, \varepsilon_\alpha)$ with centers x_α and radii ε_α, where $\varepsilon_\alpha < \varepsilon$ for any α. Since the numbers $h^i_\varepsilon A$ do not decrease as $\varepsilon \to 0$, we can define the *outer Hausdorff measure* $h^i A$ by $h^i A = \lim_{\varepsilon \to +0} h^i_\varepsilon A$.

We say that *a set* $A \subset M$ *is* h^i*-measurable* if for any subset $B \subset M$ we have $h^i B = h^i(B \cap A) + h^i(B \cap (M \setminus A))$.

We observe that if $h^p A < +\infty$, then $h^q A = 0$ for every $q > p$. Any set A is h^0-measurable, and $h^0 A$ is equal to the number of points in A.

Now let μ be Radon measure on M, let $A \subset M$ be a subset, and $z \in M$ an arbitrary point. Then we define the *upper i-dimensional μ-density* $\overline{\Theta}^i(\mu, A, z)$ and the *lower i-dimensional μ-density* $\underline{\Theta}^i(\mu, A, z)$ of the set A at the point z respectively, as the upper and lower limits as $\varepsilon \to 0$ of the expression $\alpha_i^{-1} \varepsilon^{-i} \mu(A \cap B(z, \varepsilon))$, where $B(z, \varepsilon)$ is the open ball with center z and radius ε.

If the upper and lower μ-densities are equal, we call their common value the *i-dimensional μ-density of A at z* and denote it simply by $\Theta^i(\mu, A, z)$. If $A = M$, we write for brevity $\Theta^i(\mu, A, z) = \Theta^i(\mu, z)$.

Suppose that $M = \mathbf{R}^m$, and that μ is a positive measure. We mention the following important facts.

1. If $A \subset \mathbf{R}^m$, $B \subset \mathbf{R}^m$, $t > 0$, $\overline{\Theta}^i(\mu, A, z) > t$ for $z \in B$, then $\mu(A) \geq t \cdot h^i B$.

2. If $A \subset \mathbf{R}^m$, $t > 0$, $\overline{\Theta}^i(\mu, A, z) < t$ for $z \in A$, then $\mu(A) \leq 2^i t \cdot h^i A$.

3. If B is a Borel subset and $\mu(B) < \infty$, then $\Theta^i(\mu, B, z) = 0$ for almost all $z \in \mathbf{R}^m \backslash B$.

4. If $A \subset \mathbf{R}^m$ and $h^i A < \infty$, then $2^{-i} \leq \overline{\Theta}^i(h^i, A, z) \leq 1$ for almost all $z \in A$.

Next, suppose that for μ we consider the i-dimensional Hausdorff measure h^i. In this case the upper i-dimensional h^i-density of each compactum $S \subset M$ defines the standard spherical density function $\Psi_i(x, S) = \overline{\Theta}^i(h^i, S, x)$, $0 \leq i \leq n$.

Let k be the smallest integer for which $\Psi_{k+1}(x, S) = 0$ for all $x \in M$. We construct the stratification $S = S^0 \cup S^1 \cup \cdots \cup S^k$ by setting

$$S^i = \left\{ x \in S \backslash \bigcup_{i+1 \leq j \leq k} S^j : \Psi_i(x, S) > 0 \right\}.$$

If every set S^i is h^i-measurable, we call S a *k-dimensional stratified surface*, and S^i its *i-dimensional stratum*. In this case the number $h^i S^i$ is well defined; we call it the *i-dimensional volume of* S and denote it by $\mathrm{vol}_i S$. We call the system of $k+1$ numbers $(\mathrm{vol}_0 S, \ldots, \mathrm{vol}_k S)$ the *stratified volume of* S (see [119], [129], and [232]).

Let S be a k-dimensional stratified surface in M, and let $G \subset S$. We denote an embedding of G in S by i. Then Plateau's problem A admits an equivalent reformulation in the language of bordisms.

PROBLEM A*. Among all k-dimensional stratified surfaces S in M containing a submanifold G and having the property that the singular bordism (G, i) is equivalent to zero in S (that is, there is a k-dimensional compact manifold W with boundary $\partial W = G$ and a continuous map $f: W \to S$ such that $f|_G = i$), find a stratified surface S_0 that minimizes the k-dimensional volume.

To solve this problem we could minimize only the k-dimensional volume of the stratified surface S, which is concentrated on the stratum S^k; we then preserve from the small-dimensional strata $S^0, \ldots, S^{k-1}$ only their topological functions. However, it turns out that there is a natural (and in the framework of the method developed by A. T. Fomenko, even a necessary) simultaneous minimization of the volumes of strata in all dimensions, namely minimization of the stratified volume with respect to lexicographical order. In this approach Plateau's problem for the case of spectral bordisms was solved by Fomenko [119], [232], and [129]; namely, he proved a theorem on the existence of minimal stratified surfaces (of variable topological type) with a given boundary that minimize the stratified volume with respect to lexicographical order. We note that in the formulation of Problem A*, we can require of maps f of the original Problem A only the continuity property and minimize not maps and the k-dimensional volume functional $\mathrm{vol}_k(f)$ defined on them, but the k-dimensional Hausdorff measure of the image-films $X = f(W)$ themselves (that is, without taking account of multiplicity).

The existence theorem mentioned above was proved by A. T. Fomenko for Plateau's problem in an arbitrary spectral class of surfaces whose topological type is determined by any extraordinary homology and cohomology theory. In particular, if for the extraordinary homology theory we choose the stable homotopy theory, then we obtain a theorem on the existence of stratified surfaces of least stratified volume in each spectral stable homotopy class of maps of the sphere into M.

Developing Fomenko's methods and his concept of stratified volume and stratified surface, Đào Trọng Thi formulated *the new concept of a multivarifold, which is the functional analog of a geometrical stratified surface and enables us to solve Plateau's problem in a homotopy class.* These results of Đào Trong Thi are presented in this and later chapters of the book.

2.4. Functionals of k-dimensional volume type. Let $f\colon W \to M$ be a piecewise smooth map of a k-dimensional compact manifold W into a Riemannian manifold M. Suppose there are local coordinates $x_1, \dots, x_k$ on W. We set $\partial_i f = \partial f(x)/\partial x_i$. Consider an arbitrary continuous function $l(y, u)$ of the variable $y \in M$ and the k-vector variable u, positively homogeneous of the first degree in u. We set

$$J(f) = \int_W l\left(f, \frac{\partial f}{\partial x_1}, \dots, \frac{\partial f}{\partial x_k}\right) dx_1 \wedge \dots \wedge dx_k. \tag{2.1}$$

The functional $J(f)$ defined by (2.1) is called the *functional of k-dimensional volume type specified by the Lagrangian l.* In particular, setting $l(y, u) = |u|$, the length of the k-vector u, we obtain the classical k-dimensional volume functional

$$\mathrm{vol}_k(f) = \int_W \left|\frac{\partial f}{\partial x_1} \wedge \dots \wedge \frac{\partial f}{\partial x_k}\right| dx_1 \wedge \dots \wedge dx_k.$$

The k-dimensional volume functional vol_k can also be defined on the image-films $S = f(W)$ by considering for $\mathrm{vol}_k S$ the k-dimensional Hausdorff measure $h^k S$ of the compactum S. Clearly, if S is a submanifold, then $\mathrm{vol}_k S$ coincides with the usual Riemannian volume of S. In general, let $\Gamma_k M$ (resp., $\Gamma_k^0 M$) denote the usual Grassmann bundle with base M whose fiber over a point $x \in M$ is the Grassmann manifold of k-dimensional subspaces (resp. oriented subspaces) of the tangent space $T_x M$. Let $l\colon \Gamma_k M \to \mathbf{R}$ (resp. $l\colon \Gamma_k^0 M \to \mathbf{R}$) be an arbitrary continuous function on $\Gamma_k M$ (resp. $\Gamma_k^0 M$). Then the functional of k-dimensional volume type J specified by the Lagrangian l can be defined on k-dimensional compact surfaces (resp. oriented surfaces) S by

$$J(S) = \int_S l(\vec{S}_x)\, d\sigma, \tag{2.2}$$

where $\vec{S}_x$ denotes the tangent space of S at the point x, and $d\sigma$ is the standard k-dimensional volume form on S.

Apart from classical volume functionals, there are many well-known functionals of volume type that are important in geometry, mechanics, and physics. For example, if for the Lagrangian l we take an arbitrary differential form ω of degree k on M, then the corresponding functional of k-dimensional volume type is none other than the operation of integrating the form ω:

$$J(S) = \int_S \omega(\vec{S}_x)\, d\sigma = \int_S \omega. \tag{2.3}$$

In particular, in the case $k = \dim M - 1$ (that is, S is a hypersurface), instead of the tangent space $\vec{S}_x$ we can consider the normal n^S and instead of the form ω a vector field ξ. Then (2.3) can be rewritten in the form

$$J(S) = \int_S (\xi, n^S)\, d\sigma. \tag{2.4}$$

This is the current of the vector field ξ flowing through the hypersurface S. We note that other important geometrical, mechanical and physical quantities, such as the classical Gauss integral and the flow of heat generated by temperature, the flow of a liquid in a stationary stream and the flow of energy of an electromagnetic field, the flow of energy and components of the flow of momentum of an ideal liquid, and so on, are also currents of different vector fields in Euclidean space $\mathbf{R}^3$.

§3. The functional language of multivarifolds

In this section Đào Trọng Thi develops the functional language of multivarifolds and realizes the stratified surfaces of Fomenko in this language.

This enables us to restate Problems B, B$'$ and B$''$ in the language of multivarifolds. Necessary information (and also terminology and notation) about Radon measures and Grassmannians can be found in [27].

3.1. The definition of a multivarifold. Let M be an arbitrary n-dimensional Riemannian manifold. We denote by TM the tangent bundle of this manifold, and by $\Gamma_k M$, $\Gamma_k^0 M$, $G_k M$, $G_k^0 M$ its bundle of tangent Grassmannian spaces of types $\Gamma_{k,n}$, $\Gamma_{k,n}^0$, $G_{k,n}$, $G_{k,n}^0$.

DEFINITION 6.3.1. A *multivarifold* (resp. *oriented multivarifold*) of order k on a Riemannian manifold M is any measure with compact support on $G_k M$ (resp. $G_k^0 M$); $G_k M = \bigcup_{0 \le i \le k} \Gamma_{i,n}$.

We denote the *vector space of multivarifolds* (resp. *oriented multivarifolds*) of order k on M by $V_k M$ (resp. $V_k^0 M$). Let $p\colon G_k M \to M$ (resp. $p\colon G_k^0 M \to M$) be the projection of $G_k M$ (resp. $G_k^0 M$). For any $V \in V_k M$) (resp. $V_k^0 M$) a measure $p_* V$ with compact support on M is called a *realizing measure of the multivarifold* V (resp. oriented multivarifold V) and denoted by rV.

Since it is *a measure space with compact support on* $G_k M$ (resp. $G_k^0 M$), the space of multivarifolds $V_k M$ (resp. $V_k^0 M$) can be endowed with weak (that is, weak*, weak and compactly weak) or strong (that is, strong, strong*, ultrastrong and compactly strong) topologies.

EXAMPLE 1. Let M be an n-dimensional Riemannian manifold, x an arbitrary point of M, and μ an arbitrary measure on the fiber $G_k M_x = G_{k,n}$. Then the functional μ_x over $C(G_k M)$ defined by the formula

$$\mu_x(f) = \mu(f_x), \tag{3.1}$$

where f is an arbitrary function in $C(G_k M)$ and f_x is the restriction of f to $G_k M_x$, is a multivarifold on M, which we call the *pointwise multivarifold at the point* x *with distribution of directions* μ. For each function $g \in C(M)$ we have $p_* \mu_x(g) = \mu_x(g \circ p) = \mu(g(x)) = g(x)\mu$, so that $r\mu_x$ is the pointwise measure $\mu(1)\varepsilon_x$ on M.

Similarly we can define *oriented pointwise multivarifolds.*

EXAMPLE 2. Let M be an n-dimensional Riemannian manifold and S an arbitrary k-dimensional compact submanifold of M. The map $\vec{S}$ that takes each point $z \in S$ into the tangent space S_z to S at z is a continuous map of S into $G_k M$. We define the *multivarifold* $[S]$ by

$$[S](f) = \int_S f \circ \vec{S}\, dz, \tag{3.2}$$

where f is an arbitrary function in $C(G_kM)$. For any function $g \in C(M)$ we have $p_*[S](g) = [S](g \circ p) = \int_S g \circ p \circ \vec{S}(z)\,dz = \int_S g(z)\,dz$. Hence, the resulting measure of $[S]$ is none other than the integral of the k-dimensional volume form on S.

If S is an oriented submanifold, then in a similar way we can define the *oriented multivarifold* $[S]$ *associated with* S.

EXAMPLE 3. Let $M = \mathbf{R}^n$ and let K be a finite simplicial complex in $\mathbf{R}^n$. For any i-dimensional simplex $\Delta^i \subset K$ the map Δ^i that takes each point $z \in \Delta^i$ into the tangent space Δ^i_z to Δ^i at z is a continuous map of the simplex Δ^i into $G_k\mathbf{R}^n$. We define the multivarifold $[K]$ by

$$[K](f) = \sum_{i=0}^{n} \sum_{\Delta^i} \int_{\Delta^i} f \circ \vec{\Delta}^i(z)\,dz. \tag{3.3}$$

where f is an arbitrary function in $C(G_k\mathbf{R}^n)$ and Δ^i runs through all i-dimensional simplexes in K that do not occur in a face of any $(i+1)$-dimensional simplex. For any function $g \in C(\mathbf{R}^n)$, according to (3.3), we have

$$\begin{aligned} p_*[K](g) &= [K](g \circ p) \\ &= \sum_{i=0}^{n} \sum_{\Delta^i} \int_{\Delta^i} gp\vec{\Delta}^i(z)\,dz = \sum_{\Delta} \int_{\Delta} g(z)\,dz, \end{aligned}$$

where Δ runs through all simplexes in K that do not occur in a face of any other simplex.

DEFINITION 6.3.2. A *multivarifold* (resp. *oriented multivarifold*) $V \in V_kM$ (resp. V_k^0M) is said to be *positive* if it is a positive measure on G_kM (resp. G_k^0M).

The set of all positive multivarifolds (resp. oriented multivarifolds) on a Riemannian manifold M forms a *convex cone* with vertex 0 and is denoted by V_k^+M (resp. $V_k^{0+}M$). Clearly, if $V \in V_k^+M$ (resp. $V_k^{0+}M$), then the resulting measure rV is also positive. The converse is false. The multivarifolds defined in Examples 2 and 3 are *positive multivarifolds*. The pointwise multivarifold μ_x defined in Example 1 is positive if and only if the measure μ is positive. If μ is a measure such that $\mu(1) > 0$, then μ_x need not be a positive multivarifold, while its resulting measure $r\mu_x = \mu(1)\varepsilon_x$ is always positive.

Henceforth for simplicity of presentation, we shall consider only the theory of multivarifolds on a Riemannian manifold, since the theory of oriented multivarifolds can be developed along the same lines. Where we need to distinguish these theories, we shall make special mention of it.

3.2. The structure of a multivarifold. Let M be a Riemannian manifold and let $V_{k,i}$ denote the subset of V_kM consisting of all multivarifolds V such that the support of the measure V is contained in Γ_iM. Obviously, $V_{k,i}$ is a vector subspace for each i, $0 \leq i \leq k$.

LEMMA 6.3.1. *The vector space $V_{k,i}$ is canonically isomorphic to the vector space $M(\Gamma_iM)$ of all measures with compact support on Γ_iM, $0 \leq i \leq k$.*

PROOF. Suppose that $V \in M(\Gamma_iM)$, that is, V is a measure with compact support on Γ_iM. We define a measure ρV on G_kM as follows. For an arbitrary continuous function f on G_kM we set $\rho V(f) = V(\tilde{f})$, where $\tilde{f}$ denotes the restriction of f to Γ_iM. Obviously, the ρV defined in this way is a measure with compact support contained in Γ_iM. Therefore, $\rho V \in V_{k,i}$. We have thus constructed a map $\rho\colon M(\Gamma_iM) \to V_{k,i}$. We prove that ρ is an isomorphism. Clearly, ρ is a homomorphism. We observe that Γ_iM, $0 \leq i \leq k$, is an open-and-closed set in G_kM. Therefore, we can extend any function continuous on Γ_iM by continuity to the whole of G_kM, putting it equal to 0 on $G_kM \backslash \Gamma_iM = \bigcup_{0 \leq j \leq k,\, j \neq i} \Gamma_jM$. Thus we can identify the space $C(\Gamma_iM)$ with the subspace of $C(G_kM)$ consisting of continuous functions with support contained in Γ_iM. Suppose that V_1, $V_2 \in M(\Gamma_iM)$, $\rho(V_1) = \rho(V_2)$. This means that $V_1(\tilde{f}) = V_2(\tilde{f})$ for any function $f \in C(G_kM)$, and a fortiori $V_1(f) = V_2(f)$ for any $f \in C(\Gamma_iM)$. Consequently, $V_1 = V_2$, that is, ρ is a monomorphism. Now suppose that $W \in V_{k,i}$, that is, the measure W has compact support contained in Γ_iM. We denote the restriction of W to $C(\Gamma_iM)$ by V, so that $V \in M(\Gamma_iM)$. For any function $f \in C(G_kM)$ we have $\rho V(f) = V(\tilde{f}) = W(\tilde{f}) = W(f)$, because the support of W is contained in Γ_iM, and so $\rho V = W$. Consequently, ρ is an epimorphism. This proves the lemma.

Lemma 6.3.1 shows, in particular, that the space $V_{k,i}$ does not depend essentially on the index k of the ambient space V_kM.

THEOREM 6.3.1. *For an arbitrary n-dimensional Riemannian manifold M and an arbitrary natural number $k \leq n$ we have the direct sum decomposition*

$$V_kM = \bigoplus_{0 \leq i \leq k} V_{k,i}.$$

PROOF. The inclusion $\bigoplus V_{k,i} \subset V_kM$ is obvious. We prove that any multivarifold V of V_kM can be represented uniquely as a sum $V = V^0 + \cdots + V^k$, where $V^i \in V_{k,i}$. Consider the continuous function φ_i

on G_kM that is equal to 1 on Γ_iM and to 0 on G_kM/Γ_iM, for any i. Obviously, $\varphi_0 + \cdots + \varphi_k \equiv 1$ on G_kM. We define multivarifolds V^i on M, $0 \le i \le k$, by $V^i(f) = V(f\varphi_i)$ for an arbitrary function $f \in C(G_kM)$. Clearly, $V^i \in V_{k,i}$ and $V = V^0 + \cdots + V^k$. On the other hand, if V admits a decomposition into a sum $V = W^0 + \cdots + W^k$, where $W^i \in V_{k,i}$, then for any function f in $C(G_kM)$ we have $V(f\varphi_i) = W^i(f\varphi_i) = W^i(f)$. Hence $W^i = V^i$, which proves that the decomposition of V into a sum of elements of the subspaces $V_{k,i}$ is unique. Thus the theorem is proved.

REMARK. According to Theorem 6.3.1 and the remark made at the end of the proof of Lemma 6.3.1, the space V_kM can be regarded as the subspace of $V_{k+1}M$ consisting of all measures with compact support in G_kM, so that there is a sequence of embedded vector spaces $V_0M \subset \cdots \subset V_kM$ such that $V_{k+1}M = V_kM \oplus V_{k+1,k+1}$.

DEFINITION 6.3.3. Suppose that $V \in V_kM$ and $V = V^0 + \cdots + V^k$, where $V^i \in V_{k,i}$. Then we call V^i the *i-dimensional stratum of the multivarifold* V. If V consists of a unique stratum, we call it a *homogeneous multivarifold*.

Clearly, for the multivarifolds $V, V^0, \dots, V^k$ we have the relations $rV = rV^0 + \cdots + rV^k$; $|V| = \sum_{i=0}^k |V^i|$.

REMARK. According to Lemma 6.3.1, each homogeneous multivarifold V of $V_{k,i}$ can be regarded as a measure with compact support on Γ_iM, that is, as an i-dimensional varifold. Consequently, *each multivarifold is the sum of varifolds of different dimensions*, which justifies its name.

3.3. Masses and supports of a multivarifold. Let $V \in V_kM$ be a multivarifold of order k on M. For each i we set

$$M_iV = m(rV^i) = \sup V^i(f \circ p), \tag{3.4}$$

where $f \in C(M)$, $\|f\| \le 1$, and p is the projection of the bundle G_kM.

DEFINITION 6.3.4. The *multimass of a multivarifold* V of order k is the ordered collection of $k+1$ quantities $(M_0V, \dots, M_kV)$; M_iV is called the *i-mass* (or *i-dimensional mass*) of V; the k-mass of a multivarifold V of order k is also called its *leading mass*.

From (3.4) it follows that $M_i(V) \le m(V^i)$, so that

$$m(rV) \le \sum_{i=0}^k M_iV \le \sum_{i=0}^k m(V^i) = m(V). \tag{3.5}$$

In general, the quantities in (3.5) are not equal to one another. In particular, *a nonzero multivarifold can have zero multimass.* If V is a positive multivarifold, then all its strata V^i, and also all the resulting measures rV, rV^i, are positive. In this case the inequalities (3.5) become equalities:

$$m(rV) = \sum M_i V = \sum m(V^i) = m(V). \tag{3.6}$$

We also observe that the i-dimensional stratum V^i of a multivarifold V has "homogeneous" multimass $(0, \dots, M_i V, \dots, 0)$, that is, $M_j V^i = 0$ if $j \neq i$ and $M_i V^i = M_i V$. If a positive multivarifold V has a "homogeneous" multimass of the form $(0, \dots, M_i V, \dots, 0)$, then V is a homogeneous multivarifold with one i-dimensional stratum.

Using the remarks made above, it is easy to see that the multivarifold $[S]$ defined in Example 2 has a homogeneous multimass with leading mass $\text{vol}_k S$. Moreover, for any i the i-mass $M_i[K]$ of the multivarifold $[K]$ considered in Example 3 is the i-dimensional volume of all i-dimensional simplexes of K that do not belong to faces of any $(i+1)$-dimensional simplex of K.

DEFINITION 6.3.5. The *support of a multivarifold* $V \in V_k M$ is the smallest closed set $K \subset M$ such that $V(f \circ p) = 0$ for any function $f \in C(M)$ whose support is contained in $M \backslash K$. *The support of the i-dimensional stratum* V^i *of a multivarifold* V is called its *i-support*. The support and i-support of a multivarifold V are denoted respectively by $\text{spt}\, V$ and $\text{spt}_i V$.

From the definition it follows immediately that the support $\text{spt}\, V$ and i-support $\text{spt}_i V$ of a multivarifold V coincide with the supports of the resulting measures rV and rV^i respectively.

DEFINITION 6.3.6. The absolute support of a multivarifold $V \in V_k M$ is the image of the support of a measure V under the projection p of the bundle $G_k M$. The absolute support of the i-dimensional stratum V^i of a multivarifold V is called its *absolute i-support.* The absolute support and i-support of V are denoted by $\text{spt}^* V$ and $\text{spt}_i^* V$.

We must carefully distinguish the support and absolute support of a multivarifold V from the support of a measure V, and also distinguish between them. The first two are contained in M and the third in $G_k M$. Moreover,

$$\text{spt}\, V \subset \bigcup_{i=0}^{k} \text{spt}_i V \subset \bigcup_{i=0}^{k} \text{spt}_i^* V = \text{spt}^* V. \tag{3.7}$$

The sets in (3.7) do not coincide. In particular, nonzero multivarifolds can have empty support and i-supports, $0 \le i \le k$. If V is a positive multivarifold, then the inclusions in (3.7) become equalities:

$$\operatorname{spt} V = \bigcup_{i=0}^{k} \operatorname{spt}_i V = \bigcup_{i=0}^{k} \operatorname{spt}_i^* V = \operatorname{spt}^* V. \tag{3.8}$$

It is easy to prove that in Example 1 we have $\operatorname{spt}\mu_x = x$ if $\mu(1) \neq 0$ and $\operatorname{spt}\mu_x = \varnothing$ if $\mu(1) = 0$, and $\operatorname{spt}^* \mu_x = x$ if $\mu \neq 0$ and $\operatorname{spt}^* \mu_x = \varnothing$ if $\mu = 0$; in Example 2, $\operatorname{spt}[S] = \operatorname{spt}_k[S] = \operatorname{spt}_k^*[S] = \operatorname{spt}^*[S] = S$; in Example 3, $\operatorname{spt}[K] = \operatorname{spt}^*[K] = K$; $\operatorname{spt}_i[K] = \operatorname{spt}_i^*[K]$ is the closure of the union of all i-dimensional simplexes of K that do not occur in a face of any $(i+1)$-dimensional simplex of K.

Using the property of extension of measures with compact support (see Đào Trọng Thi [42]), we can extend each multivarifold $V \in V_k M$ to a linear functional over the vector space of all bounded V-measurable functions of $G_k M$. Let φ be a V-measurable function on $G_k M$. We define the product $V \wedge \varphi$ by the formula $V \wedge \varphi(f) = V(\varphi f)$, where f is an arbitrary function in $C(G_k M)$ (see the definition of a product of a measure and a function in [25], [42]).

We observe that convergence of a bounded sequence of V-measurable functions φ_n implies the convergence of the corresponding sequence of multivarifolds $V \wedge \varphi_n$ in the compact-weak topology.

We shall often consider a function $\varphi\colon M \to \mathbf{R}$ as a function on $G_k M$ in a natural way: if $z \in p^{-1}(x)$, then $\tilde{\varphi}(z) = \varphi(x)$, that is, $\tilde{\varphi} = \varphi \circ p$, where p is the projection of the bundle $G_k M$. In what follows, we again denote $\tilde{\varphi}$ by φ if this does not lead to misunderstanding. In particular, if A is an rV-measurable subset of M, then $V \cap A$ denotes the multivarifold $V \wedge \varphi_A$, where φ_A is the characteristic function of the set $p^{-1}(A) \subset G_k M$. Obviously, $\operatorname{spt}^*(V \cap A) = \overline{\operatorname{spt}^* V \cap A}$; $\operatorname{spt}(V \cap A) = \operatorname{spt} V \cap A$ and $V^i \cap A$ is the i-dimensional stratum of the multivarifold $V \cap A$. We note that for each i we have

$$M_i(V \cap A) = [(rV^i)^+ + (rV^i)^-](A).$$

In particular, setting $A = M$ we obtain $M_i V = [(rV^i)^+ + (rV^i)^-](M)$.

3.4. Rectifiable multivarifolds. We define an important class of subsets of a Riemannian manifold.

DEFINITION 6.3.7. A subset $S \subset M$ is said to be *uniformly i-rectifiable* if the following conditions are satisfied:

(1) S is bounded and h^i-measurable and $h^i S < \infty$;
(2) $\Psi_{i+1}(z, S) = 0$, $\Psi_i(z, S) > 0$ for any $z \in S$;

(3) for each $\varepsilon > 0$ there is an i-dimensional submanifold $N \subset M$ of class C^1 such that $h^i([S\backslash N] \cup [N\backslash S]) < \varepsilon$.

We observe that if S, $P \subset M$ are uniformly i-rectifiable subsets, then $S \cup P$ and $S\backslash P$ are also uniformly i-rectifiable subsets, and $S \cap P$ differs from a uniformly i-rectifiable subset only by a set of zero measure h^i.

DEFINITION 6.3.8. A k-dimensional stratified surface $S \subset M$ is said to be *rectifiable* if every i-dimensional stratum S^i of S is a uniformly i-rectifiable subset.

Let S be a k-dimensional rectifiable stratified surface. Every i-dimensional stratum S^i has an i-dimensional tangent space $\vec{S}^i_z$ at z for almost all $z \in S^i$ in the sense of the measure h^i. We denote it by $\vec{S}_z$ $(= \vec{S}^i_z)$. Clearly, the map $\vec{S}^i : S^i \to G_k M$, which takes a point $z \in S^i$ into a point $\vec{S}^i_z \in G_k M$, is an h^i-measurable map.

REMARK. We understand by an *orientable k-dimensional rectifiable stratified surface in M* a k-dimensional rectifiable stratified surface $S = \bigcup_{i=0}^k S^i$ together with h^i-measurable maps $\vec{S}^{0i} : S^i \to G^0_k M$ such that $v \circ \vec{S}^{0i} = \vec{S}^i$, where $v : G^0_k M \to G_k M$ denotes a two-sheeted covering "by liquidation of orientation". In particular, if S has a unique nonempty i-dimensional stratum S^i, then S^i together with an h^i-measurable map $\vec{S}^{0i} : S^i \to \Gamma^0_i M$ is an oriented uniformly i-rectifiable subset. Let λ_k: $G^0_k M \to \Lambda_* M$ denote a standard embedding which maps each oriented i-dimensional subspace π into the i-vector associated with it. We write $-S = \lambda^{-1}(-\lambda \circ \vec{S})$. If S, $P \subset M$ are oriented uniformly i-rectifiable subsets, then we define their oriented intersection as follows:

$$S \vec{\cap} P = S \cap P \cap \{x : \vec{S}_x = \vec{P}_x\}.$$

Obviously, $S \vec{\cap} P$ differs from an oriented uniformly i-rectifiable subset only by a set of zero measure h^i.

Now let $S \subset M$ be a k-dimensional rectifiable stratified surface with strata S^i. We define the positive multivarifold $[S]$ of order k associated with S by

$$[S] = \sum_{i=0}^{k} (\vec{S}^i)_* (h^i \cap S^i). \tag{3.9}$$

Obviously, $M_i[S] = h^i S^i = \mathrm{vol}_i S$, $r[S^i] = h^i \cap S^i$, and $\mathrm{spt}[S]$ and $\mathrm{spt}_i[S]$ coincide with the closures of S and S^i respectively.

DEFINITION 6.3.9. A *multivarifold* $V \in V_k M$ is said to be *rectifiable* if it can be represented as a finite or convergent (in the compact-weak topology) sum of the multivarifolds $[S]$ associated with rectifiable stratified surfaces S whose union is bounded.

We denote the space of all rectifiable multivarifolds of order k on M by R_kM.

Suppose that $V \in R_kM$, $V = \sum_\alpha [S_\alpha]$, where S_α are stratified surfaces. We observe that $\operatorname{spt} V$ is the closure of the set $\bigcup_\alpha S_\alpha$. We decompose V into a sum of i-dimensional strata $V = V^0 + \cdots + V^k$. It is easy to verify that V^i can be represented in the form

$$V^i = \sum_{j=1}^{\infty} j[S_i^j], \tag{3.10}$$

where S_i^j is a uniformly i-rectifiable subset of M and $S_p^r \cap S_q^t = \varnothing$ if $p \neq q$ or $r \neq t$; $\bigcup_{i,j} S_i^j$ is bounded and $M_iV = \sum_{j=1}^{\infty} jh^i S_i^j < \infty$. We observe that $\operatorname{spt}_i V = \operatorname{spt} V^i$ coincides with the closure of the set $\bigcup_{i,j} S_i^j$, which is a uniformly i-rectifiable subset. We denote by ψ the function on M that is equal to j on S_i^j and to 0 on $M \setminus (\bigcup_{i,j} S_i^j)$. Then by (3.10) the function ψ is measurable with respect to the measure $\sum_{i,j}[S_i^j]$, and

$$V = \sum_{i,j} [S_i^j] \wedge \psi. \tag{3.11}$$

DEFINITION 6.3.10. A *multivarifold* $W \in V_kM$ is said to be *semirectifiable* if it can be represented in the form $W = V \wedge \varphi$, where V is a rectifiable multivarifold and φ is a V-measurable function on G_kM.

Obviously, semirectifiable multivarifolds of order k on M form a vector subspace of V_kM, which we shall denote by $\widetilde{R}_kM$. Clearly, $R_kM \subset \widetilde{R}_kM$.

Let us consider $V \wedge \varphi \in \widetilde{R}_kM$, $V \in R_kM$. Let φ_i be the restriction of the function φ to Γ_iM, and let $\sum_{i=0}^k V^i$ be a decomposition of V into a sum of strata. Then

$$V \wedge \varphi = \sum_{i=0}^{k} V^i \wedge \varphi_i, \tag{3.12}$$

where $V^i \wedge \varphi_i$ is the i-dimensional stratum of the multivarifold $V \wedge \varphi$.

Taking (3.11) into consideration, we see that we can take the function φ in the representation $V \wedge \varphi$ of the semirectifiable multivarifold to be a function on M, replacing it if necessary by a V-negligible subset.

3.5. Integrands.

DEFINITION 6.3.11. Let l be an arbitrary continuous function on $G_k M$. For any multivarifold V of order k on M we set

$$J_i(V) = V^i(l), \quad 0 \leq i \leq k. \tag{3.13}$$

The functional J_i defined by (3.13) (resp. the collection of functionals $(J_0, \dots, J_k)$) is called an *i-dimensional integrand* (resp. *stratified integrand*) on M specified by the Lagrangian l.

We observe that *a stratified integrand J and its Lagrangian l are uniquely determined by each other.* On the other hand, an i-dimensional integrand J_i depends only on the restriction of l to $\Gamma_i M$, so each continuous function on $\Gamma_i M$ specifies an i-dimensional integrand on M. Moreover, two distinct such functions specify distinct integrands. Thus, there are exactly as many stratified integrands (resp. i-dimensional integrands) on M as there are distinct continuous functions on $G_k M$ (resp. on $\Gamma_i M$).

If V is associated with a k-dimensional stratified surface S, then according to (3.9) $[S^i](l) = h^i \cap S^i(l \circ \vec{S})$. In this case the right-hand side of (3.13) takes the form of an integral

$$J_i[S] = \int l(\vec{S}_x)\, dr[S^i](x), \tag{3.14}$$

such that the expression under the integral sign is the Lagrangian l. In particular, for a submanifold $S \subset M$, (3.14) has the form

$$J_i[S] = \int_S l(\vec{S}_x)\, d\sigma, \tag{3.15}$$

where $d\sigma$ denotes the standard Riemannian volume form on S. Formula (3.15) shows that the i-dimensional integrand with Lagrangian l reproduces, in the language of multivarifolds, the functional of i-dimensional volume type specified by the same Lagrangian.

DEFINITION 6.3.12. A *generalized i-dimensional integrand* (resp. *stratified integrand*) on M is the upper envelope of any family of i-dimensional integrands (resp. stratified integrands) on M.

Obviously, generalized stratified integrands represent all possible collections of i-dimensional generalized integrands $(\widetilde{J}_0, \widetilde{J}_1, \dots, \widetilde{J}_k)$. Moreover, each i-dimensional generalized integrand can be given by

$$\widetilde{J}_i(V) = \sup_{J_i \in \{J_i\}} J_i(V) = \sup_{l \in \{l\}} V^i(l)$$

for an arbitrary multivarifold V of $V_k M$, where $\{J_i\}$ denotes a family of i-dimensional integrands, and $\{l\}$ is the corresponding family of Lagrangians.

The concept of a generalized i-dimensional integrand (stratified integrand) substantially enriches the stock of minimizable functionals. Thus, for example, we now have at our disposal such important functionals as the leading mass $M_k V$ and the absolute leading mass $M_k(|V|)$, which can be defined by the formulae

$$M_k V = \sup_{f \in C(M),\, \|f\| \leq 1} V^k(f \circ p),$$

$$M_k(|V|) = \sup_{f \in C(G_k M),\, \|f\| \leq 1} V^k(f),$$

where p, as usual, denotes the projection of the bundle $G_k M$. We note also that the multimass $(M_0, \ldots, M_k)$ is a generalized stratified integrand, which reproduces, in the functional language of multivarifolds, the Fomenko stratified volume $(\mathrm{vol}_0, \ldots, \mathrm{vol}_k)$.

§4. Statement of Problems B, B$'$, and B$''$ in the language of the theory of multivarifolds

4.1. Reformulation of Problems B, B$'$, and B$''$ in the language of multivarifolds. Let us first recall some definitions. Let X and Y be two metric spaces.

DEFINITION 6.4.1. A continuous map f of X into Y is said to be *locally Lipschitz* if for each point $x \in X$ there is a neighborhood $U_x \subset X$ and a number $N_x > 0$ such that for any $y, z \in U_x$ we have

$$\rho(f(y), f(z)) \leq N_x \rho(y, z), \tag{4.1}$$

where ρ denotes a distance on X and Y. We observe that if f satisfies (4.1) for some number N_x, then a fortiori it satisfies this inequality for any number $N'_x > N_x$. We denote the lower bound of all numbers N_x for which (4.1) is satisfied by $\inf N_x$. Then we call the least upper bound $\sup_{x \in X} \inf N_x$ (finite or infinite) the *Lipschitz coefficient of* f and denote it by $\mathrm{Lip}\, f$. It is natural to assume that a continuous map that is not locally Lipschitz is a map with an infinite Lipschitz coefficient.

If (4.1) is satisfied for any y and z from a subset $A \subset X$, then f is called a *Lipschitz map on* A. In particular, if $A = X$, we simply say that f is a *Lipschitz map.*

Now let $X = M$ be a Riemannian manifold. Then for each point $x \in M$ we can choose a convex open ball with center x such that for any y and z from this ball and any locally Lipschitz map f of M into Y we have

$$\rho(f(y), f(z)) \leq \mathrm{Lip}\, f \cdot \rho(y, z). \tag{4.2}$$

In particular, since G_kM is a disjoint union of Riemannian manifolds, for each point $x \in G_kM$ there is a geodesically convex open ball with center x such that (4.2) holds for any y, z from this ball and any locally Lipschitz function f on G_kM.

Henceforth for convenience we shall by convention say that *locally Lipschitz maps are maps of class* C^0.

Let $f\colon W \to M$ be a continuous differentiable map of a k-dimensional compact manifold W into a Riemannian manifold M. Consider the function $\tilde{\varphi}_f$ specified on W by the formula

$$\tilde{\varphi}_f(x) = \overline{\lim_{\varepsilon \to +0}} \alpha_l^{-1} \varepsilon^{-l} h^k(f_x^{-1}[B(f(x), \varepsilon)]),$$

where $B(f(x), \varepsilon)$ is the ball in M with center $f(x)$ and radius ε, $f_x^{-1}[B(f(x), \varepsilon)]$ is the connected component of its inverse image under f that contains the point x, and $l = \operatorname{rank} df_x$. We set $\varphi_f = 1/(\tilde{\varphi}_f \circ p)$, where $p\colon G_kW \to W$ is the projection of the bundle. It is easy to verify that φ_f is an $r[W]$-measurable function. The map f induces a differential map $df\colon TW \to TM$, which in turn induces a map $G_f\colon G_kW \to G_kM$ that switches every subspace $\pi \in G_kW$ and its image $df(\pi)$ under the map df. We now consider the induced multivarifold $W_f = G_{f*}([W] \wedge \varphi_f)$, that is, $W_f(g) = [W](\varphi_f g \circ G_f)$ for any function g from $C(G_kM)$. In Chapter 7, §3, we show that the image $S = f(W)$ is a stratified Fomenko surface, and its k-dimensional stratum S^k is connected with the k-dimensional stratum W_f^k of the multivarifold W_f by the following relation:

$$W_f^k = [S^k] \wedge \sigma'(x), \tag{4.3}$$

where $\sigma(x)$ is the number of connected components in $f^{-1}(x) \cap \{y \in W : \operatorname{rank} df_y = k\}$. Next, let J be the functional of k-dimensional volume type specified by the Lagrangian l. Since it is positively homogeneous (of the first degree), l induces a continuous function $\tilde{l}$ on G_kM, which in turn specifies a k-dimensional integrand I on M: $I(V) = V^k(\tilde{l})$, $V \in V_kM$. Clearly, from (2.1), (3.15), and (4.3) it follows that

$$J(f) = W_f^k(\tilde{l}) = I(W_f). \tag{4.4}$$

In particular, if $l(x, u) = |u|$ (that is, $\tilde{l} \equiv 1$), then

$$\operatorname{vol}_k(f) = W_f^k(1) = M_k W_f. \tag{4.5}$$

The construction of the induced multivarifold W_f described above extends naturally to the case when f is an arbitrary locally Lipschitz map; the equalities (4.4) and (4.5) still hold.

DEFINITION 6.4.2. A *locally Lipschitz parametrization on* M is any pair (W, f), where W is a semirectifiable multivarifold on some Riemannian manifold N, and f is a locally Lipschitz map of N into M.

We can now restate Problem B as follows.

PROBLEM B_*. Let M be a Riemannian manifold, G a fixed compact $(k-1)$-dimensional submanifold of M, and W a fixed k-dimensional compact manifold with boundary $\partial W = G$. Among all locally Lipschitz parametrizations $([W], f)$, where f runs through the set of maps of W into M that are the identity on G or a given homotopy class of such maps, find a parametrization $([W], f_0)$ for which the induced multivarifold W_{f_0} minimizes the given k-dimensional integrand I.

We observe that if in this statement of Problem B_* we replace the requirement that the maps f are the identity on G by the requirement that the maps f take G into some given subset of M, then we obtain a restatement of the problem "with moving boundary" B$'$ in the language of multivarifolds. Moreover, Problem B$''$ can be rewritten in the following form.

PROBLEM B''_*. Let W be a closed compact k-dimensional manifold. Among all locally Lipschitz parametrizations $([W], f)$, where f runs through a given homotopy class of locally Lipschitz maps of W into M, find a parametrization $([W], f_0)$ for which the induced multivarifold W_{f_0} minimizes the given k-dimensional integrand I.

4.2. Formulation of the main theorems. Although the language of multivarifolds is absolutely necessary for the proof of the existence of minimal surfaces and the isoperimetric inequality, the main results in a simplified version admit a simple statement in intuitive geometrical language.

THEOREM 6.4.1 (a theorem on the existence of minimal surfaces of a fixed topological type, Đào Trọng Thi [36]). *Let M be a compact connected Riemannian manifold, and W a fixed k-dimensional compact manifold with boundary ∂W. We denote by $C^r = C^r(W, M)$ the set of all maps $f: W \to M$ of class C^r, $0 \le r \le \infty$, such that the restriction of f to ∂W coincides with a given map $g: \partial W \to M$.*

(a) *The k-dimensional volume functional* vol_k *attains a minimum on C^0, and also on every class of homotopic maps in C^0.*

(b) *Let $\eta = \{\eta_i, 0 \le i \le r\}$ be a set of positive numbers. $C^r(\eta)$ consists of those maps f in C^r for which* $\mathrm{Lip}\, J^i f \le \eta_i$, *where $J^i f$ denotes the i-jet extension of f. Then any functional of k-dimensional volume type attains a minimum on $C^r(\eta)$, and also on every class of homotopic maps in $C^r(\eta)$.*

REMARK. If M is compact, then Theorem 6.4.1 remains true when $\partial W = \varnothing$ (that is, W is a compact closed manifold). Moreover, if in the definition of the set C^r in Theorem 6.4.1 we replace the condition $f|_{\partial W} = g$ (where g is a fixed map of ∂W into M) by the condition $f(\partial W) \subset K$, where K is a given compactum in M, then this theorem again remains true. Thus we have proved an existence theorem for problems of minimizing a functional of k-dimensional volume type in various classes of surfaces of fixed topological type with a "moving boundary".

THEOREM 6.4.2 (the isoperimetric inequality for surfaces of fixed topological type, Đào Trọng Thi [37]). *Let W be a k-dimensional compact manifold with boundary ∂W, and $g\colon \partial W \to \mathbf{R}^n$ a fixed map of class C^r, $0 \le r \le \infty$. Then there is a map $f\colon W \to \mathbf{R}^n$ of class C^r such that $f|_{\partial W} = g$ and*

$$(\mathrm{vol}_k f)^{(k-1)/k} \le 2\sqrt{n}(n!/(k-1)!)^{k-1} C^n_{k-1} \,\mathrm{vol}_{k-1}(g).$$

A full proof of the main Theorems 6.4.1 and 6.4.2 will be carried out in Chapter 9 in the language of the theory of multivarifolds as a very special case of a general variational problem in classes of parametrizations and parametrized multivarifolds.

CHAPTER VII

The Space of Multivarifolds

§1. The topology of the space of multivarifolds

1.1. Weak topologies. The results of Chapters 7–10 are due to Đào Trọng Thi. The space $V_k M$ of multivarifolds on a Riemannian manifold M, like the measure space with compact support on $G_k M$, can be endowed with the weak and strong topologies considered in [42]. Apart from this, by using the Riemannian metric of the space $G_k M$ we introduce for the space $V_k M$ another, the so-called N-topology.

In the space $V_k M$ we can introduce the norm $N(V) = \sup\{V(f) : f \in C(G_k M), \|f\| \leq 1, \operatorname{Lip} f \leq 1\}$. This norm induces in $V_k M$ the so-called N-metric and n-topology.

THEOREM 7.1.1. *The compact-weak topology and the N-topology induce equivalent topologies on the subspace*

$$S = \{V \in V_k M : m(|V|) \leq \eta, \operatorname{spt}^* V \subset K\},$$

where η is an arbitrary positive number, and $K \subset M$ is an arbitrary compactum.

LEMMA 7.1.1. *The subset $\{f \in C(G_k M, \widetilde{K}) : \|f\| \leq 1, \operatorname{Lip} f \leq 1\}$, where $\widetilde{K}$ is a compact subset of $G_k M$, is compact.*

PROOF. We need to show that from any infinite sequence f_n of this subset we can pick out a convergent subsequence. Let us choose a sequence ε_n converging to zero: $\varepsilon_1 > \varepsilon_2 > \varepsilon_3 > \cdots > 0$. Since $\widetilde{K}$ is compact, for each ε_n we can cover it by a finite system of geodesic convex open balls $U(x_j^n)$ with centers at points x_j^n, $1 \leq j \leq k_n$, and radii $< \varepsilon_n$. In this case all the functions f_n are Lipschitz with respect to these balls and the constant 1. From the sequence f_n we pick out a subsequence f_n^1 convergent at points x_j^1 (this is possible in view of the condition $\|f_n\| \leq 1$ for all n); we may assume that

$$|f_p^1(x_j^1) - f_q^1(x_j^1)| < \varepsilon_1 \qquad (1.1)_1$$

for all j, p, q. From the sequence f_n^1 we pick out a subsequence f_n^2 convergent at points x_j^2 and such that

$$|f_p^2(x_j^2) - f_q^2(x_j^2)| < \varepsilon_2 \tag{1.1$_2$}$$

for all j, p, q. Continuing this process, we pick out a diagonal sequence $g_n = f_n^n$ which, according to the inequalities $(1.1)_1, (1.1)_2, \dots$, satisfies the relations

$$|g_p(x_j^n) - g_q(x_j^n)| < \varepsilon_n \tag{1.2}$$

for any p, q greater than n.

Now let ε be an arbitrary positive number. We choose n so that $\varepsilon_n < \varepsilon/3$. For any point $x \in \widetilde{K}$ and any $p, q \geq n$, by (1.2) we have $|g_p(x) - g_q(x)| \leq |g_p(x) - g_p(x_j^n)| + |g_p(x_j^n) - g_q(x_j^n)| + |g_q(x_j^n) - g_q(x)| < 3\varepsilon_n < \varepsilon$, where x_j^n is the center of the ball $U(x_j^n)$ containing x. Consequently,

$$\|g_p - g_q\|_{\widetilde{K}} < \varepsilon. \tag{1.3}$$

Since the space $C(G_kM, \widetilde{K})$ is complete, (1.3) implies that the sequence g_n converges to some function $g \in C(G_kM, \widetilde{K})$ which, by the principle of extending inequalities, satisfies the condition $\|g\| \leq 1$, $\operatorname{Lip} g \leq 1$. This proves the lemma.

Proof of Theorem 7.1.1. It is sufficient to show that for any set $G \subset V_kM$ from a fundamental system of neighborhoods of zero in the compact-weak topology there is a set H from a fundamental system of neighborhoods of zero in the N-topology such that $H \cap S \subset G$, and conversely. Let $G = \{V \in V_kM : |V(f_i)| < \varepsilon,\ f_i \in C(M),\ i = 1, \dots, r\}$ be a neighborhood of zero in the weak topology. For the functions f_i we can choose bounded locally Lipschitz functions g_i such that $\|g_i - f_i\|_{\widetilde{K}} < \varepsilon/2\eta$, where $\widetilde{K} = p^{-1}(K)$ is a compact subset of G_kM. We set $s = \max\{\|g_i\|, \operatorname{Lip} g_i\}$. We prove that $S \cap \{V \in V_kM : N(V) < \varepsilon/2s\} \subset G$. In fact, if $V \in S \cap \{V \in V_kM : N(V) < \varepsilon/2s\}$, then $m(|V|) \leq \eta$, $\operatorname{spt}^* V \subset K$ and the inequality $|V(g)| < \varepsilon/2s$ is satisfied for any function g such that $\|g\| \leq 1$ and $\operatorname{Lip} g \leq 1$. In particular, $|V(g_i/s)| < \varepsilon/2s$, or $|V(g_i)| < \varepsilon/2$. Consequently, $|V(f_i)| \leq |V(g_i)| + |V(f_i - g_i)| < \varepsilon/2 + m(|V|)\|f_i - g_i\|_{\widetilde{K}} < \varepsilon$.

Conversely, let $G = \{V \in V_kM : N(V) < \varepsilon\}$ be a neighborhood of zero in the N-topology. Since the subset $\{f \in C(G_kM, \widetilde{K}) : \|f\| \leq 1, \operatorname{Lip} f \leq 1\}$ is compact by Lemma 7.1.1, we can cover it by finitely many open balls with radii $< \varepsilon/2\eta$. We denote the centers of these balls by $g_1, \dots, g_r$. We show that $S \cap \{V \in V_kM : |V(g_i)| < \varepsilon/2\} \subset G$. In fact, for any function $f \in C(M)$, $\|f\| \leq 1$, $\operatorname{Lip} f \leq 1$, we can find an i such that

$\|f - g_i\|_{\widetilde{K}} < \varepsilon/2\eta$. Then if $V \in S \cap \{V \in V_k M : V(g_i) < \varepsilon/2\}$, we have $|V(f)| \leq |V(g_i)| + |V(f - g_i)| < \varepsilon/2 + m(|V|)\|f - g_i\|_{\widetilde{K}} = \varepsilon$, so that $V \in G$. This proves the theorem.

THEOREM 7.1.2. *In the space $V_k M$ of multivarifolds on M the mass functions $M_i V$, $0 \leq i \leq k$, are lower semicontinuous in the compact-weak topology. Moreover, the restrictions of these functions to the set $V_k^+ M$ are continuous.*

PROOF. For an arbitrary continuous function $f \in C(M)$ we have $rV^i(f) = V^i(f \circ p) = V(\varphi_i f \circ p)$, where V^i is the i-dimensional stratum of the multivarifold V, φ_i is a continuous function on $G_k M$, equal to 1 on $\Gamma_i M$ and to 0 on $G_k M \setminus \Gamma_i M$, and p is the projection of the bundle $G_k M$. Consequently, we have

$$M_i V = m(rV^i) = \sup\{V(\varphi_i f \circ p) : f \in C(M), \|f\| \leq 1\},$$

that is, the function $M_i V$ is the upper envelope of functions $V \to V(\varphi_i f \circ p)$, continuous on $V_k M$ with the compact-weak topology, when f runs through the set of functions in $C(M)$ such that $\|f\| \leq 1$. Thus, the function $V \to M_i V$ is lower semicontinuous on $V_k M$ (see [25]). But if $V \in V_k^+ M$, then $M_i V = V(\varphi_i)$, and so the function $V \to M_i V$ is continuous on the set $V_k^+ M$ in the compact-weak topology. This proves the theorem.

THEOREM 7.1.3. *Let $a_0, a_1, \dots, a_k$ be arbitrary nonnegative numbers, and $K \subset M$ an arbitrary compact subset. Then the set*

$$V_k M \cap \{V : M_i V \leq a_i, 0 \leq i \leq k, \operatorname{spt}^* V \subset K\}$$

is compact in the compact-weak topology.

PROOF. The fact that this set is closed follows directly from Theorem 7.1.2 and from [42]. Let us prove that this set is relatively compact. For this it is sufficient to prove that it is compactly bounded, that is, for any function $f \in C(G_k M)$ we have $\sup|V(f)| < \infty$, where V runs through the whole set $V_k M \cap \{V : M_i V \leq a_i, \operatorname{spt}^* V \subset K\}$. We observe that $M_i(|V|) = M_i(|V|^i)$. On the other hand, since $|V|^i$ is a positive multivarifold, we have $M_i(|V|^i) = m(|V|^i)$, so that $m(|V|^i) \leq a_i$ and therefore $m(|V|) \leq a_0 + a_1 + \cdots + a_k$, for any V from the set in question. Now let f be an arbitrary function in $C(G_k M)$. Then $\sup|V(f)| \leq m(|V|)\|f\|_{\widetilde{K}} \leq (a_0 + \cdots + a_k)\|f\|_{\widetilde{K}}$, where $\widetilde{K} = p^{-1}(K)$, p is the projection of the bundle $G_k M$, and V runs through the whole set $V_k M \cap \{V : M_i(|V|) \leq a_i, \operatorname{spt}^* V \subset K\}$. This proves the theorem.

Since V_k^+M is a closed set in V_kM with respect to the compact-weak topology (see [42]), from Theorem 7.1.3 we have the following result.

COROLLARY 7.1.1. *For arbitrary numbers $a_i \geq 0$, $0 \leq i \leq k$, and an arbitrary compact subset $K \subset M$ the set*

$$V_k^+M \cap \{V : M_iV \leq a_i, \operatorname{spt} V \subset K\}$$

is compact in the compact-weak topology.

THEOREM 7.1.4. (a) *Let $A \subset M$ be an open subset. Then the function $V \to M_i(V \cap A)$ on the space V_kM is lower semicontinuous in the compact-weak topology.*

(b) *Let $K \subset M$ be a closed subset. Then the function $V \to M_i(V \cap K)$ on the set V_k^+M is upper semicontinuous in the compact-weak topology.*

PROOF. (a) From the definition it is easy to see that

$$[(rV^i)^+ + (rV^i)^-](A) = \sup rV^i(f), \quad \text{where} \quad f \in C(M), \quad |f| \leq \varphi_A.$$

Hence we have

$$M_i(V \cap A) = \sup_{f \in C(M),\, |f| \leq \varphi_A} rV^i(f) = \sup_{f \in C(M),\, |f| \leq \varphi_A} V(\varphi_i f \circ p), \tag{1.4}$$

where p is the projection of the bundle G_kM, and φ_i is a continuous function defined on G_kM, equal to 1 on Γ_iM and to 0 on $G_kM \setminus \Gamma_iM$. Formula (1.4) implies that the function $M_i(V \cap A)$ is the upper envelope of functions $V \to V(\varphi_i f \circ p)$ on V_kM, continuous in the compact-weak topology, when f runs through the set of functions on $C(M)$ such that $|f| \leq \varphi_A$. Thus the function $M_i(V \cap A)$ is lower semicontinuous on V_kM (see [25]).

(b) Let K be a closed subset of M, and V a positive multivarifold. According to the definition we have

$$|rV^i|(K) = \inf_{f \in C(M),\, f \geq \varphi_K} |rV^i(f)|,$$

so that

$$\begin{aligned} M_i(V \cap K) &= \inf_{f \in C(M),\, f \geq \varphi_K} rV^i(f) = \inf_{f \in C(M),\, f \geq \varphi_K} V^i(f \circ p) \\ &= \inf_{f \in C(M),\, f \geq \varphi_K} V(\varphi_i f \circ p), \end{aligned} \tag{1.5}$$

where p and φ_i are defined as above. Formula (1.5) shows that the function $V \to M_i(V \cap K)$ is the lower envelope of functions $V \to V(\varphi_i f \circ p)$ on V_k^+M, continuous in the compact-weak topology, when f runs through the set of functions in $C(M)$ such that $f \geq \varphi_K$. Consequently, the function $M_i(V \cap K)$ is upper semicontinuous on V_k^+M (see [25]). This proves the theorem.

1.2. Product of multivarifolds. Let M and N be arbitrary Riemannian manifolds. We denote the projections of the product manifold $M \times N$ on M and N respectively by σ_1 and σ_2, and the projections of the product space $G_rM \times G_sN$ on its factors by q_1 and q_2 respectively. Next, let p, p_1, and p_2 denote the projections of the bundles $G_{r+s}(M \times N)$, G_rM, and G_sN respectively. We define a map i of the product $G_rM \times G_sN$ into the space $G_{r+s}(M \times N)$ as follows. If $\pi_1 \in G_rM$ is a k_1-dimensional subspace of T_xM and $\pi_2 \in G_sN$ is a k_2-dimensional subspace of T_yN, then i maps the pair (π_1, π_2) into the direct product of the subspaces π_1 and π_2, that is, into the $(k_1 + k_2)$-dimensional subspace $\pi_1 \oplus \pi_2$ of $T_{(x,y)}(M \times N)$. Obviously, i is an embedding of the space $G_rM \times G_sN$ into the space $G_{r+s}(M \times N)$. Henceforth, we shall identify $G_rM \times G_sN$ with its image under the embedding i. Direct calculation shows that the diagram

$$\begin{array}{ccccc}
G_rM & \xleftarrow{q_1} & G_rM \times G_sN & \xrightarrow{q_2} & G_sN \\
 & & \downarrow{\scriptstyle i} & & \\
\downarrow{\scriptstyle p_1} & & G_{r+s}(M \times N) & & \downarrow{\scriptstyle p_2} \\
 & & \downarrow{\scriptstyle p} & & \\
M & \xleftarrow{\sigma_1} & M \times N & \xrightarrow{\sigma_2} & N
\end{array}$$

is commutative, that is, $p_1 q_1 = \sigma_1 p i$, $p_2 q_2 = \sigma_2 p i$.

DEFINITION 7.1.1. Suppose that $V \in V_rM$, $W \in V_sN$. Then the image under the map i_* of the product-measure $V \otimes W$ is called the *product of the multivarifolds* V and W and denoted by $V \times W$.

Clearly, $V \times W$ is a multivarifold of order $r + s$ on $M \times N$. According to [42] the map i_* is a continuous homomorphism of the space $M(G_rM \times G_sN)$ into the space $V_{r+s}(M \times N)$. In addition, since i is an embedding, i_* is a continuous monomorphism. Then from [42] we have the following result.

COROLLARY 7.1.2. *The map of the space V_rM into the space $V_{r+s}(M \times N)$ that takes any multivarifold $V \in V_rM$ into the multivarifold $V \times W_0$, where W_0 is a fixed multivarifold in V_sN, is a continuous monomorphism in the compact-weak topology.*

COROLLARY 7.1.3. *The map of the product $V_rM \times V_sN$ into the space $V_{r+s}(M \times N)$ that takes the pair of multivarifolds (V, W) into the multivarifold $V \times W$ is a nondegenerate bilinear form, and its restriction to the* *product* $A \times B$ *of compactly bounded subsets* A *and* B *is continuous if* V_rM, V_sN, $V_{r+s}(M \times N)$ *are endowed with the compact-weak topology.*

THEOREM 7.1.5. *For any* $V \in V_rM$ *and any* $W \in V_sN$ *we have* $r(V \times W) = rV \otimes rW$.

PROOF. According to [42], it is sufficient to show that $r(V \times W)(f) = (rV \otimes rW)(f)$ for any function f of the form $f = \sum_j g_j \circ \sigma_1 h_j \circ \sigma_2$, $g_j \in C(M)$, $h_j \in C(N)$. By definition $r(V \times W) = r(i_*(V \otimes W)) = (pi)_*(V \otimes W)$. Consequently,

$$\begin{aligned} r(V \times W)(f) &= V \otimes W(f \circ p \circ i) \\ &= \sum_j V \otimes W(g_j \circ \sigma_1 \circ p \circ ih_j \circ \sigma_2 \circ p \circ i). \quad (1.7) \end{aligned}$$

Taking into consideration the fact that the diagram (1.6) is commutative and equality (1.7), we obtain $r(V \times W)(f) = \sum_j p_{1*}V(g_j)p_{2*}W(h_j) = \sum_j r(V)(g_j)r(W)(h_j) = r(V) \otimes r(W)(\sum_j g_j \circ \sigma_1 h_j \circ \sigma_2) = r(V) \otimes r(W)(f)$.

THEOREM 7.1.6. *Suppose that* $V \in V_rM$, $W \in V_sN$. *Then*

$$M_{r+s}(V \times W) = M_rV \cdot M_sW, \qquad (1.8)$$

$$M_0(V \times W) = M_0V \cdot M_0W, \qquad (1.9)$$

$$M_k(V \times W) \le \sum_{i+j=k} M_iV \cdot M_jW, \qquad 0 < k < r+s. \qquad (1.10)$$

If V *and* W *are positive multivarifolds, then* (1.10) *becomes an equality.*

PROOF. Let V^i be the i-dimensional stratum of the multivarifolds V, and W^j the j-dimensional stratum of the multivarifolds W. Since the product is bilinear (see Corollary 7.1.3), we have

$$V \times W = \left(\sum V^i\right)\left(\sum W^j\right) = \sum_{0 \le k \le r+s} \sum_{i+j=k} V^i \times W^j. \qquad (1.11)$$

Since the support of the measure V^i is contained in Γ_iM, and the support of the measure W^j is contained in Γ_jN, the support of the measure $V^i \otimes W^j$ is contained in $\Gamma_iM \times \Gamma_jN$. Consequently, the support of the measure $V^i \times W^j$ is contained in $\Gamma_{i+j}(M \times N)$. This together with (1.11) implies that the k-dimensional stratum of $V \times W$ is

$$(V \times W)^k = \sum_{i+j=k} V^i \times W^j. \qquad (1.12)$$

In particular,

$$(V \times W)^{r+s} = V^r \times W^s; \qquad (V \times W)^0 = V^0 \times W^0. \qquad (1.13)$$

From (1.12) we obtain

$$r(V \times W)^k = \sum_{i+j=k} rV^i \otimes rW^j \tag{1.14}$$

and according to the definition we also have

$$M_k(V \times W) \leq \sum_{i+j=k} m(rV^i \otimes rW^j) = \sum_{i+j=k} M_iV \cdot M_jW,$$

$$M_{r+s}(V \times W) = m(rV^r \otimes rW^s) = M_rV \cdot M_sW,$$

$$M_0(V \times W) = m(rV^0 \otimes rW^0) = M_0V \cdot M_0W.$$

If V and W are positive multivarifolds, then $V^i \times W^j$ and $rV^i \otimes rW^j$ are positive. Then the inequality in (1.14) becomes an equality. This proves the theorem.

THEOREM 7.1.7. *Suppose that* $V \in V_rM$, $W \in V_sN$. *Then*

(a) $\operatorname{spt}(V \times W) = \operatorname{spt} V \times \operatorname{spt} W$:

(b) $\operatorname{spt}_{r+s}(V \times W) = \operatorname{spt}_r V \times \operatorname{spt}_s W$, $\operatorname{spt}_0(V \times W) = \operatorname{spt}_0 V \times \operatorname{spt}_0 W$, $\operatorname{spt}_k(V \times W) \subset \bigcup_{i+j=k} \operatorname{spt}_i V \times \operatorname{spt}_j W$, $0 < k < r+s$;

(c) $\operatorname{spt}^*(V \times W) = \operatorname{spt}^* V \times \operatorname{spt}^* W$;

(d) $\operatorname{spt}^*_k(V \times W) = \bigcup_{i+j=k} \operatorname{spt}^*_i V \times \operatorname{spt}^*_j W$, $0 \leq k \leq r+s$;

(e) *if* V *and* W *are positive multivarifolds, then the third relation in* (b) *becomes an equality.*

PROOF. As in the proof of Theorem 7.1.6, we denote the i-dimensional stratum of V by V^i and the j-dimensional stratum of W by W^j.

Assertion (a) follows directly from Theorem 7.1.5.

Assertion (b) follows from (1.13).

Let us prove assertion (c). We denote the supports of the measures V, W, and $V \times W$ by K_1, K_2, and K respectively. By definition $\operatorname{spt}^* V = p_1(K_1)$, $\operatorname{spt}^* W = p_2(K_2)$, $\operatorname{spt}^*(V \times W) = p(K)$. On the other hand, according to the properties of the product-measures, the support of the measure $V \otimes W$ is $K_1 \times K_2$. Consequently, $K = i(K_1 \times K_2)$ and $p(K) = p \circ i(K_1 \times K_2)$. Using the fact that the diagram (1.6) commutes, we have $\sigma_1[p(K)] = \sigma_1 \circ p(K) = \sigma_1 \circ p \circ i(K_1 \times K_2) = p_1 \circ q_1(K_1 \times K_2) = p_1(K_1)$, $\sigma_2[p(K)] = p_2(K_2)$. This implies that $p(K) \subset p_1(K_1) \times p_2(K_2)$, $\operatorname{spt}^*(V \times W) \subset \operatorname{spt}^* V \times \operatorname{spt}^* W$. Conversely, suppose that $(x, y) \in \operatorname{spt}^* V \times \operatorname{spt}^* W$, that is, $x \in p_1(K_1)$ and $y \in p_2(K_2)$. Then there exist $\pi_x \in K_1$ and $\pi_y \in K_2$ such that $p_1(\pi_x) = x$, $p_2(\pi_y) = y$. The pair $(\pi_x, \pi_y) \in K_1 \times K_2$, and the direct product $\pi_x \oplus \pi_y$ belongs to K. Consequently, $(x, y) = p(\pi_x \oplus \pi_y) \in p(K) = \operatorname{spt}^*(V \times W)$. Thus, $\operatorname{spt}^* V \times \operatorname{spt}^* W \subset \operatorname{spt}^*(V \times W)$, and assertion (c) is proved.

Let us prove assertion (d). Let K denote the support of the measure $(V\times W)^k$, and K_{ij} the supports of the measure $V^i\times W^j$. We observe that the supports of the measures $V^i\times W^j$ are pairwise disjoint. In particular, the supports of all the summand-measures of the sum in (1.12) are pairwise disjoint. Therefore, $K=\bigcup_{i+j=k}K_{ij}$. Consequently,

$$p(K)=\bigcup_{i+j=k}p(K_{ij}). \tag{1.15}$$

On the other hand, $p(K)=\operatorname{spt}^*_k(V\times W)$ by definition, and $p(K_{ij})=\operatorname{spt}^*(V^i\times W^j)=\operatorname{spt}^*_i V\times\operatorname{spt}^*_j W$, according to (c). Thus, (1.15) has the form of an equality in assertion (d).

If V and W are positive multivarifolds, then the terms of the sum in (1.13) are positive measures. Then the support of the sum is the union of the supports of the summand-measures, which proves assertion (e). Theorem 7.1.7 is completely proved.

§2. Local characteristics of multivarifolds

2.1. The tangent distribution of multivarifolds.

THEOREM 7.2.1. *Let V be an arbitrary multivarifold in V_kM. Then for almost all $x\in M$ (in the sense of the measure $r|V|$) there is a limit*

$$V_x=\lim_{\varepsilon\to+0}V\cap B(x,\varepsilon)/r|V|(B(x,\varepsilon)),$$

where $B(x,\varepsilon)$ is the open ball with center x and radius ε, and the limit is taken in the compact-weak topology of the space V_kM. The multivarifolds V_x satisfy the following conditions:

(a) *V_x is a pointwise multivarifold at the point x;*

(b) *$m(V_x)\le 1$, and equality holds for almost all $x\in M$ in the sense of the measure $r|V|$;*

(c) *for any function f in $C(G_kM)$ the function $x\to V_x(f)$ is $r|V|$-measurable and*

$$V(f)=\int V_x(f)\,dr|V|(x); \tag{2.1}$$

(d) *if μ is a positive measure on M and for almost all $x\in M$ in the sense of the measure μ there are specified multivarifolds $W_x\in V_kM$ that satisfy the requirements obtained from* (a), (b), (c) *by replacing V_x by W_x and $r|V|$ by μ, we have $\mu=r|V|$, and $V_x=W_x$ almost everywhere in the sense of the measure $r|V|$.*

PROOF. We prove this theorem in several steps.

Step 1. By a theorem of Whitney we can embed M in a Euclidean space $\mathbf{R}^N$. Consequently, we can regard each measure with compact support on M as a measure with compact support on $\mathbf{R}^N$.

Let f be an arbitrary function in $C(G_kM)$. Consider the measure $r(V \wedge f) \in \boldsymbol{M}(M)$. We denote the support of the measure V by $K \subset G_kM$. For each set $A \subset M$ we have $|r(V \wedge f)(A)| \leq r|V|(A)\|f\|_K$, so that the measure $r(V \wedge f)$ is absolutely continuous with respect to the positive measure $r|V|$. By the Radon-Nikodým theorem (see [217], for example) the function

$$h_f(x) = \frac{dr(V \wedge f)}{dr|V|}(x) = \lim_{\varepsilon \to +0} \frac{r(V \wedge f)(B(x, \varepsilon))}{r|V|(B(x, \varepsilon))}, \tag{2.2}$$

where $B(x, \varepsilon) \subset M$ is the open ball with center x and radius ε, is an $r|V|$-measurable function on M, and

$$r(V \wedge f) = r|V| \wedge h_f. \tag{2.3}$$

Step 2. We observe that by Whitney's theorem every manifold $\Gamma_i M$ can be embedded in a Euclidean space. Consequently, G_kM can be regarded as a disjoint union of submanifolds in some Euclidean space $\mathbf{R}^{N'}$. By Weierstrass's theorem every continuous function on $\mathbf{R}^{N'}$ is the limit, uniform on any compact subset, of a sequence of polynomials with rational coefficients. Hence, the set of all polynomials bounded on G_kM with rational coefficients forms a countable everywhere dense subset of the space $C(G_kM)$, endowed with the compact-open topology. We denote this set by P. Obviously, P is a vector space over the field $\mathbf{Q}$ of rational numbers.

Step 3. We showed above that for every fixed function f in $C(G_kM)$ the function $h_f(x)$ is defined almost everywhere in the sense of the measure $r|V|$. Hence it follows that there is a subset $G \subset M$ of full measure with respect to $r|V|$ such that for any point $x \in G$ the values $h_g(x)$ are defined simultaneously for all g of the countable set P. In this case we set $V_x(g) = h_g(x)$. From (2.2) it is obvious that the map $g \to V_x(g)$ is linear on P, and

$$|V_x(g)| \leq \|g\|_K \tag{2.4}$$

for any function $g \in P$. Hence, V_x is a linear continuous (in the compact-open topology) functional over P. Since P is everywhere dense in $C(G_kM)$, we can uniquely extend V_x to a multivarifold of order k on M, which we again denote by V_x. We prove that the symmetric derivative of $h_f(x)$ exists and is equal to $V_x(f)$ for any function f in $C(G_kM)$. In fact, let $\tilde{\varepsilon} > 0$ be an arbitrary number. We can choose a function g in

P such that $\|f - g\|_K \leq \tilde{\varepsilon}/3$. Under this condition we have

$$|V_x(f) - V_x(g)| \leq \|f - g\|_K < \tilde{\varepsilon}/3, \tag{2.5}$$

and also

$$\left| \frac{r(V \wedge f)(B(x, \varepsilon))}{r|V|(B(x, \varepsilon))} - \frac{r(V \wedge g)(B(x, \varepsilon))}{r|V|(B(x, \varepsilon))} \right| < \frac{\tilde{\varepsilon}}{3}. \tag{2.6}$$

We now choose a number $\delta > 0$ such that for any $\varepsilon < \delta$ we have

$$\left| V_x(g) - \frac{r(V \wedge g)(B(x, \varepsilon))}{r|V|(B(x, \varepsilon))} \right| < \frac{\tilde{\varepsilon}}{3}. \tag{2.7}$$

In this case from (2.5), (2.6), (2.7) we obtain

$$\left| V_x(f) - \frac{r(V \wedge f)(B(x, \varepsilon))}{r|V|(B(x, \varepsilon))} \right| < \tilde{\varepsilon}.$$

Hence the equality

$$V_x(f) = \frac{dr(V \wedge f)}{dr|V|}(x) \tag{2.8}$$

is satisfied for any function f in $C(G_k M)$.

Step 4. We verify that for every $x \in G$ the functional V_x is a point multivarifold at the point x. For this it is sufficient to show that $V_x(f) = 0$ for any function $f \in C(G_k M)$ such that $x \notin \operatorname{spt} f$. In fact, in this case there is an open ball $B(x, \varepsilon_0)$ such that $f = 0$ in $B(x, \varepsilon_0)$. Then

$$V_x(f) = \lim_{\varepsilon \to +0} \frac{r(V \wedge f)(B(x, \varepsilon))}{r|V|(B(x, \varepsilon))} = 0.$$

Step 5. We show that $m(V_x) = 1$ almost everywhere in the sense of the measure $r|V|$. We first observe that from (2.4) it follows that $m(V_x) \leq 1$ for every point $x \in G$. Taking the value of the measures on both sides of (2.3) on the constant function $\varphi \equiv 1$ on M, we obtain the required equality $V(f) = \int V_x(f) dr|V|(x)$; also, if $\|f\|_K \leq 1$, then $V(f) \leq \int m(V_x) dr|V|(x) = m(V)$, and so $m(V_x) = 1$ almost everywhere in the sense of the measure $r|V|$.

Step 6. We now prove property (d). By the hypothesis of the theorem, W_x is a pointwise multivarifold at the point x, $m(W_x) \leq 1$, where equality holds almost everywhere in the sense of the measure μ, and for any function f in $C(G_k M)$ the function $x \to W_x(f)$ is measurable and

$$V(f) = \int W_x(f)\, d\mu. \tag{2.9}$$

First of all we observe that from (2.9) it follows that

$$|V| = \int |W_x|\, d\mu. \tag{2.10}$$

Then for any function $h \in C(M)$ we have

$$r|V|(h) = \int r|W_x|(h)\,d\mu = \int h\,d\mu = \mu(h),$$

so that $r|V| = \mu$. Next, from (2.2), (2.8), (2.9), and (2.10) it follows that

$$V_x(f) = \lim_{\varepsilon\to+0} \frac{\int W_x(f)d[\mu \cap B(x, \varepsilon)](x)}{\mu(B(x, \varepsilon))}. \tag{2.11}$$

Since the functions $x \to W_x(f)$ are μ-measurable, the limit in (2.11) is equal to $W_x(f)$ for almost all $x \in M$ in the sense of the measure $\mu = r|V|$. Hence, arguing as in Step 3, we can show that for almost all $x \in M$ in the sense of the measure $r|V|$ we have $V_x(f) = W_x(f)$ simultaneously for all functions f in P, and consequently simultaneously for all functions f in $C(G_kM)$, that is, $V_x = W_x$. This completes the proof of the theorem.

REMARK. Formula (2.1) holds for any function f in $C(G_kM)$. We can therefore rewrite it in the following form:

$$V = \int V_x\,dr|V|(x), \tag{2.12}$$

which we call the *pointwise representation of the multivarifold* V.

DEFINITION 7.2.1. The pointwise multivarifold V_x in the expansion (2.12) is called the *tangent distribution of the multivarifold* V *at the point* x.

Clearly, a tangent distribution of V exists for almost all points $x \in M$ in the sense of the measure $r|V|$. For example, for the multivarifold μ_x in Example 1 of Chapter 6, §3, a tangent distribution exists at the point x and coincides with μ_x; for the multivarifold $[S]$ in Example 2 of Chapter 6, §3, a tangent distribution exists at each point $x \in S$ and coincides with the multivarifold S_x, $S_x(f) = f(\vec{S}_x)$ for any $f \in C(G_kM)$; finally, for the multivarifold $[K]$ in Example 3 of the same section a tangent distribution exists at each point x of any i-dimensional simplex in K that does not occur in a face of any $(i+1)$-dimensional simplex in K, and it is none other than the multivarifold $\Delta_x^i(f) = f(\vec{\Delta}_x^i)$ for any $f \in C(G_kM)$. Thus the concept of a tangent distribution of a multivarifold, introduced by ĐàoTrọng Thi, is a natural generalization of the concept of the tangent space to a submanifold and a surface.

REMARK. In the theory of oriented multivarifolds the tangent distribution can be regarded as a measure on the fiber $G_k^0M_x$, embedded in the vector space Λ_kM_x. In this case we call the k-vector $\vec{V}_x = \int \pi\,dV_x \in \Lambda_kM_x$ the *resultant of the tangent distribution of the oriented multivarifold* V *at the point* x. For example, in the oriented version of Examples 2 and 3 of Chapter 6, §3, the resultants of the tangent distributions $[\vec{S}]_x$ and $[\vec{K}]_x$ are none other than the oriented tangent spaces $\vec{S}_x$ and $\vec{\Delta}^i_x$ to the submanifold S and the simplex Δ^i at the point x.

2.2. Density of a multivarifold. Consider a Riemannian manifold M, and let $V \in V_k M$ be a multivarifold of order k, and $z \in M$ an arbitrary point. As usual, we denote an open ball by $B(z, \varepsilon)$.

DEFINITION 7.2.2. The *upper* (resp. *lower*) *j-dimensional i-density of the multivarifold* V *at a point* z, $0 \le j \le \infty$, $0 \le i \le k$, is the upper (resp. lower) limit of the following quantity:

$$\alpha_j^{-1} \varepsilon^{-j} M_i(V \cap B(z, \varepsilon)) \tag{2.13}$$

as $\varepsilon \to 0$, which we denote by $\overline{\Theta}^j_i(V, z)$ (resp. $\underline{\Theta}^j_i(V, z)$).

The upper (resp. lower) densities $\overline{\Theta}^j_i$ (resp. $\underline{\Theta}^j_i$) form the matrix $(\overline{\Theta}^j_i(V, z))$ (resp. $(\underline{\Theta}^j_i(V, z))$), which we call the *upper* (resp. *lower*) *matrix of densities of* V *at* z. If $\overline{\Theta}^j_i = \underline{\Theta}^j_i$, then their common value is simply called the j*-dimensional i-density of* V *at* z and denoted by $\Theta^j_i(V, z)$. If all the densities $\Theta^j_i(V, z)$ exist, then the matrix $(\Theta^j_i(V, z))$ is called the *matrix of densities of* V *at* z.

From Definition 7.2.2 it is obvious that if $0 < \overline{\Theta}^j_i < \infty$ (resp. $0 < \underline{\Theta}^j_i < \infty$), then $\overline{\Theta}^s_i = 0$ (resp. $\underline{\Theta}^s_i = 0$) for any $s < j$, and $\overline{\Theta}^s_i = \infty$ (resp. $\underline{\Theta}^s_i = \infty$) for any $s > j$. Consequently, in each column of the upper (resp. lower) density matrix there is no more than one finite positive element.

Let V be a positive multivarifold. Then rV^i are positive measures, where V^i is the i-dimensional stratum of V. Comparing (2.13), (3.12), and (2.1) of Chapter 6, we observe that the upper and lower j-dimensional i-densities of V at the point z coincide respectively with the upper and lower j-dimensional rV^i-densities of M at z:

$$\overline{\Theta}^j_i(V, z) = \overline{\Theta}^j(rV^i, z), \tag{2.14}$$

$$\underline{\Theta}^j_i(V, z) = \underline{\Theta}^j(rV^i, z). \tag{2.15}$$

Suppose that $A \subset M$ is a Borel subset and that $\overline{\Theta}^j_i(V, z) \ge a > 0$ for any $z \in A$. Then, according to well-known results (see §§2.3 and 2.4 of Chapter 6) we have

$$r(V^i \cap A) = rV^i \cap A \ge a(h^i \cap A). \tag{2.16}$$

But if $\overline{\Theta}_i^j(V, z) \le b < \infty$ for any $z \in A$, then

$$r(V^i \cap A) \le 2^j b h^j \cap A. \tag{2.17}$$

From this inequality it follows that the measure $r(V^i \cap A)$ is absolutely continuous with respect to the measure $h^j \cap A$, so that by the Radon-Nikodým theorem we have

$$r(V^i \cap A) = (h^j \cap A) \wedge \frac{dr(V^i \cap A)}{d(h^j \cap A)}. \tag{2.18}$$

In particular, if A is a uniformly j-rectifiable subset, then

$$\Theta_i^j(V, z) = \frac{dr(V^i \cap A)}{d(h^j \cap A)}(z)$$

exists almost everywhere in the sense of the measure $h^j \cap A$, and the function Θ_i^j is $(h^j \cap A)$-measurable and (2.18) has the form

$$r(V^i \cap A) = (h^j \cap A) \wedge \Theta_i^j(V, *). \tag{2.19}$$

Let $S \subset M$ be a rectifiable stratified surface and let $S = \bigcup S^i$, where S^i is a uniformly i-rectifiable subset. By direct calculation we show that for every i, $0 \le i \le k$, $\Theta_i^j([S], z)$ exists and is equal to zero for h^i-almost all $z \notin S^i$ and they exist for h^i-almost all $z \in S^i$, where $\Theta_i^j([S], z) = 0$ when $j < i$, $\Theta_i^j([S], z) = 1$ when $j = i$, and $\Theta_i^j([S], z) = \infty$ when $j > i$. In the general case, if $V \in \widetilde{R}_k M$ is a semirectifiable multivarifold and $V = \sum [S^i] \wedge \varphi_i$, where S^i is a uniformly i-rectifiable subset, and φ_i is an $r[S^i]$-measurable function on M, then $\Theta_i^j(V, z)$ exists and is equal to zero for h^i-almost all $z \notin S^i$, and exists for h^i-almost all $z \in S^i$, where

$$\Theta_i^j(V, z) = \begin{cases} 0 & \text{for } j < i, \\ |\varphi_j(z)| & \text{for } j = i, \\ \infty & \text{for } j > i, \ \varphi_i(z) \ne 0, \\ 0 & \text{for } j > i, \ \varphi_i(z) = 0. \end{cases}$$

If in addition V is positive, then from the calculation carried out above it follows that V can be represented in the form

$$V = \sum_{i=0}^{k} [\widetilde{S}^i] \wedge \Theta_i^i(V, *),$$

where $\widetilde{S}^i$ is a uniformly i-rectifiable subset defined as follows: $\widetilde{S}^i = \{z \in M : \Theta_i^i(V, z) \ne 0\}$.

§3. Induced maps

3.1. Induced maps of type T_f. Let M and N be Riemannian manifolds, and $f: M \to N$ a continuously differentiable map. The map f induces a fiber map $G_f: G_k M \to G_k N$ as follows. Let $df_x: T_x M \to T_{f(x)} N$ be the usual differential of f at the point x. Then df_x induces a map of the fiber $G_k M_x$ over the point x of the bundle $G_k M$ into the fiber $G_k N_{f(x)}$ over the point $f(x)$ of the bundle $G_k N$ that takes an i-dimensional subspace π of $T_x M$ into its image $df_x(\pi) \subset T_{f(x)} N$ under df_x. Moreover, for any $\pi \in G_k M$ we set $G_f(\pi) = df_{p(\pi)}(\pi)$, where p denotes the projection of the bundle $G_k M$. In view of the assumption made above, the map df is continuous on TM. Consequently, G_f is a Baire map on $G_k M$.

Suppose that $\operatorname{rank} df_x|_\pi = i$. We choose an orthonormal basis $e_1, \dots, e_k$ in π such that $e_{i+1}, \dots, e_k$ form a basis of $\operatorname{Ker} df_x|_\pi$. It is obvious that $df_x(e_1), \dots, df_x(e_i)$ form a basis of $df_x(\pi)$. We set

$$\tau_f(\pi) = |df_x(e_1) \wedge \cdots \wedge df_x(e_i)|. \tag{3.1}$$

It is easy to verify that the number $\tau_f(\pi)$ defined by (3.1) does not depend on the choice of orthonormal basis $e_1, \dots, e_k$ and is equal to the volume of the parallelepiped spanned by $df_x(e_1), \dots, df_x(e_i)$. It is also easy to verify that the function τ_f is a Baire function restricted to each compact subset of $G_k M$.

We now define the induced map $T_f: V_k M \to V_k N$ as follows. If $V \in V_k M$, we set

$$T_f V = G_{f_*}(V \wedge \tau_f) \in V_k N, \tag{3.2}$$

where G_{f_*} denotes, as usual, the map of measure space $\boldsymbol{M}(G_k M)$ into the measure space $\boldsymbol{M}(G_k N)$ induced by the map G_f. In particular, if $A \subset G_k N$ is a Borel subset, then

$$G_{f_*}(V \wedge \tau_f)(A) = V \wedge \tau_f(G_f^{-1} A). \tag{3.3}$$

From (3.2) it is obvious that the map T_f is linear.

REMARK. The "nonhomogeneous" induced map T_f introduced by Đào Trọng Thi differs in principle from the classical "homogeneous" induced maps used in the "homogeneous" theories of currents and varifolds (see de Rham [90], Federer [216], Fleming [217], and Almgren [147]). It is obvious from what follows that it preserves the homotopy properties of multivarifolds under local degeneracy (lowering the dimension) of the map f. If we replace τ_f by τ'_f such that

$$\tau'_f(\pi) = \begin{cases} \tau_f(\pi) & \text{if } \operatorname{rank} df_x|_\pi = \dim \pi, \\ 0 & \text{if } \operatorname{rank} df_x|_\pi < \dim \pi \end{cases}$$

for each $\pi \in G_k M_x$, $x \in M$, we obtain the classical induced map (see [90], [140], [216], [217], and [147]).

It is easy to verify that $\operatorname{spt}^* T_f V \subset f(\operatorname{spt}^* V)$ for any $V \in V_k M$, and $\operatorname{spt}^* T_f V = f(\operatorname{spt}^* V)$ if V is positive. For every bounded Baire function φ on $G_k N$ we have $T_f V \wedge \varphi = T_f(V \wedge \varphi \circ G_f)$. In particular, if for φ we take the characteristic function of a Borel set $B \subset N$, we have $T_f V \cap B = T_f(V \cap f^{-1}(B))$. By (3.1), $|\tau_f(\pi)| \le |df|^i$, for any $\pi \in G_k M$ such that $\operatorname{rank} df|_\pi = i$. Consequently,

$$M_i(T_f V) \le \left(\sum_{j=i}^{k} r|V^j| \right) (|df|^i) \tag{3.4}$$

for every i.

REMARK. It is clear from the definition that the multivarifold $T_f V$ depends only on the behavior of f on an arbitrarily small neighborhood of $\operatorname{spt}^* V$.

Let A be an arbitrary subset of M. We define the following set of multivarifolds on M:

$$V_k A = V_k M \cap \{V : \operatorname{spt}^* V \subset A\}.$$

Obviously, $V_k A$ is a linear subspace of $V_k M$. In particular, if A is an open set, we can regard it as a manifold, and the embedding $i: A \to M$ induces an isomorphism T_i of $V_k A$ into $V_k M$, where $M_j(T_i V) = M_j V$ for any $V \in V_k A$ and any $j \le k$.

3.2. Generalization of induced maps of type T_f. In the general case, an induced map T_f is defined only for a continuously differentiable map f. For a space of semirectifiable multivarifolds this concept can be generalized to the case of an arbitrary locally Lipschitz map f.

First of all we observe that if $V \in \widetilde{R}_k M$, then by (3.11) and (3.12) of Chapter 6 V can be represented in the form

$$V = \sum_{i=0}^{k} V^i = \sum_{i=0}^{k} [S^i] \wedge \varphi_i, \tag{3.5}$$

where S^i is a uniformly i-rectifiable subset of M, and φ is an $r|V|$-measurable function, and this representation is unique in the sense that if $V = \sum_{i=0}^k [\widetilde{S}^i] \wedge \tilde{\varphi}_i$ is another such representation, then S^i and $\widetilde{S}^i$ differ only by a subset of zero measure h^i and $\varphi_i \equiv \tilde{\varphi}_i$ h^i-almost everywhere on $S^i \cap \widetilde{S}^i$. Conversely, if $h^i([S^i \setminus \widetilde{S}^i] \cup [\widetilde{S}^i \setminus S^i]) = 0$ and $\varphi_i \equiv \tilde{\varphi}_i$ h^i-almost everywhere on $S^i \cap \widetilde{S}^i$ for every i, then $\sum_{i=0}^k [S^i] \wedge \varphi_i = \sum_{i=0}^k [\widetilde{S}^i] \wedge \tilde{\varphi}_i$.

Let f be a continuously differentiable map of M into N. By the remark made above we may assume that there is a tangent space $\vec{S}^i_x$ for all $x \in S^i$. We consider the set $B_i = \{\vec{S}_x : x \in S^i\} \subset \Gamma_i M$ and, as usual, let φ_{B_i} denote the characteristic function of B_i. By (3.9) of Chapter 6 we have $V^i = V^i \wedge \varphi_{B_i}$, $0 \le i \le k$. Then from the definition of T_f (see (3.2) and (3.3)) it is easy to see that

$$T_f V^i(h) = [V^i \wedge \varphi_{B_i} \tau_f](h \circ G_f) \tag{3.6}$$

for any function h in $C(G_k N)$ or

$$T_f V^i(A) = [V^i \wedge \tau_f](B_i \cap G_f^{-1}(A)) \tag{3.7}$$

for any Borel set $A \subset G_k N$. Formulae (3.6) and (3.7) show that the multivarifold $T_f V^i$ is completely determined by the behavior of the map G_f on B_i, and consequently, by the restriction of f to S^i. If $g: M \to N$ is a map whose restriction to S^i is continuously differentiable, then the maps τ_g and G_g, although they may not be defined everywhere, can nevertheless be defined on $B_i \subset G_k M$, where the restriction of G_g to B_i is a Baire map, and the restriction of τ_g to B_i is a Baire function. Consequently, in this case it is natural to define the multivarifold $T_g V^i$ by

$$T_g V^i(h) = [V^i \wedge \varphi_{B_i} \tau_g](h \circ G_g) \tag{3.8}$$

for any arbitrary function h in $C(G_k M)$. In particular,

$$T_g V^i(A) = [V^i \wedge \tau_g](G_f^{-1} A) \tag{3.9}$$

for every Borel subset $A \subset G_k N$.

Now let $f: M \to N$ be a locally Lipschitz map. We define the induced map $T_f: \widetilde{R}_k M \to \widetilde{R}_k N$ as follows. Suppose that $V \in \widetilde{R}_k M$. We represent V in the form (3.5):

$$V = \sum_{i=0}^{k} V^i = \sum_{i=0}^{k} [S^i] \wedge \varphi_i. \tag{3.10}$$

According to Definition 3.7 of Chapter 6 and properties of locally Lipschitz maps, for every i we can choose a subset $\widetilde{S}^i \subset S^i$ of full measure h^i such that the restriction of f to $\widetilde{S}^i$ is a continuously differentiable map. Obviously, $V^i = [\widetilde{S}^i] \wedge \varphi_i$. Then the multivarifold $T_f V^i$ is defined by (3.8) and (3.9), and we set

$$T_f V = \sum_{i=0}^{k} T_f V^i. \tag{3.11}$$

Clearly, the multivarifold $T_f V$ thus defined does not depend on the choice of S^i, $\widetilde{S}^i$, or φ_i.

Let us show that $T_f V \in \widetilde{R}_k N$. For this it is sufficient to show that $T_f V^i \in \widetilde{R}_k N$ for every i. Suppose that $z \in S^i_j$ and $0 \le j \le i$. We define the following quantities:

$$\overline{\Theta}^j_f(S^i, z) = \overline{\lim_{\varepsilon \to 0}} \alpha_j^{-1} \varepsilon^{-j} h^j(f[S^i \cap B(z, \varepsilon)]),$$

where $B(z, \varepsilon)$ is the open ball with center z and radius ε. It is easy to verify that if $0 < \overline{\Theta}^l_f(S^i, z) < \infty$, then $\overline{\Theta}^j_f(S^i, z) = 0$ when $j > l$ and $\overline{\Theta}^j_f(S^i, z) = \infty$ when $j < l$. Thus, for every point $z \in S^i$ there is a unique integer l, $0 \le l \le i$, such that $0 < \overline{\Theta}^l_f(S^i, z) < \infty$. Then S^i splits into disjoint subsets S^i_j, $0 \le j \le i$, where $S^i_j = \{z \in S^i : 0 < \overline{\Theta}^j_f(S^i, z) < \infty\}$, $S = \bigcup_{j=0}^i S^i_j$, $S^i_j \cap S^i_l = \varnothing$, $j \ne l$. The subset S^i_j differs only by a set of zero measure h^i from the subset of points z of S^i for which $\dim df_z(\vec{S}^i_z) = j$, that is, the restriction of the map df_z to $\vec{S}^i_z$ has rank j. Clearly, every subset S^i_j, after a subset of zero measure h^i is removed from it if necessary, becomes uniformly i-rectifiable, and therefore, $f(S^i_j)$ is a uniformly j-rectifiable subset of N. Hence we may assume that $f(S^i) = \bigcup_{j=0}^i f(S^i_j)$ is a stratified surface in N.

We define a function $\sigma_{ij} \colon f(S^i_j) \to \mathbf{R}$ as follows. Suppose that $x \in f(S^i_j)$, $D = f^{-1}(x) \cap S^i_j$, $s = i - j$. We set

$$\sigma_{ij}(x) = \int \varphi_i d(h^s \cap D). \tag{3.12}$$

It is not difficult to verify that $\sigma_{ij} \circ p$, where p is the projection of $G_k N$, is an $[f(S^i_j)]$-measurable function and

$$T_f V^i = \sum_{j=0}^{i} [f(S^i_j)] \wedge \sigma_{ij}. \tag{3.13}$$

Formula (3.13) shows that $T_f V^i \in \widetilde{R}_k N$. Thus, the map T_f that we have constructed is linear.

Remark. The image $T_f V$ of a rectifiable multivarifold $V \in R_k M$ is not necessarily a rectifiable multivarifold on N. If V has the special form $V = [S^k] + \sum_{i=0}^{k-1} [S^i] \wedge \varphi_i$, where S^i is a uniformly i-rectifiable subset, and φ_i is an $r[S^i]$-measurable function, then from (3.12) and (3.13) it is easy to establish that its image $T_f V$ under the induced map T_f has the same form.

3.3. Induced maps of type L_f. Let $f\colon M \to N$ be a locally Lipschitz map, $V \in \widetilde{R}_k M$, $V = \sum_{i=0}^k [S^i] \wedge \varphi_i$, where S^i is a uniformly i-rectifiable subset, and φ_i is an $[S^i]$-measurable function. As we remarked above, we may assume that S^i is everywhere regular and the map f is continuously differentiable on S^i. Let $B(f(z), \varepsilon)$ be the open ball with center $f(z)$ and radius ε, and $f^{-1}[B]$ its inverse image under the map f. We say that two nonempty subsets A' and A'' of M are disjoint if there are open sets U' and U'' in M such that $A' \subset U'$, $A'' \subset U''$, $U' \cap U'' = \varnothing$. Then $f^{-1}[B]$ splits into the union of disjoint components, that is, $f^{-1}[B] = \bigcup_\alpha A$, where the subsets A_α are pairwise disjoint, and no A_α can be represented as the union of disjoint subsets of itself. We denote the component A_α that contains the point z by $f_z^{-1}[B]$. We define a function $\tilde{\eta}_f^i$ on M by

$$\tilde{\eta}_f^i(z) = \varlimsup_{\varepsilon \to 0} \alpha_l^{-1} \varepsilon^{-l} r[S^i](f_z^{-1}[B]), \tag{3.14}$$

where $l = \operatorname{rank} df_z|_{S^i}$. Next we set

$$\eta_f^i = 1/\tilde{\eta}_f^i. \tag{3.15}$$

Obviously, the function η_f^i is $r[S^i]$-measurable, and we can therefore define $L_f V$ by

$$L_f V(h) = \sum_{i=0}^k ([S^i] \wedge \varphi_{B_i} \varphi_i \eta_f^i)(h \circ G_f) \tag{3.16}$$

for an arbitrary function h in $C(G_k N)$, or

$$L_f V(A) = \sum_{i=0}^k ([S^i] \wedge \varphi_i \eta_f^i)(B_i \cap G_f^{-1}(A)) \tag{3.17}$$

for an arbitrary Borel set $A \subset G_k N$, where the B_i are defined as in (3.6) and (3.7). In particular, if it is a continuously differentiable map, then

$$L_f V = G_{f_*} \left(\sum_{i=0}^k [S^i] \wedge \varphi_i \eta_f^i \right). \tag{3.18}$$

We observe that $\operatorname{spt}^* L_f V = \operatorname{spt}^* T_f V = f(\operatorname{spt}^* V)$. If f is one-to-one on $\operatorname{spt}^* V$, then $L_f V = T_f V$. In the general case, the function $\eta_f^k \circ p$

(where p is the projection of $G_k M$) coincides with the function τ_f on B_k. Therefore, the k-dimensional strata of the multivarifolds $L_f V$ and $T_f V$ are equal.

Let us show that if $V = [S]$, where S is a k-dimensional stratified surface in M, then $L_f V \subset R_k N$. Suppose that $S = \bigcup_{i=0}^k S^i$, where S^i is a uniformly i-rectifiable subset. It is sufficient to prove the assertion for each S^i. We may suppose, as we showed in the previous subsection, that $f(S^i) = \bigcup_{j=0}^i f(S^i_j)$, where $f(S^i_j)$ is a uniformly j-rectifiable subset on N, and S^i_j is a uniformly i-rectifiable subset of M. Then we can easily verify that

$$L_f[S^i] = \sum_{j=0}^{i} [f(S^i_j)] \wedge \sigma'_{ij}, \tag{3.19}$$

where σ'_{ij} is the function on $f(S^i_j)$ defined by

$$\sigma'_{ij}(x) = (\text{the number of disjoint components in } f^{-1}(x) \cap S^i_j). \tag{3.20}$$

Formulae (3.10) and (3.20) show that $L_f[S^i] \in R_k N$. Consequently, $L_f[S] \in R_k N$.

3.4. An induced map in the pointwise representation of multivarifolds.

LEMMA 7.3.1. *Let M and N be Riemannian manifolds, and $f\colon M \to N$ a continuously differentiable map. Then for any $V \in V_k M$ we have*

$$T_f V = \int T_f V_x \, dr|V|(x).$$

PROOF. Suppose that h is an arbitrary function in $C(G_k N)$. We have

$$\begin{aligned} T_f V(h) = G_{f_*}(V \wedge \tau_f)(h) &= \int V_x(h \circ G_f \tau_f)\, dr|V|(x) \\ &= \left(\int T_f V_x dr|V|(x) \right)(h). \end{aligned}$$

THEOREM 7.3.1. *Let M and N be Riemannian manifolds and $f\colon M \to N$ a continuously differentiable map, and suppose that $V \in V_k M$. Then*

(a) *for almost all $z \in N$ in the sense of the measure $f_*(r|V|)$ there is the limit*

$$W_z = \lim_{\varepsilon \to 0} \frac{\int T_f V_x d[r|V| \cap f^{-1}(B(z, \varepsilon))](x)}{f_*(r|V|)(B(z, \varepsilon))};$$

(b) *for almost all $z \in N$ in the sense of the measure $f_*(r|V|)$*

$$[T_f V]_z = W_z / m(W_z), \qquad r|T_f V| = m(W) \wedge f_*(r|V|),$$

and so

$$T_fV = \int (W_z/m(W_z))m(W) \wedge f_*(r|V|)(z),$$

where $m(W)$ *is an* $f_*(r|V|)$*-measurable function on* N *such that* $m(W)(z) = m(W_z)$;

(c) *in particular, if* V *is positive, then* $r|T_fV| = f_*(r(V \wedge \tau_f))$.

PROOF. Let h be an arbitrary function in $C(G_kN)$. For any set $A \subset N$ we have

$$r(T_fV \wedge h)(A) \leq \sup_K |\tau_f h \circ G_f| f_*(r|V|)(A),$$

where K denotes the support of V. This means that the measure $r(T_f(V \wedge h)$ is absolutely continuous with respect to the measure $f_*(r|V|)$. Arguing on the same lines as in the proof of Theorem 7.2.1, we can prove that the limit

$$W_z = \lim_{\varepsilon \to 0} \frac{T_fV \cap B(z, \varepsilon)}{f_*(r|V|)(B(z, \varepsilon))} \in V_kN \tag{3.21}$$

exists for almost all $z \in N$ in the sense of the measure $f_*(r|V|)$, and for any function φ in $C(G_kN))$ the function $z \to W_z(\varphi)$ is $f_*(r|V|)$-measurable and

$$T_fV(\varphi) = \int W_z(\varphi) f_*(r|V|)(z). \tag{3.22}$$

On the other hand, according to Lemma 7.3.1 we have

$$T_fV \cap B(z, \varepsilon) = \int T_fV_x d[r|V| \cap f^{-1}(B(z, \varepsilon))](x).$$

This together with (3.21) proves assertion (a) of the theorem.

As in the proof of Theorem 7.2.1 it is easy to verify that W_z is a pointwise multivarifold at the point $z \in N$ and that the function $m(V)$ is $f_*(r|V|)$-measurable. Then

$$T_fV = \int (W_z/m(W_z))m(W) \wedge f_*(r|V|)(z)$$

is none other than the pointwise representation of T_fV. Consequently, from the fact that such a decomposition is unique (see Theorem 7.2.1) assertion (b) of the theorem follows.

If V is a positive multivarifold, then T_fV is also positive. Then $r|T_fV| = rT_fV = r(G_{f_*} V \wedge \tau_f)$, and assertion (c) follows immediately from the fact that the following diagram commutes:

$$\begin{array}{ccc} G_kM & \xrightarrow[G_f]{} & G_kN \\ {\scriptstyle p}\downarrow & & \downarrow{\scriptstyle p'} \\ M & \xrightarrow{f} & N \end{array}$$

CHAPTER VIII

Parametrizations and Parametrized Multivarifolds

§1. Spaces of parametrizations and parametrized multivarifolds

1.1. Definition of a parametrization. Let M be a Riemannian manifold, and r an integer.

DEFINITION 8.1.1. A *parametrization of order* k *and class* C^r, $1 \leq r \leq \infty$, on M is any pair (W, f), where $W \in V_k N$ is a multivarifold on some Riemannian manifold N, and $f: N \to M$ is a map of class C^r.

DEFINITION 8.1.2. A *locally Lipschitz parametrization* (or *parametrization of class* C^0) *of order* k *on* M is any pair (W, f), where $W \in \widetilde{R}_k N$ is a semirectifiable multivarifold on some Riemannian manifold N, and $f: N \to M$ is a locally Lipschitz map.

The set of all parametrizations of order k and class C^r on M is denoted by $P_k^r(M)$, $0 \leq r \leq \infty$. Obviously, $P_k^r(M) \subset P_k^s(M)$ if $0 \leq s < r \leq \infty$.

Let $(W, f) \in P_k^r(M)$ be a parametrization. The multivarifold W is called the *inverse image* and the multivarifold $T_f W$ the *T-image of the parametrization* (W, f). If, in addition, W is semirectifiable, then the multivarifold $L_f W$ is defined and is called the *L-image of* (W, f).

If a multivarifold V is the T-image (resp. L-image) of a parametrization (W, f), we say that V is *T-parametrized* (resp. *L-parametrized*) by the pair (W, f).

We observe that a multivarifold V can be T-parametrized (resp. L-parametrized) by several different pairs. In particular, if f and g are two maps of N into M of the same class that coincide on some neighborhood of $\operatorname{spt}^* W$, then $T_f W = T_g W$, and so the multivarifold $T_f W = T_g W$ is T-parametrized by the pairs (W, f) and (W, g) simultaneously. If, in addition, $W \in \widetilde{R}_k N$, then $L_f W = L_g W$, that is, the multivarifold $L_f W = L_g W$ is L-parametrized by the pair (W, f) and by the pair (W, g).

DEFINITION 8.1.3. The *T-multimass* (resp. *leading T-mass*) of a parametrization $(W, f) \in P_k^r(M)$ is the multimass (resp. leading mass) of its T-image $T_f W$. Similarly, the L-multimass (resp. leading L-mass) of a parametrization $(W, f) \in P_k^r(M)$, $W \in \widetilde{R}_k N$, is the multimass (resp. leading mass) of its L-image $L_f W$.

We observe that the leading T-mass and L-mass of a parametrization (W, f), $W \in \widetilde{R}_k N$, are always equal. However, its T-multimass and L-multimass are different in general.

DEFINITION 8.1.4. Two parametrizations (W_1, f_1) and (W_2, f_2) in $P_k^r(M)$ are said to be *T-equivalent* if their T-images coincide, that is, $T_{f_1} W_1 = T_{f_2} W_2$. If W_1 and W_2 are semirectifiable multivarifolds and $L_{f_1} W_1 = L_{f_2} W_2$, then (W_1, f_1) and (W_2, f_2) are called *L-equivalent parametrizations.*

According to the definition, the T-multimasses (resp. L-multimasses) of T-equivalent (resp. L-equivalent) parametrizations coincide. Obviously, T-equivalence and L-equivalence are equivalence relations in sets of parametrizations.

Of special interest are subsets of $P_k^r(M)$ consisting of parametrizations with a fixed inverse image. The statement of Plateau's classical multidimensional Problem B in the language of multivarifolds was realized by Đào Trọng Thi in Chapter 6 in sets of parametrizations of this type, where the fixed inverse image W belongs to a specially chosen class of multivarifolds. Suppose we are given a Riemannian manifold N and a fixed multivarifold W of order k on it. We denote by $P^r(M, W)$ the subset of $P_k^r(M)$ consisting of all parametrizations (W, f) of class C^r with inverse image W. Obviously, $P^r(M, W) \subset P^s(M, W)$, when $s < r$.

We observe that the set $P^r(M, W)$ depends on the manifold N not only because W is a multivarifold on N, but also because the maps f in the pairs (W, f) are maps of the whole of N into M. Replacing N by an open subset N' of it containing $\operatorname{spt}^* W$, we obtain, in general, another set of parametrizations.

The relation of T-equivalence defined above splits the set $P^r(M, W)$ into classes, the set of which we denote by $T^r(M, W)$. If W is semirectifiable, then the set of classes of L-equivalent parametrizations in $P^r(M, W)$ is denoted by $L^r(M, W)$. Clearly, $T^r(M, W)$ (resp., $L^r(M, W)$) can be identified with the set of all multivarifolds T-parametrized (resp. L-parametrized) by pairs in $P^r(M, W)$, having associated each class of T-equivalent (resp. L-equivalent) parametrizations with their common T-image (resp. L-image).

REMARK. Since in the definition of the set $P^r(M, W)$ the multivarifold W is fixed, in fact $P^r(M, W)$ can be identified with the set $C^r(N, M)$, where $C^r(N, M)$ denotes the set of all maps of N into M of class C^r (we recall that by maps of class C^0 we understand locally Lipschitz maps).

1.2. Homotopy classes. We define homotopy and the relation of being homotopic in different subsets of the sets $P^r(M, W)$, $T^r(M, W)$, and $L^r(M, W)$.

Let us consider a map $F: \mathbf{R} \times N \to M$, and let $(t^0, x^0) \in \mathbf{R} \times N$. In some neighborhood of this point, we choose a system of local coordinates $t^0, x_1, \dots, x_n$, where $x_1, \dots, x_n$ are local coordinates of N in the corresponding neighborhood of the point x^0, and t is the coordinate of the line $\mathbf{R}$. For a neighborhood of the point $F(t^0, x^0)$ in M we choose a system of local coordinates $y_1, \dots, y_m$; F is called a *map of class* C^s *with respect to* N *at the point* (t^0, x^0) if it has continuous partial derivatives

$$\partial^\alpha F_i(t, x) = \partial^{|\alpha|} F_i(t, x)/\partial x_1^{\alpha_1} \cdots \partial x_n^{\alpha_n}$$

in some neighborhood of the point (t^0, x^0) for all i, $1 \le i \le m$, and $\alpha = (0, \alpha_1, \dots, \alpha_n)$, $|\alpha| \le s$, where $F = (F_1, \cdots, F_m)$ is a representation of the map F in local coordinates $y_1, \dots, y_m$. It is easy to verify that this definition does not depend on the choice of local coordinates $x_1, \dots, x_n$ and $y_1, \dots, y_m$; F is called a *map of class* C^s *with respect to* N *on some subset of* $\mathbf{R} \times N$ if it is a map of class C^s with respect to N at every point of this subset. A continuous map F is called a *locally Lipschitz map* (or a *map of class* C^0) with respect to N if $F_t: N \to M$, $F_t(x) = F(t, x)$, is a map of class C^0 for every t.

Let P be an arbitrary subset of $P^r(M, W)$.

DEFINITION 8.1.5. A *homotopy of class* C^s in P is a family of parametrizations (W, f_t) in P, $0 \le t \le 1$, that satisfies the following condition: the map $F: [0, 1] \times N \to M$ defined by $F(t, x) = f_t(x)$ is a map of class C^s when $s \ge 1$ and a continuous map when $s = 0$, and in addition F is a map of class C^r with respect to N. A homotopy of class C^0 is simply called a *homotopy*. The corresponding family of maps f_t is called a *homotopy* of class C^s in the corresponding set $C^r(N, M) \cap \{f: (W, f) \in P\}$.

DEFINITION 8.1.6. Two parametrizations (W, f) and (W, g) in P are *homotopic in the class* C^s (or simply *homotopic* when $s = 0$) in P if there is a homotopy (W, f_t) of class C^s in P such that $(W, f_0) = (W, f)$, $(W, f_1) = (W, g)$.

Now let T and L be the quotient sets of P with respect to T-equivalence and L-equivalence respectively.

DEFINITION 8.1.7. Suppose that $W_1, W_2 \in T$ (resp. L). Then W_1 and W_2 are *T-homotopic* (resp. *L-homotopic*) in the class C^s in P if there are finitely many homotopies (W, f_t^i) of class C^s in P, $1 \le i \le q$, such that $T_{f_0^1} W = W_1$, $T_{f_1^1} W = T_{f_0^2} W, \ldots, T_{f_1^{q-1}} W = T_{f_0^q} W$, $T_{f_1^q} W = W_2$ (resp. $L_{f_0^1} W = W_1$, $L_{f_1^1} W = L_{f_0^2} W, \ldots, L_{f_1^{q-1}} W = L_{f_0^q} W$, $L_{f_1^q} W = W_2$). In the case $s = 0$ we simply say that W_1 and W_2 are *T-homotopic* (resp. *L-homotopic*).

The relations of being homotopic in Definitions 8.1.6 and 8.1.7 are equivalence relations and split the corresponding sets into homotopy classes of parametrizations or parametrized multivarifolds.

1.3. A parametric metric and topology. We observe that in the space of multivarifolds $V_k M$ the weak and strong topologies (and also the N-topology) cannot cover such an important geometric fact as the continuous deformation of a k-dimensional surface into a surface of lower dimension (as in Chapter 6, §3 all surfaces in the process of deformation are regarded as semirectifiable multivarifolds). *In the topologies mentioned, this process turns out to be discontinuous.* In this connection we need to introduce a new topology that effectively reflects the geometrical properties of parametrizations and parametrized multivarifolds.

Let us first recall the definition of a jet bundle. Let N and M be Riemannian manifolds and suppose that $x \in N$. Let $f, g: N \to M$ be maps of class C^r, $0 \le r < \infty$.

DEFINITION 8.1.8. (a) *The map f has tangency of order zero with the map g at the point x* if $f(x) = g(x)$.

(b) Suppose that $1 \le k \le r$. Then *the map f has tangency of order k with the map g at the point x* if the map $df: TN \to TM$ has tangency of order $k-1$ with the map dg at every point of $T_x N$.

Obviously, the relation "tangency of order r at a point x" is an equivalence relation.

From Definition 8.1.8 it is obvious that a map f has tangency of order 1 with a map g at a point x if and only if $f(x) = g(x) = y$, and $df_x = dg_x$ as maps $T_x N \to T_y M$.

We observe that the manifold N (resp. M) can be identified with the zero section of the tangent bundle TN (resp. TM); for any map $f: N \to M$ the restriction of the map df to N (regarded as the zero section of TN) coincides with f. Then from the definition it follows that if f has tangency of order k with g at x, then f has tangency of all orders $< k$ with g at x. In particular, f has tangency of order zero with g at x, that is, $f(x) = g(x)$.

Suppose that in a neighborhood U of x we are given a system of local coordinates $x_1, \dots, x_n$, and in a neighborhood V of $f(x)$ we are given a system of local coordinates $y_1, \dots, y_m$, where $f(U) \subset V$. In these local coordinates, the restriction of f to U can be regarded as a map of some open subset of $\mathbf{R}^n$ into $\mathbf{R}^m$, which we agree to call the *local representation of* f in the given local coordinates. In particular, we can define the partial derivatives of f in the given local representation

$$\partial^\alpha f(x) = \frac{\partial^{|\alpha|} f(x)}{\partial x_1^{\alpha_1} \cdots \partial x_n^{\alpha_n}} = \left(\frac{\partial^{|\alpha|} f_1(x)}{\partial x_1^{\alpha_1} \cdots \partial x_n^{\alpha_n}}, \dots, \frac{\partial^{|\alpha|} f_m(x)}{\partial x_1^{\alpha_1} \cdots \partial x_n^{\alpha_n}} \right),$$

where $\alpha = (\alpha_1, \dots, \alpha_n)$, $|\alpha| \le r$, and $f_1, \dots, f_m$ are coordinate functions of f in the given local coordinates.

PROPOSITION 8.1.1 (Golubitsky and Guillemin [33]). *Let $f, g: N \to M$ be maps of class C^r, $0 \le r < \infty$, satisfying the condition $f(x) = g(x) = y$. Then f has tangency of order r with g at x if and only if for any system of local coordinates $x_1, \dots, x_n$, defined in some neighborhood U of x, and a system of local coordinates $y_1, \dots, y_m$, defined in some neighborhood V of y, all the partial derivatives of f and g up to order r inclusive in the local representation in these systems of local coordinates coincide.*

We denote by $J^r(N, M)$ the set of equivalence classes with respect to "tangency of order r at x" in the space of maps $f: N \to M$ of class C^r satisfying the condition $f(x) = y$. We set

$$J^r(N, M) = \bigcup_{(x, y) \in N \times M} J^r(N, M)_{x, y}.$$

An element σ of the set $J^r(N, M)$ is called an r-jet from N to M, and if $\sigma \in J^r(N, M)_{x, y}$, then the points x and y are called the *inverse image* and the *image of the jet* σ respectively. We observe that $J^0(N, M) = N \times M$.

Let P_n^r be the vector space of polynomials in n variables $z_1, \dots, z_n$ without free term, whose degree does not exceed r. The space P_n^r can be endowed with a Euclidean structure by taking the coefficients of the polynomials as orthonormal coordinates. We set $P_{n,m}^r = \bigoplus_{1 \le i \le m} P_n^r$; then $P_{n,m}^r$ is a Euclidean space.

Let $x_1, \dots, x_n$ be a system of local coordinates defined on an open subset $U \subset N$, and $y_1, \dots, y_m$ a system of local coordinates defined on an open subset $V \subset M$. We set

$$J^r(U, V) = \bigcup_{(x, y) \in U \times V} J^r(N, M)_{x, y}.$$

For any jet $\sigma \in J^r(U, V)$ we define a polynomial $D_r\sigma \in P^r_{n,m}$ given by the formula

$$\begin{aligned} D_r\sigma &= \sum_{1\le|\alpha|\le r} (1/\alpha!)\partial^\alpha f(x) z^\alpha \\ &= \left(\sum_{1\le|\alpha|\le r} (1/\alpha!)\partial^\alpha f_1(x) z^\alpha, \dots, \sum_{1\le|\alpha|\le r} (1/\alpha!)\partial^\alpha f_m(x) z^\alpha \right), \end{aligned}$$

where x is the inverse image of σ, $f: N \to M$ is a map of class C^r representing the jet σ, $f_i: U \to \mathbf{R}$, $1 \le i \le m$, are the coordinate functions of the map f in the given local coordinates, and the partial derivatives $\partial^\alpha f(x)$ are taken in the local representation in these systems of local coordinates. In view of Proposition 6.1.1 the polynomial $D_r\sigma$ is well defined, that is, it does not depend on the choice of representing map f. Next we define a map $D_{U,V}: J^r(U, V) \to U \times V \times P^r_{n,m}$ by

$$D_{U,V}(\sigma) = (x, y, D_r\sigma),$$

where x and y are the inverse image and the image of the jet σ respectively.

According to Proposition 8.1.1 the map $D_{U,V}$ is one-to-one. It is also obvious that $D_{U,V}$ is an "onto" map. The product $U \times V \times P^r_{n,m}$ can be assumed to be an open subset of the Euclidean space $\mathbf{R}^N$, $N = n + m + \dim P^r_{n,m}$, taking as orthonormal coordinates the union of the coordinates $x_1, \dots, x_n$ in U, the coordinates $y_1, \dots, y_m$ in V, and the coordinates in $P^r_{n,m}$. We have thus constructed a system of local coordinates on $J^r(U, V)$. All the possible systems of local coordinates constructed in this way define on $J^r(N, M)$ the structure of a smooth bundle (for a verification of the compatibility of these systems of local coordinates on the intersection of their domains, see [33]). The Riemannian metric of the product $U \times V \times P^r_{n,m}$ induces a Riemannian metric on $J^r(U, V)$. The metrics defined locally in this way are "glued" by means of partitions of unity on N and M into a global Riemannian metric on $J^r(N, M)$, whose restriction to each $J^r(U, V)$ is equivalent to the original metric on it. Obviously, the distance between two jets in this metric is not less than the distance between their inverse images on N and the distance between their images on M.

We observe that for any map $f: N \to M$ of class C^r there is a canonically defined map $J^r f: N \to J^r(N, M)$ that associates with each point $x \in N$ an equivalence class of f in $J^r(N, M)_{x, f(x)}$. This map is called the *r-jet extension of* f. In the local representation in systems of local coordinates, the map $J^r f$ has the form

$$J^r f(x) = \left(x\,,\, f(x)\,,\, \sum_{1 \le |\alpha| \le r} \frac{1}{\alpha!} \partial^\alpha f(x) z^\alpha \right), \tag{1.1}$$

so that $J^r f$ is a continuous map. We also observe that the class $J^r f(x_0)$ contains a map whose local representation in systems of local coordinates has the form

$$\begin{aligned} \tilde{f}(x) &= f(x_0) + \sum_{1 \le |\alpha| \le r} \frac{1}{\alpha!} \partial^\alpha f(x_0)(x - x_0)^\alpha \\ &= \sum_{0 \le |\alpha| \le r} \frac{1}{\alpha!} \partial^\alpha f(x_0)(x - x_0)^\alpha. \end{aligned}$$

Thus, $J^r f(x_0)$ is none other than an invariant way of describing the Taylor polynomial approximation of degree r of f in a neighborhood of x_0.

Let us proceed to a definition of the metric and topology. Let M and N be Riemannian manifolds. Consider the set of parametrizations $P^r(M, W)$, $W \in V_k N$. Suppose that $V_i \in P^r(M, W)$, $V_i = (W, f_i)$, $i = 1, 2$, and S is a compact subset of N. We set

$$d_r^S(V_1, V_2) = \sup_{x \in S} d_r(J^r f_1(x), J^r f_2(x)), \tag{1.2}$$

where d_r is the distance function on $J^r(N, M)$ defined by the Riemannian metric on $J^r(N, M)$ constructed above. Let S_i be a fixed sequence of compact subsets of N such that $S_i \subset \mathring{S}_{i+1}$ and $\bigcup_{i=1}^\infty S_i = N$ ($\mathring{S}_{i+1}$ is the interior of S_{i+1}). Then in $P^r(M, W)$ we can introduce a metric by

$$\rho_r(V_1, V_2) = \sum_{i=1}^\infty 2^{-i} \frac{d_r^{S_i}(V_1, V_2)}{1 + d_r^{S_i}(V_1, V_2)}. \tag{1.3}$$

If $r = \infty$, we set

$$\rho_\infty(V_1, V_2) = \sum_{i=1}^\infty 2^{-i} \frac{d_i^{S_i}(V_1, V_2)}{1 + d_i^{S_i}(V_1, V_2)}. \tag{1.4}$$

REMARK. The function $\rho_r(V_1, V_2)$ actually *does not depend* on the inverse image multivarifold W. Therefore, the function $\rho_r(f_1, f_2) = \rho_r(V_1, V_2)$, where $V_i = (W, f_i)$, $i = 1, 2$, defines the corresponding metric in the space of maps $C^r(N, M)$.

We call the metric ρ_r introduced above the *parametric metric*, and the topology induced by it the *parametric topology*. We observe that the parametric topology *does not depend* on the choice of sequence S_i (while the parametric metric *does depend* on S_i !). The sequence $V_n = (W, f_n)$ in $P^r(M, W)$ converges to $V = (W, f)$ in the parametric topology if and only if $d_r^s(V_n, V) \to 0$ as $n \to \infty$ for any compact subset $S \subset N$ in the case $r < \infty$ and $d_i^s(V_n, V) \to 0$ as $n \to \infty$ for any i and any compact subset $S \subset N$ in the case $r = \infty$. The latter, in turn, is equivalent to the uniform convergence of $J^r f_n$ to $J^r f$ (and consequently, of $J^i f_n$ to $J^i f$ for all $i \leq r$) on every compact subset $S \subset N$ in the case $r < \infty$ and the uniform convergence of $J^i f_n$ to $J^i f$ on every compact subset $S \subset N$ for any i in the case $r = \infty$. Using the local representation (1.1) of the maps $J^r f_n$ and $J^r f$, we can conclude that the convergence of V_n to V implies the uniform convergence of $\partial^\alpha f_n$ to $\partial^\alpha f$ on every compact subset $S \subset N$ on which there is specified a system of local coordinates for all α such that $|\alpha| \leq r$.

We define the topologies of the sets $T^r(M, W)$, $L^r(M, W)$, $0 \leq r \leq \infty$, as the quotient topologies of the space $P^r(M, W)$ with respect to the relations of T-equivalence and L-equivalence respectively.

§2. The structure of spaces of parametrizations and parametrized multivarifolds

In this section we study the structures of the spaces $P^r(M, W)$, $T^r(M, W)$, $L^r(M, W)$ and subsets of them. The space $P^r(M, W)$ and subsets of it are always endowed with the parametric topology, and $T^r(M, W)$, $L^r(M, W)$, and subsets of them are endowed with the quotient topologies.

2.1. The condition of locally path connectedness for spaces of parametrizations. Let M and N be Riemannian manifolds, and $W \in V_k N$ a multivarifold on N.

LEMMA 8.2.1. *Let P be a subset of $P^r(M, W)$, $0 \leq r \leq \infty$, endowed with the parametric topology. Any homotopy in P specifies a continuous path in this space. Conversely, any continuous path in P determines a homotopy in it.*

PROOF. Let us prove the lemma for the case $1 \leq r < \infty$; the cases $r = 0$ and $r = \infty$ are proved similarly. Suppose we are given a homotopy (W, f_t) in P. Consider the map $F: [0, 1] \times N \to M$ given by the formula $F(t, x) = f_t(x)$. By definition, F is a map of class C^r with respect to N. This means that the local representation of F in certain systems of local coordinates has continuous partial derivatives $\partial^\alpha F$ for any multi-index $(0, \alpha_1, \dots, \alpha_n)$, $|\alpha| \leq r$. This is equivalent to the fact that $\partial^\alpha F(t, x) = \partial^{\alpha'} f_t(x)$, $\alpha' = (\alpha_1, \dots, \alpha_n)$ are maps continuous in t, uniformly in x on every compact coordinate subset of N. This, in turn, means that the map $t \to J^r f_t(x)$ is a continuous map, uniformly in x on every compact subset of N. Consequently, the path given by the map $t \to (W, f_t)$ is continuous in the parametric topology. Conversely, suppose that the family (W, f_t) specifies a continuous path in P. Then the map $t \to J^r f_t(x)$ is a continuous map, uniformly on every compactum in N. Hence the local representation of the map $F: [0, 1] \times N \to M$, $F(t, x) = f_t(x)$ in certain systems of local coordinates has continuous partial derivatives $\partial^\alpha F$ for any multi-index $\alpha = (0, \alpha_1, \dots, \alpha_n)$, $|\alpha| \le r$. Consequently, F is a map of class C^r with respect to N, and so (W, f_t) defines a homotopy in P. This proves the lemma.

Let A be a subset of M. We set $P^r(A, W) = P^r(M, W) \cap \{(W, f): f(N) \subset A\}$. We denote the quotient spaces of $P^r(A, W)$ with respect to T-equivalence and L-equivalence by $T^r(A, W)$ and $L^r(A, W)$ respectively.

DEFINITION 8.2.1. A subset $A \subset M$ is called a *neighborhood retract of class* C^r in M if there is a neighborhood U of A in M and a map $u: U \to A$ of class C^r such that $u(x) = x$ for any $x \in A$.

A subset $A \subset M$ is a neighborhood retract of class C^r in M if and only if A is locally compact and locally contractible by means of a family of maps of class C^r. In particular, any open subset or any submanifold is a neighborhood retract of any class C^r in M, and any locally convex compactum is a neighborhood retract of class C^0 in M.

THEOREM 8.2.1. *Suppose that* N *is deformable into a compact subset* K *of itself by means of a family of maps of class* C^r, $0 \le r \le \infty$, *that* $G \subset K$ *is a subset fixed under this deformation, and that* $A \subset M$ *is a neighborhood retract of class* C^r. *Then the spaces* $P_1^r = P^r(A, M, W) \cap \{(W, f): f|_G = \chi\}$, $P_2^r = P^r(A, M, W) \cap \{(W, f): f(G) \subset S\}$, *where* $\chi: G \to A$ *is a fixed map and* $S \subset A$ *is a compact locally convex set, are locally path connected.*

PROOF. Let $u: U \to A$ be a retraction of class C^r of a neighborhood U of a subset A into this subset, and $h_t: N \to M$, $0 \le t \le 1/3$, a homotopy that deforms N into a compact subset K and is specified by a family of maps of class C^r. Let $V_0 = (W, f_0)$ be an arbitrary parametrization in P_j^r, $j = 1, 2$. We need to prove that there is a number $\varepsilon > 0$ such that any parametrization $V = (W, f)$ satisfying the condition $\rho(V, V_0) < \varepsilon$ is connected with V_0 by some homotopy (W, f_t) in P_j^r. In fact, a compact subset $f_0(K)$ can be covered by finitely many open balls $B(x_i, r_i)$

with centers $x_i \in f_0(K)$ and radii r_i, $1 \le i \le s$, such that the open balls $B(x_i, 2r_i)$ concentric with them satisfy the following conditions: (a) $B(x_i, 2r_i) \subset U$ for every i; (b) $B(x_i, 2r_i)$ is a simple convex ball for any i. In the case of the space P_2^r we also cover a compact set S by finitely many open balls $B(y_j, \rho_j)$ with centers $y_j \in S$ and radii ρ_j such that $B(y_i, 2\rho_j) \cap S$ is convex for every j. We set $U' = \bigcup_{i=1}^s B(x_i, 2r_i) \subset U$. Clearly, for sufficiently small ε, from $\rho_r(V_0, V) < \varepsilon$ it follows that for any point $x \in K$ the points $f_0(x)$ and $f(x)$ belong to the same ball $B(x_i, 2r_i)$. In particular, from this it follows that $f(K) \subset U'$. For every point $x \in K$ we denote the geodesic joining the points $f_0(x)$ and $f(x)$ by $\alpha(t, f_0(x), f(x))$, $1/3 \le t \le 2/3$, where the parameter t satisfies the condition $\alpha_t'(t, f_0(x), f(x)) = 3d(f_0(x), f(x))$; here $d(a, b)$ denotes the distance between the points a and b. Obviously, for any $x \in G$ the geodesic $\alpha(t, f_0(x), f(x))$ degenerates to one point $\chi(x)$ in the case of the space P_1^r and lies entirely in S in the case of the space P_2^r. Clearly, α is a map $[1/3, 2/3] \times K \to U$, and α depends smoothly on $f_0(x)$ and $f(x)$, so that α is a map of class C^r with respect to x. We define a homotopy $g_t\colon N \to M$ by

$$g_t(x) = \begin{cases} f_0 h_t(x), & 0 \le t \le 1/3, \\ \alpha(t, f_0 h_{1/3}(x), f h_{1/3}(x)), & 1/3 \le t \le 2/3, \\ f h_{1-t}(x), & 2/3 \le t \le 1. \end{cases}$$

Obviously, $g_0 = f_0$ and $g_1 = f$. We now obtain the required homotopy (W, f_t) joining V_0 to V by setting $f_t = ug_t\colon N \to A$.

2.2. The topology of homotopy classes. Let $P \subset P^r(M, W)$ be an arbitrary subspace of parametrizations, endowed with the parametric topology, and let T and L be the quotient spaces of P with respect to T-equivalence and L-equivalence (with the quotient topology) respectively.

THEOREM 8.2.2. *Let P be a locally path-connected space. Then the relation of homotopy in P splits it into classes that are connected open-and-closed subsets.*

PROOF. According to Lemma 8.2.1, every homotopy class U in P is essentially a path-connected component of P. Since P is locally path-connected, every element of U has a path-connected neighborhood which, according to Lemma 8.2.1, belongs entirely to U. Consequently, U is an open subset. Similarly, $P \backslash U$ is also an open subset, and so U is closed. Finally, since U is path-connected, it is a fortiori connected. This completes the proof of the theorem.

THEOREM 8.2.3. *Let P be a locally path-connected space. Then the relation of homotopy in P splits the quotient spaces T and L into classes that are connected open-and-closed subsets.*

PROOF. Let $U \subset T$ (resp. $U \subset L$) be an arbitrary homotopy class, and $p: P \to T$ (resp. $p: P \to L$) a canonical map. The complete inverse image $p^{-1}(U)$ is obviously the union of several path-connected components of P. According to Theorem 8.2.2, $p^{-1}(U)$ is an open-and-closed subset. Consequently, U is an open-and-closed subset of T (resp. of L). It remains to show that U is connected. We even prove that U is path-connected. In fact, suppose that $W_1, W_2 \in T$ (resp. $W_1, W_2 \in L$). By definition, there are finitely many homotopies (W, f_t^i) in P such that $T_{f_0^1} W = W_1$, $T_{f_1^1} W = T_{f_0^2} W, \dots, T_{f_1^{q-1}} W = T_{f_0^q} W$, $T_{f_1^q} W = W_2$ (resp. replacing T by L). Since the canonical map p is continuous, the image $T_{f_t^i} W$ (resp. $L_{f_t^i} W$) of the continuous path (W, f_t) in P is a continuous path in T (resp. in L). Moreover, since the origin of the path $T_{f_t^{i+1}} W$ (resp. for L) coincides with the end of the path $T_{f_t^i} W$ (resp. for L) for each i, the images $T_{f_t^i} W$ (resp. for L) together specify in T (resp. in L) a continuous path joining W_1 and W_2.

2.3. The main structural theorem. Let M and N be Riemannian manifolds, and $W \in V_k N$ a multivarifold on N.

LEMMA 8.2.2. *Let M be a complete Riemannian manifold, and $S \subset N$ a compact subset. Then the subspace $J^r(S, M) \subset J^r(N, M)$ consisting of jets with inverse image in S is a complete metric space for any r, $0 \le r \le \infty$.*

PROOF. We need to prove that any fundamental sequence σ_p in $J^r(S, M)$ converges to some jet σ in $J^r(S, M)$. In fact, let the inverse image and the image of the jet σ_p be denoted respectively by x_p and y_p. From the fact that the sequence σ_p is fundamental it follows that the sequences x_p and y_p are fundamental in S and M. We denote the limits of x_p and y_p by x and y respectively. Consider a compact neighborhood U of x in N and a compact neighborhood V of y in M such that systems of local coordinates are defined on U and V. We set $U' = S \cap U$. The subspace $J^r(U', V) \subset J^r(N, M)$ consisting of jets with inverse image in U' and image in V is isometric to $U' \times V \times P^r_{n,m}$ with metric equivalent to the standard Euclidean metric, and consequently it is a complete metric space. The sequence σ_p, except for finitely many terms, belongs entirely to $J^r(U', V)$ and therefore converges to some jet

$\sigma \in J^r(U', V)$ (we observe that the inverse image and image of σ are x and y respectively).

THEOREM 8.2.4 (Đào Trọng Thi). *Let M be a complete Riemannian manifold. Then the space $P^r(M, W)$, endowed with the parametric metric, is a complete metric space for any r, $0 \le r \le \infty$.*

PROOF. Consider the case $r < \infty$. Suppose that the sequence $V_p = (W, f_p)$ is fundamental in $P^r(M, W)$, that is, $\rho_r(V_p, V_q) \to 0$ as $p, q \to \infty$, or

$$d_r(J^r f_p(x), J^r f_q(x)) \to 0 \tag{2.1}$$

as $p, q \to \infty$ uniformly on every compact subset of N. In particular, it follows from (2.1) that for any point $x \in N$, $J^r f_p(x)$ is a fundamental sequence in the subspace $J^r(x, M)$ of jets with fixed inverse image x, which, according to Lemma 6.2.2, is complete. Thus, $J^r f_p(x)$ converges to some jet $\sigma(x)$ with inverse image x and image $f(x)$. We prove that the resulting map $f: N \to M$ is differentiable of class C^r and $J^r f(x) = \sigma(x)$ for any $x \in N$. In fact, let U be a compact neighborhood of the point x_0 in N and V a neighborhood of the point $f(x_0)$ in M such that systems of local coordinates are defined on U and V and $f(U) \subset V$. Since f_p converges uniformly to f on U, the images $f_p(U)$ lie entirely in V for all $p \ge N$, where N is some natural number. Consider the local representations of the maps f and f_p, $p \ge N$, in these local coordinates, and denote their coordinate functions by $(f^1, \dots, f^m)$ and $(f_p^1, \dots, f_p^m)$ respectively. From (2.1) it follows that for any i, $1 \le i \le m$, and any α, $|\alpha| \le r$, $|\partial^\alpha f_p^i(x) - \partial^\alpha f_q^i(x)| \to 0$ as $p, q \to \infty$ uniformly on the compact subset U. Hence $\partial^\alpha f_p^i$ converges uniformly to a continuous function g_α^i on U. If $|\alpha'| \le r - 1$, then by Taylor's formula, for a point $x \in U$ sufficiently close to x_0, we have

$$\partial^{\alpha'} f_p^i(x) - \partial^{\alpha'} f_p^i(x) = \sum_{|\beta|=1} \partial^{\alpha'+\beta} f_p^i(\theta_p)(x - x_0)^\beta, \tag{2.2}$$

where θ_p is some point of the interval $[x_0, x]$. We can choose a subsequence p_j such that $\theta_{p_j} \to \theta \in [x_0, x]$. If in (2.2) we replace p by p_j and make p_j tend to infinity, we obtain

$$\begin{aligned} g_{\alpha'}^i(x) - g_{\alpha'}^i(x_0) &= \sum_{|\beta|=1} g_{\alpha'+\beta}^i(\theta)(x - x_0)^\beta \\ &= \sum_{|\beta|=1} g_{\alpha'+\beta}^i(x_0)(x - x_0)^\beta + o(|x - x_0|), \end{aligned} \tag{2.3}$$

where $o(|x - x_0|)$ is a function that tends to zero as $x \to x_0$ faster than $|x - x_0|$. This equality follows from the fact that $g^i_{\alpha'+\beta}$ is continuous. Equality (2.3) shows that $\partial^\beta g^i_{\alpha'} = g^i_\alpha$ for every α, $|\beta| = 1$, so that $\partial^\alpha g^i_0 = g^i_\alpha$ for any α and any i. On the other hand, obviously $g^i_0 = f^i$. Consequently, $\partial^\alpha f^i = g^i_\alpha$, that is, $\partial^\alpha f^i_p$ converges uniformly to $\partial^\alpha f^i$ on U. This means that f is a map of class C^r and $J^r f_p$ converges to $J^r f$ uniformly on U, and therefore, uniformly on any compact subset $S \subset N$, since S can be covered by finitely many coordinate neighborhoods. In particular, $J^r f(x) = \sigma(x)$. If $r = 0$, then from the fact that the f_p are locally Lipschitz it follows that f is also locally Lipschitz. We can now conclude that $V_p = (W, f_p)$ converges to (W, f) in the parametric metric. This proves the theorem for $r < \infty$. The case $r = \infty$ is proved similarly.

COROLLARY 8.2.1. *Let M be a complete Riemannian manifold, and A a closed subset of M. Then the space $P^r(A, W)$, endowed with the parametric metric, is a complete metric space for any r, $0 \le r \le \infty$.*

PROOF. It is sufficient to show that $P^r(A, W)$ is closed in $P^r(M, W)$, that is, if a parametrization (W, f) in $P^r(M, W)$ is the limit of a sequence (W, f_p) in $P^r(A, W)$, then $(W, f) \in P^r(A, W)$. In fact, for any point $x \in N$ the points $f_p(x)$, by hypothesis, belong to A for all p. Then since A is closed, the limit $f(x)$ of the sequence of points $f_p(x)$ also belongs to A. Consequently, $f(N) \subset A$, that is, $(W, f) \in P^r(A, W)$.

LEMMA 8.2.3. *Let M be a complete connected Riemannian manifold, K a bounded subset of M, S a compact subset of N, and x_0 a point of N. For a positive number κ we set*

$$E^r = C^r(N, M) \cap \{f\colon f(x_0) \in K, \operatorname{Lip} J^r f \le \kappa\}. \tag{2.4}$$

Then for any r, $0 \le r < \infty$, the subset $\bigcup_{f \in E^r} J^r f(S)$ is bounded.

PROOF. We observe that for any map f in E^r and any points $z_1, z_2 \in N$ we have $\rho(f(z_1), f(z_2)) \le d_r(J^r f(z_1), J^r f(z_2)) \le \eta\rho(z_1, z_2)$, where ρ denotes the distance function on N and M, and d_r is the distance function on $J^r(N, M)$, so that the map f itself is locally Lipschitz and $\operatorname{Lip} f \le \eta$.

Let x be an arbitrary point of N. We first prove that the subset $K_x = \{f(x)\colon f \in E^r\}$ is bounded in M. In fact, since M is connected, the points x_0 and x can be joined by a continuous path l. Since the image of l in N is compact, it can be covered by a finite chain of geodesically convex balls, in each of which the maps in E^r are, as is well known, globally Lipschitz with Lipschitz coefficient η. On l we choose points $y_0, y_1, \dots, y_p$ corresponding to an increasing sequence of parameters such that $y_0 = x_0$, $y_p = x$ and each pair of points x_i, x_{i+1}, $i = 0, 1, \dots, p-1$, belongs to one of the balls mentioned above. Then

$$\rho(f(x), f(x_0)) \le \sum_{i=0}^{p-1} \rho(f(y_i), f(y_{i+1})) \le s(l)\eta,$$

where $s(l)$ denotes the length of the path l. From this inequality and the fact that $\{f(x_0): f \in E^r\}$ is a subset of the bounded set K it follows that the subset K_x is bounded. Now for each point $y \in \overline{K}_x$ we consider an open ball V_y with center y and radius $2\varepsilon_y$ such that a system of local coordinates is defined on V_y, and an open ball V_y' with center y and radius ε_y. Since $\overline{K}_x$ is compact, from the cover of $\overline{K}_x$ by balls V_y' we can pick out a finite cover consisting of balls $V_1', \dots, V_q'$. We denote the radii of the balls $V_1', \dots, V_q'$ by $\varepsilon_1, \dots, \varepsilon_q$ respectively, and their concentric spheres with twice the radius by $V_1, \dots, V_q$ respectively. Let U_x be an open ball in N with center x such that a system of local coordinates is defined on U_x and the radius of U_x is less than $\min(\varepsilon_1/\eta, \dots, \varepsilon_q/\eta)$. Then it is obvious that $f(U_x) \subset V_i$ for some i. Consider the local representation of f in the given systems of local coordinates on U_x and V_i. From the condition $\operatorname{Lip} J^r f \le \eta$ it follows that $\operatorname{Lip} \partial^\alpha f \le \eta_i$ for any α, $|\alpha| \le r$, where η_i does not depend on f. This in turn means that $|\partial^\alpha f(z)| \le \eta_i$ on U_x. Thus, $J^r f(U_x)$ is contained in some bounded subset of $J^r(U_x, V_i)$. Hence we can conclude that the subset $\bigcup_{f \in E^r} J^r f(U_x)$ is contained in the union of q bounded subsets of $J^r(U_x, M)$ and is itself bounded. Finally, the compact subset S can be covered by finitely many such balls U_x, so that the subset $\bigcup_{f \in E^r} J^r f(S)$ is also bounded.

LEMMA 8.2.4. *Let M be a complete connected Riemannian manifold, and S a compact subset of N. Then from any infinite family of maps from the set E^r defined by* (2.4) *we can pick out a sequence f_p with the property that for any number $\varepsilon > 0$ there is a natural number N' such that the inequality*

$$\sup_{x \in S} d_r(J^r f_p(x), J^r f_q(x)) < \varepsilon \tag{2.5}$$

is satisfied for any p and q greater than N'.

PROOF. Let δ be an arbitrary positive number. For every point $x \in S$ we can choose a strongly convex open ball $B(x) \subset N$ with center x.

Since the ball $B(x)$ is convex, the locally Lipschitz maps $J^r f$, where $f \in E^r$, are globally Lipschitz maps with Lipschitz coefficients $\leq \eta$. Then, if $U(x) \subset B(X)$ is an open ball with center x and radius less than δ/η, we have

$$\sup_{f\in E^r} \sup_{y\in U(x)} d_r(J^r f(y), J^r f(x)) < \delta. \tag{2.6}$$

We choose a decreasing sequence of numbers $\varepsilon_1 > \varepsilon_2 > \cdots > \varepsilon_n > \cdots > 0$ converging to zero. Since S is compact, it follows from (2.6) that for any ε_n there is a finite system of points $x_1^n, \dots, x_{k_n}^n$ and corresponding open balls $U(x_1^n), \dots, U(x_{k_n}^n)$ with centers at these points such that

$$\bigcup_{j=1}^{k_n} U(x_j^n) \supset S, \tag{2.7}$$

$$\sup_{f\in E^r} \sup_{y\in U(x_j^n)} d_r(J^r f(y), J^r f(x_j^n)) < \varepsilon_n. \tag{2.8}$$

According to Lemmas 8.2.2 and 8.2.3, the subset $\bigcup J^r f(S)$, $f \in E^r$, is relatively compact, so from the given infinite family of maps in E^r we can pick out a sequence g_n^1 such that the corresponding sequence $J^r g_n^1$ converges at the points $x_1^1, \dots, x_{k_1}^1$. We may assume that

$$d_r(J^r g_p^1(x_j^1), J^r g_p^1(x_j^1)) < \varepsilon_1 \tag{2.9}$$

for all j, p, q. By the same arguments, from the sequence g_n^1 we can choose a subsequence g_n^2 converging together with $J^r g_n^2$ at points $x_1^2, \dots, x_{k_2}^2$ and such that

$$d_r(J^r g_p^2(x_j^2), J^r g_q^2(x_j^2)) < \varepsilon_2 \tag{2.10}$$

for all j, p, q. Continuing this process indefinitely and picking out the diagonal sequence, we obtain a sequence of maps $f_n = g_n^n$ converging together with $J^r f_n$ at all the points x_j^n, and from (2.9), (2.10), ... it follows that for any p and q greater than n we have

$$d_r(J^r f_p(x_j^n), J^r f_q(x_j^n)) < \varepsilon_n \tag{2.11}$$

for every point x_j^n, $1 \leq j \leq k_n$. We now suppose that ε is an arbitrary positive number. We choose n so that $\varepsilon_n < \varepsilon/3$, and let s be any point of S, and p and q any natural numbers greater than n. According to (2.7), the point x belongs to some ball $U(x_j^n)$. Then from

(2.8) and (2.11) we have $d_r(J^r f_p(x), J^r f_q(x)) \leq d_r(J^r f_p(x), J^r f_p(x_j^n)) + d_r(J^r f_p(x_j^n), J^r f_q(x_j^n)) + d_r(J^r f_q(x), J^r f_q(x_j^n)) < 3\varepsilon_n < \varepsilon$. Thus, when $p \geq n$ and $q \geq n$ we have $\sup_{x\in S} d_r(J^r f_p(x), J^r f_q(x)) < \varepsilon$, and so the lemma is proved.

THEOREM 8.2.5 (Đào Trọng Thi). *Let M be a complete connected Riemannian manifold, K a compact subset of it, and x_0 a point of N. The following assertions are true*:

(a) *the subset* $P^r = P^r(M, W) \cap \{(W, f): f(x_0) \in K, \operatorname{Lip} J^k f \leq \eta_k, 0 \leq k \leq r\}$, *where η_k are positive numbers, is compact for any r, $0 \leq r < \infty$*;

(b) *the subset* $P^\infty = P^\infty(M, W) \cap \{(W, f): f(x_0) \in K, \operatorname{Lip} J^k f \leq \eta_k, 0 \leq k < \infty\}$, *where η_k are positive numbers, is compact.*

PROOF. We need to prove that from any infinite family of parametrizations in P^r (resp. P^∞) we can pick out a convergent subsequence. Let S_i be a fixed sequence of compact subsets of N from the definition of the parametric metric ρ_r. We again consider a sequence of numbers $\varepsilon_n \to 0$ as $n \to \infty$ and $\varepsilon_i > \varepsilon_{i+1} > 0$. We prove (a) and (b) separately.

Let us prove (a). We observe that the map f in the parametrization (W, f) in P^r belongs to the set E^r defined by (2.4). Then according to Lemma 8.2.4, from any infinite family of parametrizations in P^r we can choose a sequence $V_n = (W, f_n)$ such that

$$\sup_{x\in S} d_r(J^r f_p(x), J^r f_q(x)) < \delta, \qquad \text{i.e.,} \qquad d_r^S(V_p, V_q) < \delta \tag{2.12}$$

for any p and q, where S is a preassigned compact subset of N, and δ is a preassigned positive number. Successively putting $S = S_i$ and $\delta = \varepsilon_i$, we can pick out from the given infinite family of parametrizations in P^r a sequence V_n^1 such that

$$d_r^{S_1}(V_p^1, V_q^1) < \varepsilon_1 \tag{2.13}$$

for all p and q, and from the sequence V_n^1 we can pick out a subsequence V_n^2 such that for all p and q

$$d_r^{S_2}(V_p^2, V_q^2) < \varepsilon_2 \tag{2.14}$$

and so on. Continuing this process indefinitely and picking out the diagonal sequence, we obtain a sequence of parametrizations $\widetilde{V}_n$; from (2.13) and (2.14) it follows that for any p and q greater than n we have

$$d_r^{S_n}(\widetilde{V}_p, \widetilde{V}_q) < \varepsilon_n. \tag{2.15}$$

We now show that $\widetilde{V}_n$ converges in the parametric metric. Let ε be an arbitrary positive number. We choose a natural number s such that

$$\sum_{i=s}^{\infty} 2^{-i} < \varepsilon/2, \tag{2.16}$$

and a natural number l such that $\varepsilon_l < \varepsilon/2$.

We set $n = \max(s, l)$. Since $S_i \subset S_j$ when $i < j$, according to (2.15) for any $i \leq n$ and any $p, q \geq n$ we have

$$d_r^{S_i}(\widetilde{V}_p, \widetilde{V}_q) \leq \varepsilon_n < \varepsilon/2. \tag{2.17}$$

Then for all p and q greater than n, in view of (2.16) and (2.17) we have

$$\rho_r(\widetilde{V}_p, \widetilde{V}_q) \leq \varepsilon/2 + \varepsilon/2 = \varepsilon. \tag{2.18}$$

The inequality (2.18) means that the sequence $\widetilde{V}_n$ is fundamental and therefore converges in view of Theorem 8.2.4.

Assertion (b) is proved similarly. By the hypothesis of the theorem the map f in the parametrization (W, f) in P^∞ belongs to the set E^r given by (2.4) for any r, $0 \leq r < \infty$. Applying Lemma 8.2.4 several times, we find that for a given compact subset S, a given number $\delta > 0$ and a given natural number r, from any infinite family of parametrizations in P^∞ we can pick out a sequence $V_n = (W, f_n)$ such that for any p and q and any s, $1 \leq s \leq r$, we have $\sup d_s(J^s f_p(x), J^s f_q(x)) < \delta$, where $x \in S$, that is,

$$d_s^S(V_p, V_q) < \delta. \tag{2.19}$$

Successively putting $S = S_i$, and $\delta = \varepsilon_i$ and $r = i$, $0 \leq i < \infty$, we can pick out from the given infinite family of parametrizations in P^∞ a sequence V_n^1 such that for all p and q

$$d_1^{S_1}(V_p^1, V_q^1) < \varepsilon_1, \tag{2.20}$$

and from the sequence V_n^1 a sequence V_n^2 such that for all p and q

$$d_s^{S_2}(V_p^2, V_q^2) < \varepsilon_2, \qquad s = 1, 2, \tag{2.21}$$

and so on. Continuing this process indefinitely and picking out the diagonal subsequence, we obtain a sequence $\widetilde{V}_n$ for which in view of (2.20) and (2.21) we have

$$d_s^{S_n}(\widetilde{V}_p, \widetilde{V}_q) < \varepsilon_n, \qquad 1 \leq s \leq n, \tag{2.22}$$

for any p and q greater than n. We show that $\widetilde{V}_n$ converges in the parametric metric. In fact, let ε be an arbitrary positive number. We choose a natural number m such that

$$\sum_{i=m}^{\infty} 2^{-i} < \varepsilon/2, \tag{2.23}$$

and a natural number l such that $\varepsilon_l < \varepsilon/2$. We set $n = \max(m, l)$. From (2.22) it follows that for any $i \leq n$ and any p and q greater than n we have

$$d_i^{S_i}(\widetilde{V}_p, \widetilde{V}_q) \leq \varepsilon_n < \varepsilon_l < \varepsilon/2. \tag{2.24}$$

Moreover, in view of (2.23) and (2.24) for all $p \geq n$ and $q \geq n$ we have

$$\rho_\infty(\widetilde{V}_p, \widetilde{V}_q) = \sum_{i=1}^{\infty} 2^{-i} \frac{d_i^{S_i}(\widetilde{V}_p, \widetilde{V}_q)}{1 + d_i^{S_i}(\widetilde{V}_p, \widetilde{V}_q)} \leq \frac{\varepsilon}{2} + \frac{\varepsilon}{2} = \varepsilon. \tag{2.25}$$

The inequality (2.25) shows that $\widetilde{V}_n$ is a fundamental sequence, and so it converges in view of the completeness of $P^\infty(M, W)$. This completes the proof of the theorem.

The next result follows immediately from Theorem 8.2.5.

COROLLARY 8.2.2. *Let P^r, $0 \leq r \leq \infty$, be the same as in Theorem 8.2.5, and let T^r and L^r be the quotient spaces of P^r with respect to T-equiva-*
lence and L-equivalence respectively. Then T^r and L^r are compact for any r, $0 \leq r \leq \infty$.

§3. Exact parametrizations

3.1. Definitions. As we have already observed (see §1), the parametrization $(W, f) \in P^r(M, W)$ depends on the behavior of the map f on the whole of N, and not only on $\operatorname{spt}^* W \subset N$ or even on some neighborhood of $\operatorname{spt}^* W$. In this section, for a semirectifiable multivarifold W, from among all parametrizations in $P^r(M, W)$ we pick out and study an important class of parametrizations that depend only on the behavior of the map on the support itself.

Let M and N be Riemannian manifolds, and $W \subset \widetilde{R}_k N$ a semirectifiable multivarifold on N. Suppose that $\operatorname{spt}^* W$ is a neighborhood retract of class C^r in N, and denote the corresponding retraction by $\tilde{u}\colon \widetilde{U} \to \operatorname{spt}^* W$.

Let U be another neighborhood of $\operatorname{spt}^* W$ such that $\overline{U} \subset \widetilde{U}$ is compact, and $u\colon U \to \operatorname{spt}^* W$ the restriction of $\tilde{u}$ to U. We regard W as a multivarifold on U.

DEFINITION 8.3.1. A pair $(W, f) \in P^r(M, W)$ is called an *exact parametrization of class* C^r if the map f can be represented in the form $f = gu$, where g is a map of $\operatorname{spt}^* W$ into M.

We observe that the map g in this representation coincides with the restriction of f to $\operatorname{spt}^* W$. We denote the subset of all exact parametrizations in $P^r(M, W)$ by $P_e^r(M, W)$. The set $P_e^r(M, W)$ does not essentially depend on the choice of the neighborhood U in $\widetilde{U}$.

The metric in $P_e^r(M, W)$ induced by the parametric metric in $P^r(M, W)$ is not natural in the sense that the distance between the pairs (W, f_1) and (W, f_2) depends on the behavior of the maps f_1 and f_2 not only on $\operatorname{spt}^* W$, but on the whole of U. For this reason we introduce a new metric in $P_e^r(M, W)$ as follows. If $V_1, V_2 \in P_e^r(M, W)$, $V_i = (W, f_i)$, we set

$$d_r^e(V_1, V_2) = \sup_{x \in \operatorname{spt}^* W} d_r(J^r f_1(x), J^r f_2(x)), \quad 0 \le r < \infty,$$

$$d_\infty^e(V_1, V_2) = \sum_{i=1}^{\infty} 2^{-i} \frac{d_i^e(V_1, V_2)}{1 + d_i^e(V_1, V_2)},$$

where d_r is the distance function on $J^r(N, M)$. The metric introduced in $P_e^r(M, W)$ is called the *exact metric*, and the topology induced by it is called the *exact topology*.

3.2. Comparison of the exact and parametric metrics. The aim of this subsection is to prove the equivalence of the exact and parametric metrics on every set that is bounded in some sense.

THEOREM 8.3.1. *Suppose that M is connected and complete, and that $K \subset J^r(U, M)$ is a compact subset. Then the exact and parametric metrics are equivalent on the subspace*

$$P_e^r = P_e^r(M, W) \cap \{(W, f) \colon J^r f(\operatorname{spt}^* W) \subset K\} \tag{3.1}$$

for any r, $0 \le r < \infty$.

LEMMA 8.3.1. *Suppose that M is connected and complete, and that $K \subset J^r(U, M)$, $0 \le r \le \infty$, is a compact subset. There is a constant $c_1 > 0$ such that for any jets $\sigma_1, \sigma_2 \in K$ with the same inverse image x we have*

$$d_r(\sigma_1, \sigma_2) \le d_r'(\sigma_1, \sigma_2) \le c_1 d_r(\sigma_1, \sigma_2), \tag{3.2}$$

where d_r denotes the distance function in $J^r(U, M)$, and d_r' is the distance function in $J^r(x, M)$, regarded as a Riemannian manifold with Riemannian structure induced from $J^r(U, M)$.

PROOF. For each jet $\sigma \in K$ in $J^r(U, M)$ there is a strongly convex simple open ball $B(\sigma, r_\sigma)$ with center σ and radius r_σ, and together with it an open ball $B(\sigma, r_\sigma/2)$ with center σ and radius $r_\sigma/2$. Since K is compact, from a cover of it by balls $B(\sigma, r_\sigma/2)$ we can pick out a finite cover $B(\sigma_i, r_{\sigma_i}/2)$, $1 \le i \le N$. We set $B_i' = B(\sigma_i, r_{\sigma_i}/2)$, $B_i = B(\sigma_i, r_{\sigma_i})$. Consider the sets $\Delta = \{(\sigma, \sigma): \sigma \in K\}$, $H_1 = \bigcup_{x\in U} J^r(x, M) \times J^r(x, M) \subset J^r(U, M) \times J^r(U, M)$, $H_2 = H_1 \cap (K \times K \setminus \bigcup_{i=1}^N B_i \times B_i)$, $H_3 = H_1 \cap (K \times K \setminus \Delta)$. From the fact that $K \times K$ is compact and H_1 is closed it follows that H_2 is compact. We observe that $J^r(U, M)$ is a smooth bundle with fibers $J^r(x, M)$, where x runs through the set U. Since the Riemannian metric on $J^r(x, M)$ depends continuously on x, the function $D(\sigma, \sigma') = d_r'(\sigma, \sigma')/d_r(\sigma, \sigma')$, is defined and continuous on the set H_3 and therefore attains a maximum $c_1 \ge 1$ on the compact subset $H_2 \subset H_3$. If the pair $(\sigma, \sigma') \in H_3 \setminus H_2$, then $(\sigma, \sigma') = \bigcup_{i=1}^N B_i \times B_i$, that is, σ and σ' lie in some ball B_i. From the fact that the ball B_i is strongly convex and simple, and also from the fact that $J^r(x, M)$ is totally geodesic in $J^r(U, M)$ (we can easily verify this by means of local coordinates of the bundle $J^r(U, M)$) it follows that $D(\sigma, \sigma') = 1$, and so $d_r'(\sigma, \sigma') \le c_1 d_r(\sigma, \sigma')$ on H_3. From this inequality there follows immediately the second inequality in (3.2). The first inequality in (3.2) is obvious. This proves the lemma.

LEMMA 8.3.2. *Suppose that M is connected and complete, and that $K \subset J^r(U, M)$, $0 \le r < \infty$, is a compact subset. There is a constant $c_2 > 0$ such that for any pairs (W, f_1) and (W, f_2) of the set P_e^r given by (3.1) we have*

$$\sup_{y\in U} d_r(J^r f_1(y), J^r f_2(y)) \le c_2 \sup_{x\in \operatorname{spt}^* W} d_r(J^r f_1(x), J^r f_2(x)).$$

PROOF. Let y be an arbitrary point of $\widetilde{U}$ and set $x = \tilde{u}(y) \in \operatorname{spt}^* W$. We define a map $\tilde{u}_y^*: J^r(x, M) \to J^r(y, M)$ as follows. If $\sigma \in J^r(x, M)$ and the map f is a representative of the jet, we set $\tilde{u}_y^*(\sigma) =$ (the equivalence class of the map $f\tilde{u}$). Clearly, $\tilde{u}_y^*(\sigma)$ is well defined, that is, it does not depend on the choice of a representative f of the jet σ. Under a local representation in some systems of local coordinates, the map $\tilde{u}_y^*$ looks like this: if $\sigma = (x, f(x), \sum(1/\alpha!)\partial^\alpha f(x)z^\alpha)$, then $\tilde{u}_y^*(\sigma) = (y, f(x), \sum(1/\alpha!)\partial^\alpha fu(y)z^\alpha) = (y, f(x), A(y)(\sum(1/\alpha!)\partial^\alpha f(x)z^\alpha))$, where $A(y): P_{n,m}^r \to P_{n,m}^r$ is a linear map whose matrix consists of the partial derivatives of $\tilde{u}$ up to order r inclusive. Thus, $A(y)$ depends continuously on y, and since $\overline{U}$ is compact, there is a constant c_2' such that for any $y \in U$ the length of the image under $\tilde{u}_y^*$ of any path in $J^r(x, M)$ does not exceed the length of this path multiplied by c_2'. In particular, this means that for any σ and σ' in $J^r(x, M)$ we have

$$d_r'(\tilde{u}_y^*(\sigma), \tilde{u}_y^*(\sigma')) \leq c_2' d_r'(\sigma, \sigma'). \tag{3.3}$$

Applying inequality (3.2) of Lemma 8.3.1 and (3.3), we obtain

$$d_r(J^r f_1(y), J^r f_2(y)) \leq c_2' c_1 d_r(J^r f_1(x), J^r f_2(x)),$$

where (W, f_1) and (W, f_2) are any parametrizations in P_e^r, y is any point of U, and $x = \tilde{u}(y)$. The assertion of the lemma is derived immediately from the resulting inequality when $c_2 = c_1 c_2'$.

PROOF OF THEOREM 8.3.1. Suppose that $V_1, V_2 \in P_e^r$, $V_1 = (W, f_1)$, $V_2 = (W, f_2)$. Applying Lemma 8.3.2, we have

$$\rho_r(V_1, V_2) = c_2 d_r^e(V_1, V_2). \tag{3.4}$$

On the other hand, since K is compact, there is a constant c_3' such that $d_r(\sigma, \sigma') \leq c_3'$ for any $\sigma, \sigma' \in K$. In particular, from the condition $J^r f_i(\text{spt}^* W) \subset K$ it follows that

$$\sup_{x \in \text{spt}^* W} d_r(J^r f_1(x), J^r f_2(x)) \leq c_3'. \tag{3.5}$$

Using Lemma 8.3.2 and (3.5), for any r we have

$$d_r^{S_i}(V_1, V_2) \leq c_2 \sup d_r(J^r f_1(x), J^r f_2(x)) \leq c_2 c_3'. \tag{3.6}$$

Assuming that $S_i \subset \text{spt}^* W$ when $i \geq p$ and setting $c_3 = (\sum_{i=p}^{\infty} 2^{-i})/(1 + c_2 c_3')$, we obtain

$$\rho_r(V_1, V_2) \geq c_3 \cdot \sup d_r(J^r f_1(x), J^r f_2(x)) \geq c_3 d_r^e(V_1, V_2). \tag{3.7}$$

The inequalities (3.4) and (3.7) prove the theorem.

3.3. The topology of sets of exact parametrizations. For exact parametrizations the following assertions, analogous to those proved in §2, are true.

LEMMA 8.3.3. *Let P_e be a subset of $P_e^r(M, W)$, endowed with the exact topology. Any homotopy in P_e defines a continuous path in this space. Conversely, any continuous path in P_e defines a homotopy in it.*

PROOF. This lemma can be proved on the same lines as Lemma 8.2.1.

Let A be a subset of M. We set

$$P_e^r(A, W) = P_e^r(M, W) \cap \{(W, f) \colon f(\text{spt}^* W) \subset A\}$$

and denote the quotient spaces of $P_e^r(A, W)$ with respect to T-equivalence and L-equivalence by $T_e^r(A, W)$ and $L_e^r(A, W)$ respectively.

THEOREM 8.3.2. *Let* $A \subset M$ *be a neighborhood retract of class* C^r, $0 \leq r \leq \infty$, *and* G *a subset of* $\operatorname{spt}^* W$. *Then the spaces*

$$P_{e1}^r = P_e^r(A, W) \cap \{(W, f): f|_G = \chi\},$$
$$P_{e2}^r = P_e^r(A, W) \cap \{(W, f): f(G) \subset S\},$$

where $\chi: G \to A$ *is a fixed map and* $S \subset A$ *is a compact locally convex set, are locally path-connected in the exact topology.*

PROOF. Let $v: B \to A$ be a given retraction of a neighborhood B of the set A into A itself, and let $V_0 = (W, f_0) \in P_e^r$. We need to show that there is a number $\varepsilon > 0$ such that any $V = (W, f) \in P_e^r$, $d_r^e(V_0, V) < \varepsilon$, is joined to V_0 by some homotopy (W, f_t) in P_e^r. We cover the compact subset $f_0(\operatorname{spt}^* W)$ by finitely many open balls $B(x_i, r_i)$ with centers $x_i \in f_0(\operatorname{spt}^* W)$ and radii r_i, $0 \leq i \leq s$, such that the open balls $B(x_i, 2r_i)$ concentric with them are simple strongly convex balls lying entirely in B. For sufficiently small ε, from $d_r^e(V_0, V) < \varepsilon$ it follows that for any point $x \in \operatorname{spt}^* W$ the points $f_0(x)$ and $f(x)$ belong to the same ball $B(x_i, 2r_i)$. We denote the geodesic in $B(x_i, 2r_i)$ joining $f_0(x)$ and $f(x)$ by $\alpha(t, f_0(x), f(x))$, $0 \leq t \leq 1$, where the parameter t satisfies the condition $\alpha_t'(t, f_0(x), f(x)) = d(f_0(x), f(x))$; $d(a, b)$ denotes the distance between a and b on M. As in the proof of Theorem 8.2.1, it is easy to verify that the required homotopy is $v\alpha(t, f_0(x), f(x))$. This proves the theorem.

Let $P_e \subset P_e^r(M, W)$ be an arbitrary subspace of exact parametrizations, endowed with the exact topology, and let T_e and L_e be the quotient spaces of P_e with respect to T-equivalence and L-equivalence respectively.

THEOREM 8.3.3. *Let* P_e *be a locally path-connected space. Then the relation of homotopy in* P_e *splits* P_e, T_e, *and* L_e *into classes that are connected open-and-closed subsets.*

This is proved word for word like Theorems 8.2.2 and 8.2.3.

THEOREM 8.3.4. *Suppose that* M *is connected and complete. Then the space* $P_e^r(M, W)$, *endowed with the exact metric, is complete for any* r, $0 \leq r \leq \infty$.

The proof of this theorem, using Theorem 8.3.1 and the fact that $P_e^r(M, W)$ is closed in $P^r(M, W)$ in the parametric topology, is in many respects similar to that of Theorem 8.2.4 (and when $r < \infty$, it is simply a consequence of it).

COROLLARY 8.3.1. *Suppose that* M *is connected and complete, and that* A *is a closed subset of* M. *Then the space* $P_e^r(A, W)$ *with the exact metric is complete for any* r, $0 \leq r \leq \infty$.

This is proved in exactly the same way as Corollary 8.2.1.

THEOREM 8.3.5. *Suppose that* M *is connected and complete, that* K *is a compact subset of it, and that* x_0 *is a point of* $\operatorname{spt}^* W$. *Then the following assertions are true:*

(a) *the subset* $P_e^r = P_e^r(M, W) \cap \{(W, f)\colon f(x_0) \in K,\ \operatorname{Lip}(J^k f|_{\operatorname{spt}^* W}) \leq \eta_k,\ 0 \leq k \leq r\}$, *where* η_k *are positive numbers, is compact for any* r, $0 \leq r < \infty$;

(b) *the subset* $P_e^\infty = P_e^\infty(M, W) \cap \{(W, f)\colon f(x_0) \in K,\ \operatorname{Lip}(J^k f|_{\operatorname{spt}^* W}) \leq \eta_k,\ 0 \leq k < \infty\}$, *where* η_k *are positive numbers, is compact.*

The proof of this theorem by means of Theorem 8.3.1 goes in roughly the same way as for Theorem 8.2.5; assertion (a) is even a consequence of Theorem 8.3.1 and assertion (a) of Theorem 8.2.5. The next result follows immediately from Theorem 8.3.5.

COROLLARY 8.3.2. *Let* P_e^r, $0 \leq r \leq \infty$, *be the same as in Theorem* 8.2.5, *and let* T_e^r *and* L_e^r *be the quotient spaces of* P_e^r *with respect to T-equivalence and L-equivalence respectively. Then* T_e^r *and* L_e^r *are compact for any* r.

§4. Real and integral multivarifolds

In this section we consider important classes of multivarifolds that have well-defined boundaries.

Let M be a Riemannian manifold. We define the concept of a *simplex of class* C^r, $1 \leq r \leq \infty$, by induction on the dimension.

DEFINITION 8.4.1. (a) A one-dimensional simplex $\Delta^1 \subset M$ is called a *simplex of class* C^r if Δ^1 is a submanifold of class C^r in M.

(b) Suppose that the concept of "q-dimensional simplex of class C^r" has already been defined. Then a $(q+1)$-dimensional simplex $\Delta^{q+1} \subset M$ is called a *simplex of class* C^r if the interior of Δ^{q+1} (that is, the open simplex) is a $(q+1)$-dimensional submanifold of M of class C^r and its faces are q-dimensional simplexes of class C^r in M.

Let Δ^k be a k-dimensional simplex of class C^r in M. Obviously, Δ^k is a uniformly k-rectifiable subset. The *multivarifold* $[\Delta^k]$ associated with Δ^k is also called a *simplex of class* C^r *in* M.

DEFINITION 8.4.2. A *simplicial* k*-dimensional chain of class* C^r *in* M is any finite linear combination of k-dimensional simplexes of class C^r in M.

Simplicial chains will be considered with either integer or real coefficients. Clearly, simplicial chains are semirectifiable multivarifolds, and simplicial chains with integer coefficients are rectifiable multivarifolds.

The *boundary of a simplicial chain in* M is understood to be its boundary in the sense of simplicial homology theory. The boundary of a simplicial k-dimensional chain Z of class C^r is a simplicial $(k-1)$-dimensional chain of class C^r and denoted by ∂Z. Clearly, $\partial(\partial Z) = 0$.

DEFINITION 8.4.3. A homogeneous multivarifold $Z \in \widetilde{R}_k M$ (resp. $R_k M$) is called a *k-dimensional chain* (resp. *integral k-dimensional chain*) *of class* C^r if there is a homogeneous multivarifold $Y \in \widetilde{R}_{k-1} M$ (resp. $R_{k-1} M$) and for any $\varepsilon > 0$ there is a simplicial k-dimensional chain T of class C^r in M with real (resp. integer) coefficients such that

$$M_k(Z - T) < \varepsilon, \qquad M_{k-1}(Y - \partial T) < \varepsilon. \tag{4.1}$$

We observe that if such a Y exists, then it is unique. Then we naturally call Y the *boundary of the chain* Z and denote it by ∂Z. It is easy to verify that ∂Z is a $(k-1)$-dimensional chain of class C^r and $\partial(\partial Z) = 0$.

If X is a p-dimensional chain of class C^r in a Riemannian manifold N, and Z is a q-dimensional chain of class C^r in a Riemannian manifold M, then it is easy to see that $X \times Z$ is a $(p+q)$-dimensional chain of class C^r in $N \times M$, and

$$\partial(X \times Z) = \partial X \times Z - X \times \partial Z. \tag{4.2}$$

If X and Z are integral, then $X \times Z$ is also integral. In particular, if $X = [a, b] \subset \mathbf{R}$, then

$$\partial([a, b] \times Z) - ([a, b] \times \partial Z) = \partial [a, b] \times Z = T_{i_b} Z - T_{i_a} Z = L_{i_b} Z - L_{i_a} Z, \tag{4.3}$$

where $i_a, i_b \colon M \to [a, b] \times M$, $i_a(x) = (a, x)$, $i_b(x) = (b, x)$.

DEFINITION 8.4.4. A *T_0-real* (resp. *T_0-integral*) *multivarifold* on M is any multivarifold of the form $V = T_f Z$, where Z is a chain (resp. integral chain) of class C^1 on some Riemannian manifold N, and $f \colon N \to M$ is a locally Lipschitz map. If Z is a chain (resp. integral chain) of class C^r, and f is a map of class C^r, then $T_f Z$ is called a *T_r-real* (resp. *T_r-integral*) *multivarifold.*

In this case the multivarifold $T_f(\partial Z)$ is also T_r-real (resp. T_r-integral), and we call it the *T-boundary of the multivarifold* $T_f Z$ *defined by the pair* (Z, f).

DEFINITION 8.4.5. An L_0-*real* (resp. L_0-*integral*) *multivarifold* on M is any multivarifold of the form $V = L_f Z$, where Z is a chain (resp. integral chain) of class C^1 on some Riemannian manifold N, and $f: N \to M$ is a locally Lipschitz map. If Z is a chain (resp. integral chain) of class C^r, and f is a map of class C^r, then $L_f Z$ is called an L_r-*real* (resp. L_r-*integral*) *multivarifold.*

In this case the multivarifold $L_f(\partial Z)$ is again L_r-real (resp. L_r-integral) and is called the *L-boundary of the multivarifold* $L_f Z$ *defined by the pair* (Z, f).

The sets of T_r-real, T_r-integral, L_r-real, and L_r-integral multivarifolds of order k on M are denoted by $RT_k^r(M)$, $JT_k^r(M)$, $RL_k^r(M)$, $JL_k^r(M)$, respectively.

CHAPTER IX

Problems of Minimizing Generalized Integrands in Classes of Parametrizations and Parametrized Multivarifolds. A Criterion for Global Minimality

§1. A theorem on deformation

In this section we describe constructions, due to Federer and Fleming [217] and Almgren [147], of homotopies that deform $\mathbf{R}^n$ into the skeletons of a cubic cell complex of it; we study their singularities and prove the "parametrized" analog of the Federer and Fleming deformation theorem (see [271]) obtained by Đào Trọng Thi.

1.1. Estimates of multimasses of multivarifolds under a map. In this subsection we establish estimates of multimasses of multivarifolds under a map. We shall show that under certain restrictions the induced map of type T_f acts even when the inducing map f has singularities on the support of the multivarifold.

Let $u: \mathbf{R}^n \to \mathbf{R}$ be a continuous nonnegative function. We set $U_s = \{x \in \mathbf{R}^n : u(x) > s\}$, $s \in \mathbf{R}$.

DEFINITION 9.1.1. A locally Lipschitz map $f: G \to \mathbf{R}^m$, where G is an open subset of $\mathbf{R}^n$ containing U_0, is said to be *u-admissible* if f takes bounded sets into bounded sets and $|df_x| < u(x)^{-1}$ for almost all (in the sense of the measure L_n) points x of U_0.

Let f, $g: G \to \mathbf{R}^m$ be continuous maps, $I = [0, 1]$, and let $\alpha: I \to I$ be a continuous monotonically increasing function, where $\alpha(0) = 0$, $\alpha(1) = 1$. We define a homotopy $h_\alpha(f, g): I \times G \to \mathbf{R}^m$ by $h_\alpha(f, g)(x, t) = (1-\alpha(t))f(x) + \alpha(t)g(x)$. The homotopy $h_\alpha(f, g)$ differs from a linear homotopy from f to g only by the parametrization of the interval I. We, therefore, call it a *quasilinear homotopy* from f to g.

THEOREM 9.1.1. *Let f, $g: G \to \mathbf{R}^m$ be u-admissible maps (resp. continuously differentiable u-admissible maps), and let $h = h_\alpha(f, g)$ be a* *quasilinear homotopy from* f *to* g. *Then if* $V \in \widetilde{R}_k\mathbf{R}^n$ (*resp.* $V \in V_k\mathbf{R}^n$) *satisfies the conditions*

$$r|V|(\mathbf{R}^n \setminus U_0) = 0, \tag{1.1}$$

$$r|V^j|(u^{-j}) < \infty, \qquad 0 \le j \le k, \tag{1.2}$$

where V^j *denotes the* j*-dimensional stratum of* V, *it follows that*

(a) *there are limits* (*in the compact weak topology*)

$$T_f^u V = \lim_{s \to 0} T_f(V \cap U_s),$$

$$T_h^u(I \times V) = \lim_{s \to 0} T_h(I \times [V \cap U_s]);$$

(b) *for every* i, $0 \le i \le k$, *we have the estimates*

$$M_i(T_f^u V) = \sum_{j=i}^{k} r|V^j|(u^{-i}), \tag{1.3}$$

$$M_{i+1}(T_h^u(I \times V)) \le \sum_{j=i}^{k} r|V^{j+1}|(u^{-i-1}) + \sum_{j=i}^{k} r|V^j|(|f - g|u^{-i}), \tag{1.4}$$

$$M_0(T_h^u(I \times V)) \le \sum_{j=0}^{k} M_j(|V|) = m(V); \tag{1.5}$$

(c) *if* V *is semirectifiable, then* $T_f^u V$ *and* $T_h^u(I \times V)$ *are also semirectifiable.*

PROOF. We observe that if s and t are sufficiently small and $0 < s < t$, then by (3.4) of Chapter 7 we have

$$\begin{aligned} &M_i(|T_f(V \cap U_s) - T_f(V \cap U_t)|) \\ &\quad \le \sum_{j=i}^{k} r|V^j \cap (U_s \setminus U_t)|(u^{-i}) \le \sum_{j=i}^{k} r(|V^j| \cap (\mathbf{R}^n \setminus U_t))(u^{-j}). \end{aligned} \tag{1.6}$$

Next, we set $U_s' = \{x \in U_s : f(x) = g(x)\}$, $U_s'' = \{x \in U_s : f(x) \neq g(x)\}$. By means of (3.4) of Chapter 7 it is not difficult to verify that for sufficiently small s, t, $0 < s < t$,

$$M_{i+1}(|T_h(I \times (V \cap U_s)) - T_h(V \cap U_t))|) \leq \sum_{j=i}^{k} r(|V^{j+1}| \cap (\mathbf{R}^n \backslash U_t))(u^{-j-1}) + \sum_{j=i}^{k} r(|V^j| \cap (\mathbf{R}^n \backslash U_t))(|f - g|u^{-j}), \quad (1.7)$$

and also

$$M_0(|T_h(I \times (V \cap U_s)) - T_h(I \times (V \cap U_t))|) \leq \sum_{j=0}^{k} r(|V^j| \cap (\mathbf{R}^n \backslash U_t))(1) \leq \sum_{j=0}^{k} r(|V^j| \cap (\mathbf{R}^n \backslash U_t))(u^{-j}). \quad (1.8)$$

Since the quantities on the right-hand sides of (1.6), (1.7), and (1.8), according to conditions (1.1) and (1.2), tend to zero with t, and the absolute supports of the multivarifolds $T_f(V \cap U_s)$ and $T_h(I \times (V \cap U_s))$ are contained in the closure of the convex hull of the bounded set $f(\operatorname{spt}^* V) \cup g(\operatorname{spt}^* V)$, there are limits (in the compact-weak topology)

$$T_f^u V = \lim_{s \to +0} T_f(V \cap U_s),$$

$$T_h^u(I \times V) = \lim_{s \to +0} T_h(I \times (V \cap U_s)).$$

Thus, assertion (a) is proved. Next, the estimates (1.3), (1.4), and (1.5) follow from the fact that the mass functions M_i are semicontinuous in the compact-weak topology and the inequalities

$$M_i(T_f(V \cap U_s)) \leq \sum_{j=i}^{k} r(|V^j| \cap U_s)(u^{-i}),$$

$$M_{i+1}(T_h(I \times (V \cap U_s))) \leq \sum_{j=i}^{k} r(|V^{j+1}| \cap U_s)(u^{-i-1}) + \sum_{j=i}^{k} r(|V^j| \cap U_s)(|f - g|u^{-i}),$$

$$M_0(T_h(I \times (V \cap U_s))) \leq \sum_{j=0}^{k} r(|V^j| \cap U_s)(1)$$

for every $s > 0$. Thus, (b) is true. Finally, if V is semirectifiable, then $V \cap U_s$ and $I \times (V \cap U_s)$ are also semirectifiable for any $s > 0$. Then for any $s > 0$ the multivarifolds $T_f(V \cap U_s)$ and $T_h(I \times (V \cap U_s))$ are semirectifiable. Consequently, the multivarifolds $T_f^u V$ and $T_h^u(I \times V)$ are semirectifiable. Thus, assertion (c) is proved, and so the proof of the theorem is complete.

THEOREM 9.1.2. *Let u and v be continuous nonnegative functions on $\mathbf{R}^n$, and $f, g: G \to \mathbf{R}^m$ u-admissible and v-admissible (resp. continuously differentiable u-admissible and v-admissible) maps. Then if $V \in \widetilde{R}_k\mathbf{R}^n$ (resp. $V \in V_k\mathbf{R}^n$) satisfies (1.1) and (1.2) simultaneously for u and v, it follows that $T_f^u V = T_f^v V$, $T_h^u(I \times V) = T_h^v(I \times V)$, where h is a linear homotopy from f to g.*

PROOF. We set $U_s = \{x \in \mathbf{R}^n : u(x) > s\}$, $W_t = \{x \in \mathbf{R}^n : v(x) > t\}$. For sufficiently small s and t, $s > 0$, $t > 0$, we have

$$M_i(|T_f(V \cap U_s) - T_f(V \cap W_t)|) \leq \sum_{j=i}^{k} (r(|V^j| \cap (\mathbf{R}^n \setminus W_t)) + r(|V^j| \cap (\mathbf{R}^n \setminus U_s)))(u^{-j}), \tag{1.9}$$

for every i, $0 \leq i \leq k$. Next, we set $U_s' = \{x \in U_s : f(x) = g(x)\}$, $U_s'' = \{x \in U_s : f(x) \neq g(x)\}$, $W_t' = \{x \in W_t : f(x) = g(x)\}$, $W_t'' = \{x \in W_t : f(x) \neq g(x)\}$. For sufficiently small t and s, $t > 0$, $s > 0$, we have

$$\begin{aligned} M_{i+1}(|T_h(I \times (V \cap U_s)) - T_h(I \times (V \cap W_t))|) &\leq \sum_{j=i}^{k} (r(|V^{j+1}| \cap (\mathbf{R}^n \setminus W_t)) + r(|V^{j+1}| \cap (\mathbf{R}^n \setminus U_s)))(u^{-j-1}) \\ &+ \sum_{j=i}^{k} (r(|V^j| \cap (\mathbf{R}^n \setminus W_t)) + r(|V^j| \cap (\mathbf{R}^n \setminus U_s)))(|f - g| u^{-j}) \end{aligned} \tag{1.10}$$

for every i, and also

$$M_0(|T_h(I \times (V \cap U_s)) - T_h(I \times (V \cap W_t))|) \leq \sum_{j=0}^{k} (r(|V^j| \cap (\mathbf{R}^n \setminus W_t)) + r(|V^j| \cap (\mathbf{R}^n \setminus U_s)))(u^{-j}). \tag{1.11}$$

Since the quantities on the very right-hand sides of (1.9), (1.10) and (1.11) tend to zero as $t, s \to 0$, we obtain $T_f^u V = T_f^v V$, $T_h^u(I \times V) = T_h^v(I \times V)$. Consequently, $T_f^u V$ and $T_h^u(I \times V)$ do not depend on the choice of the function u. Thus, instead of $T_f^u V$ and $T_h^u(I \times V)$ we sometimes simply write $T_f V$ and $T_h(I \times V)$.

1.2. A theorem on deformation. Let $Z^n \subset \mathbf{R}^n$ be the lattice of integer points in $\mathbf{R}^n$, and $Z_j^n \subset Z^n$ the subset of all points that have j even and $n-j$ odd coordinates. In particular, Z_0^n and Z_n^n are the subsets of points with odd coordinates and even coordinates respectively. Each point $\xi \in Z_j^n$ is the center of a unique j-dimensional cube $c'(\xi)$ with vertices in Z_0^n and a unique $(n-j)$-dimensional cube $c''(\xi)$ with vertices in Z_n^n. It is well known that $C' = \bigcup_{j=0}^n \bigcup_{\xi\in Z_j^n} c'(\xi)$, $C'' = \bigcup_{j=0}^n \bigcup_{\xi\in Z_{n-j}^n} c''(\xi)$ are dual rectilinear cubic complexes whose k-dimensional skeletons are $C_k' = \bigcup_{j=0}^k \bigcup_{\xi\in Z_j^n} c'(\xi)$, $C_k'' = \bigcup_{j=0}^k \bigcup_{\xi\in Z_{n-j}^n} c''(\xi)$ respectively. In particular, the sets of vertices of C' and C'' are Z_0^n and Z_n^n respectively.

In $\mathbf{R}^n$ we consider the metric $\theta(x, y) = \sup_{1\le i\le n} |x_i - y_i|$, which, as we know, is equivalent to the usual Euclidean metric. We denote the distance $\theta(x, C_{n-k-1}'')$ from the point x to C_{n-k-1}'' by $u_k(x)$. It is easy to show that u_k is a Lipschitz function, $u_k(x) \ge 0$, and $u_k(x) = 0$ if and only if $x \in C_{n-k-1}''$. It is also easy to prove that

$$u_k(x) = \inf_{\lambda\in S_n} \sup_{1\le i\le k+1} |x_{\lambda(i)} - \xi_{\lambda(i)}|,$$

where S_n is the group of permutations of the elements $1, 2, \ldots, n$, and $\xi \in Z_n^n$ is the center of the n-dimensional cube $c'(\xi)$ such that $x \in \overline{c'(\xi)}$. Later we shall need the following property of u_k.

PROPOSITION 9.1.1 (Federer and Fleming [217]). *Let μ be a positive Radon measure on $\mathbf{R}^n$, and let $c'(0)$ denote, as usual, the n-dimensional cube $\{a \in \mathbf{R}^n : |a_i| < 1,\ i = 1, \ldots, n\}$. Then $\mu(\{x \in \mathbf{R}^n : (u_k\tau_a)(x) = 0\}) = 0$ for almost all (in the sense of the measure L_n) a in $c'(0)$ and*

$$\int_{c'(0)} \mu((u_k\tau_a)^{-k})\, dL_n(a) = 2^n C_n^k \mu(\mathbf{R}^n). \tag{1.12}$$

REMARK. If for μ we take the pointwise measure ε_y, then from (1.12) it follows that

$$\int_{c'(0)} (u_k(x+y))^{-k}\, dL_n(x) = 2^n C_n^k. \tag{1.13}$$

Next, we define a map $P_k\colon C_{k+1}' \setminus C_{n-k-1}'' \to C_k'$, $0 \le k \le n-1$, by the following conditions: the restriction of P_k to C_k' is the identity map, and its restriction to each $(k+1)$-dimensional cell is the central projection of this cell (without the center) onto its boundary.

The map P_k is smooth everywhere except for the set of cones $P_k^{-1}(C_{k-1}')$, the inverse image of the places of "junction of k-dimensional cells", where the differentiability of P_k is violated because the surface of cubes in C'_{k+1} is not smooth. On the other hand, for multivarifolds the induced map T_f exists in the general case only when the inducing map f is continuously differentiable. For this reason we need to construct smooth approximations to P_k (see [147]).

Let $s\colon [-1, 1] \to [-1, 1]$ be a monotonically increasing function of class C^∞ satisfying the following conditions:

(a) $s(-1) = -1$, $s(0) = 0$, $s(1) = 1$;

(b) $d^j s(-1)/dt^j = d^j s(1)/dt^j = 0$, $1 \le j \le \infty$;

(c) $s(t) \ge t$ when $t > 0$ and $s(t) \le t$ when $t < 0$.

We define a map $\vec{s}\colon \mathbf{R}^n \to \mathbf{R}^n$ as follows. If $x = z + y$, where $z \in Z^n_n$, and $y \in [-1, 1]^n$, then

$$\vec{s}(x) = z + (s(y_1), \dots, s(y_n)). \tag{1.14}$$

Obviously, $\vec{s}$ is a map of class C^∞ that takes each cell of the cubic complexes C' and C'' into itself and leaves the lattice Z^n of integer points fixed.

For each k, $0 \le k \le n-1$, we set

$$P^s_k = \vec{s}P_k : C'_{k+1}\backslash C''_{n-k-1} \to C'_k;$$
$$\sigma^s_k = P^s_k \cdots P^s_{n-1} : \mathbf{R}^n \backslash C''_{n-k-1} \to C'_k.$$

Since on C'_{k-1} the map $\vec{s}$ has zero derivatives of all orders in all "directions of junction of k-dimensional cells", the map P^s_k, and hence the map σ^s_k, is infinitely differentiable.

THEOREM 9.1.3. *The map σ^s_k, $0 \le k \le n-1$, is a v_k-admissible map of class C^∞ for $v_k = k!u_k/(n!(\mathrm{Lip}\, s)^{n-k})$. Moreover, we have*

$$|\sigma^s_k(x) - x| < \sqrt{n}, \qquad x \in \mathbf{R}^n \backslash C''_{n-k-1}. \tag{1.15}$$

PROOF. Inequality (1.15) follows from the fact that the map σ^s_k takes each closed n-dimensional cube $\overline{c'(\xi)}$, $\xi \in Z^n_n$, into itself. Now to complete the proof of the theorem it remains to show that the estimate

$$|d\sigma^s_k(x)| \le n!(k!)^{-1}(\mathrm{Lip}\, s)^{n-k} u_k(x)^{-1} \tag{1.16}$$

holds for almost all $x \in \mathbf{R}^n$ in the sense of the measure L_n. To prove the last assertion we consider the subset $B_{i+p} = \{z \in \mathbf{R}^n : 0 < z_1 < \cdots < z_{i+p} < 1, z_{i+p+1} = \cdots = z_n = 1\}$ and prove by induction on p that the inequality

$$|d(P^s_i P^s_{i+1} \cdots P^s_{i+p-1})(z)| \le (i+p)!(i!)^{-1} z^{-1}_{i+1} (\mathrm{Lip}\, s)^p \tag{1.17}$$

holds for all $z \in B_{i+p}$. In fact, when $p = 1$ we have $P_i(z) = (z_1/z_{i+1}, \dots, z_i/z_{i+1}, 1, \dots, 1)$, $P_i^s(z) = (s(z_1/z_{i+1}), \dots, s(z_i/z_{i+1}), 1, \dots, 1)$ for any point $z \in B_{i+1}$. For any vector $h \in \mathbf{R}^n$ the jth coordinate of the vector $dP_i^s(z)(h)$ is equal to

$$s'(z_i/z_{i+1})[h_j - (z_j/z_{i+1})h_{i+1}]/z_{i+1}$$

when $j \leq i$ and to 0 when $j > i$. Therefore,

$$|dP_i^s(z)(h)| \leq \operatorname{Lip} s(|h| + i|h_{i+1}|)/z_{i+1} \leq \operatorname{Lip} s(i+1)|h|/z_{i+1}, \tag{1.18}$$

so that (1.17) is verified for $p = 1$. Suppose that this inequality is true for $p - 1$ and suppose that $z \in B_{i+p}$. We set $P_{i+p-1}(z) = z'$. Clearly, $z' \in B_{i+p-1}$, $z_i' = s(z_{i+1}/z_{i+p}) \geq z_{i+1}/z_{i+p}$.

From the inductive hypothesis we obtain

$$|d(P_i^s \cdots P_{i+p-2}^s)(z')| \leq (i+p-1)!(i!)^{-1}(z_{i-1})^{-1}(\operatorname{Lip} s)^{p-1} z_{p+i}. \tag{1.19}$$

On the other hand, by (1.18) we have

$$|dP_{i+p-1}^s(z)| \leq z_{p+i}^{-1} \operatorname{Lip} s(i+p). \tag{1.20}$$

From (1.19) and (1.20) it follows that

$$|d(P_i^s \cdots P_{i+p-1}^s)(z)| \leq \frac{(i+p-1)!(\operatorname{Lip} s)^{p-1} z_{p+i}}{i! z_{i+1}} \cdot \frac{\operatorname{Lip} s(i+p)}{z_{p+i}} = \frac{(i+p)!(\operatorname{Lip} s)^p}{i! z_{i+1}}.$$

Thus, (1.17) is proved. Since $u_k(x) = x_{k+1}$ for any $x \in B_n$, the inequality (1.16) when $x \in B_n$ is a special case of (1.17) if for i and p we take k and $n - k$ respectively. Finally, by means of suitable isometric maps this inequality can be extended from B_n to almost all (in the sense of the measure L_n) the space $\mathbf{R}^n$. Hence the theorem is completely proved.

REMARK. The function s can be chosen so that P_k^s arbitrarily uniformly closely approximates P_k and $\operatorname{Lip} s$ is arbitrarily close to 1. We also observe that the coefficient in the estimate (1.16) is far from the best possible values.

Let us agree to denote by σ_n the identity map of $\mathbf{R}^n$ onto itself. It is easy to verify that the quasilinear homotopy $h_\alpha(\sigma_n, \sigma_k)$, where α is an arbitrary monotonically increasing map of class C^∞ of the interval $[0, 1]$ onto itself, is a smooth homotopy that deforms $\mathbf{R}^n \setminus \mathbf{C}''_{n-k-1}$ into the k-dimensional skeleton C_k' of the complex C'.

THEOREM 9.1.4. *Suppose that* $V \in RT_k^r(\mathbf{R}^n)$ *with* T*-boundary* ∂V *and that* $\varepsilon > 0$. *Then there are multivarifolds* P, $Q \in RT_k^r(\mathbf{R}^n)$, $S \in RT_{k+1}^r(\mathbf{R}^n)$ *and their* T*-boundaries* ∂P, ∂Q, ∂S *satisfying the following requirements*:

(a) $V = P + Q + \partial S$;

(b) $\operatorname{spt}^* P \cup \operatorname{spt}^* S \subset \{x \in \mathbf{R}^n : \rho(x, \operatorname{spt}^* V) \leq 2\sqrt{n}\varepsilon\}$, $\operatorname{spt}^* \partial P \cup \operatorname{spt}^* Q \subset \{x \in \mathbf{R}^n : \rho(x, \operatorname{spt}^* \partial V) \leq 2\sqrt{n}\varepsilon\}$;

(c) $M_k P \leq c_1 M_k V + \sqrt{n} c_2 \varepsilon M_{k-1}(\partial V)$, $M_{k-1}(\partial P) \leq c_2 M_{k-1}(\partial V)$, $M_k Q \leq 3\sqrt{n} c_2 \varepsilon M_{k-1}(\partial V)$, $M_{k+1} S \leq 2\sqrt{n} c_1 \varepsilon M_k V$, *where* $c_1 = 2(n!(\operatorname{Lip} s)^{n-k}/k!)^k C_k^n$, $c_2 = 2(n!(\operatorname{Lip} s)^{n-k+1}/k!)^{k-1} C_{k-1}^n$;

(d) *if* $V \in JT_k^r(\mathbf{R}^n)$, *then* P, $Q \in JT_k^r(\mathbf{R}^n)$ *and* $S \in JT_{k+1}^r(\mathbf{R}^n)$;

(e) *the* k*-dimensional stratum of the multivarifold* P *is a* k*-dimensional polyhedral chain of cubes belonging to* $\mu_\varepsilon(C')$, *and if* $V \in JT_k^r(\mathbf{R}^n)$, *then this polyhedral chain is integral.*

PROOF. We shall prove the theorem for the case $\varepsilon = 1$. The general case can be reduced to this by means of a homothety μ_ε.

Suppose that $r \geq 1$. The case $r = 0$ can be considered in exactly the same way. Let $W \in \widetilde{R}_k N$ be a k-dimensional chain of class C^r on N and $f\colon N \to \mathbf{R}^n$ a map of class C^r such that $V = T_f W$ and $\partial V = T_f(\partial W)$. The supports $\operatorname{spt}^* W$ and $\operatorname{spt}^* \partial W$ are uniformly k-rectifiable and $(k-1)$-rectifiable subsets of N, respectively. We show that for almost all $a \in \mathbf{R}^n$ (in the sense of the measure L_n) we have

$$f(\operatorname{spt}^* W) \cap \tau_a(C''_{n-k-1}) = \varnothing, \tag{1.21}$$

$$f(\operatorname{spt}^* \partial W) \cap \tau_a(C''_{n-k}) = \varnothing. \tag{1.22}$$

In fact, the set of points $a \in \mathbf{R}^n$ not satisfying (1.21) is the image of the subset $\operatorname{spt}^* W \times C''_{n-k-1}$ under the map $\Phi\colon N \times \mathbf{R}^n \to \mathbf{R}^n$; $\Phi(x, y) = f(x) - y$, where $x \in N$ and $y \in \mathbf{R}^n$; $\operatorname{spt}^* W \times C''_{n-k-1}$ is obviously an $(n-1)$-rectifiable subset of $N \times \mathbf{R}^n$, and Φ is a map of class C^∞. Consequently, $\Phi(\operatorname{spt}^* W \times C''_{n-k-1})$ has measure zero (with respect to the measure L_n). We can prove similarly that the set of points $a \in \mathbf{R}^n$ that do not satisfy (1.22) has measure zero, and so the set of points $a \in \mathbf{R}^n$ that satisfy (1.21) and (1.22) simultaneously is a subset of full measure with respect to the measure L_n in $\mathbf{R}^n$.

Next we prove that the set of points a in $c'(0)$ for which

$$\begin{aligned} r|V^k|((u_k\tau_a)^{-k}) &\leq 2C_k^n M_k V, \\ r|(\partial V)^{k-1}|((u_{k-1}\tau_a)^{-k+1}) &\leq 2C_{k-1}^n M_{k-1}(\partial V), \end{aligned} \tag{1.23}$$

where V^j and $(\partial V)^j$ denote the j-dimensional strata of V and ∂V respectively, has positive measure L_n. In fact, according to Proposition 9.1.1 we have

$$\int_{c'(0)} r|V^k|((u_k\tau_a)^{-k})\,dL_n(a) = 2^n C_k^n M_k V,$$

$$\int_{c'(0)} r|(\partial V)^{k-1}|((u_{k-1}\tau_a)^{-k+1})\,dL_n(a) = 2^n C_{k-1}^n M_{k-1}(\partial V).$$

If the set of points a in $c'(0)$ for which (1.23) holds had zero measure, we should have

$$\begin{aligned} L_n(c'(0)) &= \int_{c'(0)} \left(\frac{r|V^k|((u_k\tau_a)^{-k})}{2C_k^n M_k V} + \frac{r|(\partial V)^{k-1}|((u_{k-1}\tau_a)^{-k+1})}{2C_{k-1}^n M_{k-1}(\partial V)} \right) dL_n \\ &> \int_{c'(0)} 1\, dL_n = 2^n = L_n(c'(0)), \end{aligned}$$

which is impossible. From the assertions proved above it follows, in particular, that we can choose $a \in c'(0)$ so that (1.21), (1.22), and (1.23) hold simultaneously.

We observe that $\{0\} \times W + [0, 1] \times \partial W$ is a k-dimensional chain of class C^r in $[0, 1] \times N$, and $\partial(\{0\} \times W + [0, 1] \times \partial W) = \{1\} \times \partial W$. Let α be a monotonically increasing function of class C^∞ on $[0, 1]$, where $\alpha(1) = 1$, $\alpha(0) = 0$, $d^j\alpha(0)/dt^j = d^j\alpha(1)/dt^j = 0$, $1 \le j < \infty$. Consider a map of class C^r, $\overline{h}\colon [0, 2] \times N \to \mathbf{R}^n$, given by the formulas $\overline{h}|_{[0,1]\times N} = h_\alpha(f, \sigma_k^s\tau_a f)$, $\overline{h}|_{[1,2]\times N} = h_\alpha(\sigma_k^s\tau_a f, \sigma_{k-1}^s\tau_a f)\beta$, where $\beta\colon [1, 2] \times N \to [0, 1] \times N$, $\beta(t, x) = (t-1, x)$. We set $P = T_{\overline{h}}(\{1\} \times W + [1, 2] \times \partial W)$, $S = -T_{\overline{h}}([0, 1] \times W)$, $Q = -T_{\overline{h}}([0, 2] \times \partial W)$, $\partial P = T_{\overline{h}}(\partial(\{1\} \times W + [1, 2] \times \partial W)) = T_{\overline{h}}(\{2\} \times \partial W)$, $\partial S = -T_{\overline{h}}(\partial([0, 1] \times W))$, $\partial Q = -T_{\overline{h}}(\partial([0, 2] \times \partial W))$. Clearly, $S \in RT_{k+1}^r(\mathbf{R}^n)$, P, Q, $\partial S \in RT_k^r(\mathbf{R}^n)$; ∂P, $\partial Q \in RT_{k-1}^r(\mathbf{R}^n)$. We have $P = P_1 + T_{\overline{h}}([1, 2] \times \partial W)$, $Q = Q_1 - T_{\overline{h}}([1, 2] \times \partial W)$, where $P_1 = T_{\overline{h}}(\{1\} \times W) = T_{\sigma_k^s\tau_a f} W$, $Q_1 = -T_{\overline{h}}([0, 1] \times \partial W) = -T_{h_\alpha(f, \sigma_k^s\tau_a f)}([0, 1] \times \partial W)$.

On the other hand, $\partial S = -T_{\overline{h}}(\partial([0, 1] \times W)) = T_f W - T_{\sigma_k^s\tau_a f} W - Q_1 = V - P_1 - Q_1 = V - P - Q$, from which we obtain $V = P + Q + \partial S$, so assertion (a) is proved. Next we have

$$|\sigma_k^s\tau_a(x) - x| < |\sigma_k^s\tau_a(x) - \tau_a(x)| + |\tau_a(x) - x| \le 2\sqrt{n}$$

for any point $x \in \mathbf{R}^n \setminus \tau_a^{-1}(C''_{n-k-1})$, and

$$|\sigma_k^s\tau_a(x) - \sigma_{k-1}^s\tau_a(x)| < \sqrt{n}$$

for any point $x \in \mathbf{R}^n \backslash \tau_a^{-1}(C''_{n-k})$. Assertion (b) follows immediately from these inequalities and (1.21) and (1.22). To prove (c) we observe that

$$P_1^k = (T_{\sigma_k^s \tau_a} V)^k, \qquad Q_1^k = (T_{h_\alpha(\sigma_n, \sigma_k^s \tau_a)}([0, 1] \times \partial V))^k,$$

$$S^{k+1} = (-T_{h_\alpha(\sigma_k^s \tau_a)}([0, 1] \times V))^{k+1}, \qquad (\partial P)^{k-1} = (T_{\sigma_{k-1}^s \tau_a}(\partial V))^{k-1},$$

$$P^k - P_1^k = Q^k - Q_1^k = (T_{h_\alpha(\sigma_k^s \tau_a, \sigma_{k-1}^s \tau_a)}([0, 1] \times \partial V))^k.$$

On the other hand, observing that $\operatorname{Lip} s \geq 1$, we see that $v_k \geq v_{k-1}$ and σ_n, $\sigma_k^s \tau_a$ are v_k-admissible, and $\sigma_k^s \tau_a$, $\sigma_{k-1}^s \tau_a$ are $v_{k-1} \tau_a$-admissible. Now using Theorems 9.1.1 and 9.1.3 and (1.23) we obtain the estimates

$$M_k P_1 \leq \left(\frac{n!(\operatorname{Lip} s)^{n-k}}{k!}\right)^k \cdot 2C_k^n M_k V,$$

$$M_k Q_1 \leq \left(\frac{n!(\operatorname{Lip} s)^{n-k}}{k!}\right)^{k-1} \cdot 4\sqrt{n} C_{k-1}^n M_{k-1}(\partial V),$$

$$M_{k+1} S \leq 2\sqrt{n} \left(\frac{n!(\operatorname{Lip} s)^{n-k}}{k!}\right)^k \cdot 2C_k^n M_k V,$$

$$M_{k+1}(\partial P) \leq \left(\frac{n!(\operatorname{Lip} s)^{n-k+1}}{(k-1)!}\right)^{k-1} \cdot 2C_{k-1}^n M_{k-1}(\partial V),$$

$$\begin{aligned} M_k(P - P_1) &= M_k(Q - Q_1) \\ &\leq 2\sqrt{n} \left(\frac{n!(\operatorname{Lip} s)^{n-k+1}}{(k-1)!}\right)^{k-1} \cdot C_{k-1}^n M_{k-1}(\partial V), \end{aligned}$$

from which assertion (c) follows.

If $V \in JT_k^r(\mathbf{R}^n)$, that is, W is an integral chain, then obviously P, $Q \in JT_k^r(\mathbf{R}^n)$, $S \in JT_{k+1}^r(\mathbf{R}^n)$, so assertion (d) is proved. Finally we prove assertion (e). Since the images of the maps σ_k^s and σ_{k-1}^s coincide with C'_k and C'_{k-1} respectively, and for every point $x \in \mathbf{R}^n \backslash C''_{n-k}$ the segment joining $\sigma_k^s(x)$ and $\sigma_{k-1}^s(x)$ belongs to C'_k, we see that $\operatorname{spt}^* P' \subset C'_k$, $\operatorname{spt}^* \partial P \subset C'_{k-1}$.

It is not difficult to verify that by means of a suitable smooth deformation of $\operatorname{spt}^*(\{0\} \times W + [0, 1] \times \partial W)$ in itself we can construct a homotopy that takes P, Q, S, ∂P, ∂Q, and ∂S into the multivarifolds $\widetilde{P}$, $\widetilde{Q}$, $\widetilde{S}$, $\partial \widetilde{P}$, $\partial \widetilde{Q}$, and $\partial \widetilde{S}$, which again satisfy all the requirements (a), (b),

(c) and (d); the k-dimensional stratum of the multivarifold $\widetilde{P}$ is a k-dimensional polyhedral chain of cubes belonging to C'. Clearly, in this case $\widetilde{P}$ satisfies requirement (e). This proves the theorem.

§2. Isoperimetric inequalities

DEFINITION 9.2.1. Let A and B be subsets of $\mathbf{R}^n$ with $B \subset A$. We shall say that A is *contractible to* B *by means of a homotopy of class* C^r if there is a map $F\colon [0, 1] \times \mathbf{R}^n \to \mathbf{R}^n$ of class C^r such that $F(0, x) = x$ when $x \in \mathbf{R}^n$, $F(t, x) = x$ when $x \in B$ and $t \in [0, 1]$, and $F([0, 1] \times A) \subset A$, $F(\{1\} \times A) \subset B$. In this case, we call B a *strong deformation retract of class* C^r *of the set* A. If B consists of a single point, then A is said to be *contractible by means of a homotopy of class* C^r.

THEOREM 9.2.1. *Let $B \subset A \subset \mathbf{R}^n$ be subsets, where A is contractible to B by means of a homotopy of class C^r, $0 \le r < \infty$. Let U_a and U_b be neighborhoods of A and B in $\mathbf{R}^n$, let $\varphi_a\colon U_a \to A$ and $\varphi_b\colon U_b \to B$ be retractions of class C^r, and let C be a compact subset of A, where a and b are positive numbers; suppose that φ_a satisfies a Lipschitz condition with constant ξ on the set $\{x \in \mathbf{R}^n : \rho(x, C) \le a\}$ and φ_b satisfies a Lipschitz condition with constant η on the set $\{x \in \mathbf{R}^n : \rho(x, B \cap C) \le b\}$.*

Then if $V \in JT_k^r(\mathbf{R}^n)$ with T-boundary ∂V, $\operatorname{spt}^* V \subset C$, $\operatorname{spt}^* \partial V \subset B$, $2\sqrt{n}\varepsilon \le \inf\{b, (\eta + 3)^{-1} a\}$, *where $\varepsilon > 0$ is defined by the equation*

$$c_1' M_k V + \sqrt{n} c_2 \varepsilon M_{k-1}(\partial V) = 2^k \varepsilon^k, \qquad c_1' > c_1,$$

it follows that there is a multivarifold $X \in JT_{k+1}^r(\mathbf{R}^n)$ and its T-boundary ∂X such that $\operatorname{spt}^* X \subset A$, $\operatorname{spt}^*(V - \partial X) \subset B$,

$$M_{k+1} X \le \xi^{k+1} (\eta + 3)^{k+1} 2\sqrt{n}(1 + c_1)\varepsilon M_k V,$$

$$M_k(\partial X) \le M_k V + \xi^k (\eta + 3)^k 3\sqrt{n} c_2 \varepsilon M_{k-1}(\partial V).$$

PROOF. Consider the cubic cell complex $\mu_e(C')$. Applying Theorem 9.1.4 to the multivarifold V, we obtain $V = P + Q + \partial S$, where $V = T_{\bar{h}}(\{0\} \times W)$, $\partial V = T_{\bar{h}}(\{0\} \times \partial W)$, $P = T_{\bar{h}}(\{1\} \times W + [1, 2] \times \partial W)$, $S = -T_{\bar{h}}([0, 1] \times W)$, $Q = -T_{\bar{h}}([0, 2] \times \partial W)$, $\partial P = T_{\bar{h}}(\{2\} \times \partial W)$, $\partial S = -T_{\bar{h}}(\partial([0, 1] \times W))$, $\partial Q = -T_{\bar{h}}(\partial([0, 2] \times \partial W))$, and $W \in R_k N$ is an integral k-dimensional chain of class C^r, and $\bar{h}\colon [0, 2] \times N \to \mathbf{R}^n$ is a homotopy defined as in the proof of Theorem 9.1.4. According to this theorem

$$\operatorname{spt}^* Q \cup \operatorname{spt}^* \partial P \subset \{x \in \mathbf{R}^n : \rho(x, B \cap C) \le 2\sqrt{n}\varepsilon\},$$

$$\operatorname{spt}^* S \cup \operatorname{spt}^* P \subset \{x \in \mathbf{R}^n : \rho(x, C) \le 2\sqrt{n}\varepsilon\},$$

$$M_k P < c_1' M_k V + c_2 \sqrt{n} \varepsilon M_{k-1}(\partial V) = 2^k \varepsilon^k.$$

But by Theorem 9.1.4 the k-dimensional stratum P^k is an integral linear combination of k-dimensional cubes in $\mu_\varepsilon(C')$. Therefore, $P^k = 0$. Thus, P is essentially a multivarifold of order $k-1$. We observe that if $y \in B \cap C$, then $\varphi_b(y) = y$, so that for any point x such that $\rho(x, B \cap C) \le 2\sqrt{n}\varepsilon \le b$, we have $|\varphi_b(x) - x| \le |\varphi_b(x) - \varphi_b(y)| + |y - x| \le (\eta+1)2\sqrt{n}\varepsilon$, where $y \in B \cap C$ is chosen so that $|y - x| \le 2\sqrt{n}\varepsilon$.

We construct a map $\varphi_b' \colon \mathbf{R}^n \to \mathbf{R}^n$ of class C^r such that $\operatorname{Lip}\varphi_b' \le \eta$ and $\varphi_b'(x) = \varphi_b(x)$ if $\rho(x, B \cap C) \le 2\sqrt{n}\varepsilon$. Next we construct a map of class C^∞:

$$\gamma \colon \mathbf{R}^n \to \mathbf{R}^n \cap \{y \in \mathbf{R}^n : |y| \le (\eta + 2)2\sqrt{n}\varepsilon\},$$

that satisfies the following requirements: the restriction of γ to the ball $\{y \in \mathbf{R}^n : |y| \le (\eta+1)2\sqrt{n}\varepsilon\}$ is the identity map, and on the complement to the ball $\{y \in \mathbf{R}^n : |y| < (\eta+2)2\sqrt{n}\varepsilon\}$ it is a contraction along radii, and $\operatorname{Lip}\gamma \le (\eta+2)/(\eta+1)$. We now define a map $\varphi_b'' \colon \mathbf{R}^n \to \mathbf{R}^n$ by $\varphi_b''(x) = x + \gamma(\varphi_b'(x) - x)$, $x \in \mathbf{R}^n$. Clearly, φ_b'' is a map of class C^r, $\varphi_b''(x) = \varphi_b(x)$, if $\rho(x, B \cap C) \le 2\sqrt{n}\varepsilon$, $|\varphi_b''(x) - x| \le (\eta+2)2\sqrt{n}\varepsilon$ for all $x \in \mathbf{R}^n$, and $|\varphi_b''(x) - \varphi_b''(y)| \le |x - y| + (\eta+1)^{-1}(\eta+2)(\varphi_b'(x) - \varphi_b'(y) + (x - y)) \le (\eta+3)|x - y|$, so that $\operatorname{Lip}\varphi_b'' \le \eta + 3$.

Consider the map $\overline{h}' \colon [-1, 2] \times N \to \mathbf{R}^n$ defined by

$$\overline{h}_1|_{[-1,0]\times N} = h_\alpha(f, \varphi_b'' f)\beta, \qquad \overline{h}_1|_{[0,2]\times N} = \varphi_b''\overline{h},$$

where $h_\alpha(f, \varphi_b'' f)$ is a quasilinear homotopy from f to $\varphi_b'' f$ with function α, defined as in the proof of Theorem 7.1.4, and $\beta \colon [-1, 0] \times N \to [0, 1] \times N$ is given by the formula $\beta(t, x) = (t+1, x)$. Then $\overline{h}_1([0, 1] \times \operatorname{spt}^* W) \subset \{x \in \mathbf{R}^n : \rho(x, C) \le (\eta+3)2\sqrt{n}\varepsilon\}$. Thus,

$$\overline{h}_1([-1, 1] \times \operatorname{spt}^* W) \subset \{x \in \mathbf{R}^n : \rho(x, A) \le a\} \tag{2.1}$$

and we can consider the integral multivarifolds $X_1 = T_{\varphi_a \overline{h}_1}(-[-1, 1] \times W)$, $\partial X_1 = T_{\varphi_a \overline{h}_1}(\{-1\} \times W - \{1\} \times W + [-1, 1] \times \partial W)$. Observing that $\overline{h}_1([1, 2] \times \operatorname{spt}^* \partial W) \subset \{x \in \mathbf{R}^n : \rho(x, A) \le a\}$, we can transform ∂X_1 to the following form: $\partial X_1 = T_{\varphi_a \overline{h}_1}(\{-1\} \times W) - T_{\varphi_a \overline{h}_1}(\{1\} \times W + [1, 2] \times \partial W) + T_{\varphi_a \overline{h}_1}([-1, 2] \times \partial W)$. Moreover, $T_{\varphi_a \overline{h}_1}(\{-1\} \times W) = T_{\varphi_a f}(W) = V$, since $f(\operatorname{spt}^* W) = \operatorname{spt}^* V \subset C$, and φ_a is the identity map on A, and $T_{\varphi_a \overline{h}_1}([-1, 2] \times \partial W) = T_{\varphi_a \overline{h}_1}([0, 2] \times \partial W) + Y_1$, where $\operatorname{spt}^* Y_1 = \operatorname{spt}^* T_{\varphi_a \varphi_b' f}(\partial W) = \operatorname{spt}^* V \subset B \cap C$, $M_k Y_1 = 0$, since φ_b'' is the identity on

$B \cap C$ and φ_a is the identity on A. Moreover, $\varphi_a \overline{h}_1([0, 2] \times \operatorname{spt}^* \partial W) = \varphi_a \varphi_b''(\operatorname{spt}^* Q) = \varphi_a \varphi_b(\operatorname{spt}^* Q) \subset B$, so that $\operatorname{spt}^* T_{\varphi_a \overline{h}_1}([0, 2] \times \partial W) \subset B$, and consequently $\operatorname{spt}^* T_{\varphi_a \overline{h}_1}([-1, 2] \times \partial W) \subset B$. We thus obtain

$$\partial X_1 = V - T_{\varphi_a \overline{h}_1}(\{1\} \times W + [1, 2] \times \partial W) + T_{\varphi_a \overline{h}_1}([0, 2] \times \partial W) + Y_1, \quad (2.2)$$

where $\operatorname{spt}^*(T_{\varphi_a \overline{h}_1}([0, 2] \times \partial W) + Y_1) \subset B$, $M_k Y_1 = 0$.

Now let $F: [0, 1] \times \mathbf{R}^n \to \mathbf{R}^n$ be a strong deformation retraction of class C^r of A onto B. We set $A' = \overline{h}_1^{-1}(\{x \in \mathbf{R}^n : \rho(x, A) \leq a\}) \subset [-1, 2] \times N$ and consider the map $\psi: [0, 1] \times A' \to \mathbf{R}^n$ defined by

$$\psi(s, z) = F(s, \varphi_a \overline{h}_1(z)).$$

According to (2.1) and (2.2), we have $[-1, 1] \times \operatorname{spt}^* W \subset A'$, $\operatorname{spt}^*(\{1\} \times W + [1, 2] \times \partial W) \subset A'$. We can, therefore, consider the integral multivarifolds

$$X = X_1 - T_\psi([0, 1] \times (\{1\} \times W + [1, 2] \times \partial W)),$$
$$\partial X = \partial X_1 - T_\psi(\partial([0, 1] \times (\{1\} \times W + [1, 2] \times \partial W))).$$

The second term in this decomposition of ∂X can be transformed as follows:

$$- T_\psi(\{1\} \times (\{1\} \times W + [1, 2] \times \partial W))$$
$$+ T_{\varphi_a \overline{h}_1}(\{1\} \times W + [1, 2] \times \partial W) + T_\psi([0, 1] \times (\{2\} \times \partial W)). \quad (2.3)$$

We set $Y_2 = T_\psi([0, 1] \times (\{2\} \times \partial W))$. From the fact that $\varphi_a \overline{h}_1(\{2\} \times \operatorname{spt}^* \partial W) = \varphi_a \varphi_b''(\operatorname{spt}^* \partial P) \subset B \cap C$, $\psi(s, x) = x$ when $x \in B$ it follows that

$$\operatorname{spt}^* Y_2 \subset B, \qquad M_k Y_2 = 0. \quad (2.4)$$

On the other hand, in view of the fact that $\psi(1, z) \in B$ for any point $z \in A'$ we have

$$\operatorname{spt}^* T_\psi(\{1\} \times W + [1, 2] \times \partial W) \subset B. \quad (2.5)$$

Using (2.2), (2.3), (2.4), and (2.5), we obtain $\partial X = V + Y_1 + Y_2 + T_{\varphi_a \overline{h}_1}([0, 2] \times \partial W) + T_\psi(\{1\} + (\{1\} \times W + [1, 2] \times \partial W))$, and $\operatorname{spt}^*(\partial X - V) \subset B$.

We now estimate the leading masses of the multivarifolds X and ∂X. Taking into consideration the fact that the equality

$$(T_{fg} Z)^k = (T_f T_g(Z))^k$$

holds for any multivarifold Z of order k and any maps f and g, by means of Theorem 7.1.4 and the inequality (3.4) of Chapter 7 we can establish the following estimates:

$$M_{k+1}(T_{\varphi_a \bar{h}_1}(-[0, 1] \times W)) \leq 2\xi^{k+1}(\eta + 3)^{k+1}\sqrt{n}c_1 \varepsilon M_k V,$$
$$M_{k+1}(T_{\varphi_a \bar{h}_1}(-[-1, 0] \times W)) \leq \xi^{k+1}(\eta + 2)2\sqrt{n}\varepsilon(\eta + 3)^k M_k V,$$
$$M_k(T_{\varphi_a \bar{h}_1}([0, 2] \times \partial W)) \leq (\eta + 3)^k \xi^k 3\sqrt{n}c_2 \varepsilon M_{k-1}(\partial V).$$

Moreover, since $P^k = 0$, we have $(T_{\varphi_a \bar{h}_1}(\{1\} \times W + [1, 2] \times \partial W))^k = (T_{\varphi_a} T_{\varphi_b''} P)^k = 0$. Hence, it follows that $(T_\psi([0, 1] \times (\{1\} \times W + [1, 2] \times \partial W)))^{k+1} = 0$, $(T_\psi(\{1\} \times (\{1\} \times W + [1, 2] \times \partial W)))^k = 0$ or $M_{k+1}(T_\psi([0, 1] \times (\{1\} \times W + [1, 2] \times \partial W))) = 0$, $M_k(T_\psi(\{1\} \times (\{1\} \times W + [1, 2] \times \partial W))) = 0$. We finally obtain the following estimates:

$$M_{k+1}(X) \leq \xi^{k+1}(\eta + 3)^{k+1} 2\sqrt{n}(1 + c_1)\varepsilon M_k V,$$
$$M_k(\partial X) \leq M_k(V) + \xi^k(\eta + 3)^k 3\sqrt{n}c_2 \varepsilon M_{k-1}(\partial V).$$

The theorem is thus completely proved.

THEOREM 9.2.2. *Let $A \subset \mathbf{R}^n$ be a set contractible by means of a homotopy of class C^r, $0 \leq r < \infty$, let C be a compact subset, U_a a neighborhood of A in $\mathbf{R}^n$, and $\varphi_a \colon U_a \to A$ a retraction of class C^r defined and satisfying a Lipschitz condition with constant ξ on the set $\{x \in \mathbf{R}^n : \rho(x, C) \leq a\}$. Then if $\partial V \in JT_k^r(\mathbf{R}^n)$ is the T-boundary of some $V \in JT_{k+1}^r(\mathbf{R}^n)$, $\operatorname{spt}^* \partial V \subset C$, $c_1' \sqrt{n}^k M_k(\partial V) \leq a^k$, $c_1' > c_1$, it follows that there is a multivarifold $X \in JT_{k+1}^r(\mathbf{R}^n)$ with T-boundary ∂V such that $\operatorname{spt}^* X \subset A$ and*

$$(M_{k+1}(X))^{k/k+1} \leq \xi^k \sqrt{n} c_1' M_k(\partial V). \tag{2.6}$$

PROOF. We choose a positive number ε such that $c_1' M_k(\partial V) = 2^k \varepsilon^k$. Consider the cubic cell complex $\mu_\varepsilon(C')$. Applying Theorem 9.1.4 to ∂V, we obtain $\partial V = P + \partial S$, and $\partial V = T_{\bar{h}}(\{0\} \times \partial W)$, $V = T_{\bar{h}}(\{0\} \times W)$, $P = T_{\bar{h}}(\{1\} \times \partial W)$, $S = -T_{\bar{h}}([0, 1] \times \partial W)$, $\partial S = -T_{\bar{h}}(\partial([0, 1] \times \partial W))$, where $W \in R_{k+1}N$ is an integral $(k + 1)$-dimensional chain of class C^r on the Riemannian manifold N, and $\bar{h} = \mu_\varepsilon h_\alpha(\mu_{1/\varepsilon} f, \sigma_k^s \tau_a \mu_{1/\varepsilon} f)$. From the same theorem it follows that $\operatorname{spt}^* S \cup \operatorname{spt}^* P \subset \{x \in \mathbf{R}^n : \rho(x, C) \leq 2\sqrt{n}\varepsilon\}$, $M_k P < c_1' M_k(\partial V) = 2^k \varepsilon^k$. However, P^k is an integral linear combination of k-dimensional cubes with edge 2ε. Therefore, $P^k = 0$.

From the requirement $a^k \geq c_1' \sqrt{n}^k M_k(\partial V)$ it follows that $2\sqrt{n}\varepsilon \leq a$. Therefore, $\overline{h}([0, 1] \times \operatorname{spt}^* W) = \operatorname{spt}^* S \subset \{x \in \mathbf{R}^n : \rho(x, A) \leq a\}$. We denote by $F: [0, 1] \times \mathbf{R}^n \to \mathbf{R}^n$ a strong deformation retraction of class C^r of the set A, and by $\varphi_a': \mathbf{R}^n \to \mathbf{R}^n$ a map of class C^r that coincides with φ_a on $\{x \in \mathbf{R}^n : \rho(x, C) \leq a\}$. We define a map $\psi: [0, 2] \times N \to \mathbf{R}^n$ by

$$\psi(t, x) = \varphi_a' \overline{h}(t, x) \qquad 0 \leq t \leq 1,$$

$$\psi(t, x) = F(\alpha(t-1), \varphi_a' \overline{h}(1, x)), \qquad 1 \leq t \leq 2.$$

Clearly, ψ is a map of class C^r. From the fact that $\operatorname{spt}^* P = \overline{h}(\{1\} \times \partial W) \subset \{x \in \mathbf{R}^n : \rho(x, C) \leq a\}$, $F(1, A) = \{x\}$, it follows that $\psi(\{2\} \times \operatorname{spt}^* \partial W) = \{x_0\}$. We can therefore construct a map $\psi': [0, 2] \times N \to \mathbf{R}^n$ of class C^r that coincides with ψ on $[0, 2] \times \operatorname{spt}^* \partial W$ and maps $\{2\} \times \operatorname{spt}^* W$ into the point x_0. We set

$$X = -T_{\psi'}([0, 2] \times \partial W - \{2\} \times W) \in JT_{k+1}^r(\mathbf{R}^n),$$
$$\partial X = -T_{\psi'}(\partial([0, 2] \times \partial W - \{2\} \times W)) \in JT_k^r(\mathbf{R}^n).$$

These multivarifolds can be transformed as follows:

$$X = -T_{\psi}([0, 2] \times \partial W) + T_{\psi'}(\{2\} \times W), \qquad \partial X = T_{\varphi_a' f}(\partial W) = \partial V, \tag{2.7}$$

since $f(\operatorname{spt}^* \partial W) = \operatorname{spt}^* \partial V \subset A$, and φ_a' is the identity map on A.

Let us estimate the leading mass of the multivarifold X. According to Theorem 9.1.4, we have

$$M_{k+1}(T_{\psi'}([0, 1] \times \partial W) \leq 2\xi^{k+1} \sqrt{n} c_1 \varepsilon M_k(\partial V). \tag{2.8}$$

Moreover, $[T_{\psi'}([1, 2] \times \partial W)]^{k+1} = 0$ (since $P^k = 0$) and $[T_{\psi'}(\{2\} \times W)]^{k+1} = 0$, so that

$$M_{k+1}(T_{\psi'}([1, 2] \times \partial W)) = 0, \qquad M_{k+1}(T_{\psi'}(\{2\} \times W)) = 0. \tag{2.9}$$

From (2.6), (2.8), and (2.9) it follows that

$$M_{k+1} X \leq \xi^{k+1} \sqrt{n} (c_1' M_k(\partial V))^{k+1/k}.$$

Consequently,

$$(M_{k+1} X)^{k/k+1} < \xi^k \sqrt{n} c_1' M_k(\partial V). \tag{2.10}$$

The equalities (2.7) and the inequality (2.10) show that X is the desired multivarifold. This proves the theorem.

REMARK. If the set A is convex, then we can choose a arbitrarily large and $\zeta = 1$ and $\xi = 1$.

§3. Statement of variational problems in classes of parametrizations and parametrized multivarifolds

The solution of the multidimensional variational Problems B_*, B'_*, B''_*, stated in Chapter 6, §4 in the language of multivarifolds, will be carried out in §4 as a consequence of general theorems of Đào Trọng Thi on the structure of sets of parametrizations and parametrized multivarifolds (see Chapter 8, §§2 and 3). However, it turns out that a more general point of view is the most natural. In fact there is no need to restrict ourselves to these problems. By the general method we can solve a wide class of variational problems with parametrizations and parametrized multivarifolds; we now turn to a statement of these problems.

3.1. Boundary conditions. We recall that in the statements of the Problems B_*, B'_*, B''_* we considered a compact k-dimensional manifold W as a multivarifold $[W]$ of order k on W itself (in addition, the multivarifold $[W]$ is a k-dimensional chain of class C^∞). We can treat the boundary ∂W of the manifold W as the support of the boundary of the chain $[W]$, and also as the boundary of the support of $[W]$. We now consider parametrizations of the form (W, f) on M, where W is an arbitrary fixed multivarifold on some Riemannian manifold N, and f is a map from N to M. By analogs of the statements of the Problems B_*, B'_* and B''_* the boundary conditions of the general problem must consist in the requirement that the restrictions of the maps f in the parametrization (W, f) to the support of the boundary ∂W (if ∂W exists!) or to the boundary of $\operatorname{spt}^* W$ are either a given fixed map (for problems "with fixed boundary") or have an image in some given subset of M (for problems "with moving boundary"). As we have seen, not every multivarifold has a boundary (we can introduce the natural concept of boundary only for integral and real multivarifolds). The question of the boundary of an arbitrary compactum is completely different, namely: the homological boundary of a compactum X in dimension k is, according to the definition of Reifenberg and Adams (see [318] and [365]) any compactum $A \subset X$ such that $H_{k-1}(A) \subset \operatorname{Ker} i_*$, where i_* is the homomorphism induced by the embedding $i: A \to X$. Geometrically the condition $H_{k-1}(A) \subset \operatorname{Ker} i_*$ means that the $(k-1)$-dimensional homological cycles in A are completely glued by the compactum X, so that homological boundaries exist for each compactum X, and there are infinitely many of them. Next, for every pair of compacta $A \subset X$, Reifenberg and Adams (see [318], [365]) suggested defining the homological boundary of the pair (X, A) as the kernel of the homomorphism $i_*: H_*(A) \to H_*(X)$ induced by the embedding i.

Thus, the compactum A can be interpreted as the homological boundary of the incomplete gluing of X (the compactum X glues only the subgroup $\operatorname{Ker} i_*$ of the group $H_*(A)$). Fomenko [119], [232], [129], and [134] substantially generalized the concept of the homological boundary of a pair of compacta (X, A) by using instead of the usual homology theory an arbitrary generalized homology or cohomology theory. The concept of the algebraic boundary and coboundary of a pair of compacta (X, A) that he introduced makes it possible to treat a compactum A in a very diverse way as the "boundary" of a compactum X containing A. Thus, as the boundary condition of the general problem, there naturally appears either the requirement that the restriction of the maps f in the parametrizations (W, f) to the set G is a given fixed map (for problems "with fixed boundary") or the requirement that these maps f take the set G into some given set $S \subset M$ (for problems "with moving boundary"), where G is an arbitrary compact subset of $\operatorname{spt}^* W$.

3.2. Parametrization-solutions and multivarifold-solutions. We recall that we wish to solve variational problems in the classical formulations (that is, problems of minimizing some functional in classes of parametrized films). Therefore, according to the analysis carried out in Chapter 6, §2, we can admit to the class of parametrization-solutions those parametrizations (W, f) for which the map f has film-images that are nonhomogeneous in dimension. In addition, depending on the practical aspect of the specific problem, we can additionally require of the maps f some of the following properties.

1. The images of the maps f are contained in a given closed subset K of M.
2. The maps f have smoothness of class C^r.
3. The maps f, and also their jet extensions $J^k f$ (the "derivatives" of f), have uniformly bounded extensibility (that is, their rate of change, rate of change of velocity, and so on, are uniformly bounded).

If we solve variational problems not in classes of parametrizations, but in classes of their T-images (or L-images), then the admissible multivarifold-solutions must be T-parametrized (or L-parametrized) by a fixed multivarifold W and maps f with the properties picked out above; here we mention a very important feature: the possible rise of the low-dimensional strata of image-films of the maps f leads to the fact that we should admit to the class of multivarifold-solutions not only homogeneous multivarifolds but also nonhomogeneous multivarifolds.

3.3. Variational classes. Let M be a Riemannian manifold, and let W be a fixed multivarifold on some Riemannian manifold N.

DEFINITION 9.3.1. A *basic variational class of type* (F) in $P^r(M, W)$ (resp. in $P_e^r(M, W)$) is any family of parametrizations (W, f) in $P^r(M, W)$ (resp. in $P_e^r(M, W)$) such that

(1) $f(N) \subset K$ (resp. $f(\operatorname{spt} W) \subset K$), where K is a closed subset of M;

(2) $f|_G = g$, where G is a compact subset of $\operatorname{spt}^* W$, and $g: G \to M$ is a given map;

(3) $\operatorname{Lip} J^i f \leq \eta_i$ (resp. $\operatorname{Lip}(J^i f|_{\operatorname{spt}^* W}) \leq \eta_i$), $0 \leq i < r+1$, where η_i are certain positive numbers (the η_i may be infinite), which we call the *expansion coefficients of the class.*

DEFINITION 9.3.2. A *basic variational class of type* (F) in $T^r(M, W)$ or $L^r(M, W)$ (resp. in $T_e^r(M, W)$ or $L_e^r(M, W)$) is the quotient space of any basic variational class of type (F) in $P^r(M, W)$ (resp. in $P_e^r(M, W)$) with respect to T-equivalence or L-equivalence.

DEFINITION 9.3.3. A *basic variational class of type* (M) in $P^r(M, W)$ (resp. in $P_e^r(M, W)$) is any family of parametrizations (W, f) in $P^r(M, W)$ (resp. in $P_e^r(M, W)$) such that

(1) $f(N) \subset K$ (resp. $f(\operatorname{spt}^* W) \subset K$), where K is a closed subset;

(2) $f(G) \subset S$, where G is a compact subset of $\operatorname{spt}^* W$, and S is a given compact subset of K;

(3) $\operatorname{Lip} J^i f \leq \eta_i$ (resp. $\operatorname{Lip}(J^i f|_{\operatorname{spt}^* W}) \leq \eta_i$), $0 \leq i < r+1$, where η_i are certain positive numbers (the η_i may be infinite), which we call the *expansion coefficients of the class.*

DEFINITION 9.3.4. A *basic variational class of type* (M) in $T^r(M, W)$ or $L^r(M, W)$ (resp. in $T_e^r(M, W)$ or $L_e^r(M, W)$) is the quotient space of any basic variational class of type (M) in $P^r(M, W)$ (resp. in $P_e^r(M, W)$) with respect to T-equivalence or L-equivalence.

Finally, we understand by a *variational class of the general problem* an arbitrary subset of some basic variational class listed in Definitions 9.3.1–9.3.4. In particular, all the homotopy classes of any basic variational class are variational classes of the general problem. We also observe that basic variational classes of type (F) (resp. of type (M)), and also certain subclasses of them (for example, their homotopy classes) serve as classes of admissible variations for different parametrized and homotopic problems "with fixed boundary" (resp. "with moving boundary").

3.4. Formulation of the general variational problem. For the definitive statement of the general problem in the language of the theory of multivarifolds we need to define the concept of minimality. This concept, as usual, is specified by some functional over a space of parametrizations (or parametrized multivarifolds), which must satisfy the following natural requirements:

1. The value of the functional on some parametrization depends only on its multivarifold-image, in other words, the functional is invariant with respect to the relation of T-equivalence (or L-equivalence).

2. The value of the functional depends only on the stratum of maximal dimension of the multivarifold-image measured by the parametrization, that is, the value of the functional on the degenerate part of the parametrization is zero.

3. The functional has a definite continuity property.

All these requirements are satisfied by generalized integrands, which were introduced over the space of multivarifolds $V_k M$ in Ch. 6, §3 and over the space of parametrizations $P_k^r(M)$ in the following definition.

DEFINITION 9.3.5. An *integrand* (resp. *generalized integrand*) over a space $P_k^r(M)$ of parametrizations of order k, $0 \leq r \leq \infty$, on a Riemannian manifold M is any functional $\tilde{J}$ over $P_k^r(M)$ defined by the formula

$$\tilde{J}(W, f) = J(T_f W) = J(L_f W)$$

for an arbitrary parametrization (W, f) in $P_k^r(M)$, where J is an integrand (resp. generalized integrand) over $V_k M$.

Let us proceed to the definitive statement of the general problem in classes of parametrizations and parametrized multivarifolds.

Let E be an arbitrary variational class of parametrizations (resp. parametrized multivarifolds), and J a generalized integrand over the space of parametrizations (resp. parametrized multivarifolds). Then the general problem with parametrizations (resp. with parametrized multivarifolds) can be stated as follows: to find an element of E that minimizes the generalized integrand in the given class E.

The next section will be devoted to a solution of this problem.

§4. Existence and properties of minimal parametrizations and parametrized multivarifolds

4.1. Semicontinuity of generalized integrands. Theorem 7.1.2 shows that the multimass of multivarifolds is lower semicontinuous in the compact-weak topology. This means that all the mass functions $M_0 V, \dots, M_k V$ are lower semicontinuous in this topology. Under transition to the parametric topology, the multimass ceases to be lower semicontinuous because of possible local degenerations in the process of continuous deformation, causing the appearance of the additional part of the i-mass on account of the degenerations of masses of higher dimension. Despite this, we show below that generalized integrands, in particular the leading mass, are lower semicontinuous in the parametric topology.

LEMMA 9.4.1. *Let M be an arbitrary Riemannian manifold. Then any integrand (resp. generalized integrand) J over the space V_kM of multivarifolds is continuous (resp. lower semicontinuous) in the compact-weak topology.*

PROOF. We first suppose that J is an integrand. According to the definition there is a continuous function l on G_kM such that $J(V) = V^k(l)$ for an arbitrary multivarifold V, where V^k denotes the k-dimensional stratum on V. Consider the continuous function $\tilde{l}$ such that $\tilde{l} = 0$ on $G_kM \setminus \Gamma_kM$ and $\tilde{l} \equiv l$ on Γ_kM. Clearly, $J(V) = V^k(l) = V(\tilde{l})$. Hence, it follows that J is continuous in the compact-weak topology. Next, if J is a generalized integrand, then by definition J is the upper envelope of some family of integrands continuous in the compact-weak topology, so it is a lower semicontinuous function in the compact-weak topology (see [25]).

THEOREM 9.4.1 (Đào Trọng Thi). *Let M and N be arbitrary Riemannian manifolds, W a multivarifold of order k on N, and r an integer, $1 \leq r \leq \infty$. Then any integrand (resp. generalized integrand) J over $P_k^r(M)$ is continuous (resp. lower semicontinuous) on $P^r(M, W)$ in the parametric topology.*

PROOF. We assume that a sequence (W, f_n) in $P^r(M, W)$ converges to (W, f) in the parametric topology. We set $V_n = T_{f_n}W$, $V = T_fW$ and denote the k-dimensional strata of the multivarifolds W, V, $V_1, \dots,$ $V_n, \dots$ by W^k, V^k, $V_1^k, \dots, V_n^k, \dots$, respectively. According to Ch. 7, (3.2), for an arbitrary continuous function φ on G_kM we have $V(\varphi) = W(\tau_f \varphi G_f)$, $V_n(\varphi) = W(\tau_{f_n} \varphi G_{f_n})$, where τ_f and τ_{f_n} are functions on G_kN, and G_f and G_{f_n} are the maps of G_kN into G_kM induced by df and df_n. Consider the following new functions on G_kN:

$$\tau_g^k(\pi) = \begin{cases} \tau_g(\pi) & \text{if } \pi \in \Gamma_kN,\ \text{rank}\, dg_x|_\pi = k, \\ 0 & \text{for the remaining } \pi \in G_kN, \end{cases}$$

where $g = f, f_1, \dots, f_n, \dots$. It is easy to verify that τ_g^k are continuous functions on G_kN and moreover $\tau_g^k \varphi G_g$ are also continuous functions on G_kN. In addition, since f_n converges to f and df_n converges to df uniformly on every compact subset of N, the functions $\tau_{f_n}^k \varphi G_{f_n}$ converge to the function $\tau_f^k \varphi G_f$ uniformly on every compact subset of G_kN.

Consequently,

$$W(\tau_{f_n}^k \varphi G_{f_n}) \to W(\tau_f^k \varphi G_f)$$

since the functional W is continuous on $C(G_k N)$. On the other hand, it is obvious that

$$V_n^k(\varphi) = W^k(\tau_{f_n}^k \varphi G_{f_n}) = W(\tau_{f_n}^k \varphi G_{f_n}),$$
$$V^k(\varphi) = W^k(\tau_f^k \varphi G_f) = W(\tau_f^k \varphi G_f).$$

Thus, $V_n^k(\varphi) \to V^k(\varphi)$ for an arbitrary function $\varphi \in C(G_k M)$, so V_n^k converges to V^k in the compact-weak topology. Now suppose that $J(W, f) = \widetilde{J}(T_f W)$, $(W, f) \in P^r(M, W)$. Observing that $\widetilde{J}(V) = \widetilde{J}(V^k)$ and applying Lemma 9.4.1 to the integrand (resp. generalized integrand) $\widetilde{J}$, we find that the integrand (resp. generalized integrand) J is continuous (resp. lower semicontinuous) in the parametric topology. This proves the theorem.

COROLLARY 9.4.1. *Let M, N, W be chosen as in Theorem 9.4.1. Then any integrand (resp. generalized integrand) J over the space $V_k M$ is continuous (resp. lower semicontinuous) on the spaces $T^r(M, W)$ and $L^r(M, W)$, $1 \le r \le \infty$, in the parametric topology.*

REMARK. The fact that an arbitrary integrand (resp. generalized integrand) is continuous (resp. lower semicontinuous) on the spaces $P_e^r(M, W)$, $T_e^r(M, W)$, $L_e^r(M, W)$, $1 \le r \le \infty$, in the exact topology is proved in exactly the same way.

THEOREM 9.4.2 (Đào Trọng Thi). *Let M and N be arbitrary Riemannian manifolds, W a k-dimensional chain of class C^2 on N, and η, η' positive numbers. Let P denote the set of all parametrizations (W, f) in $P^0(M, W)$ such that f is a map of class C^2 and $\operatorname{Lip} f \le \eta$, $\operatorname{Lip} J^1 f \le \eta'$. Then any integrand (resp. generalized integrand) over the space $P_k^0(M)$ is continuous (resp. lower semicontinuous) on the closure $\overline{P}$ of the set P in the space $P^0(M, W)$ with the parametric topology.*

To prove this theorem we need the following lemma.

LEMMA 9.4.2. *Suppose (W, f), $(W, g) \in \overline{P}$, and*

$$\sup |f(x) - g(x)| < \varepsilon, \qquad x \in \operatorname{spt}^* W. \tag{4.1}$$

Then there is a constant c such that for sufficiently small ε we have

$$N([T_f W]^k - [T_g W]^k) \le c\varepsilon(M_k W + M_{k-1}(\partial W)),$$

where $[T_f W]^k$ *and* $[T_g W]^k$ *denote the* k*-dimensional strata of the multivarifolds* $T_f W$ *and* $T_g W$ *respectively.*

PROOF. First of all we observe that if Δ is a $(k+1)$-dimensional simplicial chain of class C^2 on M, then by Ch. 6, (3.3) we have

$$\partial[\Delta](\varphi) = \int_{\partial\Delta} \varphi \overrightarrow{\partial\Delta}(z)\, dz \tag{4.2}$$

for any locally Lipschitz function φ on $\Gamma_k M$. The continuously differentiable section $\overrightarrow{\partial\Delta}$ can be extended to a continuously differentiable section $\overline{\Delta}\colon \Delta \to \Gamma_k M$ defined on Δ. Now applying Stokes's theorem to the right-hand side of (4.2) we obtain

$$|\partial|\Delta|(\varphi)| \le \operatorname{Lip}\overline{\Delta} \operatorname{Lip}\varphi M_{k+1}(|\Delta|). \tag{4.3}$$

The inequality (4.3) can be generalized in a natural way to an arbitrary $(k+1)$-dimensional chain Z of class C^2 on M, namely

$$|\partial Z(\varphi)| \le c' \operatorname{Lip}\varphi M_{k+1} Z, \tag{4.4}$$

where the number c' depends only on Z.

Suppose that ε is sufficiently small. Then for every point $x \in \operatorname{spt}^* W$ the points $f(x)$ and $g(x)$ can be joined by a unique minimal geodesic $\alpha(t, f(x), g(x))$, $0 \le t \le 1$. Consider the homotopy $h\colon [0, 1]\times \operatorname{spt}^* W \to M$, $h(t, x) = \alpha(t, f(x), g(x))$. It is easy to verify that

$$[T_g W]^k - [T_f W]^k = \partial[T_h([0, 1] \times W)]^{k+1} + [T_h([0, 1] \times \partial W)]^k. \tag{4.5}$$

Suppose that (W, f), $(W, g) \in P$ and that (4.1) is satisfied. Then h is a map of class C^2; $\operatorname{Lip} h \le \eta$, $\operatorname{Lip} J^1 h \le \eta'$. Consequently, $[T_h([0, 1] \times W)]^{k+1}$ is a $(k+1)$-dimensional chain of class C^2 on $[0, 1] \times N$. For every function $\varphi \in C(\Gamma_k M)$, $\|\varphi\| \le 1$, $\operatorname{Lip}\varphi \le 1$, it follows from (4.4) that

$$|\partial[T_h([0, 1] \times W)]^{k+1}(\varphi)| \le c' \eta^k \varepsilon M_k W, \tag{4.6}$$

$$|[T_h([0, 1] \times \partial W)]^k(\varphi)| \le \varepsilon \eta^{k-1} M_{k-1}(|\partial W|), \tag{4.7}$$

where since $\operatorname{Lip} J^1 h \le \eta'$, the constant c' can be chosen to be common for all f and g in P. From (4.5), (4.6), and (4.7) we obtain

$$|[T_g W]^k(\varphi) - [T_f W]^k(\varphi)| \le c\varepsilon(M_k(|W|) + M_{k-1}(|\partial W|)), \tag{4.8}$$

where $c = \max\{c'\eta^k, \eta^{k-1}\}$. Hence we have

$$N([T_g W]^k - [T_f W]^k) \le c\varepsilon(M_k(|W|) + M_{k-1}(|\partial W|)). \tag{4.9}$$

If (W, f), $(W, g) \in P$, then we can choose sequences (W, f_n) and

(W, g_n) in P that converge in the parametric topology of $P^0(M, W)$ to (W, f) and (W, g) respectively. The estimate (4.9) together with Theorem 7.1.1 shows that $[T_{f_n} W]^k$ and $[T_{g_n} W]^k$ tend to $[T_f W]^k$ and $[T_g W]^k$ as $n \to \infty$, and the inequality remains true for f and g in this general case. This proves the lemma.

PROOF OF THEOREM 9.4.2. Suppose that the sequence (W, f_n) in $\overline{P}$ converges to (W, f) in the parametric topology. From Lemma 9.4.2 it follows that the corresponding sequence of k-dimensional strata $[T_{f_n} W]^k$ converges to $[T_f W]^k$ in the N-topology. On the other hand, obviously $\operatorname{Lip} f_n \leq \eta$, and from this it follows that the $M_k(|[T_{f_n} W]^k|)$ are uniformly bounded and the $\operatorname{spt}^*[T_{f_n} W]^k$ are contained in some compact subset. According to Theorem 7.1.1, the sequence $[T_{f_n} W]^k$ also converges to $[T_f W]^k$ in the compact-weak topology. Now the assertion of the theorem follows from the continuity (resp. lower semicontinuity) of the integrands (resp. generalized integrands) over the space of multivarifolds in the compact-weak topology.

COROLLARY 9.4.2. *Let M, N, W, P, $\overline{P}$ be chosen as in Theorem 9.4.2. Then any integrand (resp. generalized integrand) J over $V_k M$ is continuous (resp. lower semicontinuous) on the quotient spaces of $\overline{P}$ with respect to T-equivalence and L-equivalence.*

4.2. Theorems on the existence of minimal solutions. Consider an arbitrary Riemannian manifold M.

DEFINITION 9.4.1. Let E be a variational class in $P^r(M, W)$ (resp. $T^r(M, W)$ or $L^r(M, W)$), $r \geq 1$. Then the problem of minimizing a generalized integrand J in the class E is said to be *consistent* if the following conditions are satisfied:

(1) there are constants η_i, $0 \leq i < r+1$, such that for any $V \in E$ there is a $V' \in E$, $V' = (W, f)$ (resp. V' is T-parametrized or L-parametrized by the pair (W, f)) for which $\operatorname{Lip} J^i f \leq \eta_i$, $J(V') \leq J(V)$;

(2) the set of all V' in E, $V' = (W, f)$ (resp. V' can be T-parametrized or L-parametrized by the pair (W, f)) such that $\operatorname{Lip} J^i f \leq \eta_i$ is compact.

DEFINITION 9.4.2. Let E be a variational class in $P^0(M, W)$ (resp. $T^0(M, W)$ or $L^0(M, W)$). The problem of minimizing a generalized integrand J in the class E is said to be *consistent* if the following conditions are satisfied:

(1) there are constants $\eta > 0$ and $\eta' > 0$ such that for any $V \in E$ there is a $V' \in E$, $V' \in \overline{P}$ (resp. V' is T-parametrized or L-parametrized by a pair in $\overline{P}$) for which $J(V') \le J(V)$, where P denotes the set of all parametrizations (W, f) such that f is a map of class C^2, $\operatorname{Lip} f \le \eta$, $\operatorname{Lip} J^1 f \le \eta'$;

(2) the set of all V' in E, $V' \in \overline{P}$ (resp. V' can be T-parametrized or L-parametrized by a pair in $\overline{P}$) is compact.

DEFINITION 9.4.3. Let E be a variational class in $P_e^r(M, W)$ (resp. $T_e^r(M, W)$ or $L_e^r(M, W)$), $r \ge 1$. The problem of minimizing a generalized integrand J in the class E is said to be *consistent* if the following conditions are satisfied:

(1) there are constants $\eta_i > 0$, $0 \le i < r+1$, such that for any $V \in E$ there is a $V' \in E$, $V' = (W, f)$ (resp. V' is exactly T-parametrized or L-parametrized by the pair (W, f)) for which $\operatorname{Lip}(J^i f|_{\mathrm{spt}^* W}) \le \eta_i$, $0 \le i < r+1$, $J(V') \le J(V)$;

(2) the set of all V' in E, $V' = (W, f)$ (resp. V' can be exactly T-parametrized or L-parametrized by the pair (W, f)) such that $\operatorname{Lip}(J^i f|_{\mathrm{spt}^* W}) \le \eta_i$ is compact.

DEFINITION 9.4.4. Let E be a variational class in $P_e^0(M, W)$ (resp. $T_e^0(M, W)$ or $L_e^0(M, W)$). The problem of minimizing a generalized integrand J in the class E is said to be *consistent* if the following conditions are satisfied:

(1) there are constants $\eta > 0$ and $\eta' > 0$ such that for any $V \in E$ there is a $V' \in E$, $V' \in \overline{P}_e$ (resp. V' is exactly T-parametrized or L-parametrized by a pair in $\overline{P}_e$) for which $J(V') \le J(V)$, where P_e denotes the set of all exact parametrizations $(W, f) \in P_e^0(M, W)$ such that f is a map of class C^2, $\operatorname{Lip}(f|_{\mathrm{spt}^* W}) \le \eta$, $\operatorname{Lip}(J^1 f|_{\mathrm{spt}^* W}) \le \eta'$;

(2) the set of all V' in E, $V' \in \overline{P}_e$ (resp. V' can be exactly T-parametrized or L-parametrized by a pair in $\overline{P}_e$) is compact.

THEOREM 9.4.3. *Suppose that the problem of minimizing a generalized integrand J in the variational class E is consistent. Then this problem has a solution.*

PROOF. We consider, for example, the case where E is a variational class in $P^r(M, W)$, $r \ge 1$; all the other cases can be proved in exactly the same way.

Condition (1) in Definition 9.4.1 means that if the given problem has solutions, then at least one of them can be found in the set E' of all

$V' = (W, f) \in E$ such that $\operatorname{Lip} J^i f \leq \eta_i$. At the same time, condition (2) guarantees that this set E' is compact. Now the attainment of a minimum by the generalized integrand J is conditioned by the classical Weierstrass-Tonelli principle, since J is lower semicontinuous, as we proved in Theorem 9.4.1. This proves the theorem.

REMARK. When $J = M_k$ Definitions 9.4.1 and 9.4.3 can be extended to the case $r = 0$, and Theorem 9.4.3 remains true because M_k is lower semicontinuous (see Đào Trọng Thi [42]). Furthermore, condition (2) in Definitions 9.4.1, 9.4.2, 9.4.3, and 9.4.4 has been investigated in some detail in Chapter 8, §§2–3. In particular, from the results of these sections it follows that under natural assumptions this condition is satisfied for all the basic variational classes listed in §3 and their homotopy classes. As for condition (1), it is satisfied automatically for those variational classes for which the expansion coefficients η_i are finite. In particular, assertion (b) of the fundamental Theorem 6.4.1 of Chapter 6 follows from this fact and Theorem 9.4.3. The case where the η_i are infinite has still not been examined satisfactorily. We mention only the following important case.

THEOREM 9.4.4. *Let E be a fundamental variational class listed in* § 3 (*or any homotopy class of it*) *for which M is an arbitrary compact connected Riemannian manifold, N is a k-dimensional compact manifold with boundary G, $W = [N]$, $K = M$ and $r = 0$. Then the problem of minimizing M_k in the class E has a solution. In particular, assertion* (a) *of the fundamental Theorem* 6.4.1 *of Chapter* 6 *follows from this.*

PROOF. According to Theorem 8.2.5 of Chapter 8, condition (2) in Definition 9.4.1 is satisfied for E. In view of Theorem 9.4.3 (see also the remark to it), it remains to show that the problem of minimizing the leading mass M_k in E satisfies condition (1) in Definition 9.4.1. For simplicity, we consider the class E of parametrizations $\{(W, f)\}$ and denote by $\widetilde{E}$ the corresponding class of locally Lipschitz maps $\{f\}$. In this case, condition (1) implies the existence of a constant η such that for any $f \in \widetilde{E}$ there is an $f' \in \widetilde{E}$ satisfying the inequalities $\operatorname{vol}_k(f') \leq \operatorname{vol}_k(f)$ and $\operatorname{Lip} f' \leq \eta$. Let $X_0 \in W$ be an interior point. Consider a smooth Morse function φ on W such that $\varphi = 0$ on G and $0 < \varphi \leq 1$ on $W \backslash G$, and φ has a unique maximum x_0 and finitely many other critical points, among which there are no local minima and maxima. We set $W(t) = \{x \in W : \varphi(x) = t\}$, $W(t, t') = \{x \in W : t < \varphi(x) < t'\}$, $W[t, t'] = \{x \in W : t \leq \varphi(x) \leq t'\}$. Let $\lambda_1 < \lambda_2 < \cdots < \lambda_{p+1} = 1$ be the critical values of φ, and let $\lambda_0 = 0$. The open submanifold $W(\lambda_i, \lambda_{i+1})$, $0 \leq i \leq p$, is a cylinder, parametrized by the level surfaces of φ and trajectories of the field $\operatorname{grad} \varphi$. Let d denote the infimum of k-dimensional volume in the class $\widetilde{E}$, and suppose we are given $f \in \widetilde{E}$. We shall reconstruct f on the k-dimensional stratum and on lower-dimensional strata of its image separately.

Let us choose $f' \in \widetilde{E}$ so that $\mathrm{vol}_k(f') \leq \max\{\mathrm{vol}_k(f), d+1\}$. For simplicity, we again denote f' by f. Let S_k denote the k-dimensional stratum of $f(W)$, and set $W_k = f^{-1}(S_k)$. Since M is compact, there is a constant $c_0 > 0$ such that every point $x \in M$ has a spherical neighborhood $B(x, c_0)$ with center x and radius c_0. We choose and fix some constant $c < c_0$. For each subset $X \subset M$ we denote its c-neighborhood $\bigcup_{x\in X} B(x, c)$ by X^c. Let $Y = \bigcup_{0\leq t\leq 1} Y_t \subset S_k$ be a cylinder-cone for which the horizontal sections Y_t $(0 < t < 1)$ are closed $(k-1)$-dimensional surfaces, and the bases Y_0 and Y_1 may degenerate into surfaces of dimension $\leq k-2$. We say that Y is *thin* if there is a cylinder-cone $X = \bigcup_{0\leq t\leq 1} X_t \subset Y$, $X_t \subset Y_t$ $(0 \leq t \leq 1)$, $\dim X \leq k-1$, and a locally Lipschitz homotopy $h\colon Y^c \times [0, 1] \to Y^c$ such that

(a) $h(x, 0) = x$ for any $x \in Y^c$,
(b) $h(Y_t \times \{1\}) = X_t$ for any $t \in [0, 1]$,
(c) $h(x, \tau) \in B(h(x, 1), c)$ for any $x \in Y$.

Next, we define restrictions that enable us, in particular, to condense thin cylinder-cones into low-dimensional pieces.

Reconstructions of type I. Consider cylinder-cones Y, X $(X \subset Y)$ as above and suppose that there is a locally Lipschitz homotopy $h\colon [Y_0^c \cup Y_1^c] \times [0, 1] \to [Y_0^c \cup Y_1^c]$ such that

(a′) $h(x, 0) = x$ for any $x \in Y_0^c \cup Y_1^c$,
(b′) $h(Y_0 \times \{1\}) = X_0$, $h(Y_1 \times \{1\}) = X_1$,
(c′) $h(x, \tau) \in B(h(x, 1), c)$ for any $x \in Y_0 \cup Y_1$,
(d′) $\mathrm{vol}_k(h([Y_0 \cup Y_1] \times [0, 1])) \leq \frac{1}{2}\mathrm{vol}_k(Y)$.

(For the case of the homotopy class $\widetilde{E}$ we also require

(e′) $Y' = X \cup h([Y_0 \cup Y_1] \times [0, 1])$ and Y are homotopic.)

Then we reconstruct S_k by replacing Y by Y'. We observe that conditions (a′)–(c′) imply that Y has thin bases, while condition (d′) shows that Y is long in relation to its boundary.

After all possible reconstructions of this type, the resulting k-dimensional stratum S_k does not contain "long thin pieces". On the other hand, under such reconstructions $\mathrm{vol}_k(S_k)$ decreases, although some new low-dimensional pieces may appear. Now under the condition $\mathrm{vol}_k(S_k) \leq d+1$, we can reparametrize S_k so that $\mathrm{Lip}\, f$ on W_k is bounded by some constant μ_1 depending only on the class $\widetilde{E}$, since a strong expansion of S_k in some directions would imply a contraction of it in other directions, and this would cause the appearance of long thin cylinder-cones.

Let us turn to low-dimensional strata of the image $f(W)$. Henceforth, we use only these, leaving the k-dimensional stratum S_k untouched. Since we are minimizing only the k-dimensional volume, the low-dimensional strata carry out only their topological functions here. Thus, this part of the proof has a purely topological character. Let $\widetilde{W}(t, t')$ (resp. $\widetilde{W}[t, t']$) denote low-dimensional strata of $f(W(t, t'))$ (resp. $f(W[t, t'])$) and set $\widetilde{W}(t) = f(W(t) \cap \widetilde{W}[0, 1])$. Low-dimensional strata are due to the degeneracy of the map f, and to the i-dimensional pieces there correspond the parts of W where $\operatorname{rank} df = i$. We note that the value of rank df in some sufficiently small neighborhood of the point $x \in W$ is always at least equal to its value at this point itself. Therefore, by means of a deformation and reparametrization we can arrange that the degeneracy of f on different parts of W takes one of the following forms:

1. The degeneracy of f takes place on some subcylinder $A[\tau, \tau'] \subset W[\tau, \tau']$ (formed by some of the trajectories in $W[\tau, \tau']$) along the level surfaces. Correspondingly, the image $\widetilde{A}[\tau, \tau'] = f(A[\tau, \tau'])$ is a mixed cylinder-cone such that the inverse image of its generators consists of trajectories in $W[\tau, \tau']$. The cylinder-cones $\widetilde{A}[\tau, \tau']$ are of two kinds. For the first kind, the lateral boundary $\bigcup_{\tau \leq t \leq \tau'} \partial A(t)$ of the subcylinder $A[\tau, \tau']$ is empty, and for the second kind this boundary belongs to $\partial W_k \backslash G$. In the latter case, the bases of $\widetilde{A}[\tau, \tau']$ are pasted either to the k-dimensional stratum or to cylinder-cones of the first kind. The parts of $\widetilde{W}[t, t']$ that belong to cylinder-cones of the first and second kinds are denoted respectively by $\widetilde{W}_1[t, t']$ and $\widetilde{W}_2[t, t']$. We set $\widetilde{W}_1(t) = \widetilde{W}(t) \cap \widetilde{W}[0, 1]$ and $\widetilde{W}_2(t) = \widetilde{W}(t) \cap \widetilde{W}_2[0, 1]$. Clearly, we can arrange that $\widetilde{W}_1(t)$ and $\widetilde{W}_2(t)$ do not contain trivial cycles.

2. The degeneracy of f forms a "horizontal branch". Namely, there is a $(k-1)$-dimensional surface $X \subset W$ such that ∂X is either empty or belongs to $\partial W_k \backslash G$. The surface X does not lie entirely on any level surface, and the map f takes it into $f(X \cap W(\tau))$ for some $\tau \in (0, 1)$. This situation can also be described as follows. The region $H \subset W \backslash W_k$ bounded by the surface X is represented as a cylinder-cone $H = \bigcup_{0 \leq \tau \leq 1} H_\tau$, for which $H_0 = X$, $H_1 \subset W(s)$ and the lateral boundary $\bigcup_{0 \leq \tau \leq 1} \partial H_\tau$ is either empty or belongs to $\partial W_k \backslash G$. The horizontal section H_τ $(0 \leq \tau < 1)$ does not lie entirely on any level surface, and under the map f the cycle of H_0 that distinguishes it from H_1 is annihilated. Furthermore, the degeneracy of f either takes place on H along horizontal sections H_τ and

forms cylinder-cones of the first and second kinds, as in part 1, or it gives rise to "horizontal branches".

Henceforth for simplicity we present only reconstructions of the cylinder-cones $\widetilde{W}_1[\tau, \tau']$ and $\widetilde{W}_2[\tau, \tau']$. Similar operations are carried out for each horizontal branch $\widetilde{H} = f(H)$, $(\widetilde{H}_\tau = f(H_\tau))$. We start by considering cylinder-cones of the first kind. $\widetilde{W}_1[0, 1]$ splits into a chain of cylinder-cones $\{\widetilde{W}_1[\gamma_j, \gamma_{j+1}]\}$, $\{\lambda_0, \lambda_1, \dots, \lambda_{p+1}\} \subset \{\gamma_0, \gamma_1, \dots, \gamma_{q+1}\}$ (some of the $\widetilde{W}_1[\gamma_j, \gamma_{j+1}]$ may be empty). A topological function of the image $f(W)$ is the pasting of nontrivial cycles of the boundary $f(G)$ and a realization of definite nontrivial cycles in M. Each cylinder-cone $\widetilde{W}_1[\gamma_j, \gamma_{j+1}]$ realizes either a contraction or an expansion (blowing-up) of a definite nontrivial cycle in W, or both these operations simultaneously. In the last situation, $\widetilde{W}_1[\gamma_j, \gamma_{j+1}]$ must be replaced by two cylinder-cones, which carry out these operations separately. We then carry out all possible reconstructions of the following types:

Reconstructions of type II. If there are two regions in $W \backslash W_k$ that realize mutually exclusive cycles under f, then they must be joined and contracted together to a point. These reconstructions are necessary only for the homotopic case.

Reconstructions of type III. Let $\widetilde{A}[\gamma_u, \gamma_v] \subset \widetilde{W}_1[\gamma_u, \gamma_v]$ be a subcylinder formed by the part of the generators of the cylinder-cones $\widetilde{W}_1[\gamma_j, \gamma_{j+1}]$ in $\widetilde{W}_1[\gamma_u, \gamma_v]$ $(\gamma_u \le j \le \gamma_v - 1)$. Suppose that some nontrivial cycle in $\widetilde{A}(t) = \widetilde{W}(t) \cap \widetilde{A}[\gamma_u, \gamma_v]$ $(\gamma_u < t < \gamma_v)$ is contractible to a point in the two bases $\widetilde{A}(\gamma_u)$ and $\widetilde{A}(\gamma_v)$ simultaneously (this means that $\widetilde{A}[\gamma_u, \gamma_v]$ first expands and then contracts this nontrivial cycle). Let $\widetilde{A}'[\gamma_u, \gamma_v] \subset \widetilde{A}[\gamma_u, \gamma_v]$ denote the subcylinder obtained from $\widetilde{A}[\gamma_u, \gamma_v]$ by a simultaneous contraction of this cycle in all sections $\widetilde{A}(t)$ $(\gamma_u \le t \le \gamma_v)$. We then reconstruct $\widetilde{W}_1[0, 1]$ by replacing $\widetilde{A}[\gamma_u, \gamma_v]$ by $\widetilde{A}'[\gamma_u, \gamma_v]$ (for the homotopic case this replacement will be carried out only on condition that $\widetilde{A}'[\gamma_u, \gamma_v]$ and $\widetilde{A}[\gamma_u, \gamma_v]$ are homotopic). Under this operation, cylinders may appear, each of which simply joins two contracting-expanding cylinder-cones.

Consider a connected chain of contracting and "joining" cylinder-cones $\{\widetilde{W}_1[\gamma_j, \gamma_{j+1}], s \le j \le s' - 1\}$, that is, all the $\widetilde{W}_1[\gamma_j, \gamma_{j+1}]$ in the chain are nonempty contracting or joining cylinder-cones. Since the number of nontrivial cycles in $\widetilde{W}_1(t)$ $(s \le t \le s')$ gradually decreases as t increases, and each contracting cylinder-cone is followed by no more than one joining cylinder, the number of cylinder-cones in a given chain must be restricted by the structures of W and M. A similar assertion is true for an arbitrary connected chain of expanding and joining cylinder-cones $\{\widetilde{W}_1[\gamma_j, \gamma_{j+1}], s \leq j \leq s' - 1\}$ (that is, all the $\widetilde{W}_1[\gamma_j, \gamma_{j+1}]$ in the chain are nonempty and are expanding or joining cylinder-cones). On the other hand, after reconstructions of type III, each connected chain of expanding and joining cylinder-cones either ends in $f(\partial W_k \backslash G)$ (that is, it is joined to some "k-dimensional expanding cylinder-cone"), or it realizes a nontrivial cycle in M (for the homotopic case). In view of the absence of thin pieces in S_k, each k-dimensional expanding cylinder-cone takes up a k-dimensional volume greater than a certain number. Now from the fact that $\mathrm{vol}_k(S_k) \leq d + 1$ and the homotopy class is fixed (in the homotopic case), we can conclude that the number of connected chains of expanding and joining cylinder-cones in $\widetilde{W}_1[0, 1]$, and consequently the number of all cylinder-cones of the first kind $\widetilde{W}_1[\gamma_j, \gamma_{j+1}]$, is bounded by the class $\widetilde{E}$.

Consider cylinder-cones of the second kind $\widetilde{W}_2[\tau, \tau']$. Their lateral boundaries are pasted to the k-dimensional stratum S_k. Since there are no thin pieces in S_k and $\mathrm{vol}_k(S_k) \leq d + 1$, it is easy to verify that the number of these cylinder-cones is also bounded by the class $\widetilde{E}$.

Finally, let us consider horizontal branches. Suppose that the horizontal branch $\widetilde{H}$ contains only cylinder-cones of the first kind, and among them there is no expanding piece. Obviously, the restriction of f to H can be represented as a composition of locally Lipschitz maps $f_1 \circ f_2$, where $H' = f_2(H)$ is a cylinder with horizontal sections $H'_\tau = f_2(H_\tau) \cong H'_1 = H_1$. Then we can replace f by a map homotopic to it that coincides with f on $W \backslash H$ and takes H into $\widetilde{H}_0$. We again denote this new map by f. Thus, the horizontal branch $\widetilde{H}$ is reducible to its base $\widetilde{H}_0$, which lies in the union S^* of the subsets S_k, $\widetilde{H}_1[0, 1]$, $\widetilde{W}_2[0, 1]$ and irreducible horizontal branches. Consequently, the necessary boundedness of $\mathrm{Lip}\, f$ on $\widetilde{H}$ will follow from the bound of $\mathrm{Lip}\, f$ on S^*. It remains to consider irreducible horizontal branches. As we have seen, an obstruction to the reduction of the horizontal branch $\widetilde{H}$ is either the fact that its lateral boundary is pasted to S_k, or the fact that it contains expanding cylinder-cones and realizes a nontrivial cycle in M (for the homotopic case). Therefore, in view of reconstructions of type I and the fact that $\mathrm{vol}_k(S_k) \leq d+1$ and the homotopy class is fixed (for the homotopic case), we can conclude that the number of irreducible horizontal branches is bounded by the class $\widetilde{E}$. Now, as we did for $\widetilde{W}_1[0, 1]$ and $\widetilde{W}_2[0, 1]$, we can show that the number of cylinder-cones of both kids in each irreducible horizontal branch is bounded by the class $\widetilde{E}$. Hence we have proved that the total number of cylinder-cones in $\widetilde{W}_1[0, 1]$, $\widetilde{W}_2[0, 1]$ and in all irreducible horizontal branches is bounded by the class $\widetilde{E}$.

Let us choose and fix a triangulation $\{\Delta_\alpha\}$ on M such that each simplex Δ_α lies entirely in some ball $B(x, c)$. Consider an arbitrary cylinder-cone $\widetilde{W}_1[\gamma_j, \gamma_{j+1}]$. For each $t \in [\gamma_j, \gamma_{j+1}]$ we denote by $\Delta(t)$ the simplicial neighborhood of the section $\widetilde{W}_1(t)$, that is, $\Delta(t) = \{\bigcup \Delta_\beta : \Delta_\beta \cap \widetilde{W}_1(t) \neq \varnothing\}$. By means of deformations we can arrange that $\Delta(t)$ is contractible in itself to $\widetilde{W}_1(t)$ so long as $\widetilde{W}_1(t) \not\subset \Delta(\gamma_j)$ and $\widetilde{W}_1(t) \not\subset \Delta(\gamma_{j+1})$.

Reconstructions of type IV. Let $t, t \in (\gamma_j, \gamma_{j+1})$ be such that $\widetilde{W}_1(t) \not\subset \Delta(\gamma_j)$, $\widetilde{W}_1(t) \not\subset \Delta(\gamma_{j+1})$, $\widetilde{W}_1(t') \not\subset \Delta(\gamma_j)$, and $\widetilde{W}_1(t') \not\subset \Delta(\gamma_{j+1})$. Suppose that $\Delta(t) = \Delta(t')$ and $\widetilde{W}_1[t, t'] \not\subset \Delta(t)$, that is, $\widetilde{W}_1[\gamma_j, \gamma_{j+1}]$ passes through the simplicial neighborhood $\Delta(t) = \Delta(t')$ for t and t'. Then we reconstruct $\widetilde{W}_1[\gamma_j, \gamma_{j+1}]$ by replacing $\widetilde{W}_1[t, t']$ by some cylinder $[\widetilde{W}_1(t), \widetilde{W}_1(t')] \subset \Delta(t)$ joining $\widetilde{W}_1(t)$ and $\widetilde{W}_1(t')$ (for the homotopic case this replacement will be carried out only on condition that $\widetilde{W}_1[t, t']$ and $[\widetilde{W}_1(t), \widetilde{W}_1(t')]$ are homotopic).

After all possible reconstructions of type IV, the resulting cylinder-cone, which we again denote by $\widetilde{W}_1[\gamma_1, \gamma_{j+1}]$, passes through each simplicial neighborhood $\Delta(t)$ exactly once in the nonhomotopic case and at most a certain number of times in the homotopic case (this "certain number" depends only on the class $\widetilde{E}$). Since the total number of combinations of simplexes in a given triangulation is already determined, the number of simplicial neighborhoods $\Delta(t)$ is bounded (by the class $\widetilde{E}$). Now we can deform $\widetilde{W}_1[\gamma_j, \gamma_{j+1}]$ inside each simplicial neighborhood $\Delta(t)$ and reparametrize it so that $\operatorname{Lip} f \leq \mu_2$ on $f^{-1}(\widetilde{W}_1[\gamma_j, \gamma_{j+1}])$, where μ_2 is a constant that depends only on the class $\widetilde{E}$. Carrying out similar operations for all cylinder-cones in $\widetilde{W}_1[0, 1]$, $\widetilde{W}_2[0, 1]$ and in all irreducible horizontal branches $\widetilde{H}$, we obtain $\operatorname{Lip} f \leq \mu_2'$ on $W \backslash W_k$ for some constant μ_2'. Finally, reparametrizing f, we arrange that $\operatorname{Lip} f \leq \mu$, where μ is a constant that depends only on the class $\widetilde{E}$. This completes the proof of the theorem.

4.3. The structure of the set of minimal solutions.

THEOREM 9.4.5 (*Đào Trong Thi*). *The set of all solutions of the problem of minimizing an arbitrary generalized integrand J in an arbitrary variational class E in $P^r(M, W)$ (resp. $P_e^r(M, W)$, $T^r(M, W)$, $L^r(M, W)$, $T_e^r(M, W)$, $L_e^r(M, W)$) when $r \geq 1$, is closed in E.*

PROOF. Consider the case $E \subset P^r(M, W)$; the remaining cases can be proved in exactly the same way. According to Theorem 9.4.1, the integrand J is lower semicontinuous. Suppose that the sequence of solutions V_n converges to V (in the parametric topology). We need to prove that V is again a solution. Indeed, from the fact that J is semicontinuous it follows that $J(V) \leq \lim_{n\to\infty} J(V_n) = J(V_n)$. On the other hand, $J(V) \geq J(V_n)$, since V_n minimizes J, so $J(V) = J(V_n)$, that is, V is a solution. This proves the theorem.

THEOREM 9.4.6. *Let* $P = P^2(M, W) \cap \{(W, f)\colon \operatorname{Lip} f \leq \eta,\ \operatorname{Lip} J^1 f \leq \eta'\}$, $P_e = P_e^2(M, W) \cap \{(W, f)\colon \operatorname{Lip}(f|_{\operatorname{spt}^* W}) \leq \eta,\ \operatorname{Lip}(J^1 f|_{\operatorname{spt}^* W}) \leq \eta'\}$, *and let* $\overline{P}$ (*resp.,* $\overline{P}_e$) *denote the closure of* P (*resp.* P_e) *in* $P^0(M, W)$ (*resp.* $P_e^0(M, W)$). *Also, let* $\overline{T}$ *and* $\overline{L}$ (*resp.* $\overline{T}_e$ *and* $\overline{L}_e$) *denote the quotient spaces of* $\overline{P}$ (*resp.* $\overline{P}_e$) *with respect to the* T-*equivalence and* L-*equivalence. Then the set of solutions of the problem of minimizing an arbitrary generalized integrand* J *is an arbitrary variational class* E *in* $\overline{P}$ (*resp.* $\overline{P}_e$, $\overline{T}$, $\overline{L}$, $\overline{T}_e$, $\overline{L}_e$) *is closed in* E.

The proof of this theorem is carried out on the same lines as the proof of Theorem 9.4.5.

CHAPTER X

Criteria for Global Minimality

§1. Statement of the problem in the functional language of currents

Let M be a Riemannian manifold, and let J be a functional over the space of k-dimensional compact surfaces in M. In this chapter we shall be concerned with the question of effective criteria for a given k-dimensional compact surface S to minimize J in the class of k-dimensional compact surfaces of variable topological type with a given boundary (that is, for it to be a solution of Problem A; see Chapter 6, §2). We observe that if S minimizes J in the class of k-dimensional compact surfaces of variable topological type with a given boundary, then a fortiori it minimizes J in the narrower class of k-dimensional compact surfaces of fixed topological type with a given boundary (that is, it is a solution of Problem B; see Chapter 6, §2).

It is well known that every k-dimensional compact surface S can be regarded as the k-current $[S]$ obtained by integrating along S. The basic concepts connected with surfaces (boundary, tangent space, volume, homology, and so on) can be carried over to functional language. Under this approach, the question of criteria for global minimality of surfaces can be carried over, as far as the solution is concerned, to functional spaces of currents. On the other hand, since the space of currents on a manifold M well reflects its topology, the statement and investigation of variational problems in different classes of currents have independent interest.

1.1. Concepts of globally minimal currents. Let M be a Riemannian manifold, and let J be a functional over the space of currents $E_k M$.

DEFINITION 10.1.1. A current $S \in E_k M$ (with or without boundary) is said to be *absolutely* (resp. *homologically*) *minimal with respect to* J if $J(S) \leq J(S')$ for any $S' \in E_k M$ such that the current $S - S'$ is closed (resp. exact).

Among all possible functionals over $E_k M$ we pick out the important class of functionals specified by Lagrangians.

DEFINITION 10.1.2. A *Lagrangian of degree* k *on* M is any map l: $\Lambda_k M \to \mathbf{R}$ such that its restriction to every fiber $\Lambda_k M_x$ of the bundle $\Lambda_k M$ is *positively homogeneous*.

Every Lagrangian l of degree k on M specifies a positively homogeneous functional over $E_k M$ by the formula

$$J(S) = \int l(\vec{S}_x)\, d\|S\|(x), \qquad S \in E_k M,$$

which is called an *integrand over currents*. Obviously, if ξ is a point current, then $J(\xi) = l(\xi)$. Thus, l is completely determined by J.

Clearly, absolutely (resp. homologically) minimal currents with respect to J are solutions of the problem of minimizing J in the class of currents with a given common boundary (resp. in the class of homological currents with a given common boundary). Like any variational problem, the investigation of this problem is carried out in two stages. At the first stage there is the question of the existence of solutions, which has been solved successfully (see [217] and [219], for example). At the second stage it is required to establish necessary and sufficient conditions for the global minimality of currents. The present chapter is devoted to this question.

1.2. Modern analysis of the classical Huyghens algorithm. In this subsection we present briefly the analysis carried out by Young [140] of the classical Huyghens algorithm concerning a family of light rays in a nonhomogeneous medium. This remarkable algorithm of geometrical optics will be the basis of the minimization method created by Đào Trọng Thi.

According to Fermat's principle the light paths are solutions of the classical one-dimensional problem of minimizing the integral

$$J(x) = \int l(x(t), \dot{x}(t))\, dt \tag{1.1}$$

in the class of curves $x = x(t)$, $a \le t \le b$, in Euclidean space $\mathbf{R}^3$ joining two given points, where the Lagrangian l is determined by the properties of the medium in question. Since the Lagrangian l is positively homogeneous, it follows that

$$\dot{x} l_{\dot{x}} = l. \tag{1.2}$$

Let $p = p(x)$ be a function with values in $\mathbf{R}^3$. We call p a *geodesic slope* if there is a function $S(x)$ such that for any pair $(x, \dot{x})$ we have

$$l(x, \dot{x}) \ge \dot{x} S_x(x), \tag{1.3}$$

and equality holds when $\dot{x} = p$. The function of two vector variables $\Phi(x, \dot{x}) = \dot{x} S_x(x)$ is called the "*exact derivative*". We observe that in this case Φ is completely determined by p. In fact the difference $l - \Phi$, regarded as a function of $\dot{x}$, attains a minimum when $\dot{x} = p$, and so its derivative with respect to $\dot{x}$ vanishes when $\dot{x} = p$, that is,

$$S_x(x) = l_{\dot{x}}(x, p(x)). \tag{1.4}$$

If p is a geodesic slope, then solutions of the vector differential equation $\dot{x} = p$ are called *curves of geodesic slope.* The Huyghens algorithm can now be stated thus: curves of geodesic slope minimize the functional J. From (1.2) and (1.4) it follows that

$$\Phi(x, \dot{x}) = l(x, p) + (\dot{x} - p)l_{\dot{x}}(x, p),$$

so that the integral

$$\int (l(x, p) + (\dot{x} - p)l_{\dot{x}}(x, p))\, dt \tag{1.5}$$

along any curve depends only on the ends of this curve (it is obviously equal to $S(x_1) - S(x_0)$, where x_0 and x_1 are the beginning and end of the path of integration). The expression (1.5) is called the *invariant Hilbert integral.* If we rewrite (1.3) in the form $l - \Phi \geq 0$ and substitute for Φ the integrand in (1.5), we obtain the Weierstrass condition

$$l(x, \dot{x}) - l(x, p) - (\dot{x} - p)l_{\dot{x}}(x, p) \geq 0. \tag{1.6}$$

Inequality (1.6) means that the function l is convex with respect to $\dot{x}$ at the point $\dot{x} = p$, which is the real geometrical meaning of the Weierstrass condition. It is more complicated to discover the connection between the condition that the Hilbert integral is invariant and the classical results. A theorem of Mályusz, which is not easy to prove, asserts that this condition is equivalent to the requirement that the family of integral curves of the equation $\dot{x} = p$ satisfies Euler's equation.

1.3. Convex functionals and the Hahn-Banach theorems.

DEFINITION 10.1.3. A functional J, defined on a vector space E, is said to be *convex at the point* p if for every convex combination $\sum \lambda_i z_i$, $z_i \in E$, $\lambda_i \geq 0$, $\sum \lambda_i = 1$, equal to p we have $J(p) \leq \sum \lambda_i J(z_i)$. The functional J is said to be *convex* if it is convex at all points of E.

PROPOSITION 10.1.1 (Young [140]). *A functional J, defined on a vector space E, is convex at a point $p \in E$ if and only if there is an affine functional l over E such that $l(z) \leq J(z)$ for all $z \in E$, and $l(p) = J(p)$.*

In this case the functional l is called the *support functional for* J *at* p.

PROPOSITION 10.1.2 (see [32]). *If J_i are convex functionals and $\lambda_i \geq 0$, then $\sum_{i=1}^n \lambda_i J_i$ is also a convex functional.*

If J_α, $\alpha \in I$, are convex functionals (I is any set of indices), then $\sup_{\alpha \in I} J_\alpha$ is also convex. If J is a convex functional, then the sets $\{z : J(z) \leq \lambda\}$ and $\{z : J(z) < \lambda\}$ are convex for any λ.

Suppose that a functional J over a vector space E is differentiable at p in any direction. We denote the derivative of J at p in the direction z by $J_p(z)$. It is easy to see that J_p is a positively homogeneous functional over E.

PROPOSITION 10.1.3 (see [32]). *Let J be a convex functional over a vector space E. Then for any point $p \in E$ the functional J_p exists, it is convex, and it satisfies the inequality*

$$J_p(z) \leq J(p+z) - J(p) \tag{1.7}$$

for all $z \in E$.

We now state the Hahn-Banach theorem on extending a linear functional, which will play an important part in our later investigation.

THEOREM 10.1.1 (Young [140]). *Let J be a convex positively homogeneous functional, defined on a vector space E. Let $\bar{\omega}$ be a linear functional, defined on a subspace L of E, and suppose that $\overline{\omega}(z) \leq J(z)$ for any $z \in L$. Then there is a linear functional ω, defined on the whole space E, such that $\omega(z) = \bar{\omega}(z)$ when $z \in L$ and $\omega(z) \leq J(z)$ for all $z \in E$.*

§2. Generalized forms and their properties

DEFINITION 10.2.1. A *generalized k-form* on a Riemannian manifold M is any continuous linear functional over $E_k M$.

We denote the vector space of all generalized k-forms on M by $\widetilde{E}^k M$. Clearly, every differential k-form $\varphi \in E^k M$ can be regarded as a generalized k-form defined by the formula $\varphi(S) = S(\varphi)$, $S \in E_k M$.

As D. V. Anosov remarked, *generalized forms are in fact ordinary differential forms.* This can be established with no difficulty by means of a well-known property that the space of differential forms is reflexive. However, it is technically convenient to regard ordinary forms as generalized, so henceforth we shall use the language of generalized forms.

DEFINITION 10.2.2. A generalized k-form ω is said to be *closed* if $\omega(S) = 0$ for any exact k-current S, and *exact* if $\omega(S) = 0$ for any closed k-current S.

Clearly, if φ is a closed (resp. exact) ordinary differential form in the sense of Definition 10.2.2, then it is closed (resp. exact) in the classical sense.

DEFINITION 10.2.3. The *product of a k-current S and a generalized i-form ω* is the $(k-i)$-current $S \wedge \omega$ defined by

$$S \wedge \omega(\varphi) = (-1)^{i(k-i)} \omega(S \wedge \varphi),$$

where φ is an arbitrary differential $(k-i)$-form.

THEOREM 10.2.1. *Let ω be an arbitrary generalized k-form on M. For every $x \in M$ we denote by ω_x the k-covector in $\Lambda^k M_x$ defined by*

$$\omega_x(\xi) = \omega(\xi) \tag{2.1}$$

for all k-vectors ξ in $\Lambda_k M_x$, also regarded as pointwise k-currents. Then for any current $S \in E_k M$ we have

$$\omega(S) = \int \omega_x(\vec{S}_x)\, d\|S\|(x). \tag{2.2}$$

REMARK. Theorem 10.2.1 shows that generalized forms on M can be regarded as sections of the Grassmann bundle $\Lambda^k M$. We also observe that under this correspondence ordinary differential forms correspond to sections of class C^∞.

PROOF. Clearly, $\vec{S}_x = \lim_{\varepsilon \to 0} (S \cap B(x, \varepsilon))/\|S\|(B(x, \varepsilon))$. Therefore, from the fact that ω is continuous it follows that

$$\lim_{\varepsilon \to 0} \frac{\omega(S \cap B(x, \varepsilon))}{\|S\|(B(x, \varepsilon))} = \omega(\vec{S}_x) = \frac{d(S \wedge \omega)}{d\|S\|},$$

from which we obtain $\omega(S) = \int \omega_x(\vec{S}_x) d\|S\|(x)$.

REMARK. Equality (2.2) can be obtained by starting from the pointwise representation of the current $S : \omega(S) = \int \omega(\vec{S}_x)\, d\|S\|(x)$.

DEFINITION 10.2.4. The *outer product of a generalized i-form ω_1 and a generalized j-form ω_2* is the generalized $(i+j)$-form $\omega_1 \wedge \omega_2$ defined by

$$\omega_1 \wedge \omega_2(S) = \omega_2(S \wedge \omega_1) \tag{2.3}$$

for an arbitrary $(i+j)$-current S.

It is easy to verify that for any $x \in M$ we have

$$(\omega_1 \wedge \omega_2)_x = (\omega_1)_x \wedge (\omega_2)_x, \tag{2.4}$$

that is, the outer product of generalized forms introduced above is a natural generalization of the concept of outer product of ordinary differential forms. Now, by means of (2.3) and (2.4), from the basic properties of the operation of outer product of covectors we can immediately derive the associativity and antisymmetry of the outer product of generalized forms, and also the distributivity, with respect to addition.

DEFINITION 10.2.5. The *support* $\operatorname{spt}\omega$ *of a generalized k-form* $\omega \in \widetilde{E}^k M$ is the smallest closed subset K of M such that $\omega(S) = 0$ for any k-current S for which $\operatorname{spt} S \subset M \backslash K$.

DEFINITION 10.2.6. The *comass of a generalized k-form* ω is the following quantity:

$$M^*(\omega) = \sup\{\omega(S) : S \in E_k M, MS \leq 1\}.$$

By simple arguments we can show that $\operatorname{spt}\omega$ coincides with the closure of the set $\{x \in M : \omega_x \neq 0\}$, and the comass $M^*(\omega)$ is equal to $\sup\{\|\omega_x\|^2 : x \in M\}$, so these concepts are the natural generalizations of the corresponding concepts for ordinary differential forms.

§3. Conditions for global minimality of currents

3.1. The modern Euler "equation" and Huyghens algorithm. Consider a functional J over the space $E_k M$ of all k-currents on a Riemannian manifold M.

THEOREM 10.3.1 (Đào Trọng Thi). (a) *Suppose that the functional* J *is differentiable at the point* $S \in E_k M$ *in any direction. Then if the current* S *is absolutely* (*resp. homologically*) *minimal with respect to* J, *we have*

$$J_S(T) \geq 0 \tag{3.1}$$

for any closed (*resp. exact*) *current* $T \in E_k M$.

(b) *If, in addition,* J_S *is a convex functional, bounded above in a neighborhood of some point, then there is an exact* (*resp. closed*) *generalized form* $\omega \in \widetilde{E}^k M$ *such that*

$$J_S(T) - \omega(T) \geq 0 \tag{3.2}$$

for any current $T \in E_k M$.

(c) *If* J_S *is a homogeneous functional, then the inequality* (3.1) *becomes the equality*

$$J_S(T) = 0 \tag{3.3}$$

for every closed (*resp. exact*) *current* $T \in E_k M$. *In particular, if* J_S *is a generalized form, then it is exact* (*resp. closed*).

PROOF. (a) Let S be an absolutely (resp. homologically) minimal current with respect to J, and let T be any closed (resp. exact) current in $E_k M$. According to Definition 10.2.1,

$$J(S + \varepsilon T) - J(S) \geq 0$$

for arbitrary $\varepsilon \geq 0$, since $(S+\varepsilon T)-S=\varepsilon T$ is closed (resp. exact). Hence by the principle of extending inequalities we obtain

$$J_S(T)=\lim_{\varepsilon\to 0}\varepsilon^{-1}(J(S+\varepsilon T)-J(S))\geq 0.$$

(b) If J_S is a convex positively homogeneous functional, then from (3.1) we can assert, according to the Hahn-Banach theorem, that there is a linear functional ω over E_kM that vanishes on the subspace of closed (resp. exact) currents, and

$$J_S(T)\geq \omega(T) \tag{3.4}$$

for any current $T\in E_kM$. Let us prove that ω is a generalized form. For this it is sufficient to show that ω is bounded in some neighborhood of zero. In fact, by the assumption of the theorem there is a neighborhood U of the point $P\in E_kM$ and a constant $c>0$ such that

$$J_S(T)\leq c \tag{3.5}$$

for any $T\in U$. Then there is a symmetric neighborhood U_0 of zero such that $P+U_0\subset U$. From (3.4) and (3.5) it follows that $\omega(P+T)\leq c$ and $\omega(P-T)\leq c$ for every $T\in U_0$. Consequently,

$$|\omega(T)|\leq c-\omega(P), \qquad T\in U_0. \tag{3.6}$$

Inequality (3.6) implies that ω is continuous, that is, it is a generalized form. Finally, since $\omega(T)=0$ for any closed (resp. exact) current T, the generalized form ω is exact (resp. closed).

(c) If J_S is homogeneous, then according to (3.1) $J_S(T)\geq 0$ and $-J_S(T)=J_S(T)\geq 0$. Consequently, $J_S(T)=0$ for any closed (resp. exact) current $T\in E_kM$. This proves the theorem.

The relations (3.1), (3.2), and (3.3) were derived from the condition with respect to "first variation". By tradition we call them the *Euler relations.*

THEOREM 10.3.2. *Let J be an arbitrary functional over E_kM, and suppose that for the current $S\in E_kM$ there is an exact (resp. closed) generalized form $\omega\in\widetilde{E}^kM$ such that*

$$J(T)-\omega(T)\geq J(S)-\omega(S) \tag{3.7}$$

for any $T\in E_kM$. We make the following assertions:

(a) *The current S is absolutely (resp. homologically) minimal with respect to J.*

(b) *If $P\in E_kM$ and $P-S$ is closed (resp. exact), then P is absolutely (resp. homologically) minimal with respect to J if and only if*

$$J(P)-\omega(P)=J(S)-\omega(S). \tag{3.8}$$

(c) *Suppose additionally that* J *is a positively homogeneous functional. Then* (3.7) *is equivalent to the following requirements*:

$$J(S) = \omega(S), \qquad J(T) - \omega(T) \geq 0 \tag{3.9}$$

for any $T \in E_k M$.

PROOF. (a) Suppose that $T - S$ is a closed (resp. exact) current. From the fact that ω is exact (resp. closed) it follows that $\omega(T - S) = 0$ or

$$\omega(T) = \omega(S). \tag{3.10}$$

Then from (3.7) and (3.10) we obtain $J(T) \geq J(S)$, that is, S is absolutely (resp. homologically) minimal with respect to J.

(b) Suppose that $P \in E_k M$ and that $P - S$ is closed (resp. exact). According to (3.10), $\omega(P) = \omega(S)$. Then (3.8) is equivalent to the equality $J(P) = J(S)$. The latter equality implies that P is an absolutely (resp. homologically) minimal current with respect to J.

(c) Now let J be a positively homogeneous functional, and suppose that (3.7) holds for any $T \in E_k M$. Substituting, in particular, $T = \lambda S$, where $\lambda > 0$, in (3.7), we obtain

$$\lambda(J(S) - \omega(S)) = J(\lambda S) - \omega(\lambda S) \geq J(S) - \omega(S).$$

This inequality is possible for all $\lambda > 0$ only when $J(S) - \omega(S) = 0$, that is, $J(S) = \omega(S)$, and then (3.7) has the form $J(T) - \omega(T) \geq 0$ for any $T \in E_k M$. Conversely, if (3.9) is satisfied for any $T \in E_k M$, then obviously (3.7) is also true for any $T \in E_k M$. This proves the theorem.

Setting $l(T) = J(S) - \omega(S) + \omega(T)$, we can rewrite (3.7) in the form $l(T) \leq J(T)$ for any $T \in E_k M$ and $l(S) = J(S)$, so that l is the support functional of J at the point S. This is equivalent to J being convex at S. In particular, if J is a positively homogeneous functional, then l coincides with the homogeneous part of ω. Thus, in this case the generalized form ω itself is the support of J at S.

For every current $S \in E_k M$ we denote by $F^*(S)$ (resp. $\widetilde{F}^*(S)$) the set of all exact (resp. closed) generalized forms ω for which (3.7) is satisfied (this means that ω is the homogeneous part of some support functional J at the point S). Then assertion (a) of Theorem 10.3.2 can be stated thus: the current S is absolutely (resp. homologically) minimal with respect to J if the set $F^*(S)$ (resp. $\widetilde{F}^*(S)$) is not empty.

THEOREM 10.3.3. *Let* J *be an arbitrary functional over* $E_k M$. *Suppose that the currents* S *and* P *are both absolutely* (*resp. homologically*) *minimal with respect to* J, *and that* $S - P$ *is closed* (*resp. exact*). *Then* $F^*(S) = F^*(P)$ (*resp.* $\widetilde{F}^*(S) = \widetilde{F}^*(P)$).

PROOF. Suppose that $\omega \in F^*(S)$ (resp. $\widetilde{F}^*(S)$). Then ω is exact (resp. closed) and $J(T) - \omega(T) \geq J(S) - \omega(S)$ for any $T \in E_k M$. We prove that $\omega \in F^*(P)$ (resp. $\widetilde{F}^*(P)$). For this it is sufficient to show that $J(P) - \omega(P) = J(S) - \omega(S)$. Indeed, from the fact that $S - P$ is closed (resp. exact) it follows that $\omega(S-P) = 0$ or $\omega(S) = \omega(P)$. Hence, $J(P) - \omega(P) = J(S) - \omega(P) = J(S) - \omega(S)$. Thus, $F^*(S) \subset F^*(P)$ (resp. $\widetilde{F}^*(S) \subset \widetilde{F}^*(P)$). Since S and P are completely equivalent in our argument, the reverse inclusion $F^*(P) \subset F^*(S)$ (resp. $\widetilde{F}^*(P) \subset \widetilde{F}^*(S)$) is true. This completes the proof of the theorem.

Theorem 10.3.3 shows that the set $F^*(S)$ (resp. $\widetilde{F}^*(S)$) depends not on the current S itself, but on a given class of currents with given boundary (resp. on a given class of homological currents with given boundary).

3.2. The convex case. It turns out that if the functional J is convex, the necessary condition and the sufficient condition in the previous subsection coincide, which gives a general necessary and sufficient condition.

THEOREM 10.3.4 (Đào Trọng Thi). *Let J be a convex functional over $E_k M$, bounded above in a neighborhood of some point. A necessary and sufficient condition for S to be absolutely* (*resp. homologically*) *minimal with respect to J is the existence of an exact* (*resp. closed*) *generalized form $\omega \in E^k M$ such that*

$$J(T) - \omega(T) \geq J(S) - \omega(S) \tag{3.11}$$

for any current $T \in E_k M$.

PROOF. Necessity. Let J be a convex functional. According to Proposition 10.1.3, the functional J_S exists and is convex, and

$$J_S(T) \leq J(S+T) - J(S) \tag{3.12}$$

for all $T \in E_k M$. From (3.12) and the assumption that J is bounded above in a neighborhood of some point $P \in E_k M$ it follows that J_S is itself bounded above in the corresponding neighborhood of the point $P - S$, so that we can apply assertion (b) of Theorem 10.3.1 to J_S: if S is absolutely (resp. homologically) minimal with respect to J, then there is an exact (resp. closed) generalized form $\omega \in \widetilde{E}^k M$ such that

$$J(S+T') - J(S) \geq J_S(T') \geq \omega(T') \tag{3.13}$$

for any current $T' \in E_k M$. Replacing T' by $T + S$ in (3.13), we obtain the inequality $J(T) - J(S) \geq \omega(T) - \omega(S)$, from which (3.11) follows.

The sufficiency follows immediately from Theorem 10.3.2 as a special case.

If J is a positively homogeneous functional, then from Theorem 10.3.4 and assertion (c) of Theorem 10.3.2 we obtain the following result.

THEOREM 10.3.5 (Đào Trọng Thi). *Let J be a positively homogeneous convex functional over $E_k M$, bounded above in a neighborhood of some point. Then the current S is absolutely (resp. homologically) minimal with respect to J if and only if there is an exact (resp. closed) generalized form $\omega \in \widetilde{E}^k M$ such that $J(T) - \omega(T) \geq 0$ for any current $T \in E_k M$, and $J(S) = \omega(S)$.*

3.3. The case of integrands. We now co: sider the case where J is an integrand. It turns out that in this case we can reduce the global minimality condition to families of "pointwise" requirements.

LEMMA 10.3.1. *Let l be a Lagrangian of degree k, let ω be a generalized form, and let S be a k-current on a Riemannian manifold M. Let J be the integrand over $E_k M$ determined by l. Then the following assertions are equivalent:*

$$\text{(a)} \quad J(T) - \omega(T) \geq 0 \tag{3.14}$$

for any current $T \in E_k M$, and

$$J(S) = \omega(S); \tag{3.15}$$

$$\text{(b)} \quad l(\xi) - \omega(\xi) \geq 0 \tag{3.16}$$

for any $\xi \in \Lambda_k M$, and

$$l(\vec{S}_x) = \omega(\vec{S}_x) \tag{3.17}$$

for almost all $x \in M$ in the sense of the measure $\|S\|$.

PROOF. Suppose that (3.16) and (3.17) are satisfied. Then for any current $T \in E_k M$ we have

$$J(T) \geq \int \omega(\vec{T}_x)\, d\|T\|(x) = \omega(T),$$

$$J(S) = \int \omega(\vec{S}_x)\, d\|S\|(x) = \omega(S),$$

so that (3.14) and (3.15) are also satisfied. Conversely, suppose that (3.14) and (3.15) are satisfied. For any $\xi \in \Lambda_k M$ we consider the pointwise current $T = \xi$:

$$J(T) = \int l(\xi/\|\xi\|)\, d\|T\| = l(\xi),$$

$$\omega(T) = \int \omega(\xi/\|\xi\|)\, d\|T\| = \omega(\xi).$$

When $T = \xi$, (3.14) gives

$$l(\xi) \geq \omega(\xi). \tag{3.18}$$

On the other hand, (3.15) can be rewritten in the form

$$\int l(\vec{S}_x)\, d\|S\|(x) = \int \omega(\vec{S}_x)\, d\|S\|(x). \tag{3.19}$$

Obviously, (3.18) and (3.19) together give (3.17). This proves the lemma.

THEOREM 10.3.6. *Let J be the convex integrand over $E_k M$ defined by the Lagrangian l, where J is bounded above in a neighborhood of some point. Then the current S is absolutely (resp. homologically) minimal with respect to J if and only if there is an exact (resp. closed) generalized form $\omega \in \widetilde{E}^k M$ such that $l(\xi) - \omega(\xi) \geq 0$ for any $\xi \in \Lambda_k M$, and $l(\vec{S}_x) = \omega(\vec{S}_x)$ for almost all $x \in M$ in the sense of the measure $\|S\|$.*

PROOF. This theorem is a direct consequence of Theorem 10.3.5 and Lemma 10.3.1.

Let J be an arbitrary integrand over $E_k M$ defined by the Lagrangian l, and let F^* (resp. $\widetilde{F}^*$) be a nonempty set of exact (resp. closed) generalized forms ω in $\widetilde{E}_k M$ such that $\omega(T) \leq J(T)$ for any $T \in E_k M$. We set

$$F_x = \{\xi \in \Lambda_k M_x : l(\xi) = \omega(\xi) \text{ for any } \omega \in F^*\},$$
$$\widetilde{F}^* = \{\xi \in \Lambda_k M_x : l(\xi) = \omega(\xi) \text{ for any } \omega \in \widetilde{F}^*\}.$$

THEOREM 10.3.7. *Let J be the integrand defined by the Lagrangian l, and let $S \in E_k M$ be a current such that $F^*(S)$ (resp. $\widetilde{F}^*(S)$) is not empty. Let $F^* \subset F^*(S)$ (resp. $\widetilde{F}^* \subset \widetilde{F}^*(S)$) be an arbitrary nonempty subset. Then S is absolutely (resp. homologically) minimal with respect to J, and $P \in E_k M$, where $P - S$ is closed (resp. exact), is absolutely (resp. homologically) minimal with respect to J if and only if $\vec{P}_x \in F_x$ (resp. $\vec{P}_x \in \widetilde{F}_x$) for almost all $x \in M$ in the sense of the measure $\|P\|$.*

PROOF. First of all we observe that the fact that the set $F^*(S)$ (resp. $\widetilde{F}^*(S)$) is nonempty implies, by the remark to Theorem 10.3.2, that S is absolutely (resp. homologically) minimal with respect to J. Next we observe that for the current P, where the current $P - S$ is closed (resp. exact), the assertion "$\vec{P}_x \in F_x$ (resp. $\widetilde{F}_x$) almost everywhere in the sense of the measure $\|P\|$" is equivalent, by Lemma 10.3.1, to the requirements

$$J(P) = \omega(P), \qquad J(T) \geq \omega(T) \tag{3.20}$$

for any $T \in E_k M$ and any $\omega \in F^*$ (resp. $\widetilde{F}^*$). By definition the requirements (3.20) imply that

$$F^* \subset F^*(P) \qquad (\text{resp. } \widetilde{F}^* \subset \widetilde{F}^*(P)). \tag{3.21}$$

According to Theorem 10.3.3 and the remark to Theorem 10.3.2, the inclusion (3.21) is equivalent to the fact that P is absolutely (resp. homologically) minimal with respect to J. This completes the proof of the theorem.

To apply the criteria for absolute (resp. homological) minimality given in the section we need to define "pencils of minimal directions" F_x (resp. $\widetilde{F}_x$); to calculate these it is necessary to know at least one exact (resp. closed) support form ω of the functional J in $F^*(S)$ (resp. $\widetilde{F}^*(S)$). Since the set $F^*(S)$ (resp. $\widetilde{F}^*(S)$) does not depend on the choice of minimal current S, for a symmetric (in some sense) functional J it is reasonable to choose the initial current S among "symmetric" currents, and the supporting generalized form ω among "symmetric" generalized forms. We shall be concerned with this question in the following sections.

§4. Globally minimal currents in symmetric problems

4.1. Problems with invariant functionals. We shall say that the (left) action of a Lie group G is defined on a Riemannian manifold M if there is defined a map $\pi: G \times M \to M$ of class C^∞ such that for every $g \in G$ the map $\pi_g: M \to M$, $\pi_g x = \pi(g, x)$, is a diffeomorphism of M onto itself, and

$$\pi_{gh} = \pi_g \pi_h \tag{4.1}$$

for any $g, h \in G$. It is easy to see that

$$\pi_{g^{-1}} = \pi_g^{-1} \tag{4.2}$$

for any $g \in G$. In view of (4.1) and (4.2), henceforth we shall denote $\pi_g x$ simply by gx.

The set $H_x = \{h \in G : hx = x\}$ is a closed subgroup of G and is called the *stationary subgroup of the point* x.

As we know, for every $g \in G$ the diffeomorphism $g: M \to M$ induces an automorphism $g^*: E^k M \to E^k M$, and

$$(gh)^* = h^* g^*, \qquad (g^{-1})^* = (g^*)^{-1}. \tag{4.3}$$

We define a map $g_*: E_k M \to E_k M$ by

$$g_* S(\varphi) = S(g^* \varphi), \qquad \varphi \in E^k M.$$

Clearly, g_* is an automorphism, and

$$(gh)_* = g_* h_*, \qquad (g^{-1})_* = g_*^{-1}. \tag{4.4}$$

Finally, for every generalized k-form $\omega \in \widetilde{E}^k M$ we set $\tilde{g}^*\omega(S) = \omega(g_*S)$ for any $s \in E_k M$. It is easy to verify that $\tilde{g}^*\omega$ is again a generalized k-form and the correspondence $\omega \to \tilde{g}^*\omega$ defines an automorphism of the space $\widetilde{E}^k M$ that extends an automorphism g^* of the space $E^k M$, and the relations (4.3) are preserved. For this reason we again denote $\tilde{g}^*$ simply by g^*.

DEFINITION 10.4.1. A current $S \in E_k M$ (resp. a generalized form $\omega \in \widetilde{E}^k M$) is said to be *invariant with respect to the action of the group* G, or simply *G-invariant*, if $g_*S = S$ (resp. $g^*\omega = \omega$) for any g.

Obviously, G-invariance of the current S (resp. of the generalized form ω) is equivalent to the equality $S(\varphi) = S(g^*\varphi)$ for any $g \in G$ and any $\varphi \in E^k M$ (resp. $\omega(g_*S) = \omega(S)$ for any g and any $S \in E_k M$).

DEFINITION 10.4.2. A functional J over $E_k M$ is said to be *invariant with respect to the action of the group* G, or simply *G-invariant*, if $J(g_*S) = J(S)$ for any $g \in G$ and any $S \in E_k M$.

LEMMA 10.4.1. *For every* $S \in E_k M$ *the map* $\rho\colon G \to E_k M$, $\rho(g) = g_*S$, *is continuous.*

We now assume that G is a compact Lie group. For G there is a unique two-sided invariant Haar measure, and the measure of the whole group is equal to unity. For every current $S \in E_k M$ the map $g \to g_*S$ is continuous, since the integral $\pi_G S = \int_G g_*S\,dg$ exists. If $\omega \in \widetilde{E}^k M$, then from the fact that the functional ω and the function $g \to g_*S$ are continuous it follows that the map $g \to \omega(g_*S) = g^*\omega(S)$ is continuous, so that for every $S \in E_k M$ we can consider the integral $\int_G g^*\omega(S)\,dg$. We have thus obtained a functional $\pi_G^*\omega$ over $E_k M$ defined by the equality $\pi_G^*\omega(S) = \int_G g^*\omega(S)\,dg$, for any $S \in E_k M$. It is easy to verify that $\pi_G^*\omega$ is a linear functional; we call $\pi_G S$ and $\pi_G^*\omega$ the *averages with respect to G of the current S and the generalized form ω* respectively.

LEMMA 10.4.2. π_G *and* π_G^* *are linear operators, and* π_G^* *is the adjoint operator for* π_G.

PROOF. The fact that π_G and π_G^* are linear is obvious. Also, for any $S \in E_k M$ and any $\omega \in \widetilde{E}^k M$ we have

$$\pi_G^*\omega(S) = \int_G \omega(g_*S)\,dg. \tag{4.5}$$

On the other hand,

$$\int_G \omega(g_*S)\,dg = \omega(\pi_G S). \tag{4.6}$$

From (4.5) and (4.6) it follows that $\pi_G^*\omega(S) = \omega(\pi_G S)$.

LEMMA 10.4.3. (a) *For any current* S *and any generalized form* ω *the currents* $\pi_G S$ *and* $\pi_G^* \omega$ *are G-invariant.*

(b) *For any G-invariant current* S *and any G-invariant generalized form* ω *we have* $\pi_G S = S$, $\pi_G^* \omega = \omega$.

PROOF. (a) For every $g \in G$ we have $g_*(\pi_G S) = g_*(\int_G h_* S\, dh) = \int_G (gh)_* S d(gh) = \pi_G S$ and $g^*(\pi_G^* \omega)(T) = \int_G (hg)^* \omega(T)\, d(hg) = \pi_G^* \omega(T)$ for any $T \in E_k M$, so $g^*(\pi_G^* \omega) = \pi_G^* \omega$. Thus, $\pi_G S$ and $\pi_G^* \omega$ are G-invariant.

(b) If S is a G-invariant current, then $g_* S = S$ for any $g \in G$. Consequently, $\pi_G S = S$. Similarly we can prove that $\pi_G^* \omega = \omega$ for a G-invariant generalized form ω. This proves the lemma.

LEMMA 10.4.4. *Let* J *be a convex lower semicontinuous G-invariant functional over* $E_k M$. *Then for any* $S \in E_k M$ *we have* $J(\pi_G S) \leq J(S)$.

PROOF. The integral $\pi_G S = \int_G g_* S\, dg$ can be treated as the limit of some sequence of finite sums of the form $\sum \lambda_i g_{i*} S$, where each sum $\sum \lambda_i g_{i*} S$ is formed by means of a finite partition of G into disjoint domains U_i, where $g_i \in U_i$ and λ_i is equal to the Haar measure of the domain U_i for any i. From the fact that J is convex and G-invariant we obtain

$$J\left(\sum_i \lambda_i g_{i*} S\right) \leq \sum_i \lambda_i J(S) = J(S).$$

On the other hand, from the fact that J is lower semicontinuous it follows that $J(\pi_G S) \leq J(S)$.

THEOREM 10.4.1 (Đào Trọng Thi). *Suppose we are given the action of a compact Lie group* G *on a Riemannian manifold* M *(in the case of homological minimality we also require that* G *is connected). Let* J *be a convex lower semicontinuous G-invariant functional over* $E_k M$, *bounded above in a neighborhood of some point. Then if a current* $S \in E_k M$ *is absolutely (resp. homologically) minimal with respect to* J, *and its boundary* ∂S *is G-invariant, it follows that*

(a) *the current* $\pi_G S - S$ *is closed (resp. exact), and* $\pi_G S$ *is absolutely (resp. homologically) minimal with respect to* J;

(b) *the set of G-invariant generalized forms in* $F^*(S)$ *(resp.* $\widetilde{F}^*(S)$*) is not empty and coincides with the image of* $F^*(S)$ *(resp.* $\widetilde{F}^*(S)$*) under* π_G^*.

PROOF. (a) For any form $\varphi \in E^k M$, from the fact that ∂S is G-invariant we have

$$\partial \pi_G S(\varphi) = \int_G \partial S(\varphi)\, dg = \partial S(\varphi).$$

Thus, $\pi_G S - S$ is closed. In the case of homological minimality, from the fact that G is path-connected and the homotopy axiom of homology theory it follows that $g_* S$ is homologous to S for any g. Consequently, $\pi_G S$ is homologous to S. This means that $\pi_G S - S$ is exact. Also, $J(S) \leq J(\pi_G S)$, since S is absolutely (resp. homologically) minimal with respect to J. On the other hand, according to Lemma 10.4.4, $J(\pi_G S) \leq J(S)$, so that $J(\pi_G S) = J(S)$ and $\pi_G S$ is also absolutely (resp. homologically) minimal with respect to J.

(b) According to part (a) (already proved) and Theorem 10.3.3 we have $F^*(S) = F^*(\pi_G S)$ (resp. $\widetilde{F}^*(S) = \widetilde{F}^*(\pi_G S)$). Let $\omega \in F^*(\pi_G S)$ (resp. $\widetilde{F}^*(\pi_G S)$). Then $J(T) - \omega(T) \geq J(\pi_G S) - \omega(\pi_G S)$ for any $T \in E_k M$. In particular,

$$J(\pi_G T) - \omega(\pi_G T) \geq J(\pi_G S) - \omega(\pi_G S) \tag{4.7}$$

for any $T \in E_k M$. According to Lemma 10.4.3, $\pi_G S$ is G-invariant and $\pi_G(\pi_G S) = \pi_G S$. Now according to Lemma 10.4.2, we have

$$\omega(\pi_G S) = \pi_G^* \omega(\pi_G S), \tag{4.8}$$

$$\omega(\pi_G T) = \pi_G^* \omega(T). \tag{4.9}$$

On the other hand, by Lemma 10.4.4 we have

$$J(\pi_G T) \leq J(T). \tag{4.10}$$

Substituting (4.8), (4.9), and (4.10) in (4.7), we obtain

$$J(T) - \pi_G^* \omega(T) \geq J(\pi_G S) - \pi_G^* \omega(\pi_G S) \tag{4.11}$$

for any $T \in E_k M$. Let us prove that $\pi_G^* \omega$ is a generalized form. For this it is sufficient to show that $\pi_G^* \omega$ is bounded in some neighborhood of zero. In fact, by the hypothesis of the theorem there is a neighborhood U of a point $P \in E_k M$ and a constant $c > 0$ such that

$$J(T) \leq c \tag{4.12}$$

for any $T \in U$. Then there is a symmetric neighborhood U_0 of zero such that $P + U_0 \subset U$. From (4.11) and (4.12) it follows that $\pi_G^* \omega(P+T) \leq c'$ and $\pi_G^* \omega(P-T) \leq c'$ for any $T \in U_0$, where $c' = c - J(\pi_G S) + \pi_G^* \omega(\pi_G S)$. Therefore, $|\pi_G^* \omega(T)| \leq c' - \pi_G^* \omega(P)$, $T \in U_0$, so that $\pi_G^* \omega$ is a generalized form. Let us prove that $\pi_G^* \omega$ is exact (resp. closed). Suppose that $T \in E_k M$ is closed (resp. exact, that is, $T = \partial T'$). Then for any form $\varphi \in E^k M$ we have $\partial \pi_G T(\varphi) = \int_G \partial g_* T(\varphi)\, dg = \int_G g_* \partial T(\varphi)\, dg = 0$ (resp. $\pi_G T(\varphi) = \pi_G \partial T'(\varphi) = \partial \pi_G T'(\varphi)$), which implies that $\partial \pi_G T = 0$ (resp. $\pi_G T = \partial \pi_G T'$), that is, $\pi_G T$ is closed (resp. exact). Therefore,

$\pi_G^*\omega(T) = \omega(\pi_G T) = 0$, since ω is exact (resp. closed). We have thus proved that the generalized form $\pi_G^*\omega$ is exact (resp. closed). Next, the inequality (4.11) shows that $\pi_G^*\omega \in F^*(\pi_G S)$ (resp. $\widetilde{F}^*(\pi_G S)$). Since, according to Lemma 10.4.3, $\pi_G^*\omega$ is G-invariant, the image of $F^*(\pi_G S)$ (resp. $\widetilde{F}^*(\pi_G S)$) under π_G^* is included in the set of G-invariant generalized forms in $F^*(\pi_G S)$ (resp. $\widetilde{F}^*(\pi_G S)$). Conversely, if $\omega \in F^*(\pi_G S)$ (resp. $\widetilde{F}^*(\pi_G S)$) is a G-invariant generalized form, then, again by Lemma 10.4.3, $\omega = \pi_G^*\omega$, that is, it is in the image of $F^*(\pi_G S)$ (resp. $\widetilde{F}^*(\pi_G S)$) under π_G^*. Thus the set of G-invariant elements of $F^*(\pi_G S)$ (resp. $\widetilde{F}^*(\pi_G S)$) coincides with the image of $F^*(\pi_G S)$ (resp. $\widetilde{F}^*(\pi_G S)$) under π_G^*, which completes the proof of the theorem.

Now let J be an integrand defined by a Lagrangian l.

LEMMA 10.4.5. *An integrand J over $E_k M$ is invariant with respect to a given action of a Lie group G on M if and only if the corresponding Lagrangian l satisfies the condition*

$$l(g_*\xi) = l(\xi) \tag{4.12$'$}$$

for any $\xi \in \Lambda_k M$.

PROOF. Suppose that the integrand J is G-invariant. Then for any $g \in G$ and any $S \in E_k M$ we have

$$J(g_*S) = J(S). \tag{4.13}$$

In particular, if for S we take the point current ξ, then $J(g_*\xi) = J(\xi)$, which is equivalent to $(4.12)'$. Conversely, suppose that $(4.12)'$ is satisfied for any $\xi \in \Lambda_k M$. For any $g \in G$ and any $S \in E_k M$ we have $J(g_*S) = \int l(\vec{S}_{g^{-1}x})\, d\|S\|(g^{-1}x) = \int l(\vec{S}_x)\, d\|S\|(x) = J(S)$, that is, J is a G-invariant integrand.

THEOREM 10.4.2. *Suppose that an integrand J over $E_k M$ is invariant with respect to the given action of a Lie group G on M, and that F^* (resp. $\widetilde{F}^*$) is a nonempty set of G-invariant exact (resp. closed) generalized forms ω in $\widetilde{E}^k M$ such that $\omega(T) \le J(T)$ for any $T \in E_k M$. Then for any $g \in G$ and any $x \in M$ we have*

$$F_{gx} = g_*(F_x) \qquad (\textit{resp. } \widetilde{F}_{gx} = g_*(\widetilde{F}_x)). \tag{4.14}$$

In other words, the family F_x (resp. $\widetilde{F}_x$) is invariant under the action of the group G. In particular, every subset of F_x (resp. $\widetilde{F}_x$) is invariant under the action of the stationary subgroup H_x at the point x.

PROOF. We first prove that

$$g_*(F_x) \subset F_{gx} \qquad (\text{resp. } g_*(\widetilde{F}_x) \subset \widetilde{F}_{gx}) \tag{4.15}$$

for any $g \in G$ and any $x \in M$. In fact, suppose that $\xi \in F_x$ (resp. $\widetilde{F}_x$), that is, $l(\xi) = \omega(\xi)$ for every form $\omega \in F^*$ (resp. $\widetilde{F}^*$). Then from the fact that J and ω are G-invariant we have, by Lemma 8.4.5, $l(g_*\xi) = (l(\xi) = \omega(\xi)) = \omega(g_*\xi)$, which means that $g_*\xi \in F_{gx}$ (resp. $\widetilde{F}_{gx}$). Also, if in (4.13) we take g^{-1} for g, and gx for x, we obtain

$$g_*^{-1}(F_{gx}) \subset F_x \qquad (\text{resp. } g_*^{-1}(\widetilde{F}_{gx}) \subset \widetilde{F}_x);$$

from which it follows that

$$F_{gx} = g_*g_*^{-1}(F_{gx}) \subset g_*(F_x) \qquad (\text{resp. } \widetilde{F}_{gx} \subset g_*(\widetilde{F}_x)). \tag{4.16}$$

The inclusions (4.15) and (4.16) give us the equality (4.14). Finally we observe that if $h \in H_x$, then $hx = x$. Thus, (4.14) has the form $h_*(F_x) = F_x$ (resp. $h_*(\widetilde{F}_x) = \widetilde{F}_x$) for any $h \in H_x$, that is, F_x (resp. $\widetilde{F}_x$) withstands the action of the group H_x. This completes the proof of the theorem.

4.2. Problems with covariantly constant Lagrangians. Consider a compact simply-connected Riemannian manifold M. By the de Rham-Hodge theorem, in every cohomology class of M there is a unique harmonic differential form (see [63]). Among the harmonic forms we pick out the covariantly constant forms φ, that is, $\nabla\varphi = 0$, where ∇ denotes covariant differentiation on M. As Lichnerowicz showed in [63], a k-form φ is covariantly constant if and only if it is invariant with respect to any parallel translation (along any path), that is,

$$\varphi(\xi) = \varphi(t_*\xi) \tag{4.17}$$

for any $\xi \in \Lambda_k M$ and any parallel translation t. Let Ψ_x denote the holonomy group of M at the point x (that is, the group of all parallel translations along loops passing through x). Clearly, if φ is a covariantly constant k-form, then the k-covector $\varphi_x \in \Lambda^k M_x$ is invariant under the action of the group Ψ_x for every point $x \in M$. The converse is also true: any Ψ_x-invariant k-covector $\varphi_x \in \Lambda^k M_x$ generates by means of parallel translations covariantly constant k-form φ defined on the whole of M. Thus, covariantly constant forms are completely defined by their behavior at one arbitrary point.

Let us fix a point e on M. In what follows we identify the set of all covariantly constant k-forms on M with the set Λ_I^k of all Ψ_e-invariant

k-covectors in $\Lambda^k M_e$. It is known (see [63]) that for a compact simply-connected manifold M its holonomy group at the point e is a closed subgroup of the orthogonal group $O(n)$. Thus, Ψ_e is a compact Lie group. In view of the compactness we can consider on Ψ_e a two-sided invariant Haar measure such that the measure of the whole group Ψ_e is equal to unity. For each covector $\eta \in \Lambda_k M_e$ (we recall that $\Lambda_k M_e$ and $\Lambda^k M_e$ are identified with respect to scalar product) we consider the integral $\pi\eta = \int_{\Psi_e} t_* \eta \, dt$.

Lemma 10.4.6. *The map $\eta \to \pi\eta$ is a linear selfadjoint operator of the space $\Lambda_k M_e$ into itself.*

Proof. It is obvious that this map is linear. Let ξ and η be arbitrary k-vectors in $\Lambda_k M_e$. Since $t \in \Psi_e$ is an orthogonal transformation, $t^* = t^{-1}$. Then $(\pi\xi, \eta) = (\int_{\Psi_e} t_* \xi \, dt, \eta) = (\xi, \int_{\Psi_e} t_*^{-1} \eta \, dt^{-1}) = (\xi, \pi\eta)$. Thus π is a linear selfadjoint operator, and the lemma is proved.

Lemma 10.4.7. (a) *For any $\eta \in \Lambda_k M_e$, $\pi\eta$ belongs to Λ_I^k.*
(b) *For any $\eta \in \Lambda_I^k$ we have $\pi\eta = \eta$.*

Proof. For every $s \in \Psi_e$ we have

$$s_*(\pi\eta) = s_* \left(\int_{\Psi_e} t_* \eta \, dt \right) = \int_{\Psi_e} (st)_* \eta \, d(st) = \pi\eta,$$

so that $\pi\eta \in \Lambda_I^k$. Also, if $\eta \in \Lambda_I^k$, then $t_*\eta = \eta$ for any $t \in \Psi_e$. Consequently, $\pi\eta = \eta$. This proves the lemma.

For every $\xi \in \Lambda_I^k$ we define the k-current S_ξ by

$$S_\xi(\varphi) = \int_M (\xi, \varphi) \, dx, \qquad \varphi \in E^k M. \tag{4.18}$$

Theorem 10.4.3. *Suppose that $\xi \in \Lambda_I^k$. Then the current S_ξ given by* (4.18) *is closed, and if we denote the homology class containing S_ξ by $[S_\xi]$, then the map $\xi \to [S_\xi]$ is a monomorphism of Λ_I^k into $H_k(N_* M)$.*

Proof. For any form $\varphi \in E^k M$ we have

$$\partial S_\xi(\varphi) = \int_M (\xi, d\varphi) \, dx. \tag{4.19}$$

Since ξ is a harmonic form, and $d\varphi$ is an exact form, the right-hand side of (4.19) is equal to zero, by the de Rham-Hodge theorem. Thus, $\partial S_\xi = 0$, that is, S_ξ is a closed current. Next, the fact that the map $\xi \to [S_\xi]$ is linear is obvious, so it remains to show that the kernel of this linear map is trivial. In fact, suppose that $[S_\xi] = 0$. Then $S_\xi(\varphi) = 0$ for any closed form $\varphi \in E^k M$. In particular, $S_\xi(\xi) = 0$, which implies that $(\xi, \xi) = 0$ or $\xi = 0$. This proves the theorem.

DEFINITION 10.4.3. A Lagrangian l of degree k on a Riemannian manifold M is said to be *covariantly constant* if for any $\xi \in \Lambda_k M$ and any parallel translation t (along any path) we have $l(\xi) = l(t_* \xi)$.

THEOREM 10.4.4. *Let J be the integrand defined by a covariantly constant Lagrangian l, where the restriction of l to the fiber $\Lambda_k M_e$ is a convex function. Then for any $\xi \in \Lambda_I^k$ the current S_ξ is homologically minimal with respect to J, and $F^*(S_\xi) \cap \Lambda_I^k$ is not empty.*

LEMMA 10.4.8. *Let l be a covariantly constant Lagrangian of degree k on M, and suppose that its restriction to $\Lambda_k M_e$ is a convex function. Then for any $\eta \in \Lambda_k M_e$ we have $l(\pi\eta) \le l(\eta)$.*

The proof of this lemma is completely analogous to that of Lemma 10.4.4.

PROOF OF THEOREM 10.4.4. First of all we observe that $(\vec{S}_\xi)_e = \xi$. Next, from the fact that the restriction of l to $\Lambda_k M_e$ is convex and positively homogeneous it follows, by the Hahn-Banach theorem, that there is a k-vector $\zeta \in \Lambda_k M_e$ such that

$$l(\eta) \le (\zeta, \eta), \qquad l(\xi) = (\zeta, \xi) \tag{4.20}$$

for any $\eta \in \Lambda_k M_e$. According to Lemma 10.4.7, $\pi\xi = \xi$. Then from Lemma 10.4.6 we have

$$l(\xi) = (\zeta, \xi) = (\pi\xi, \xi). \tag{4.21}$$

On the other hand, substituting $\eta = \pi\gamma$ into the inequality in (4.20) and applying Lemmas 10.4.8 and 10.4.9, we obtain

$$l(\gamma) \ge l(\pi\gamma) \ge (\zeta, \pi\gamma) = (\pi\zeta, \gamma), \qquad \gamma \in \Lambda_k M_e. \tag{4.22}$$

By Lemma 10.4.7, $\pi\zeta \in \Lambda_I^k$, so that we can regard $\pi\zeta$ (by means of parallel translations) as a covariantly constant form. We also observe that, for any point $x \in M$, $(\vec{S}_\xi)_x$ is obtained from $(\vec{S}_\xi)_e$ by a parallel translation along an arbitrary path joining x to e. Now from (4.22) and (4.21) and the fact that l and $\pi\zeta$ are invariant under all possible parallel translations we obtain

$$l(\gamma) - \pi\zeta(\gamma) \ge 0 \tag{4.23}$$

for any $\gamma \in \Lambda_k M$, and

$$l(\vec{S}_\xi)_x = \pi\zeta(\vec{S}_\xi)_x \tag{4.24}$$

for all $x \in M$. The relations (4.23) and (4.24) show that $\pi\zeta \in F^*(S_\xi) \cap \Lambda_I^k$, and in particular, the current S_ξ is homologically minimal. This proves the theorem.

THEOREM 10.4.5. *Let J be an integrand over $E_k M$ defined by a covariantly constant Lagrangian l, and let $\widetilde{F}^*$ be a nonempty set of covariantly constant forms ω in Λ_I^k such that $\omega(T) \le J(T)$ for any $T \in E_k M$. Then for any parallel translation t along any path joining the points x and y we have*

$$\widetilde{F}_y = t_*(\widetilde{F}_x).$$

In other words, the family $\widetilde{F}_x$ is invariant under all possible parallel translations. In particular, every subset of $\widetilde{F}_x$ withstands the action of the holonomy group Ψ_x.

This theorem is proved in exactly the same way as Theorem 10.4.2.

The theorem shows that for any point $x \in M$ the set $\widetilde{F}_x$ is obtained from $\widetilde{F}_e$ by a parallel translation along an arbitrary path joining x and e.

§5. Specific examples of globally minimal currents and surfaces

5.1. Minimal currents on Kähler manifolds. Consider a connected Kähler manifold M with fundamental 2-form Ω. It is known that a 2-form Ω, and consequently all its exterior powers Ω^k, are covariantly constant. We denote the structure orthogonal transformation of the Kähler structure at the point x by $J_x\colon T_xM \to T_xM$. Then Ω is expressed in terms of J_x by the formula

$$\Omega_x(v_1, v_2) = (J_x v_1, v_2), \qquad v_1, v_2 \in T_xM; \tag{5.1}$$

a $2k$-vector $\xi \in \Lambda_{2k}M$ is called a *positive complex $2k$-vector* if it can be represented as a finite sum of $2k$-vectors of the form

$$v_1 \wedge J_x v_1 \wedge v_2 \wedge J_x v_2 \wedge \cdots \wedge v_k \wedge J_x v_k,$$

where $v_1, \dots, v_k$ are arbitrary vectors in T_xM.

LEMMA 10.5.1 (Federer [216], Lawson [286]). *For any $\xi \in \Lambda_{2k}M$ we have $(k!)^{-1}\Omega^k(\xi) \le \|\xi\|$, where equality holds if and only if ξ is a positive complex $2k$-vector.*

Consider an integrand of mass M specified by a covariantly constant Lagrangian of mass $\|\ \|$. From Theorem 10.3.6 and Lemma 10.5.1 there follows a well-known result of Federer [222]: any complex current (in particular, a complex surface) on a Kähler manifold is homologically minimal with respect to M.

We now suppose that the Kähler manifold M is compact and simply-connected. By Theorems 10.4.3 and 10.4.4 the $2k$-current S_{Ω^k} is closed and homologically minimal. We choose an orthonormal basis $e_1, \dots, e_{2n}$ of the space T_eM, where $\dim M = n$, and nonnegative numbers $\lambda_1, \dots, \lambda_n$ such that

$$\Omega_e = \lambda_1 e_1 \wedge e_2 + \dots + \lambda_n e_{2n-1} \wedge e_{2n}. \tag{5.2}$$

According to Lemma 10.5.1, $\lambda_i \leq 1$ for every i, $1 \leq i \leq n$. Let $J_e e_i = \sum_{j=1}^{2n} \alpha_{ij} e_j$. Clearly, $|\alpha_{ij}| \leq 1$, and $|\alpha_{ij}| = 1$ if and only if $J_e e_i = \pm e_j$. For every i by (5.2) we have

$$\Omega_e(e_{2i-1}, J_e e_{2i-1}) = \lambda_i \alpha_{2i-1,2i}. \tag{5.3}$$

On the other hand, according to (5.1)

$$\Omega_e(e_{2i-1}, J_e e_{2i-1}) = (e_{2i-1}, e_{2i-1}) = 1. \tag{5.4}$$

From (5.3) and (5.4) it follows that $\lambda_i \alpha_{2i-1,2i} = 1$. Consequently, $\lambda_i = 1$ and $\alpha_{2i-1,2i} = 1$. These equalities together with (5.2) show that Ω_e, and consequently all its exterior powers, are positive complex $2k$-vectors, $k = 1, 2, \dots, n$. This together with Lemma 5.1 implies that $(k!)^{-1}\Omega^k \in \widetilde{F}^*(S_{\Omega^k})$. We set

$$\widetilde{F}^* = \{(k!)^{-1}\Omega^k\}. \tag{5.5}$$

Then from Lemma 10.5.1 and Theorem 10.4.5 it follows that $\widetilde{F}_x$ is the set of all positive complex $2k$-vectors in $\Lambda_{2k}M_x$.

Summarizing the results, we obtain the following. For any k, $1 \leq k \leq n$, the current S_{Ω^k} is closed and homologically minimal with respect to an integrand specified by any covariantly constant convex Lagrangian. Moreover, a closed current S in the homology class $[S_{\Omega^k}]$ is homologically minimal with respect to M if and only if $\vec{S}_x$ is a positive complex $2k$-vector almost everywhere in the sense of the measure $\|S\|$.

5.2. Minimal currents on symmetric spaces. Consider a compact Riemannian symmetric space M. It is known that M admits a representation in the form G/H, where $G = I(M)_0$ is a connected group of isometries of M, H is a stationary subgroup, and on G there is specified an involutory automorphism σ, the set of all fixed points of which coincides with the connected component of the identity of H. For a compact symmetric space M every G-invariant generalized k-form is a closed differential form and is identified with the H-invariant k-covector determined by it at an arbitrary fixed point. Moreover, in every cohomology class there is a unique G-invariant form, so that the group of G-invariant forms is isomorphic to the cohomology group. There are close connections between the group H and the holonomy group Ψ_e (if M is simply-connected, then Ψ_e simply coincides with the connected component of the identity of the group H) (see Lawson [286], E. Cartan [55]). For this reason, for a compact symmetric space M all the results of Subsection 10.4.2 are reproduced word for word, where instead of the holonomy group Ψ_e we have the stationary subgroup H, and instead of covariantly constant forms and Lagrangians we have G-invariant forms and Lagrangians. For the functional $M(\ )$ all that we have said is presented in detail in the papers [38] and [43] of Đào Trọng Thi.

Below we apply the results obtained in §§3 and 4 to the investigation of global minimality of some important classes of surfaces in symmetric spaces. For this the following assertion will be useful.

THEOREM 10.5.1 (Đào Trọng Thi). *Let $M = G/H$ be a compact symmetric space, and V a closed compact surface in M. Suppose there is a point $e \in M$ and a k-vector $\xi \in \Lambda_k M_e$ such that for almost all $x \in V$ the equality*

$$g_* \vec{V}_x = \xi \tag{5.6}$$

holds for some $g \in G$. Then the k-current $[V]$ is homologically minimal with respect to $M(\ )$ if and only if there is an H-invariant k-covector $\omega_e \in \Lambda^k M_e$ such that $\|\omega_e\|^* = 1$ and $\omega_e(\xi) = 1$.

PROOF. First of all we observe that $M(\ )$ is the convex G-invariant integrand specified by the Lagrangian $l(\xi) = \|\xi\|$. Suppose that $[V]$ is homologically minimal with respect to M. According to Theorems 10.3.6 and 10.4.1 there is at least one G-invariant form ω such that

$$\|\eta\| \geq \omega(\eta) \tag{5.7}$$

for any $\eta \in \Lambda_k M$; the equality

$$\omega_x(\vec{V}_x) = 1 \tag{5.8}$$

holds for almost all $x \in V$. In view of the fact that ω is G-invariant and condition (5.6) we have

$$\omega_e(\xi) = 1\,. \tag{5.9}$$

The relations (5.7) and (5.9) obviously imply that $\|\omega_e\|^* = 1$ and $\omega_e(\xi) = 1$. In addition, from the fact that ω is G-invariant it follows that ω_e is H-invariant. Conversely, suppose there is an H-invariant k-covector

$\omega_e \in \Lambda^k M_e$ such that $\|\omega_e\|^* = 1$ and $\omega_e(\xi) = 1$; the k-covector ω_e generates a k-form ω by the formula

$$\omega(g_*\xi) = \omega_e(\xi) \tag{5.10}$$

for any $\xi \in \Lambda_k M_e$ and any $g \in G$. Clearly, ω is G-invariant and therefore closed. From the equalities $\|\omega_e\|^* = 1$ and $\omega_e(\xi) = 1$ and the fact that ω is G-invariant it follows that (5.7) is true for any $\eta \in \Lambda_k M$, and (5.8) is true for almost all $x \in V$. This means, according to Theorem 10.3.6, that the current $[V]$ is homologically minimal with respect to M. This proves the theorem.

If V is a *totally geodesic submanifold of the symmetric space* G/H, then V itself is a symmetric space, and its connected isometry group $I(V)_0$ is a subgroup of G. Since $I(V_0)$ acts transitively on V, it is clear that V satisfies (5.6).

EXAMPLE 1. Let $M = SU(n)$, $V = SU(2)$, where $SU(2)$ is embedded in the standard way in $SU(n)$. Obviously, the trilinear function $\omega(X, Y, Z) = \mathrm{Re}(\mathrm{Tr}\, XYZ)$, where $X, Y, Z \in T_e M$, is a 3-covector in $\Lambda^3 M_e$ invariant with respect to the adjoint action of the group $SU(n)$. By direct calculation we can show that $\|\omega\|^* = 1$ and $\omega(\vec{V}_e) = 1$. Hence, according to Theorem 10.5.1, it follows that the current $[V]$ is homologically minimal with respect to the mass $M(\ )$.

Consider a compact symmetric space $M = G/H$, and let σ be an involutory automorphism of the group G, the set of fixed points of which coincides with H. The manifold M admits a Cartan embedding into G of the form $\{g\sigma(g^{-1})\}$, where g runs through the whole group G. The action of G on M has the form $g(x) = gx\sigma(g^{-1})$, where $x \in M$, $g \in G$. If $h \in H$, then $h(x) = hxh^{-1}$. We denote the Lie algebras of G and H respectively by $\mathbf{G}$ and $\mathbf{H}$. We set $B = T_e M$. The subspaces $\mathbf{H}$ and B correspond to the eigenvalues 1 and -1 respectively.

EXAMPLE 2. The symmetric space $M = Sp(2n)/U(n)$. In this case $G = Sp(2n)$, $H = U(n)$, $n \geq 2$, and $\mathbf{G} = sp(2n)$. We describe the embedding of the subgroup $U(n)$ in the group $Sp(2n)$. The group $Sp(2n)$ is regarded as the subgroup of $SU(2n)$ consisting of the matrices g such that $gJ = J\bar{g}$, where

$$J = \begin{pmatrix} 0 & E \\ -E & 0 \end{pmatrix}, \qquad E = \begin{pmatrix} 1 & & 0 \\ & \ddots & \\ 0 & & 1 \end{pmatrix}.$$

Let $SO(2n)$ be the subgroup of $SU(2n)$ consisting of all real matrices, that is, matrices g such that $\bar{g} = g$. Then $U(n) = Sp(2n) \cap SO(2n)$. We observe that if $g \in U(n)$, then $gJ = Jg$. The isomorphism $g \to JgJ^{-1}$ is an involutory automorphism of the group $Sp(2n)$, the subgroup of fixed points of which is $U(n)$. The automorphism σ defines a symmetric space M. Consequently, if $X \in T_eM$, then $JXJ^{-1} = d\sigma(X) = -X$, that is, $JX = -XJ$. Consider the 2-form $\omega \in \Lambda^2 M_e$ given by $\omega(X, Y) = \operatorname{Re}(\operatorname{Tr} XJY)$. If $h \in H$, then since J commutes with h, we have

$$\omega(hXh^{-1}, hYh^{-1}) = \operatorname{Re}(\operatorname{Tr}(hXh^{-1}JhYh^{-1})) = \omega(X, Y),$$

so that ω is invariant under the action of H. The Euclidean metric on T_eM induced by the Riemannian metric of M has the form $(X, Y) = \operatorname{Re}(\operatorname{Tr} XY^*)$. Then for any $X, Y \in T_eM$ we have $\omega(X, Y) = \operatorname{Re}(\operatorname{Tr} XJY) = -\operatorname{Re}(\operatorname{Tr} X(JY)^*) = -(X, JY)$. Hence it follows that $|\omega(X, Y)| = |(X, JY)| \leq |X| \cdot |JY|$, and equality holds if and only if $JY = -X$.

Consider the H-invariant $2k$-forms $\theta_k = (k!)^{-1}\omega^k \in \Lambda^{2k} M_e$. Let $X_1, \dots, X_{2k}$ be an oriented orthonormal system of vectors in T_eM. Let K denote the $2k$-dimensional linear hull $[X_1, \dots, X_{2k}]$, and let f denote the embedding of K in T_eM as a subspace. We choose dual orthonormal bases $X'_1, \dots, X'_{2k}$ and $\omega_1, \dots, \omega_{2k}$ of the spaces K and $\Lambda^1 K$ and nonnegative numbers $\lambda_1, \dots, \lambda_k$ such that $f^*\omega = \sum_{j=1}^k \lambda_j \omega_{2j-1} \wedge \omega_{2j}$. We see that $\lambda_j = \omega(X'_{2j-1}, X'_{2j}) = 1$ for every j and $f^*\theta_k = (k!)^{-1}(f^*\omega)^k = \lambda_1 \cdots \lambda_k \omega_1 \wedge \cdots \wedge \omega_{2k}$. Consequently, $\theta_k(X_1, \dots, X_{2k}) = \theta_k(X'_1, \dots, X'_{2k}) = \lambda_1 \cdots \lambda_k \leq 1$; equality holds if and only if all the $\lambda_j = 1$, that is, $JX'_{2j} = -X'_{2j-1}$. In particular, $\|\theta_k\|^* = 1$. We now consider a submanifold of the form $V = Sp(2n')/U(n')$, $2 \leq n' < n$, where $Sp(2n')$ and $U(n')$ are embedded in $Sp(2n)$ and $U(n)$ respectively as follows. If $Q \in Sp(2n')$, then

$$i(Q) = \begin{pmatrix} E & 0 & 0 \\ 0 & Q & 0 \\ 0 & 0 & E \end{pmatrix}, \qquad E = E_{n-n'},$$

where $i\colon Sp(2n') \to Sp(2n)$ denotes the embedding, and $iU(n') = iSp(2n') \cap U(n)$. It is easy to define an orthonormal basis $X_1, \dots, X_{2k}$ in T_eV for which $X_{2j} = JX_{2j-1}$. Then $\theta_k(X_1, \dots, X_{2k}) = 1$, that is, $\theta_k(\vec{V}_e) = 1$. Hence, according to Theorem 10.5.1, the current $[V]$ is homologically minimal with respect to the mass $M(\ )$.

EXAMPLE 3. The symmetric space $M = SO(2n)/U(n)$. In this case $G = SO(2n)$, $H = U(n)$, $n \geq 2$, $\mathbf{G} = so(2n)$, and $\mathbf{H} = u(n)$. The group $U(n)$ is embedded in $SO(2n)$ as in the case of the symmetric space $Sp(2n)/U(n)$, that is, $U(n)$ consists of all matrices g in $SO(2n)$ such that $gJ = Jg$. The isomorphism $g \to JgJ^{-1}$ is an involutory automorphism of the group $SO(2n)$, the subgroup of fixed points of which coincides with $U(n)$. Consider a submanifold of the type $V = SO(2n')/U(n')$, $2 \le n' < n$, where $SO(2n')$ and $U(n')$ are embedded in $SO(2n)$ and $U(n)$ respectively as follows. If $Q \in SO(2n')$, then

$$i(Q) = \begin{pmatrix} E & 0 & 0 \\ 0 & Q & 0 \\ 0 & 0 & E \end{pmatrix},$$

where $i\colon SO(2n') \to SO(2n)$ denotes the embedding. Then $iU(n') = iSO(2n') \cap U(n)$. Next, arguing in exactly the same way as in the case of the symmetric space $Sp(2n)/U(n)$, we arrive at the following conclusion: $[V]$ is homologically minimal with respect to the mass $M(\ )$.

EXAMPLE 4. Consider the Lie group $U(n)$. For the Lie algebra $u(n)$ we choose an orthonormal basis of matrices $e_{ij} = \frac{1}{\sqrt{2}}(E_{ij} - E_{ji})$, $f_{ij} = \frac{i}{\sqrt{2}}(E_{ij} - E_{ji})$, $f_{jj} = iE_{jj}$, $1 \le i < j \le n$, where E_{ij} denotes the matrix in which the element at the intersection of the ith row and jth column is 1 and all the other elements are 0. The algebra $u(n)$ splits into the direct sum $u(n) = u^*(1) \oplus su(n)$, where $u^*(1)$ consists of scalar matrices, and $su(n)$ consists of all matrices with zero trace in $u(n)$. In the space $\Lambda^1(u(n))$, we choose a basis ω^{ij}, $1 \le i < j \le n$; θ^{ij}, $1 \le i \le j \le n$, dual to the basis e_{ij}, $1 \le i < j \le n$; f_{ij}, $1 \le i \le j \le n$. Obviously the 1-covector $\omega = n^{-1/2}\sum_{i=1}^n \theta^{ii}$ is invariant under the adjoint action of the group $U(n)$, $\omega(\eta) \le |\eta|$ for any $\eta \in u(n)$, and equality holds only in the case where $\eta = \eta_0 = n^{-1/2}\sum_{i=1}^n f_{ii}$. Consider the standard embedding of $U(k)$ in $U(n)$; $U(k)$ splits into the semidirect product $U(k) = U^{**}(1) \times SU(k)$, where $U^{**}(1)$ consists of scalar matrices, and $SU(k)$ consists of all matrices with unit determinant in $U(k)$. We set $U = U(k)$ and $V = SU(k)$. We observe that any invariant (with respect to the adjoint action of $U(n)$) k^2-covector in $\Lambda^{k^2} u(n)$ has the form $\omega \wedge \Omega$, where Ω is an invariant (k^2-1)-covector in $\Lambda^{k^2-1} su(n)$. We set $\eta_0' = k^{-1/2}\sum_{i=1}^k f_{ii}$. Obviously, $\omega(\eta_0) > \omega(\eta_0')$ and $\vec{U}_e = \eta_0' \wedge \vec{V}_e$. Hence, we have $\omega \wedge \Omega(\eta_0 \wedge \vec{V}_e) > \omega \wedge \Omega(\vec{U}_e)$. This shows, according to Theorem 10.5.1, that $[U(k)]$ is not a homologically minimal current with respect to the mass $M(\ )$.

5.3. Pontryagin cycles in Lie groups. Let G be a compact Lie group, and let K be a closed one-dimensional subgroup of G. Let $Z(K)$ denote the connected component of the identity in the centralizer of the subgroup

K. We denote the space of left cosets of G with respect to $Z(K)$ by W. Consider the map $\varphi: W \times K \to G$ given by $\varphi(X, h) = xhx^{-1}$, where $X \in W$, $h \in K$ and x is the arbitrary element of G that occurs in the class X. It was shown in [49] that $V = \varphi(W \times K)$ realizes a primitive element in the homology algebra $H_*(G, \mathbf{R})$. The surface V has a unique singular point e (e is the identity of G). We observe that K is contained in the center of the subgroup $Z(K)$. Let H and S denote the Lie algebras of $Z(K)$ and K respectively. Suppose that the subalgebra H' is the orthogonal complement of S in H. Let Z' denote the subgroup of $Z(K)$ corresponding to H'.

For the Lie groups $U(n)$, $SO(2n+1)$, and $Sp(2n)$ the construction we have described enables us to obtain all the primitive cycles constructed by Pontryagin [89]; the subgroup Z' coincides with the subgroups $U(n-1)$, $SO(2n-1)$, and $Sp(2n-2)$ respectively, embedded in the standard way in $U(n)$, $SO(2n+1)$, and $Sp(2n)$. We denote the cycles constructed in this way by P_n, Σ_{2n} and T_{2n} respectively; $\dim P_n = 2n-1$, $\dim \Sigma_{2n} = \dim T_{2n} = 4n-1$.

THEOREM 10.5.2 (Đào Trọng Thi). *The current $[V]$ is homologically minimal with respect to the mass M if and only if the current $[Z']$ is homologically minimal with respect to M.*

LEMMA 10.5.2. *Consider the isomorphism $D_k: \Lambda_{k,n} \to \Lambda_{n-k,n}$, given by the formula $(D_k\xi, \eta) = (\theta, \xi \wedge \eta)$, for any $\xi \in \Lambda_{k,n}$ and any $\eta \in \Lambda_{n-k,n}$, where θ denotes the unit n-vector in $\Lambda_{n,n}$. The following assertions are true:*

(a) *D_k takes simple k-vectors into simple $(n-k)$-vectors;*
(b) *D_k commutes with the action of the group* $\operatorname{Ad} G$;
(c) *D_k is an isometric isomorphism;*
(d) $D_k^{-1} = D_{n-k}$.

PROOF. (a) Let ξ be a simple k-vector. In $\mathbf{R}^n$ we choose an oriented orthonormal basis $v_1, \dots, v_n$ such that $\xi = \lambda_1 v_1 \wedge \cdots \wedge v_k$. Then it is easy to show that $D_k\xi = \lambda v_{k+1} \wedge \cdots \wedge v_n$, that is, $D_k\xi$ is a simple $(n-k)$-vector.

(b) Suppose that $\xi \in \Lambda_{k,n}$, $g \in \operatorname{Ad} G$. Then for any $\eta \in \Lambda_{n-k,n}$ we have $(g_* D_k\xi, \eta) = (D_k\xi, g_*^{-1}\eta) = (\theta, \xi \wedge g_*^{-1}\eta) = (\theta, g_*^{-1}(g_*\xi \wedge \eta)) = (g_*\theta, g_*\xi \wedge \eta) = (\theta, g_*\xi \wedge \eta) = (D_k g_*\xi, \eta)$. Hence it follows that $g_*(D_k\xi) = D_k(g_*\eta)$. This proves (b).

(c) This property means that D_k takes an orthonormal basis of the space $\Lambda_{k,n}$ into an orthonormal basis of the space $\Lambda_{n-k,n}$. This follows immediately from the explicit form of D_k for simple k-vectors (see the proof of (a)).

(d) This assertion is obvious.

PROOF OF THEOREM 10.5.2. Let X_0 denote the class of W that contains the identity e of the group G. For any $X \in W$ and $h \in K$ we have $\varphi(X, h) = c(X)hc(X)^{-1}$ where $c(X)$ is an arbitrary element of the class X. We observe that $\varphi(X_0, h) = h \in K$. We set $c(X_0) = e$, and for X in some sufficiently small neighborhood of X_0 in the space W we choose $c(X)$ so that c is a differentiable map of this neighborhood into G. We prove that the tangent space to the surface $V = \varphi(W \times K)$ at an arbitrary point a of $K \backslash e$, moved to the point e by the shift $L_{a^{-1}}$, is orthogonal to H'. Differentiating both sides of the equality $\varphi(X, h) = c(X)hc(X)^{-1}$, setting $X = X_0$, $h = a$ and making the shift $L_{a^{-1}}$, we obtain

$$L_{a^{-1}}(d\varphi|_{X_0 \times a}) = L_{a^{-1}}R_a(dc|_{X=X_0}) + dh|_{h=e} + dc^{-1}|_{X=X_0}. \tag{5.11}$$

From the fact that $c(X)c(X)^{-1} = e$ and $c(X_0) = e$ it follows that

$$dc|_{X=X_0} + dc^{-1}|_{X=X_0} = 0. \tag{5.12}$$

The action of the adjoint group $\operatorname{Ad} Z(K)$ on the Lie algebra $\mathbf{G}$ of the group G leaves invariant the subspace H and its orthogonal complement B; $\mathbf{G} = H \oplus B$. Suppose that $dc|_{X=X_0}$ splits into the sum $u + v$, where $u \in H$, $v \in B$. Since $a \in Z(K)$, it follows that $L_{a^{-1}}R_a u = \operatorname{Ad}_a u = u$, $u \in H$, so that we have $L_{a^{-1}}R_a(dc|_{X=X_0}) = L_{a^{-1}}R_a v + u$. Taking (5.12) into consideration, we obtain

$$L_{a^{-1}}R_a(dc|_{X=X_0}) + dc^{-1}|_{X=X_0} = L_{a^{-1}}R_a v - v \in B. \tag{5.13}$$

From (5.11) and (5.13) it follows that $L_{a^{-1}}(d\varphi|_{X_0 \times a}) \in B \oplus S$, that is, it is orthogonal to H'. We also observe that the surface V is invariant under the action of the group $\operatorname{Int} G$, so that V satisfies the condition (5.6) in Theorem 10.5.2 with k-vector ξ associated with the orthogonal complement to H'. On the other hand, it is obvious that Z' satisfies (5.6) with $(n-k)$-vector $D_k\xi$, which is associated with H', $k = \dim V$, $n = \dim M$. Now, applying Lemma 10.5.2, it is easy to see that if ω is an $\operatorname{Ad} G$-invariant k-covector in $\Lambda^k G_e$ and $\|\omega\|^* = 1$, $\omega(\xi) = 1$, then $D_k\omega$ is also $\operatorname{Ad} G$-invariant and $\|D_k\omega\|^* = 1$, $D_k\omega(D_k\xi) = 1$. According to Theorem 10.5.1, this proves Theorem 10.5.2.

The following examples are a consequence of Theorem 10.5.2.

EXAMPLE 5. As we showed in Example 1, $Sp(2) = SU(2)$, embedded in $SU(4)$ in the standard way, is homologically minimal with respect to M

in $SU(4)$. Consequently, $Sp(2)$ is homologically minimal with respect to M in $Sp(4)$. Then, according to Theorem 10.5.2, the Pontryagin cycle T_4 is also homologically minimal with respect to M in $Sp(4)$.

EXAMPLE 6. As we showed in Example 4, the subgroup $U(n-1)$, embedded in $U(n)$ in the standard way, is homologically not minimal with respect to M. Then, according to Theorem 10.5.2, the Pontryagin cycle P_n is also homologically not minimal with respect to M in $U(n)$.

Let $G = SU_n$, and let Z' be a canonically embedded subgroup of SU_{n-1}. For K we take the centralizer of the group SU_{n-1} in SU_n. We denote the primitive Pontryagin cycle $V(K)$ by P_{2n-1}. The canonical embedding $SU_n \to SU_m$ maps P_{2n-1} into SU_m. We also denote this image by P_{2n-1}. By the method of relative calibrations (see Chapter 4, Subsection 4.5) Le Hong Van proved the following theorem.

THEOREM 10.5.3. *The primitive Pontryagin cycle* P_{2k-1} *is a stably minimal surface in the group* SU_n *for any* $2k-1 \leq n$.

CHAPTER XI

Globally Minimal Surfaces in Regular Orbits of the Adjoint Representation of the Classical Lie Groups

§1. Statement of the problem. Formulation of the main theorem

In this chapter we briefly present results obtained by Le Hong Van.

In Chapter 5 (§5, Theorem 5.3.3) we showed that every globally minimal surface in a homogeneous compact Riemannian space has an invariant absolute calibration. In view of this there arises the question of describing and classifying globally minimal surfaces in important classes of compact homogeneous spaces. This question has been solved for one remarkable class of homogeneous spaces. Namely, we consider the class $\mathscr{C}$ of spaces realized as orbits of the adjoint representation of the compact Lie groups. This class attracts our attention because of the following properties:

(1) $\mathscr{C}$ coincides with the class of strictly homogeneous symplectic spaces with a compact group of motions (Kirillov [448]);

(2) $\mathscr{C}$ often arises in Hamiltonian mechanics: the orbits of the adjoint representation support many integrable systems (Fomenko [459]);

(3) $\mathscr{C}$ coincides with the class of complex homogeneous manifolds (M^{2n}, J) with negative canonical Hermitian form $h_J(X, Y) = -2\pi\Omega_J(X, JY)$, where Ω_J is the first Chern form on M^{2n} ([443]);

(4) additional invariant structures (complex, symplectic) on a homogeneous Riemannian manifold enable us to discover hidden symmetries of its minimal submanifolds.

Harvey and Lawson [253] began to study calibration geometry systematically in the space $\mathbf{R}^{2n}$ with canonical complex structure and obtained striking results.

We note that the space $G/C(t) \in \mathscr{C}$ has many invariant Riemannian structures [443]; among them we pick out the Einstein-Kähler metric and the Killing metric, which generates a canonical connection of the second kind [449]. Any suborbit of $G/C(t)$ is a complex submanifold, and therefore a globally minimal submanifold of $G/C(t)$ endowed with the Kähler metric [216]. However, when $G/C(t)$ is endowed with the Killing metric, the picture abruptly changes. We restrict ourselves to a consideration of the case of regular orbits, that is, when $C(t) = T_G$ is a maximal torus in G.

THEOREM (Le Hong Van). *Let N be a globally minimal submanifold of the homogeneous space G/T_G, endowed with the Killing metric. Then N is the image $\rho_*(H/T_H)$ of a homogeneous manifold H/T_H, where T_H is a maximal torus in the group H, and ρ_* is induced by the representation $\rho: H \to G$ so that the following diagram is commutative:*

$$\begin{array}{ccc} H & \xrightarrow{\rho} & G \\ \downarrow & & \downarrow \\ H/T_H & \xrightarrow{\rho_*} & G/T_G \end{array}$$

If G and H are classical Lie groups, and $\rho(H) \neq \mathrm{Spin}_n$, $n \geq 7$, then ρ is one of the admissible representations in Table 1, where π_i are the ith fundamental representations of the group H.

TABLE 1. All admissible representations

$H \xrightarrow{\rho} G$	SU_{m+1}	SO_m	Sp_m
SU_{n+1}	$\oplus\pi_i$ $(0 \leq i \leq n)$	$\oplus\pi_i$ $(i = 0$ *or* $i + i' = n + 1)$	$\oplus\pi_i$ $(i = 0$ *or* $i + i' = n + 1)$
SO_{2n+1}, $n \neq 1$	—	$\oplus\pi_i$ $(i = 0, 1)$	— $(n = 2 \downarrow)$
Sp_n	$\oplus\pi_i$ $(i = 0, 1)$	$(i = 2 \uparrow)$ $\oplus\pi_i$ $(i = 0, 1)$	$\oplus\pi_i$ $(i = 0, 1)$
SO_{2n}, $n \geq 4$	$\oplus\pi_i$ $(i = 0, 1)$	$\oplus\pi_i$ $(i = 0, 1)$	$\oplus\pi_i$ $(i = 0, 1)$

We describe briefly the fundamental representations of the classical Lie groups. The representation π_0 is the trivial representation of dimension 1, and the representation π_i of the group SU_{n+1} is the outer ith power of π_1. (There is a similar correspondence for other groups: see [446], [447], or Subsection 2.2 of this chapter.)

The main theorem of Le Hong Van demonstrates the deep connection between algebra, topology, and geometry. Namely, minimal (in the sense of dimension) representations of the classical Lie groups correspond one-to-one with minimal (in the sense of volume) models of nontrivial cycles in homogeneous spaces of the class $\mathscr{C}$.

§2. Necessary information from the theory of representations of the compact Lie groups

2.1. The weight of an irreducible representation. It is easier to construct a theory of representations of compact Lie groups by starting from the theory of representations of complex semisimple Lie groups (and algebras). The fact is that any compact Lie group is the direct product of a torus (a torus is a connected compact Abelian group) and a compact semisimple Lie group. Any complex representation of a torus is the direct sum of one-dimensional irreducible representations. A representation of a compact semisimple Lie group is uniquely defined up to equivalence by a representation of its complexification; the latter is determined by a representation of its Lie algebra. We first write out explicitly an embedding of a compact semisimple Lie algebra in its complexification. Let lG be a complex semisimple Lie algebra, and lT_G a Cartan subalgebra of it. We denote its root system by $\Delta(lG)$ and the subsystem of positive roots by $\Delta^+(lG)$. We normalize the root vectors $E_\alpha \in lG$, $\alpha \in \Delta$, so that for any $\alpha, \beta \in \Delta$ we have $N_{\alpha,\beta} = N_{-\alpha,-\beta}$, where $N_{\alpha\beta}$ (here $N_{\alpha,\beta} = N_{\alpha\beta}$) is defined by $[E_\alpha, E_\beta] = N_{\alpha\beta}E_{\alpha+\beta}$. The real compact form lG_u of the algebra lG is the direct sum $\operatorname{span}_{\mathbf{R}}\{H_\alpha\}$ $\oplus\operatorname{span}_{\mathbf{R}} V^+$, where H_α is an element of lT_G such that for any $h \in lT_G$ we have $\alpha(h) = \langle H_\alpha, h\rangle$, where α runs through $\Delta(lG)$ and

$$V^+ = \{(E_\alpha + E_{-\alpha}), (iE_\alpha - iE_{-\alpha}), \alpha \in \Delta(lG)\}.$$

Theorem on the weight decomposition of the space of a representation of a complex semisimple Lie algebra. Let ρ be a complex representation of a semisimple Lie algebra lG in the space V. Let lT_G be the Cartan subalgebra of lG. Then V splits into a direct sum of weight subspaces $V = \oplus_\alpha V_\alpha$, where V_α are invariant subspaces of the algebra lT_G, and for any $h \in lT_G$ we have $\rho(h)v_\alpha = \alpha(h)v_\alpha$. Here $v_\alpha \in V_\alpha$, and α is a linear functional on lT_G. We shall call α the weight of the representation ρ. Let E_β be a root vector of the algebra lG; then we have $\rho(E_\beta)V_\alpha \subset V_{\alpha+\beta}$. A weight vector $v_\alpha \in V_\alpha$ is called a highest vector if for any positive root $\beta \in \Delta^+(lG)$ we have $\rho(E_\beta)V_\alpha = 0$, and in this case α is called the highest weight of the representation ρ. The representation ρ is irreducible if and only if it has a unique (up to multiplication by scalars) highest vector. Any other weight vector of an irreducible representation can be obtained from the highest vector by successive application of the operators $\rho(E_{-\beta_i})$ and linear combinations of them, where $\beta_i \in \Delta^+(lG)$. Similarly, the lowest vector of the representation ρ is defined as the weight vector $v_\gamma \in V_\gamma$ for which $\rho(E_{-\beta})v_\gamma = 0$ for any positive root $\beta \in \Delta^+(lG)$, and in this case γ is called the lowest weight.

2.2. Fundamental representations of a compact semisimple Lie group. Thus, an irreducible representation is uniquely determined by its highest weight. A linear functional Λ on lT_G is the highest weight of some irreducible representation if and only if for any simple root $\alpha \in \Delta(lG)$

$$\Lambda_\alpha = 2\frac{\langle \Lambda , \alpha \rangle}{\langle \alpha , \alpha \rangle}$$

is a nonnegative integer. Here $\langle\ ,\ \rangle$ denotes the induced Killing form on lT_G^*.

Let $\Pi = \{\alpha_1, \dots, \alpha_r\}$ be a system of simple roots. Then the irreducible representation π_i, defined by the highest weight $\Lambda_{\alpha_i} = \delta_{ij}$, is called the ith fundamental representation of the algebra lG. In other words, the fundamental representations of the algebra lG form a basis of the integral lattice of linear functionals on lT_G, which are the highest weights of irreducible representations of lG. Addition in this lattice corresponds to the Cartan composition of irreducible representations. We recall that the Cartan composition of irreducible representations φ and ψ of lG is the irreducible representation of lG with highest weight equal to the sum of the highest weights of φ and ψ. The Cartan composition of φ and ψ is denoted by $\varphi \dotplus \psi$ (in contrast to the direct sum of representations $\varphi \oplus \psi$). If $\psi = \varphi$, we denote $\varphi \dotplus \varphi$ by 2φ (and $\varphi \oplus \varphi$ by $\oplus_{2\varphi}$) and so on. The trivial representation of lG is denoted by $\pi_0 \colon \pi_0(lG) = e \in \operatorname{End} V_0$, $\dim V_0 = I$.

REMARK. A representation of a Lie group is determined by a representation of its Lie algebra: $\rho(\exp x) = \exp \rho(x)$. The whole terminology of its Lie algebra (highest vector, fundamental representations, etc.) also goes over to the group. Similarly, the ith fundamental representation of a compact semisimple group G is defined as the restriction to G of the ith fundamental representation of the complexification $G_{\otimes \mathbf{C}}$.

2.3. Decomposition of the tensor product of representations of a compact Lie group. The following result is well known.

PROPOSITION 11.2.1 ([463], Theorem 3.8). (i) *Let ρ be a representation of the group G in the group SU_{n+1}. Then the representation* $\operatorname{Ad}\rho$ *in the algebra $gl_n(\mathbf{C}) \simeq u_{n+1 \otimes \mathbf{C}}$ is equivalent to the representation $\rho \otimes \rho^*$, where ρ^* is the representation contragredient to ρ.*

(ii) *Let* ρ *be a representation of the group* G *in the group* SO_n. *Then the representation* $\operatorname{Ad}\rho$ *in the algebra* $so_n(\mathbf{C})$ *is equivalent to the representation* $\Lambda^2(\rho)$.

(iii) *Let* ρ *be a representation of the group* G *in the group* Sp_n. *Then the representation* $\operatorname{Ad}\rho$ *in the algebra* $sp_n(\mathbf{C})$ *is equivalent to the representation* $S^2(\rho)$.

It is well known that in the decomposition of a product of irreducible representations $\varphi \otimes \psi$ there occurs the Cartan composition $\psi \dotplus \varphi$. Moreover, $\varphi \otimes \varphi = S^2(\varphi) \oplus \Lambda^2(\varphi)$. The representation $\psi \dotplus \varphi$ is called the highest component of irreducibility of the product $\psi \oplus \varphi$, since its highest weight is the largest of all weights of the representation of the latter (with respect to lexicographic order in $\Delta(lG)$). Theorem 11.2.1 below, obtained by Le Hong Van, enables us to find "small" components of irreducibility of the representation $\psi \otimes \psi^*$.

Let φ and ψ be irreducible representations of the semisimple algebra lG with highest weights Λ and M respectively. Following Dynkin, we shall say that ψ is subordinate to φ if for any simple root $\alpha_i \in \Delta(G)$ we have $\Lambda_{\alpha_i} \geq M_{\alpha_i}$.

THEOREM 11.2.1. *Suppose that an irreducible representation* ψ *is subordinate to an irreducible representation* φ.

(i) *Then in the decomposition of the tensor product* $\varphi \otimes \varphi^*$ *there occurs a component* $\psi \otimes \psi^*$.

(ii) *Suppose that the representations* ψ *and* φ *are selfadjoint. Then in the decomposition of the representation* $\Lambda^2(\varphi)$ *(resp.* $S^2(\varphi)$*) there occurs a component* $\Lambda^2(\psi)$ *(resp.* $S^2(\psi)$*) if* ψ *and* φ *are simultaneously orthogonal or symplectic. Otherwise we have the inclusions* $\Lambda^2(\psi) \subset S^2(\varphi)$; $S^2(\psi) \subset \Lambda^2(\varphi)$.

PROOF. Let $h(\psi)$ and $h(\varphi)$ denote the highest weights of the representations ψ and φ respectively. Then $h(\varphi) - h(\psi)$ is the highest weight of some irreducible representation γ. Let $E(\psi)$ (resp. $E(\varphi)$, $E(\gamma)$) denote the space of the representation ψ (resp. φ, γ) and $E^*(\psi)$ (resp. $E^*(\varphi)$, $E^*(\gamma)$) its adjoint space. We define a convolution map from $E(\psi) \otimes E(\gamma) \otimes E^*(\psi) \otimes E^*(\gamma)$ to $E(\psi) \otimes E^*(\psi)$ as follows:

$$\sigma(x \otimes y \otimes x^* \otimes y^*) = y^*(y) x \otimes x^*. \tag{2.1}$$

Clearly, σ is an lG-invariant linear operator. Since $h(\psi) + h(\gamma) = h(\varphi)$, we can embed the space $E(\varphi)$ (resp. $E^*(\varphi)$) in the tensor product $E(\psi) \otimes E(\gamma)$ (resp. $E^*(\psi) \otimes E^*(\gamma)$) as the highest component of irreducibility of the latter. Let $U(lG)$ be the covering algebra of the Lie algebra lG. It is well known that $E(\psi) \otimes E^*(\psi)$ is generated, as a $U(lG)$-module, by the vector $v_h(\psi) \otimes v_l(\psi^*)$, where $v_h(\psi)$ and $v_l(\psi^*)$ are the highest and lowest vectors in $E(\psi)$ and $E^*(\psi)$. Taking account of the fact that $v_h(\psi) \otimes v_l(\psi^*) \in \operatorname{Im} \sigma(E(\varphi) \otimes E^*(\varphi))$, the restriction of σ to the subspace $E(\varphi) \otimes E^*(\varphi)$ is a surjective map onto $E(\psi) \otimes E^*(\psi)$. Hence we immediately obtain the first assertion of Theorem 11.2.1. To prove assertion (ii), it is sufficient to observe that if φ and ψ are selfadjoint representations, then γ is also a selfadjoint representation. Moreover, γ is an orthogonal representation if φ and ψ are simultaneously either orthogonal or symplectic; otherwise γ is symplectic. Correspondingly, the map in (2.1) takes the form

$$\sigma(x \otimes y \otimes x' \otimes y') = \sigma_\gamma(y, y')x \otimes x', \tag{2.2}$$

where σ_γ is a nondegenerate bilinear lG-invariant form on $E(\gamma)$, and σ_γ is symmetric if γ is orthogonal and skew-symmetric if γ is symplectic. Regarding $S^2(E(\varphi))$ and $\Lambda^2(E(\varphi))$ as subspaces of symmetric and skew-symmetric tensors in $E(\varphi) \otimes E(\varphi)$, we easily derive assertion (ii) from (i) by taking account of (2.2). This proves the theorem.

REMARK. In a similar way we can prove a more general assertion. Namely, the representation $\varphi_1 \otimes \varphi_2^*$ contains all irreducible components of the representation $\psi_1 \otimes \psi_2^*$ if $h(\varphi_1) - h(\varphi_2) = h(\psi_1) - h(\psi_2)$ and ψ_1 is subordinate to φ_1.

By means of Theorem 11.2.1 and using the technique of its proof we obtain the following propositions.

PROPOSITION 11.2.2. (a) *If an irreducible representation ψ of the algebra su_{n+1} differs from the fundamental representations π_i, then in the decomposition of the representation $\psi \otimes \psi^*$ there occurs the component $2\pi_1 \dotplus 2\pi_n$ $(\approx 2\pi_1 \dotplus 2\pi_1^*)$.*

(b) *In the decomposition of the representation $\psi \otimes \psi^*$, where ψ is any irreducible representation of the algebra so_{2n+1} $(n \neq 1)$, there occurs the component $2\pi_1$.*

PROPOSITION 11.2.3. (A) *Let ψ be an irreducible orthogonal representation of the group SU_{n+1}, distinct from fundamental representations π_i. Then in the decomposition of the representation $\Lambda^2(\psi)$ there occurs the component $2\pi_1 \dotplus \pi_{n-1}$.*

(B+D) *Let ψ be an irreducible orthogonal representation of the group SO_n, distinct from fundamental representations π_i. Then in the decomposition of the representation $\Lambda^2(\psi)$ there occurs the component $2\pi_1 \dotplus \pi_2$.*

(C) *Let* ψ *be an irreducible orthogonal representation of the group* Sp_n. *Then in the decomposition of the representation* $\Lambda^2(\psi)$ *there occurs either the component* $\pi_1 \dotplus \pi_3$ *or the component* $2\pi_1 \dotplus \pi_2$.

PROPOSITION 11.2.4. (a) *Let* Ψ *be an irreducible symplectic representation of the group* SU_{n+1}, *distinct from fundamental representations* π_i. *Then in the decomposition of the representation* $S^2(\psi)$ *there occurs the component* $2\pi_1 \dotplus 2\pi_n$.

(b) *Let* ψ *be a symplectic representation of the group* Sp_n, *distinct from the fundamental representation* π_1 *of least dimension. Then in the decomposition of the representation* $S^2(\psi)$ *there occurs either the component* $\pi_1 \dotplus \pi_2$ *or the component* $2\pi_1 \dotplus \pi_2$.

§3. Topological structure of the space G/T_G

3.1. Representation of the cohomology algebra $H^*(G/T_G, \mathbf{R})$ as the algebra $\Lambda(G/T_G)$ of outer closed differential forms. It is known that an orbit in general position of the adjoint representation of a compact semisimple group G is diffeomorphic to the quotient G/T_G, where T_G is a maximal torus of G. Moreover, we have the isomorphism (see [444])

$$H^*(G/T_G; \mathbf{R}) \cong S^I_{T_G}/\rho(S^I_G), \tag{3.1}$$

where $S^I_{T_G}$ is the algebra of all polynomials on the Cartan subalgebra lT_G (of the algebra lG), and $\rho(S^I_G)$ is the ideal generated by the set of all restrictions to lT_G of G-invariant polynomials of positive degree on lG. Therefore, the Poincaré polynomial for the space G/T_G has the form

$$P(G/T) = \frac{(1-x^{2\nu_1})\cdots(1-x^{2\nu_r})}{(1-x^2)^r}. \tag{3.2}$$

In (3.2), $r = \operatorname{rk} G = \dim T$, and $2\nu_k - 1$ is the degree of the primitive kth generating element of the cohomology algebra $H^*(G, \mathbf{R})$ of the group G. Formulae (3.1) and (3.2) show that the cohomology algebra $H^*(G/T_G, \mathbf{R})$ has generating elements of degree 2; moreover, we have $\operatorname{rk} H^2(G/T, \mathbf{R}) = \operatorname{rk} G = \dim T$.

Let V be the orthogonal complement of the subalgebra lT_G in the algebra lG. Then we can identify V with the tangent space to G/T_G at the point $\{eT_G\}/T_G$. Let $\langle\ ,\ \rangle$ denote the Killing form on lG. For each $v \in lT_G$ we set

$$\omega_v(X, Y) = \langle v, [X, Y]\rangle, \qquad X, Y \in V. \tag{3.3}$$

Let $\{v_i, i = 1, \ldots, r\}$ denote a basis in lT_G.

THEOREM 11.3.1 (Le Hong Van). *For any* $v \in lT_G$ *the form* ω_v *can be extended to a* G*-invariant closed differential* 2*-form* $\hat{\omega}_v$ *on the orbit* G/T_G. *The set* $\{[\hat{\omega}_{v_i}],\ i = 1, \dots, r\}$ *is a set of generators of the group* $H^2(G/T_G, \mathbf{R})$, *where* $[\hat{\omega}_{v_i}]$ *denotes the cohomology class of the form* $\hat{\omega}_{v_i}$. *Thus, a map from the algebra of outer differential forms* $\Lambda(G/T_G) = \Lambda(\hat{\omega}_{v_i})$ *to the cohomology algebra* $H^*(G/T_G, \mathbf{R})$ *is an epimorphism.*

REMARK. By means of the Killing metric we define an isomorphism from the algebra lG to its dual space $(lG)^*$:

$$X \to X^*: X^*(Y) = \langle X, Y\rangle \text{ for any } X, Y \in lG.$$

Obviously, the 2-form ω_i is the differential of the 1-form v_i^* on lG. The forms v_i^* can be regarded as polynomials of degree 1 on lG (their restrictions to lT_G generate the algebra of polynomials on lT_G). Since the forms $\hat{\omega}_i$ and the polynomials v_i are generators of the commutative algebra $\Lambda(G/T_G)$ and the algebra of polynomials $S^I_{T_G}$ respectively, we can map $\Lambda(G/T_G)$ into $S^I_{T_G}$ as follows: $\sigma(\hat{\omega}_i) = v_i^*$. Consider the composition

$$j\sigma: \Lambda(G/T_G) \to S^I_{T_G} \to S^I_{T_G}/\rho(S^I_{T_G}) \cong H^*(G/T_G, \mathbf{R}).$$

It is easy to verify that $j\sigma$ is the map $\Lambda(G/T_G) \to H(G/T_G, \mathbf{R})$, defined in Theorem 11.3.1. Consequently, $j\sigma$ is an epimorphism, and $\operatorname{Ker} j\sigma = \{\omega, \sigma(\omega) \in \rho(S^I_{T_G})\}$. In other words we have $H^*(G/T_G, \mathbf{R}) = \Lambda(\hat{\omega}_i)/\rho(S^I_{T_G})(\hat{\omega}_i)$.

3.2. Realization of cycles in $H_*(G/T_G, \mathbf{R})$ by homogeneous subspaces.

THEOREM 11.3.2 (Le Hong Van). *Let* G *be a compact semisimple Lie group,* T_G *a maximal torus of* G, *and* H *a compact Lie subgroup of* G. *Then the quotient subspace* $H/(H \cap T_G)$ *realizes a nontrivial cycle in* $H_*(G/T_G, \mathbf{R})$ *if and only if* $H \cap T_G$ *is a maximal torus* T_H *in* H.

PROOF. Clearly, $H \cap T_G$ is a torus T in the subgroup H. We first prove that if T is not a maximal torus in H, then $H/(H \cap T_G)$ does not realize any cycle in $H^*(G/T_G, \mathbf{R})$. For this it is sufficient to show that the restriction of any cocycle $\alpha \in H(G/T_G, \mathbf{R})$, where $\deg \alpha = \dim \rho(H/T)$, to the cycle $\rho[H/T]$ is equal to zero. Here ρ is an embedding $H \to G$. It is well known [444] that the algebra $H^*(G/T_G, \mathbf{R})$ is characteristic and it has two-dimensional generators $\{\alpha_i^2,\ i = 1, \dots, \operatorname{rk} G\}$. Therefore, our assertion will be proved if we show that for any set $\{r_i, r_i \geq 0, \sum_{i=1}^{k} r_i = 2\dim(H/T)\}$ we have

$$[(\alpha_1^2)^{r_1} \cdot (\alpha_2^2)^{r_2} \cdot \cdots \cdot (\alpha_k^2)^{r_k}][\rho(H/T)] = 0. \tag{3.4}$$

Here $k = \operatorname{rk} G$.

In the previous section we showed that the classes α_i^2 have representatives ω_i^2 in the class of G-invariant 2-forms on G/T_G. Namely, the value of ω_i^2 in the tangent space $T_e(G/T_G)$ has the form

$$\omega_i^2(X, Y) = \langle r_i, [X, Y]\rangle \tag{3.5}$$

where X and Y are tangent vectors in $T_e(G/T_G) = V \subset lG$, r_i is an element in the Cartan algebra lT_G, and $\langle\ ,\ \rangle$ is the Killing form on g (see (3.3)). This means that the integral of the form $[(\omega_1^2)^{r_1} \wedge \cdots \wedge (\omega_k^2)^{r_k}]$ on the submanifold $\rho(H/T)$ is nonzero. We observe that the form $\rho^*[(\omega_1^2)^{r_1} \wedge \cdots \wedge (\omega_k^2)^{r_k}]$ is H-invariant since $\rho(H)$ is a subgroup of G. We now calculate the value of the form ω_k^2 on the subspace $\rho(H/T)$.

LEMMA 11.3.1. *The restriction of the form* ω_k^2 *to the subspace* $\rho(H/T)$ *is cohomologous to the restriction of the form* $\overline{\omega}_k^2$, *defined as in* (3.3) *when* $\bar{r}_k = \pi_{lT}(r_k)$, *the orthogonal projection of the vector* r_k *from* lT_G *to* lT_G.

CONTINUATION OF THE PROOF OF THEOREM 11.3.1. By hypothesis, T is not a maximal torus in H, so $Z(T) \cap W \neq 0$. Suppose that $W_0 \in Z(T) \cap W$. Then we have $(W_0 \rfloor \rho^*(\overline{\omega}_k^2))_W = 0$, for any 2-form $\overline{\omega}_k^2$, so that

$$\rho^*((\overline{\omega}_1^2)^{r_1} \wedge \cdots \wedge (\overline{\omega}_k^2)^{r_k})(W_0, W_1, \ldots, W_s) = 0$$

for the orthonormal basis $\{w_i\} \in W$. By Lemma 11.3.1 this equality implies that

$$\rho^*[(\alpha_1^2)^{r_1} \cdot \cdots \cdot (\alpha_k^2)^{r_k}]([H/T]) = 0.$$

We have thus proved the first part of the theorem.

We now prove that the quotient subspace H/T_H, where T_H is a maximal torus in H, realizes a nontrivial cycle in G/T_G. We choose a symplectic H-invariant form ω_2 on H/T_H, that is, $(\omega_2)^n$ is the volume form on H/T_H. Such a form ω_2 always exists, since H/T_H is realized as an orbit of the adjoint representation in lH. It is easy to see that ω_2 has the form $\omega_2(X, Y) = \langle t, [X, Y]\rangle$, $t \in lT_H$. Clearly, ω_2 is the restriction of the 2-form $\overline{\omega}_2$ to V such that $\overline{\omega}_2(X, Y) = \langle t, [X, Y]\rangle$. As we observed in Subsection 3.2, the form $\overline{\omega}_2$ generates a cocycle α_2 in $H^*(G/T_G, \mathbf{R})$. Clearly, the form $(\overline{\omega}_2)^n$ is also closed and its restriction to $T_e(H/T_H)$ is equal to $(\omega_2)^n$. Consequently, the value of the cocycle α_2^n on $\rho_*(H/T_H)$ is nonzero. Hence, H/T_H realizes a nontrivial cycle in $H_*(G/T_G, \mathbf{R})$. This proves the theorem.

§4. A brief outline of the proof of the main theorem

Theorem 3.2.1 reduces the question of classification of globally minimal homogeneous spaces in regular orbits G/T_G to the problem of the global minimality of the embedding $\rho_*:(H/T_H) \to G/T_G$.

4.1. The complete geodesic property of the embedding $\rho_*:(H/T_H) \to G/T_G$.

THEOREM 11.4.1. *Suppose that G/T_G is endowed with the G-invariant metric induced by the Killing form on the algebra g. Then H/T_H is a totally geodesic submanifold of G/T_G and the embedding $\rho_*:(H/T_H) \to G/T_G$ is defined uniquely up to an isometry by the embedding $\rho: H \to G$.*

PROOF. Consider the Cartan decomposition $lG = lT_G + V$. Here we identify V with the tangent space to G/T_G at the point $\{eT_G\}$, and accordingly $H = lT_H + W$, where $lT_H \subset lT_G$. Clearly, lT_H acts on W by the adjoint representation and we have the direct sum decomposition $W = \bigoplus_{\alpha\in\Delta^+(H)} W_\alpha$, where W_α is a (real) irreducible two-dimensional space, that is, $[lT_H, W_\alpha] = W_\alpha$, and $\Delta^+(H)$ is the system of positive roots of the subgroup H. Obviously, for any $t \in lT_G$ and for any $\alpha \in \Delta^+(H)$ we have

$$\langle t, W_\alpha\rangle = \langle t, [lT_H, W_\alpha]\rangle = \langle [t, lT_H], W_\alpha\rangle = 0.$$

This means that the plane $W = \bigoplus_\alpha W_\alpha$ is orthogonal to the Cartan subalgebra lT_G. In other words, W belongs to V. Since the Riemannian structure on G/T_G is induced by the Killing form on lG, every geodesic $\mathrm{Exp}(tw)$, $w \in W$, on the manifold H/T_H is also geodesic on G/T_G (see also [449]). Consequently, H/T_H is a totally geodesic submanifold. Obviously, a totally geodesic submanifold is a minimal submanifold.

4.2. Construction of variations that decrease the volume of "sufficiently complicated" embeddings ρ_*. Since $\rho_*(H/T_H)$ is a totally geodesic submanifold of G/T_G, we can apply the previous results to it (Corollary 3.5.1). We choose an orthonormal basis in $P = T_e(H/T_H)$ as follows: $p_\alpha^+ = (E_\alpha + E_{-\alpha})/\sqrt{2}$, $p_\alpha^- = (iE_\alpha - iE_{-\alpha})/\sqrt{2}$, $\alpha \in \Delta(H)$, where E_α are normalized as in Subsection 2.1. Then Corollary 3.5.1 can be rewritten as follows:

$$\overline{I\psi} = \sum_{\alpha\in\Delta(H)} \{-X_{-\alpha}X_\alpha(\psi) - X_{-\alpha}[E_\alpha, \psi]_W + [E_{-\alpha}[E_\alpha, \psi]_{lT_G}]_W\} \quad (4.1)$$

where X_α is a left-invariant differential operator on H with value E_α at unity, and W is normal to $\rho(H/T_H)$ at the point e.

Let $t^\perp$ be the orthogonal complement to $\rho_*(T_H)$ in T_G. Clearly, $l G = \rho(lH) = (W + t^\perp)$. We set $\sigma(\rho) = \mathrm{Ad}_G\, \rho|(W + t^\perp)$; then $\sigma(\rho) = \mathrm{Ad}_G \rho - \mathrm{Ad}_H$. We also denote the action of $\sigma(\rho)$ on $(W + t^\perp)_{\otimes \mathbf{C}}$ by $\sigma(\rho)$. The space $(W + t^\perp)_{\otimes\mathbf{C}}$ splits into the direct sum of irreducible subspaces m_i of the representation $\sigma(\rho)$. Consequently, $W_{\otimes\mathbf{C}}$ is the sum (not necessarily direct) of its subspaces $\pi_{W_{\otimes\mathbf{C}}}(m_i) = W_i$. It is easy to see that W_i is an invariant subspace of the restriction of the action of $\sigma(\rho)$ to the torus T_H. Accordingly, the space $C^\infty(H, W_{\otimes\mathbf{C}})_{T_H}$ splits into the sum of its subspaces $\sum_i \sum_{\lambda\in D(H)} \tilde{A}(V_\lambda \otimes \mathrm{Hom}_{T_H}(V_\lambda, W_i))$. Here $\tilde{A}$ is defined by

$$\tilde{A}(v_\lambda \otimes L)(h) = L(\rho(h^{-1})v_\lambda)$$

(ρ is the representation of the group H in the complex irreducible space V_λ).

By Theorem 3.5.1 and Corollary 3.5.1 $\tilde{A}(V_\lambda \otimes \mathrm{Hom}_{T_H}(V_\lambda, W_i))$ is an invariant subspace of the action of the Jacobi operator $\overline{I}$, whose restriction has the form

$$\begin{aligned}(\overline{I}_* L)(v_\lambda) = \sum_{\alpha\in\Delta(H)} (&-L[E_{-\alpha}[E_\alpha, v_\lambda]] + \pi_{W_i}([E_\alpha, L[E_{-\alpha}, v_\lambda]]) \\ &+ \pi_{W_i}[E_{-\alpha}, \pi_{lT_G}[E_\alpha, Lv_\lambda]].\end{aligned} \tag{4.2}$$

Consequently, the eigenvalues of the Jacobi operator coincide with the eigenvalues of the induced operator $\overline{I}_*$ on $\mathrm{Hom}_{T_H}(V_\lambda, W_i)$.

Formula (4.2) is our working formula. Henceforth, for each series of "sufficiently complicated" embeddings ρ_* of the space V_λ, W_i (and consequently also W_i) we give an operator $L \in \mathrm{Hom}_{T_H}(V_\lambda, W_i)$ for which $\langle \overline{I}_* L, L\rangle < 0$. One of the two functions $\mathrm{Re}\,\tilde{A}(v_\lambda \otimes L)$, $\mathrm{Im}\,\tilde{A}(v_\lambda \otimes L)$ is nonzero. Consequently, the variation corresponding to the section $\tau\mathrm{Re}(\mathrm{Im})\tilde{A}(v_\lambda \otimes L)$ where τ is the isomorphism $C^\infty(H, W)_{T_H} \simeq \Gamma(n(H/T_H))$, decreases the volume of $\rho_*(H/T_H)$ in G/T_G. Moreover, $\mathrm{ind}\,\rho_*(H/T_H) \geq \dim_{\mathbf{C}} V_\lambda$.

A. The case where $G = SU_{m+1}$. By Proposition 2.3.1 the restriction of the representation $\mathrm{Ad}_G\rho$ to W is equivalent to the representation $\rho\otimes\rho^* - \mathrm{Ad}_H = \sigma(\rho)$.

A.1. $H = SU_{n+1}$. Let ρ be the direct sum of irreducible representations, among which there is a nonfundamental representation. Since $(\sum \rho_i)\otimes(\sum \rho_i^*) = \sum_{ij} \rho_i \otimes \rho_j^*$, taking account of Proposition 11.2.2, we see that the representation $2\pi_1 \dotplus 2\pi_n$ is one of the irreducible components of the representation $\sigma(\rho)$.

We choose a canonical basis in $V(\pi_1) \cong \mathbf{C}^{n+1}$ from the normalized weight vectors v_{x_i}. We denote the dual basis in $V(\pi_n)$ by $v^*_{-x_i}$. We embed the space $V(2\pi_1 \dotplus 2\pi_n)$ in the space $V(2\pi_1 \otimes 2\pi_n)$. For convenience we redenote $v_{x_i} \otimes v^*_{-x_j}$ by $v_{x_i - x_j}$, $v_{x_i} \cdot v_{x_j} \in V(S^2(\pi_1))$ by $v_{x_i + x_j}$, and so on.

LEMMA 11.4.1. *The linear operator* $\hat{L}$ *defined by*

$$\hat{L}(v_{x_i - x_j}) = c_{i1} c_{j1} v_{x_i + x_1} \otimes v^*_{-x_1 - x_j} - c_{i2} c_{j2} v_{x_i + x_2} \otimes v^*_{-x_2 - x_j},$$

$$\hat{L}(v_{x_1 - x_1}) = (1/2) v_{2x_1} \otimes v^*_{-2x_1} - v_{x_1 + x_2} \otimes v^*_{-x_2 - x_1},$$

$$\hat{L}(v_{x_2 - x_2}) = v_{x_2 + x_1} \otimes v^*_{-x_1 - x_2} - (1/2) v_{2x_2} \otimes v^*_{-2x_2},$$

where $(2 - \delta_{ij} - \delta_{i1} - \delta_{i2})(2 - \delta_{ij} - \delta_{j1} - \delta_{j2}) \neq 0$, $c_{kl} = 1$ *if* $k \neq l$, $c_{ll} = 1/2$, *takes the space* $V(\pi_1 \dotplus \pi_n)$ *into the space* $V(2\pi_1 \dotplus 2\pi_n)$.

PROOF. We define the convolution operator $\tau : V(2\pi_1 \otimes 2\pi_n) \to V(\pi_1 \otimes \pi_n)$ as in the proof of Theorem 11.2.1. It is easy to verify that $\operatorname{Ker} \tau = V(2\pi_1 \dotplus 2\pi_n)$ and $\tau \hat{L}(V(\pi_1 \dotplus \pi_n)) = 0$, which implies our assertion.

We embed the space $V(2\pi_1 \dotplus 2\pi_n)$ in the space $W_{\otimes \mathbf{C}}$. Let π_W denote the orthogonal projection on $W_{\otimes \mathbf{C}}$. Then from Lemma 11.4.1, taking account of the fact that $\hat{L}$ and π_W take weight vectors into weight vectors, preserving the weight, it follows that the operator $L = \pi_W \hat{L}$ belongs to the space $\operatorname{Hom}_{T_n}(V(\pi_1 \dotplus \pi_n), W_{\otimes \mathbf{C}})$. On the space $\operatorname{Hom}_{T_n}(V, W)$ there is a natural metric induced by the metrics on V and W. Direct verification, using (4.2), shows that

$$\begin{aligned}\langle I_* L, L\rangle &= \sum_{\alpha \in \Delta(SU_{n+1})} \langle I_* L v_\alpha, L v_\alpha \rangle \\ &= \sum_{j=1}^{n+1} (|L v_{x_1 - x_j}|^2 + |L v_{x_j - x_1}|^2 + |L v_{x_2 - x_j}|^2 + |L v_{x_j - x_2}| \\ &\quad - 3(|L v_{x_1 - x_2}|^2 + |L v_{x_2 - x_1}|^2) < 0.\end{aligned}$$

Consequently, the minimal submanifold $\rho_*(SU_{n+1}/T_n)$ is not stable in SU_{m+1}/T_m.

A.2. $H = SO_{2n+1}$, $n > 1$. From Proposition 11.2.2, the representation $2\pi_1$ is one of the components of irreducibility of the representation $\rho \otimes \rho^*$. If $n \neq 2$, then $\sigma(\rho) = \rho \otimes \rho^* - \mathrm{Ad}_H$ contains the component $2\pi_1$. For V_λ

we choose $V(\pi_1) \cong \mathbf{C}^{2n+1}$. Next, we choose a Witt basis in $\mathbf{C}^{2n+1}$ from the weight vectors $v_{\pm x_i}$, v_0, $|v_{\pm x_i}| = 1 = (1/2)|v_{x_0}$ of the representation π_1 of the group SO_{2n+1}. We define the operator $\hat{L}\colon V(\pi_1) \to V(S^2(\pi_1))$ as follows:

$$\hat{L}(v_{\pm x_i}) = v_{\pm x_i} \cdot v_0,$$
$$\hat{L}(v_0) = (v_0 \cdot v_0) + (2/n) \sum_{i=1}^{n} v_{x_i} \cdot v_{-x_i}.$$

It is easy to verify that $\hat{L}(V(\pi_1)) \subset V(2\pi_1) \subset V(S^2(\pi_1))$. Let π_W denote the orthogonal projection on $W_{\otimes \mathbf{C}}$. Then, arguing as in A.1, we see that the operator $L = \pi_\omega \hat{L}$ belongs to the subspace $\mathrm{Hom}_{T_n}(V(\pi_1), W_{\otimes \mathbf{C}})$. Direct calculation, using (4.2), shows that $\langle I_* L, L\rangle < 0$. Consequently, the minimal submanifold $\rho_*(SO_{2n+1}/T_n)$ is not stable in SU_{m+1}/T_m.

A.3. $H = Sp_n$. It is well known that any representation of the group Sp_n is selfadjoint, that is, the embedding $\rho : Sp_n \to SU_m$ is either a composition of $\rho_1 : Sp_n \to SO_m$, $\rho_2 : SO_m \to SU_m$, or a composition of $\rho_1 : Sp_n \to Sp_{m/2}$, $\rho_2 : Sp_{m/2} \to SU_m$. Therefore, the instability of the embedding $\rho_* : (Sp_n/T_n) \to SU_{m+1}/T_m$ follows from the instability of either the embedding $\rho_* : Sp_n/T_n \to SO_m/T_{[m/2]}$ or the embedding $\rho_* : Sp_n/T_n \to Sp_m/T_m$.

A.4. $H = SO_{2n}$. This case can be considered as in A.1, A.2, A.3.

B. $G = SO_m$. By Proposition 11.2.1, the restriction of the representation $\mathrm{Ad}_G \rho$ to W is equivalent to the representation $\Lambda^2(\rho) - \mathrm{Ad}_H = \sigma(\rho)$. We consider only the model case where $H = SO_{2n}$.

PROPOSITION 11.4.1 (a) *Suppose that the representation* $\rho : SO_{2n} \to SO_m$ *is the sum of irreducible representations, among which there is an irreducible component distinct from the fundamental representations* π_i. *Then the induced minimal embedding* $\rho_* : SO_{2n}/T_n \to SO_m/T_{[m/2]}$ *is not stable.*

(b) Suppose that the representation $\rho : SO_{2n} \to SO_m$ is the sum of fundamental representations π_i, among which there is an irreducible component distinct from π_1. Then the minimal embedding $\rho_* : SO_{2n}/T_n \to SO_m/T_{[m/2]}$ is not stable.

PROOF. (a) By Proposition 11.2.3 (B+D), in the case under consideration in the decomposition of the representation $\Lambda^2(\rho)$ there occurs the irreducible component $\pi_2 \dotplus 2\pi_1$. Since the latter is distinct from $\pi_2 = \mathrm{Ad}_{SO_{2n}}$, $2\pi_1 \dotplus \pi_2$ occurs in the decomposition of the representation

$\sigma(\rho) = \Lambda^2(\rho) - \mathrm{Ad}_{SO_{2n}}$. We realize the space $V(2\pi_1 \dotplus \pi_2)$ as a subspace of the representation $V(\Lambda^2(S^2(\pi_1))) = V(\Lambda^2(2\pi_1))$ (see Proposition 11.2.3 and Table 5 of [446]). We define the operator $\hat{L} : V(\pi_2) \to V(\Lambda^2(2\pi_1))$ as follows:

$$\begin{aligned}
\hat{L}(v_{x_1 - x_1}) &= 0, \\
\hat{L}(v_{\pm x_i \pm x_j}) &= v_{\pm i+1} \wedge v_{\pm j-1} - v_{\pm i-1} \wedge v_{\pm j+1}, \\
\hat{L}(v_{\pm x_i + x_1}) &= \tfrac{1}{2} v_{\pm i+1} \wedge v_{1-1} - v_{\pm i-1} \wedge v_{-1+1}, \\
\hat{L}(v_{\pm x_i + x_1}) &= v_{\pm i+1} \wedge v_{1-1} - \tfrac{1}{2} v_{\pm i-1} \wedge v_{1+1}, \\
\hat{L}(v_{x_i - x_i}) &= v_{i+1} \wedge v_{-1-i} - v_{i-1} \wedge v_{1-i} - \tfrac{1}{2} v_{1+1} \wedge v_{-1-1},
\end{aligned}$$

where $1 \neq i \neq j \neq 1$, $v_{\pm x_i \pm x_j} = v_{\pm x_i} \wedge v_{\pm x_j} \in V(\pi_2)$, $v_{\pm i \pm j} = v_{\pm x_i} \cdot v_{\pm x_j}$ $\in V(S^2(\pi_1))$. Then the operator $\hat{L}$ takes the space $V(\pi_2)$ into $V(2\pi_1 \dotplus \pi_2)$
$\to V(\Lambda^2(S^2(\pi_1))$. Direct verification, using (4.2), shows that the operator $L = \pi_W \hat{L}$ belongs to $\mathrm{Hom}_{T_n}(V(\pi_2), W_{\otimes \mathbf{C}})$ and $\langle I_* L, L\rangle < 0$. Consequently, the minimal embedding $\rho_* : SO_{2n}/T_n \to SO_m/T_{[m/2]}$ is not stable. This proves part (a).

(b) Let ρ be the sum of fundamental representations of the groups SO_{2n}, $i \neq 1$, $n \neq 4$. Then from the data of Table 5 of [446], among the components of irreducibility $\sigma(\rho) = \Lambda^2(\rho) - \mathrm{Ad}_{SO_{2n}}$ there is a representation $\pi_1 \dotplus \pi_3$. We realize the component $\pi_1 \dotplus \pi_3$ as a subspace of $V(\Lambda^2(\pi_2))$. Next, we define an operator $\hat{L} : V(\pi_2) \to V(\Lambda^2(\pi_2))$ as follows:

$$\hat{L}(v_{\pm i \pm j}) = v_{\pm i \pm j} \wedge v_{1-1} \qquad \text{for } i \neq 1 \neq j,$$

$$\hat{L}(v_{1 \pm i}) = v_{1 \pm i} \wedge v_{1-1} + 1/[2(n-2)] \sum_{i \pm k \neq 1} (v_{1-k} \wedge v_{k \pm i} + v_{1+k} \wedge v_{k \pm i})$$
$$\text{for } i \neq 1,$$

$$\hat{L}(v_{-1 \pm i} = v_{-1 \pm i} \wedge v_{1-1} - 1/[2(n-2)] \sum_{i \pm k \pm 1} (v_{-1-k} \wedge v_{k \pm i} + v_{-1+k} \wedge v_{-k \pm i})$$
$$\text{for } i \neq 1,$$

$$\hat{L}(v_{i-i}) = v_{i-i} \wedge v_{1-1} \qquad \text{for any } i.$$

Here for simplicity we redenote $v_{\pm x_i} \wedge v_{\pm x_j}$ by $v_{\pm i \pm j}$.

It is easy to verify that the vector $v_{\pm i \pm k} \wedge v_{\mp k \mp i} \in \Lambda^2(\pi_2)$, realized in $\Lambda^2(\pi_2)$, is an element in the canonical Cartan subalgebra of $so(V(\pi_i)) \subset so(V(\rho))$. Let π_W denote the projection of $V(\pi_1 \dotplus \pi_3)$ on $W_{\otimes\mathbf{C}}$. Then $\pi_{W_0}(\sum c_{ik} v_{\pm i \pm k} \wedge v_{\mp k \mp i}) = 0$. By means of this observation we easily obtain the following assertion.

ASSERTION. *The operator* $\hat{L}$ *belongs to the subspace* $V(\pi_1 \dotplus \pi_3) \subset V(\Lambda^2(\pi_2))$. *The operator* $L = \pi_{W_0}\hat{L}$ *belongs to the space* $\mathrm{Hom}_{T_n}(V(\pi_2), W_0)$, *and we have* $\langle I_* L, L\rangle < 0$.

Hence it follows immediately that the minimal embedding $\rho_* : SO_{2n}/T_n \to SO_m/T_{[m/2]}$ is not stable. This proves the proposition.

The case where $G = Sp_m$ and $\rho : H \to Sp_m$ is proved along the same lines.

4.3. Invariant calibrations on G/T_G. We define a complex structure J on V as follows:

$$J(E_\alpha + E_{-\alpha}) = i(E_\alpha - E_{-\alpha}), \qquad \alpha \in \Delta^+(G),$$
$$J(iE_\alpha - iE_{-\alpha}) = -(E_\alpha + E_{-\alpha}), \qquad \alpha \in \Delta^+(G).$$

LEMMA 11.4.2. *Every metric on* V *that is invariant under the action of* $\mathrm{Ad}\, T_G$ *is invariant under the action of the operator* J.

From Lemma 11.4.2 it follows that the space V, endowed with the T_G-invariant metric and the operator J, is a Hermitian space.

DEFINITION 11.4.1 Suppose that the set $\{(v_i, Jv_i) : i = 1, \dots, n\}$ is a unitary basis in the Hermitian space V^{2n}. Then a form of degree $2k$ in $\Lambda^{2k}(V^{2n})$ is called a *real form of bidegree* (k, k) if it can be written as a sum $\sum_{i\in I} c_i \xi^i$, where ξ^i is a basis complex form in $\Lambda^{2k}(V^{2n})$, $\xi^i = v_{i_1}^* \wedge Jv_{i_1}^* \wedge \cdots \wedge v_{i_k}^* \wedge Jv_{i_k}^*$.

EXAMPLES OF REAL FORMS OF BIDEGREE (k, k). Let G/T_G be endowed with a G-invariant metric. Then its restriction to the tangent space V is a T_G-invariant form. By Lemma 3.2.2 the complex structure J preserves the T_G-invariant metric on V. Then it is not difficult to verify that any form ω_v (see (2.3)), and consequently any form $\omega \in \Lambda(V)$, is of real bidegree (k, k) with respect to the given Hermitian structure on V.

The next theorem was obtained independently by Le Hong Van [451] and Dadok, Harvey, and Morgan [462].

THEOREM 11.4.1. *Suppose that a real form of bidegree* (k, k) *has the expansion* $\psi = \sum_{i\in I} c_i \xi^i$. *Then its comass is equal to the largest absolute value of the coefficients* $|c_i|$.

REMARK. Substituting the value $c_i = 1$ in Theorem 11.4.1, we obtain a well-known inequality of Wirtinger [216].

4.4. Global minimality of an embedding "of small dimension".

PROPOSITION 11.4.2. *Let $\rho: H \to G$ be one of the representations listed in Table 1. Let H_α, $\alpha \in \Delta^+(H)$, denote the vector in lT_H corresponding to the root α (see Subsection 2.1). Let φ_α be the 2-form on $T_e(G/T_G)$ associated with the vector H_α (see (3.3)) such that*

$$\varphi_\alpha(X, Y) = \langle -\rho(H_\alpha), [X, Y]\rangle. \tag{4.3}$$

We set $\psi = \prod_{\alpha\in\Delta^+(H)} \varphi_\alpha$ and $\hat{\psi} = (|\psi|^*)^{-1}\psi$. Then $\hat{\psi}$ can be extended to an invariant absolute calibration on G/T_G. Moreover, $\rho_*(H/T_H)$ is a $\hat{\psi}$-submanifold.

PROOF. The first assertion of the proposition is a consequence of Theorem 11.3.1. As we remarked in Subsection 4.3, ψ is a real form of bidegree (N, N), where $N = \dim(H/T_H)$, and so its comass is equal to the maximal value of ψ in the basis complex vectors in $T_e(G/T_G)$ (see Theorem 11.4.1).

We observe that if $V \in \operatorname{span}_{\mathbf{R}}\{iE_\beta \wedge E_{-\beta}\}$, $W \in \operatorname{span}_{\mathbf{R}}\{iE_\gamma \wedge E_{-\gamma}\}$, where $\beta \neq \gamma$; $\beta, \gamma \in \Delta^+(SU_{m+1})$, then we have $\varphi_\alpha(V, W) = 0$ for any $\alpha \in \Delta^+(H)$. From the explicitly given formula (4.3) and Theorem 11.4.1, taking account of the fact that all the vectors $E_\beta + E_{-\beta}$, $E_\beta - E_{-\beta}$, $\beta \in \Delta^+(SU_{m+1})$ have identical lengths, we obtain

$$\|\psi\|^* = \max_{\{\beta_i\}} \left\{ \left(\prod_{\alpha\in\Delta^+(H)} t_\alpha^* \right) (H_{\beta_1} \cdots H_{\beta_N}) \right\}, \tag{4.4}$$
$$\beta_i \neq \beta_j, \qquad \beta_i \in \Delta^+(SU_{m+1}),$$

where $H_\beta = i[E_\beta, E_{-\beta}]$, and t_α^* is the linear function (polynomial of the first degree) associated with the vector H_α on the Cartan subalgebra lT_G. Therefore, the second assertion of Proposition 11.4.2 follows from the equality

$$\|\psi\|^* = \psi(\rho_*(\overrightarrow{H/T_H})) \tag{4.5}$$

where $\rho_*(\overrightarrow{H/T_H})$ is the unit N-vector associated with the tangent space to $\rho_*(H/T_H)$ at the point $\rho(\{eT_H\}/T_H)$. Let $\overline{v}$ denote the normalized vector $v/|v|$. Then we have

$$[\overline{\rho_* E_\alpha}, \overline{\rho_* \sqrt{-1} E_{-\alpha}}] = \frac{-2\rho_*(H_\alpha)}{|\rho_* H_\alpha|^2}.$$

Consequently, for any α, $\beta \in \Delta^+(H)$ we have

$$\varphi_\beta(\overline{\rho_* E_\alpha}, \overline{\rho_* \sqrt{-1} E_\alpha}) = \langle H_\beta, H_\alpha \rangle = d_{\beta\alpha}. \tag{4.6}$$

Thus, the value of $\psi(\rho_*\overrightarrow{(H/T_H)})$ is a constant $d(H)$ which does not depend on ρ.

The rest of the proof of (4.5) is carried out separately for each group H with the corresponding representation ρ.

Let us consider the model case where $G = SU_{mH}$, $H = SU_{n+1}$, $\rho = \bigoplus_i \pi_i$. Since the weight of the fundamental representation π_k of the group SU_{n+1} runs through the set $\{X_{i_1 \cdots i_k} = x_{i_1} + \cdots + x_{i_k}$, where $x_{i_l} \neq x_{i_j}$, $1 \le x_{i_l} \le n+1$, we have $\pi_k(H_{x_i - x_j}) = \sum_{I_{k-1}} \{H_{i_{k-1}} \cup x_i - H_{I_{k-1}} \cup x_j\}$, where k_{k-1} runs through the set of $k-1$ distinct elements of $\{x_1, \dots, \hat{x}_i, \dots, \hat{x}_j, \dots, x_{n+1}\}$. Suppose that ρ is the sum of the fundamental representations $\{\pi_i^j\}$, $\pi_i^j \simeq \pi_i$, where j is the number of the representation π_i. Then $\rho(H_{x_i - x_j}) = \sum \pi_j^i(H_{x_i - x_j})$. Accordingly, the roots of the algebra su_{m+1} are $X_I - X_J$, where I is the multi-index corresponding to one of the representation spaces π_i^j. Suppose that $I = i_1 i_2 \cdots i_k$, $J = j_1 \cdots j_l$, and $\alpha = x_p - x_q$. Then from the above we have

$$\varphi_\alpha(H_{X_I - X_J}) = \langle \rho_*(H_\alpha), H_{X_I} - H_{X_J} \rangle = \sum_{s=1}^{k} \delta_{i_s} p - \delta_{j_r} q. \tag{4.7}$$

In particular, by means of Theorem 11.4.1 we deduce from (4.7) that $\|\varphi_\alpha\|^* = 2$.

LEMMA 11.4.3. *We have the expansion*

$$\psi = \sum_{\sigma \in S_{n+1}} \varepsilon_\sigma (\varphi_{\sigma(1)})^n (\varphi_{\sigma(2)})^{n-1} \cdots \varphi_{\sigma(n)}$$

where $\varphi_i(H_\beta) = \langle \rho_* H_{x_i}, H_\beta \rangle$.

Lemma 11.4.3 and formula (4.7) give a simple prescription for calculating the coefficients of the polynomial $\overline{\psi} = \Pi t_\alpha^*$. In particular, we find that all these coefficients of the expansion of ψ are multiples of the number $\alpha = n!(n-1)! \cdots 1$ and we have

$$\psi(H_{\beta_1} \cdots H_{\beta_N}) \le d(SU_{n+1}) = (n+1)!n! \cdots 1, \tag{4.8}$$

where $\beta_i = X_I - X_J$ is a root of the algebra su_{m+1}.

From (4.8) there immediately follows (4.5) for $H = SU_{n+1}$, where ρ is the sum of fundamental representations.

§5. Globally minimal currents of codimension 2 on the orbit G/T_G

There is a close connection between invariant complex structures on the orbit G/T_G, where G is a compact semisimple Lie group, and its globally minimal currents of codimension 2. We recall that a current S^{2k} on a Hermitian manifold M^{2n} is said to be positive if it is a φ-current, where φ is the kth power of a fundamental (Kähler) 2-form on M^{2n}.

THEOREM 11.5.1 (Le Hong Van). *A current S of codimension 2 in G/T_G is globally minimal if and only if it is a positive current with respect to some invariant complex structure J on G/T_G.*

Borel [444] showed that G/T_G is a complex manifold, since it is diffeomorphic to the quotient $G^{\mathbf{C}}/B$, where $G^{\mathbf{C}}$ is the complexification of the group G, and B is a maximal solvable subgroup of G. In fact, a stronger assertion is true.

PROPOSITION 11.5.1 ([443], [451]). *There is a one-to-one correspondence between invariant (integrable) complex structures on G/T_G and partitions of the root system $\Delta(G)$ into positive and negative roots.*

With each closed differential 2-form ω on G/T_G we associate a current $S_{D\omega}$ of codimension 2 on G/T_G as follows:

$$S_{D\omega}(\varphi) = \int_{G/T_G} \varphi \wedge \omega \, dx. \tag{5.1}$$

Using the fact that every invariant $(n-2)$-form, where $n = \dim G/T_G$, is a closed form, and by means of Proposition 11.5.1, we obtain the following proposition.

PROPOSITION 11.5.2. *Suppose that the 2-form ω is invariant and closed. Then the current $S_{D\omega}$ is globally minimal. Moreover, the current $S_{D\omega}$ is positive with respect to some invariant structure on G/T_G.*

Since the map $H^2(G/T_G, \mathbf{R}) \to H_{n-2}(G/T_G, \mathbf{R})$, given by (5.1), is the Poincaré duality isomorphism, we easily obtain Theorem 11.5.1 from Proposition 11.5.2.

APPENDIX

Volumes of Closed Minimal Surfaces and the Connection with the Tensor Curvature of the Ambient Riemannian Space

The topology of a Riemannian manifold is closely connected with its Riemannian curvature. It is intuitively obvious that topologically nontrivial cycles (realized by minimal surfaces) tend to take up a position in the ambient Riemannian manifold corresponding in a certain sense to the "maximal curvature" of this manifold. We now present new results obtained by Le Hong Van and A. T. Fomenko in the course of developing Fomenko's method presented above in Chapter 4, §4.

Let $B_r(x)$ be a ball of radius r in the tangent space T_xM. The injective radius of the manifold M^n at the point x is the quantity $R(x) = \sup\{r$, where $\exp\colon B_r(x) \to M$ is a diffeomorphism$\}$. Let $R(M) = \inf R(x)$, where $x \in M$. We call $R(M)$ the injective radius of the manifold. Let $\eta_k(v, x)$ be the k-dimensional deformation coefficient of the vector field v at the point x, introduced by Fomenko (see Chapter 4, §4). Next, we recall that the quantity

$$\Omega_k^0(M) = \inf \Omega_k^0(x_0), \qquad x_0 \in M,$$

where

$$\Omega_k^0(x_0) = \gamma_k q_{x_0}(R(x_0)), \qquad q_{x_0}(r) = \exp \int_0^r \left(\max_{x \in \{f=t\}} \eta_k(v, x)\right)^{-1} dt,$$

$\gamma_k = \mathrm{vol}_k B_1^k$ is the volume of a k-dimensional unit ball, is called the geodesic deficiency of M^n if $f(x)$ is the function of distance from x to x_0 on the manifold, and $v = \mathrm{grad} f$. For the details see Chapter 4, §4. We also recall that the volume of a k-dimensional nontrivial cycle in M is not less than the number $\Omega_k^0(M)$ (see Chapter 4, Theorems 4.4.1 and 4.4.2).

Suppose that the sectional curvature of M in any two-dimensional direction does not exceed a^2, where a is real or purely imaginary. Let the injective radius of M be equal to R.

THEOREM 1 (a lower bound of the geodesic deficiency of a manifold; Le Hong Van and A. T. Fomenko).

(1) *If* $a^2 > 0$ *and* $Ra \leq \pi$, *then* $\Omega_k^0(M) \geq k\gamma_k a^{-k} \int_0^{Ra} \sin^{k-1} t\, dt$.

(2) *If* $a^2 > 0$ *and* $Ra > \pi$, *then* $\Omega_k^0(M) \geq \mathrm{vol}_k S_{1/a}^k$, *where* $S_{1/a}^k$ *is a Euclidean sphere of dimension* k *and radius* $1/a$.

(3) *If* $a = 0$, *then* $\Omega_k^0(M) \geq \gamma_k R^k$.

(4) *If* $a^2 < 0$, *then* $\Omega_k^0(M) \geq k\gamma_k |a|^{-k} \int_0^{R|a|} \sinh^{k-1} t\, dt$.

THEOREM 2 (isoperimetric inequalities; Le Hong Van and A. T. Fomenko). *Let* X^k *be a globally minimal surface passing through a point* $x \in M$, *let* $B_x(r)$ *be the geodesic ball of radius* r *and center* x *in* M, *and let* A_r^{k-1} *be the boundary of intersection of the surface with the ball* $X^k \cap B_x(r) = X_r^k$.

(1) *If* $a^2 > 0$ *and* $r \leq \min(R, \pi/a)$, *then we have* $\mathrm{vol}_{k-1} A_r \geq k\gamma_k a^{1-k} (\sin ar)^{k-1}$. *Consequently,*

$$(\mathrm{vol}_{k-1} A_r)/(\mathrm{vol}_k X_r) \geq (\sin ar)^{k-1} \Big/ \left(\int_0^r (\sin at)^{k-1}\, dt \right).$$

(2) *If* $a = 0$ *and* $r \leq R$, *then* $\mathrm{vol}_{k-1} A_r \geq k\gamma_k r^{k-1}$ *(which is equal to the volume of the standard* $(k-1)$*-dimensional sphere of radius* r). *Consequently,*

$$\mathrm{vol}\, X_r \leq kr\, \mathrm{vol}\, A_r \quad \textit{and} \quad \mathrm{vol}\, X_r \leq (\mathrm{vol}\, A_r)^{k/(k-1)} k^{-1} (k\gamma_k)^{1/(1-k)}.$$

(3) *If* $a^2 < 0$ *and* $r \leq R$, *then* $\mathrm{vol}_{k-1} A_r \geq k\gamma_k |a|^{1-k} (\sinh |a| r)^{k-1}$. *Consequently,*

$$(\mathrm{vol}\, A_r)/(\mathrm{vol}\, X_r) \geq (\sinh |a| r)^{k-1} \Big/ \left(\int_0^r (\sinh |a| t)^{k-1}\, dt \right).$$

In the general case, the estimates in these theorems are the best possible, that is, in many cases equalities are attained. The theorems confirm the conjecture that in a certain sense absolutely minimal surfaces tend to take up a position corresponding to "maximal curvature" in the ambient manifold. Nevertheless, it follows that the curvature of globally minimal surfaces "senses" the curvature of the ambient manifold.

COROLLARY 1. *Suppose that a complete noncompact Riemannian manifold* M *has nonpositive curvature, and that* X^k *is a globally minimal surface in* M. *Then the function* $V(r) = \mathrm{Vol}_k B_X(r)$, *where* $B_X(r)$ *is a geodesic ball of radius* r *in* X^k, *increases no more slowly than a polynomial of degree* k *in* r. *If* M *has negative curvature bounded below, then*

$V(r)$ increases no more slowly than the exponential function of r. In particular, it follows that at each point $x \in X^k$ there is a $v \in T_x X$ such that $R_X(v, v) < 0$, where R_X is the Ricci tensor on X at the point x.

COROLLARY 2. *If M is a compact simply-connected symmetric space with sectional curvature everywhere not greater than a, then the volume of any nontrivial k-dimensional cycle (that is, corresponding to a closed minimal surface) is not less than the volume of a k-dimensional standard sphere of curvature a.*

COROLLARY 3. *The length of a homologically nontrivial loop in a manifold M is not less than twice the injective radius of the manifold.*

COROLLARY 4 (lower bound of the volume of a manifold).
(1) *If $a^2 > 0$, then* $\mathrm{vol}_n M^n \geq n\gamma_n a^{1-n} \int_0^R (\sin at)^{n-1}\, dt$.
(2) *If $a = 0$, then* $\mathrm{vol}_n M^n \geq \gamma_n R^n$.
(3) *If $a^2 < 0$, then* $\mathrm{vol}_n M^n \geq n\gamma_n |a|^{1-n} \int_0^R (\sinh |a|t)^{n-1}\, dt$.

These estimates of the volume coincide with those that can be obtained on the basis of the Rauch-Bishop comparison theorem. Let us consider the case of symmetric spaces.

THEOREM 3 (Le Hong Van and A. T. Fomenko). *Let B be a Cartan subspace of the tangent space $T_e M$, where M is a symmetric space, $x = \exp v$, $v \in B$. We denote by α_i the root system of M with respect to B. Suppose that the roots are ordered as follows: $\alpha_1^2(v) \geq \alpha_2^2(v) \geq \cdots > \alpha_p^2(v) = 0 = \alpha_{p+1}(v) = \cdots$. Then*
(1) *for $k < p$ we have*

$$\eta_k(x) = \frac{\int_0^1 (\sin \alpha_1(v)t \cdots \sin \alpha_{k-1}(v)t)\, dt}{\sin \alpha_1(v) \cdots \sin \alpha_{k-1}(v)};$$

(2) *for $k \geq p$ we have*

$$\eta_k(x) = \frac{\int_0^1 (\sin \alpha_1(v)t \cdots \sin \alpha_{k-1}(v)t)(|v|t)^{k-p}\, dt}{(\sin \alpha_1(v) \cdots \sin \alpha_{k-1}(v))|v|^{k-p}}.$$

Thus, for example, for symmetric spaces of rank 1 we can calculate the functions $\Omega_k^0(M)$ explicitly, and hence the volumes of these spaces and the volumes of their minimal cycles.

It is well known [257], [258] that in a simply-connected irreducible compact symmetric space M there is a totally geodesic sphere of curvature a^2, where a^2 is the least upper bound of two-dimensional curvature on M. Moreover, each such sphere lies in some totally geodesic Helgason sphere of maximal dimension $i(M)$. All the Helgason spheres go into one another under the action of the group $\mathrm{Iso}(M)$, and they also have curvature a^2. Obviously, the following assertion follows immediately from Corollary 2.

PROPOSITION. *If the Helgason sphere $S(M)$ realizes a nontrivial cycle in the homology group $H_*(M, \mathbf{R})$, then it is a globally minimal surface in M.*

Below we give a list of Helgason spheres that realize nontrivial cycles in (compact) irreducible symmetric spaces.

(1) M is a simple compact group, $\dim S(M) = 3$,
(2) $M = SU_{l+m}/SU_l \times SU_m$, $\dim S(M) = 2$,
(3) $M = SO_{l+2}/SO_l \times SO_2$, $\dim S(M) = 2$,
(4) $M = SU_{2n}/Sp_n$, $\dim S(M) = 5$,
(5) $M = Sp_{m+n}/Sp_m \times Sp_n$, $\dim S(M) = 4$,
(6) $M = SO_{2n}/U_n$, $\dim S(M) = 2$,
(7) $M = F_4/\mathrm{Spin}\,9$, $\dim S(M) = 8$,
(8) $M = \mathrm{Ad}\,E_6/T^1\,\mathrm{Spin}\,10$, $\dim S(M) = 2$,
(9) $M = \mathrm{Ad}\,E_7/T^1 E_6$, $\dim S(M) = 2$.

The explicit embedding $S(M) \to M$ is described in [257].

In conclusion we give one more corollary of Theorem 1 for noncompact symmetric spaces. Clearly [258], the upper sectional curvature of such spaces is equal to zero.

COROLLARY 5. *Let N be a flat totally geodesic subspace of a noncompact symmetric space M. Then N is a globally minimal surface.*

Bibliography

1. S. I. Al′ber, *Spaces of mappings in a manifold of negative curvature*, Dokl. Akad. Nauk SSSR **178** (1968), 13–16; English transl. in Soviet Math. Dokl. **9** (1968).

2. ——, *The topology of functional manifolds and the calculus of variations in the large*, Uspekhi Mat. Nauk **25**, no. 4 (1970), 57–123; English transl. in Russian Math. Surveys **25**, no. 4 (1970).

3. ——, *Multidimensional problems in the calculus of variations in the large*, Dokl. Akad. Nauk SSSR **156** (1964), 727–730 ; English transl. in Soviet Math. Dokl. **5** (1964).

4. Yu. A. Aminov, *An analogue of the intrinsic Ricci condition for a minimal manifold in a Riemannian space*, Ukrain. Geom. Sb. No. 17 (1975), 15–22.

5. ——, *Minimal surfaces*, Izdat. Kharkov Univ., Kharkov, 1978. (Russian)

6. ——, *The metric of approximately minimal surfaces*, Sibirsk. Mat. Zh. **7** (1967), 483–493; English transl. in Siberian Math. J. **7** (1967).

7. ——, *On the stability of a minimal surface in an n-dimensional Riemannian space of positive curvature*, Mat. Sb. **100** (1976), 400–419; English transl. in Math. USSR-Sb. **29** (1976), 359–375.

8. ——, *Defining a surface in 4-dimensional Euclidean space by means of its Grassmann image*, Mat. Sb. **117** (1983), 147–160; English transl. in Math. USSR.-Sb. **45** (1983), 155–168.

9. ——, *On the problem of stability of a minimal surface in a Riemannian space of positive curvature*, Dokl. Akad. Nauk SSSR **224** (1975), 745–747; English transl. in Soviet Math. Dokl. **16** (1975).

10. ——, *The outer diameter of an immersed Riemannian manifold*, Mat. Sb. **92** (1973), 456–460; English transl. in Math. USSR-Sb. **21** (1973).

11. ——, *On the unboundedness of a minimal surface in a Riemannian space of nonpositive curvature*, Ukrain. Geom. Sb. No. 19 (1976), 3–9.

12. C. V. Anosov, *Complexes and Whitehead's theorem. Poincaré duality and the gluing of handles*, Appendix to the Russian transl. of J. W. Milnor, *Morse theory*, "Mir", Moscow, 1965.

13. V. I. Arnol′d, *Mathematical methods in classical mechanics*, "Nauka", Moscow, 1974; English transl., Springer-Verlag, Berlin and New York, 1978.

14. V. I. Arnol′d, A. N. Varchenko, and S. M. Husein-Zade, *Singularities of differentiable maps. Classification of critical points, caustics and wave fronts*, "Nauka", Moscow, 1982; English transl., Birkhäuser, Basel, 1985, 1988.

15. I. Ya. Bakel′man, *Mean curvature and quasilinear elliptic equations*, Sibirsk. Mat. Zh. **9** (1968), 1014–1040; English transl. in Siberian Math. J. **9** (1968).

16. D. V. Beklemishev, *On strongly minimal surfaces of a Riemannian space*, Dokl. Akad. Nauk SSSR **114** (1957), 156–158. (Russian)

17. ——, *Strongly minimal surfaces*, Izv. VUZ Mat. **1958**, no. 3, 13–23. (Russian)

18. S. N. Bernstein, *On surfaces defined by means of their mean or total curvature*, Collected works, vol. 3, Izdat. Akad. Nauk SSSR, Moscow, 1960, 122–140. (Russian)

19. ——, *On a geometrical theorem and its applications to partial differential equations of elliptic type*, Collected works, vol. 3, Izdat. Akad. Nauk SSSR, Moscow, 1960, 251–258.

20. R. L. Bishop and R. J. Crittenden, *Geometry of manifolds*, Academic Press, New York and London, 1964.

21. A. N. Bogolyubov, *Gaspard Monge*, "Nauka", Moscow, 1978. (Russian)

22. Le Hong Van, *Relative calibrations and the stability problem of minimal surfaces*, in: Topological and geometrical methods of analysis, Izdat. Voronezh Univ., Voronezh, 1989, 122–136.

23. A. Yu. Borisovich, *The Lyapunov-Schmidt method and singularity type of critical points in the bifurcation problem of minimal surfaces*, in: Topological and geometrical methods of analysis, Izdat. Voronezh Univ, Voronezh, 1989, 138–141.

24. Yu. G. Borisovich, N. M. Bliznyakov, Ya. A. Izraelevich, and T. N. Fomenko, *Introduction to topology*, "Vysš. Škola", Moscow, 1980; English transl.: "Mir", Moscow, 1985.

25. N. Bourbaki, *Éléments de mathématique*, Fasc. III, Livre III, *Topologie générale*, Ch. IV–VIII, 3rd ed., Hermann, Paris, 1942, 1947.

26. ——, *Éléments de mathématique*, Fasc. XV, Livre V, *Espaces vectoriels topologiques*, 2nd ed., Hermann, Paris, 1966.

27. ——, *Éléments de mathématique*, Fasc. XIII, Livre VI; Fasc. XXI, Livre VI, *Intégration*, Ch. I–V, 2nd ed., Hermann, Paris, 1965, 1967.

28. A. Ya. Vikaruk, *The principle of the argument and the integral curvature of minimal surfaces*, Mat. Zametki **15** (1974), 691–700; English transl. in Math. Notes **15** (1974).

29. ——, Analogues of the Nevanlinna theorems for minimal surfaces, Mat. Sb. **100** (1976), 555–579; English transl. in Math. USSR-Sb. **29** (1976).

30. M. L. Gavril′chenko, *Geometrical properties of the indicatrix of rotations*, Deposited at VINITI, 4 Feb. 1981, no. 5553-81.

31. *Geometrical theory of integration*, Collection of articles translated from English, Izdat. Inostr. Lit., Moscow, 1960.

32. I. V. Girsanov, *Lectures on mathematical theory of extremal problems*, Izdat. Moskov. Gos. Univ., Moscow, 1970. (Russian)

33. M. Golubitsky and V. Guillemin, *Stable mappings and their singularities*, Springer-Verlag, New York and Heidelberg, 1973.

34. V. P. Gorokh, *On hypersurfaces containing minimal submanifolds*, Ukrain. Geom. Sb., No. 24 (1981), 18–26.

35. M. L. Gromov and V. A. Rokhlin, *Embeddings and immersions in Riemannian geometry*, Uspekhi Mat. Nauk **25**, no. 5 (1970), 3–62; English transl. in Russian Math. Surveys **25**, no. 5 (1970).

36. Đào Trọng Thi, *Multivarifolds and minimization problems for functionals of multidimensional volume type*, Dokl Akad. Nauk SSSR **276** (1984), 1042–1045; English transl. in Soviet Math. Dokl. **29** (1984).

37. ——, *Isoperimetric inequalities for multivarifolds*, Izv. Akad. Nauk Ser. Mat. **48** (1984), 309–325; English transl. in Math. USSR-Izv. **26** (1986).

38. ——, *Minimal real currents on compact Riemannian manifolds*, Izv. Akad. Nauk Ser. Mat. **41** (1977), 853–867; English transl. in Math. USSR-Izv. **11** (1977).

39. ——, *On the stability of the homology of compact Riemannian manifolds*, Izv. Akad. Nauk SSSR Ser. Mat. **42** (1978), 500–505; English transl. in Math. USSR-Izv. **12** (1978).

40. ——, *Minimal surfaces in compact Lie groups*, Uspekhi Mat. Nauk **33**, no. 3 (1978), 163–164; English transl. in Russian Math. Surveys **33**, no. 3 (1978).

41. ——, *Real minimal currents in compact Lie groups*, Trudy Sem. Vektor. Tenzor. Anal. No. 19, Izdat. Mosk. Univ., Moscow, 1979, 112–129. (Russian)

42. ——, *Multivarifolds and classical multidimensional Plateau problems*, Izv. Akad. Nauk SSSR Ser. Mat. **44** (1980), 1031–1065; English transl. in Math. USSR-Izv. **17** (1981).

43. ——, *A multidimensional variational problem in symmetric spaces*, Funktsional. Anal. i Prilozhen. **12**, no. 1 (1978), 72–73; English transl. in Functional Anal. Appl. **12** (1978).

44. ——, *On minimal currents and surfaces in Riemannian manifolds*, Dokl. Akad. Nauk SSSR **233** (1977), 21–22; English transl. in Soviet Math. Dokl. **18** (1977).

45. ——, *Spaces of parametrizations and parametrized multivarifolds*, Trudy Sem. Vektor. Tenzor. Anal. No. 22 Izdat. Mosk. Univ., Moscow, 1985, 31–59. (Russian)

46. ——, *Algebraic questions on the realization of cycles in symmetric spaces*, Vestnik Moskov. Univ. Ser. I Mat. Mekh. **31**, no. 2 (1976), 62–66; English transl. in Moscow Univ. Math. Bull. **31** (1976).

47. Đào Trọng Thi and A. T. Fomenko, *The topology of absolute minima of functionals of volume type and the Dirichlet functional*, in: Modern questions of real and complex analysis, Inst. Mat. Akad. Nauk UkrSSR, Kiev, 1984, 40–64. (Russian)

48. B. A. Dubrovin, S. P. Novikov, and A. T. Fomenko, *Modern geometry*, "Nauka", Moscow, parts 1 and 2, 1979; part 3, 1984; English transl., Parts I, II, Springer-Verlag, Berlin and New York, 1984, 1985.

49. E. B. Dynkin, *The homology of compact Lie groups*, Uspekhi Mat. Nauk **8**, no. 5 (1953), 73–120.

50. N. V. Efimov, *Surfaces with a slowly changing negative curvature*, Uspekhi Mat. Nauk **21**, no. 5 (1966), 3–58; English transl. in Russian Math. Surveys **21**, no. 5 (1966).

51. ——, *Impossibility of a complete regular surface in three-dimensional Euclidean space whose Gaussian curvature has a negative upper bound*, Dokl. Akad. Nauk SSSR **150** (1963), 1206–1290; English transl. in Soviet Math. Dokl. **4** (1963).

52. N. V. Efimov and E. G. Poznyak, *Generalization of Hilbert's theorem on surfaces of constant negative curvature*, Dokl. Akad. Nauk SSSR **137** (1961), 509–512; English transl. in Soviet Math. Dokl. **2** (1961).

53. H. Seifert and W. Threlfall, *Variationsrechnung im Grossen*, Chelsea, New York, 1951.

54. A. D. Ioffe and V. M. Tikhomirov, *Extension of variational problems*, Trudy Moskov. Mat. Obshch. **18** (1968), 187–246; English transl. in Trans. Moscow Math. Soc. **18** (1969).

55. E. Cartan, *The geometry of Lie groups and symmetric spaces*, IL, Moscow, 1949.([1])

56. V. E. Katznelson and L. I. Ronkin, *On the minimal volume of an analytic set*, Sibirsk. Mat. Zh. **15** (1974), 516–528; English transl. in Siberian Math. J. **15** (1974).

57. V. M. Kesel′man, *On Bernstein's theorem for surfaces with a quasiconformal map*, Mat. Zametki **35** (1984), 445–453; English transl. in Math. Notes **35** (1984).

58. W. Klingenberg, *Lectures on closed geodesics*, Springer-Verlag, Berlin and New York, 1978.

59. E. A. Korolev and T. N. Fomina, *Minimal Peterson surfaces*, Ukrain. Geom. Sb. No. 22 (1979), 92–96.

60. O. A. Ladyzhenskaya, *Boundary-value problems of mathematical physics*, "Nauka", Moscow, 1973; English transl., Springer-Verlag, Berlin and New York, 1985.

([1])This volume consists of Russian translations of the following papers:

a) *La géométrie des groupes de transformations*, J. Math. Pures Appl. (9) **6** (1927), 1–119;

b) *Sur une classe remarquable d'espaces de Riemann*, Bull. Soc. Math. France **54** (1926), 214–264;

c) *Groupes simples clos et ouverts et géométrie riemannienne*, J. Math. Pures. Appl. (9) **8** (1929), 1–33;

d) *La géométrie des groupes simples*, Ann. Mat. Pura Appl. (4) **4** (1927), 209–256; *Complément*, ibid. **5** (1928), 253–260;

e) *La théorie des groupes finis et continus et l'analysis situs*, Mém. Sci. Math. fasc. **4** (1930).

61. O. A. Ladyzhenskaya and N. N. Ural′tseva, *Linear and quasilinear equations of elliptic type*, "Nauka", Moscow, 1964; English transl., Academic Press, New York, 1968.

62. L. D. Landau and E. M. Lifshits, *Mechanics of continuous media*, Gostekhizdat, Moscow, 1954; English transl., Addison-Wesley, Reading, Mass., 1959.

63. L. Lichnerowicz, *Théorie globale des connexions et des groupes d'holonomie*, Edizioni Cremonese, Rome, 1955.

64. O. V. Manturov, *On some cohomology properties of Cartan models of symmetric spaces*, Trudy Vektor. Tenzor. Anal. No. 18, Izdat. Moskov. Univ., Moscow, 1978, 169–175. (Russian)

65. L. A. Masal′tsev, *Instability of minimal cones in Lobachevsky space*, Ukrain. Geom. Sb. No. 21 (1978), 72–81.

66. ——, *Minimal surfaces in* $\mathbf{R}^5$ *whose Gaussian image has constant curvature*, Mat. Zametki **35** (1984), 927–932; English transl. in Math. Notes **35** (1984).

67. ——, *Dimensions of stable domains on certain minimal submanifolds of a sphere*, Ukrain. Geom. Sb. No. 22 (1979), 103–108.

68. V. M. Miklyukov, *On some properties of tubular minimal surfaces in* $\mathbf{R}^n$, Dokl. Akad. Nauk SSSR **247** (1979), 549–552; English transl. in Soviet Math. Dokl. **20** (1979).

69. ——, *An estimate of the modulus of a family of curves on a minimal surface and applications*, Uspekhi Mat. Nauk **34**, no. 3 (1979), 207–208; English transl. in Russian Math. Surveys **34**, no. 3 (1979).

70. ——, *Some peculiarities of the behavior of solutions of minimal surface type equations in unbounded domains*, Mat. Sb. **116** (1981), 72–86; English transl. in Math. USSR-Sb. **41** (1982).

71. ——, *Capacity and a generalized maximum principle for quasilinear equations of elliptic type*, Dokl. Akad. Nauk SSSR **250** (1980), 1318–1320; English transl. in Soviet Math. Dokl. **21** (1980).

72. J. W. Milnor, *Morse theory*, Princeton Univ. Press, Princeton, N. J., 1963.

73. ——, *Lectures on the h-cobordism theorem*, Princeton Univ. Press, Princeton, NJ, 1965.

74. A. S. Mishchenko and A. T. Fomenko, *A course of differential geometry and topology*, Izdat. Moskov. Gos. Univ., Moscow, 1980. (Russian)

75. R. Narasimhan, *Analysis on real and complex manifolds*, North-Holland Publ. Co., Amsterdam, 1968.

76. J. C. C. Nitsche, *On new results in the theory of minimal surfaces*, Bull. Amer. Math. Soc. **71** (1965), 195–270.

77. R. Osserman, *Minimal surfaces*, Uspekhi Mat. Nauk **22**, no. 4 (1967), 55–136; English transl. in Russian Math. Surveys **22** (1967).

78. A. I. Pluzhnikov, *Some properties of harmonic mappings in the case of spheres and Lie groups*, Dokl. Akad. Nauk SSSR **268** (1983), 1300–1302; English transl. in Soviet Math. Dokl. **27** (1983).

79. ——, *Harmonic mappings of Riemann surfaces and foliated manifolds*, Mat. Sb. **113** (1980), 339–347; English transl. in Math. USSR-Sb. **41** (1982).

80. ——, *Indices of harmonic maps of spheres*, in: Geometry and topology in global nonlinear problems, Izdat. Voronezh. Univ., Voronezh, 1984, 162–166. (Russian)

81. ——, *The problem of minimizing the energy functional*, Inst. of Control Problems, Akad. Nauk SSSR, Deposited at VINITI, no. 5584-84, 1984.

82. ——, *A topological criterion for the unattainability of global minima of the energy functional*, in: Analysis on manifolds and differential equations, Izdat. Voronezh. Univ., Voronezh, 1986, 149–156. (Russian)

83. ——, *On minima of the Dirichlet functional*, Dokl. Akad. Nauk SSSR **290** (1986), 289–293; English transl. in Soviet Math. Dokl. **34** (1987).

84. ——, *Some geometrical properties of harmonic maps*, Trudy Sem. Vektor. Tenzor. Anal. No. 22, Izdat. Moskov. Univ., Moscow, 1986, 132–147. (Russian)

85. A. V. Pogorelov, *Extrinsic geometry of convex surfaces*, "Nauka", Moscow, 1969; English transl., Amer. Math. Soc., Providence, R.I., 1973.

86. ——, *On the stability of minimal surfaces*, Dokl. Akad. Nauk SSSR **260** (1981), 293–295; English transl. in Soviet Math. Dokl. **24** (1981).

87. ——, *On minimal hypersurfaces in spherical space*, Dokl. Akad. Nauk SSSR **206** (1972), 291–292; English transl. in Soviet Math. Dokl. **13** (1972).

88. E. G. Poznyak, *Isometric immersions of two-dimensional Riemannian metrics in Euclidean spaces*, Uspekhi Mat. Nauk **28**, no. 4 (1973), 47–76; English transl. in Russian Math. Surveys **28**, no. 4 (1973).

89. L. S. Pontryagin, *Homologies in compact Lie groups*, Mat. Sb. **6** (1938), 389–422 (in English).

90. G. De Rham, *Variétés différentiables*, Hermann, Paris, 1955.

91. P. K. Raševskiĭ, *Riemannian geometry and tensor analysis*, "Nauka", Moscow, 1967; German transl., VEB Deutscher Verlag, Berlin, 1959.

92. Yu. G. Reshetnyak, *A new proof of the theorem on the existence of an absolute minimum for two-dimensional problems of the calculus of variations in parametric form*, Sibirsk. Mat. Zh. **3** (1962), 744–768.

93. E. R. Rozendorn, *Weakly irregular surfaces of negative curvature*, Uspekhi Mat. Nauk **21**, no. 5 (1966), 59–116; English transl. in Russian Math. Surveys **21**, no. 5 (1966).

94. ——, *Investigation of the basic equations of the theory of surfaces in asymptotic coordinates*, Mat. Sb. **70** (1966), 490–507.

95. L. I. Ronkin, *Discrete uniqueness sets for entire functions of exponential type in several variables*, Sibirsk. Mat. Zh. **19** (1978), 142–152; English transl. in Siberian Math. J. **19** (1978).

96. K. A. Rybnikov, *The history of mathematics*, Izdat. Moskov. Univ., Moscow, 1974. (Russian)

97. I. Kh. Sabitov, *Formal solutions of the Hilbert problem for an annulus*, Mat. Zametki **12** (1972), 221–232; English transl. in Math. Notes **12** (1972).

98. ——, *Minimal surfaces with certain boundary conditions*, Mat. Sb. **76** (1968), 368–389; English transl. in Math. USSR-Sb. **5** (1968).

99. ——, *A minimal surface as the rotation graph of a sphere*, Mat. Zametki **2** (1967), 645–656; English transl. in Math. Notes **2** (1967).

100. ——, *Local structure of Darboux surfaces*, Dokl. Akad. Nauk SSSR **162** (1965), 1001–1004; English transl. in Soviet Math. Dokl. **6** (1965).

101. L. I. Sedov, *Continuum mechanics*, "Nauka", Moscow, 1976; French transl., "Mir", Moscow, 1975.

102. A. G. Sigalov, *Two-dimensional problems of calculus of variations in nonparametric form*, Trudy Moskov. Mat. Obshch. **2** (1953), 201–233.

103. ——, *Variational problems with admissible surfaces of arbitrary topological types*, Uspekhi Mat. Nauk **12**, no 1 (1957), 53–93.

104. ——, *Two-dimensional problems of the calculus of variations*, Uspekhi Mat. Nauk **6**, no. 2 (1951), 16–101.

105. S. Smale, *Generalized Poincáre's conjecture in dimensions greater than four*, Ann. of Math. (2) **74** (1961), 391–406.

106. ——, *On the structure of manifolds*, Amer. J. Math. **84** (1962), 387–399.

107. Russian translation of [212].

108. D. J. Struik, *A concise history of mathematics* (2 vols.), Dover, New York, 1948.

109. O. V. Titov, *Minimal hypersurfaces stretched over soft obstacles*, Dokl. Akad. Nauk SSSR **211** (1973), 293–296; English transl. in Soviet Math. Dokl. **14** (1973).

110. ——, *Minimal hypersurfaces over soft obstacles*, Izv. Akad. Nauk SSSR Ser. Mat. **38** (1974), 374–417; English transl. in Math. USSR-Izv. **8** (1974).

111. A. A. Tuzhilin and A. T. Fomenko, *Many-valued maps, minimal surfaces and soap films*, Vestnik Moskov. Univ. Mat. Mekh. **1986**, no. 3, 3–12; English transl. in Moscow Univ. Math. Bull. **41** (1986).

112. A. V. Tyrin, *Critical points of the multidimensional Dirichlet functional*, Mat. Sb. **124** (1984), 146–158; English transl. in Math. USSR - Sb. **52** (1985), 141–153.

113. ——, *On the absence of local minima of the multidimensional Dirichlet functional*, Uspekhi Mat. Nauk **39**, no. 2 (1984), 193–194; English transl. in Russian Math. Surveys **39**, no. 2 (1984).

114. ——, *The problem of minimizing functionals of Dirichlet type*, in: Geometry, differential equations and mechanics, Izdat. Moskov. Univ., Moscow, 1986, 146–150. (Russian)

115. Russian translation of [234].

116. A. T. Fomenko, *Indices of minimal and harmonic surfaces*, in: Methods of topology and Riemannian geometry in mathematical physics, Izdat. Vilnius Univ., Vilnius, 1984, 95–108. (Russian)

117. ——, *On certain properties of extremals of variational problems*, in: Geometry and topology in global nonlinear problems, Izdat. Voronezh. Univ., Voronezh, 1984, 114–122; English transl. in: Lecture Notes in Math. **1108** (1984), 209–217.

118. ——, *Differential geometry and topology. Additional chapters*, Izdat. Moskov. Univ., Moscow, 1983. (Russian)

119. ——, *Variational methods in topology*, "Nauka", Moscow, 1982. (Russian)

120. ——, *The existence and almost everywhere regularity of minimal compacta with given homology properties*, Dokl. Akad. Nauk SSSR **187** (1969), 747–749; English transl. in Soviet Math. Dokl. **10** (1969).

121. ——, *Some cases when elements of homotopy groups of homogeneous spaces are realized by totally geodesic spheres*, Dokl. Akad. Nauk SSSR **190** (1970), 792–795; English transl. in Soviet Math. Dokl. **11** (1970).

122. ——, *The multidimensional Plateau problem and singular points of minimal compacta*, Dokl. Akad. Nauk SSSR **192** (1970), 293–296; English transl. in Soviet Math. Dokl. **11** (1970).

123. ——, *Homological properties of minimal compacta in the multidimensional Plateau problem*, Dokl. Akad. Nauk SSSR **192** (1970), 38–41; English transl. in Soviet Math. Dokl. **11** (1970).

124. ——, *Realization of cycles in compact symmetric spaces by totally geodesic submanifolds*, Dokl. Akad. Nauk SSSR **195** (1970), 789–792; English transl. in Soviet Math. Dokl. **11** (1970).

125. ——, *Totally geodesic models of cycles*, Trudy Sem. Vektor. Tenzor. Anal. No. 16, Izdat. Moskov. Univ., Moscow, 1972, 14–98. (Russian)

126. ——, *The multidimensional Plateau problem in extraordinary homology and cohomology theories*, Dokl. Akad. Nauk SSSR **200** (1971), 797–800; English transl. in Soviet Math. Dokl. **12** (1971).

127. ——, *Bott periodicity from the point of view of an n-dimensional Dirichlet functional*, Izv. Akad. Nauk SSSR Ser. Mat. **35** (1971), 667–681; English transl. in Math. USSR-Izv. **5** (1971).

128. ——, *Minimal compacta in Riemannian manifolds and a conjecture of Reifenberg*, Izv. Akad. Nauk SSSR Ser. Mat. **36** (1972), 1049–1080; English transl. in Math. USSR-Izv. **6** (1972).

129. ——, *The multidimensional Plateau problem in Riemannian manifolds*, Mat. Sb. **89** (1972), 475–519; English transl. in Math. USSR-Sb. **18** (1972).

130. ——, *Multidimensional Plateau problems on Riemannian manifolds and extraordinary homology and cohomology theories*. I, Trudy Sem. Vektor. Tenzor. Anal. No. 17, Izdat. Moskov. Univ., Moscow, 1974, 3–176. (Russian)

131. ——, *Multidimensional Plateau problems on Riemannian manifolds and extraordinary homology and cohomology theories*. II, Trudy Sem. Vektor. Tenzor. Anal. No. 18, Izdat. Moskov. Univ., Moscow, 1978, 4–93. (Russian)

132. ——, *Geometrical variational problems*, Sovremennye Problemy Matematyki, Vol. 1, VINITI, Moscow, 1973, 39–60. (Russian)

133. ——, *Minimal volumes of topological globally minimal surfaces in cobordisms*, Izv. Akad. Nauk SSSR Ser. Mat. **45** (1981), 187–212; English transl. in Math. USSR-Izv. **18** (1982).

134. ——, *Multidimensional variational problems in the topology of extremals*, Uspekhi Mat. Nauk **36**, no. 6 (1981), 105–135; English transl. in Russian Math. Surveys **36**, no. 6 (1981).

135. ——, *A universal lower bound on the rate of growth of globally minimal solutions*, Dokl. Akad. Nauk SSSR **251** (1980), 295–299; English transl. in Soviet Math. Dokl. **21** (1980).

136. ——, *Topological variational problems*, Izdat. Moskov. Univ., Moscow, 1985. (Russian)

137. D. B. Fuks, A. T. Fomenko, and V. L. Gutenmakher, *Homotopic topology*, Izdat. Moskov. Univ., Moscow, 1969; English transl., Acad. Kiadó, Budapest, 1986.

138. N. G. Khor′kova, *Minimal cones in Riemannian manifolds*, Application of topology to modern analysis, Izdat. Voronezh. Univ., Voronezh, 1985, 167–171. (Russian)

139. S.-S. Chern, *Complex manifolds*, Universidade do Recife, 1959.

140. L. C. Young, *Lectures on the calculus of variations and optimal control theory*, Saunders, Philadelphia, London, and Toronto, 1969.

141. H. Alexander and R. Osserman, *Area bounds for various classes of surfaces*, Amer. J. Math. **97** (1975), 753–769.

142. W. K. Allard, *On the first variation of a varifold*, Ann. of Math. (2) **95** (1972), 417–491.

143. ——, *On the first variation of a varifold. Boundary behavior*, Ann. of Math. (2) **101** (1975), 418–446.

144. ——, *On boundary regularity for Plateau's problem*, Bull. Amer. Math. Soc. **75** (1969), 522–523.

145. W. K. Allard and F. J. Almgren, *On the radial behaviour of minimal surfaces and the uniqueness of their tangent cones*, Ann. of Math. (2) **113** (1981), 215–265.

146. ——, *The structure of stationary one-dimensional varifolds with positive density*, Invent. Math. **34** (1976), 83–97.

147. F. J. Almgren, *Existence and regularity almost everywhere of solutions to elliptic variational problems among surfaces of varying topological type and singularity structure*, Ann. of Math. (2) **87** (1968), 321–391.

148. F. J. Almgren and J. E. Taylor, *The geometry of soap films and soap bubbles*, Scientific American, July 1976, 82–93.

149. F. J. Almgren, *The homotopy groups of the integral cycle groups*, Topology **1** (1962), 257–299.

150. ——, *Plateau's problem. An introduction to varifold geometry*, Benjamin, New York, 1966.

151. ——, *Some interior regularity theorems for minimal surfaces and an extension of Bernstein's theorem*, Ann. of Math. (2) **84** (1966), 277–293.

152. ——, *Measure theoretic geometry and elliptic variational problems*, Bull. Amer. Math. Soc. **75** (1969), 285–304.

153. ——, *Q valued functions minimizing Dirichlet's integral and regularity of area minimizing rectifiable currents up to codimension two*, Bull. Amer. Math. Soc. (N.S.) **8** (1983), 327–328.

154. F. J. Almgren, I. R. Schoen, and L. Simon, *Regularity and singularity estimates on hypersurfaces minimizing parametric elliptic variational integrals*, Acta Math. **139** (1977), 217–265.

155. F. J. Almgren and W. P. Thurston, *Examples of unknotted curves which bound only surfaces of high genus within their convex hulls*, Ann. of Math. (2) **105** (1977), 527–538.

156. F. J. Almgren and L. Simon, *Existence of embedded solutions of Plateau's problem*, Ann. Scuola Norm. Sup. Pisa (4) **6** (1979), 447–495.

157. F. J. Almgren, *Existence and regularity almost everywhere of solutions to elliptic variational problems with constraints*, Mem. Amer. Math. Soc. **169** (1976), 1–199.

158. H. W. Alt, *Verzweigungspunkte von H-Flächen*. I, Math. Z. **127** (1972), 333–362.

159. ——, *Verzweigungspunkte von H-Flächen*. II, Math. Ann. **201** (1973), 33–55.

160. F. Arago, *Biographie de Gaspard Monge*, Mem. Acad. Sci. V. XXIV, Paris, 1854.

161. P. V. Aubry, *Monge le savant ami de Napoléo Bonaparte, 1746–1818*, Paris, 1954.

162. J. L. M. Barbosa and M. P. do Carmo, *On the size of a stable minimal surface in* $\mathbf{R}^3$, Amer. J. Math. **98** (1976), 515–528.

163. A. Beer, *Theorie der Elastisität und Capillarität*, Einl. in die Math., Leipzig, 1869.

164. M. J. Beeson and A. J. Tromba, *The cusp catastrophe of Thom in the bifurcation of minimal surfaces*, Manuscripta Math. **46** (1984), 273–308.

165. M. J. Beeson, *Some results on finiteness in Plateau's problem*. I, II, Math. Z. **175** (1980), 103–123; **181** (1982), 1–30.

166. B. Berndtsson, *Zeros of analytic functions of several variables*, Arkiv Mat. **16** (1978), 251–262.

167. L. Bers, *Isolated singularities of minimal surfaces*, Ann. of Math. (2) **53** (1951), 364–368.

168. ——, *Abelian minimal surfaces*, J. Analyse Math. **1** (1951), 43–58.

169. D. Bindschadler, *Invariant solutions to the oriented Plateau problem of maximal codimension*, Trans. Amer. Math. Soc. **261** (1980), 439–462.

170. E. Bombieri, E. de Giorgi, and E. Giusti, *Minimal cones and the Bernstein problem*, Invent. Math. **7** (1969), 243–268.

171. E. Bombieri and M. Miranda, *Una maggiorazione a priori relativa alle ipersuperfici minimali non parametriche*, Arch. Rational Mech. Anal. **32** (1969), 255–267.

172. R. Böhme and A. Tromba, *The index theorem for classical minimal surfaces*, Ann. of Math. (2) **113** (1981), 447–499.

173. R. Böhme, *Stability of minimal surfaces*, Lecture Notes in Math. **561** (1976), 123–137.

174. R. Böhme, S. Hildebrandt, and E. Tausch, *The two-dimensional analogue of the catenary*, Pacific J. Math. **88** (1980), 247–278.

175. A. Borel, *Sur la cohomologie des espaces fibrés principaux et des espaces homogènes de groupes de Lie compacts*, Ann. of Math. (2) **57** (1953), 115–207.

176. R. Bott, *Nondegenerate critical manifolds*, Ann. of Math. (2) **60** (1954), 248–261.

177. J.-P. Bourguignon, *Harmonic curvature for gravitational and Yang-Mills fields*, Lecture Notes in Math. **949** (1982), 35–47.

178. J. E. Brothers, *Invariance of solutions to invariant parametric variational problems*, Trans. Amer. Math. Soc. **262** (1980), 159–179.

179. R. Caccioppoli, *Misura e integrazione sugli insiemi dimensionalmente orientati*. I, II, Atti Accad. Naz. Lincei Rend (8) **12** (1952), 3–11, 137–146.

180. E. Calabi, *Minimal immersions of surfaces in Euclidean spheres*, J. Diff. Geom. **1** (1967), 111–125.

181. ——, *Quelques applications de l'analyse complexe aux surfaces d'aire minimal* (*together with topics in complex manifolds by Rossi*), Montreal Univ. Press, Montreal, 1968.

182. T. Carleman, *Zur Theorie der Minimalflächen*, Math. Z. **9** (1921), 154–160.

183. M. P. Do Carmo and C. K. Peng, *Stable minimal surfaces in* $\mathbf{R}^3$ *are planes*, Bull. Amer. Math. Soc. (N.S.) **1** (1979), 903–906.

184. H. Cartan, *Transgression dans un groupe de Lie et dans un espace fibré principal*, Colloque de topologie algébrique, Brussels, 1950, 57–71.

185. Y. W. Chen, *Branch points, poles and planar points of minimal surfaces in* $\mathbf{R}^3$, Ann. of Math. (2) **49** (1948), 790–806.

186. S.-S. Chern, *Minimal surfaces in a Euclidean space of* N *dimensions*, Differential and combinatorial topology, Princeton Univ. Press, Princeton, N.J., 1965, 187–199.

187. ——, *Complex manifolds*, Univ. of Chicago, Chicago, IL, 1955–1956.

188. ——, *Brief survey of minimal submanifolds*, Tagungsbericht, Oberwolfach, 1969, 43–60.

189. ——, *Differential geometry: its past and future*, Actes Congrès Internat. Math., Nice, 1970, vol. 1, 41–53.

190. S.-S. Chern and R. Osserman, *Complete surfaces in Euclidean* n*-space*, J. Analyse Math. **19** (1967), 15–34.

191. R. Courant, *Plateau's problem and Dirichlet's principle*, Ann. of Math. (2) **38** (1937), 679–724.

192. R. Courant and N. Davids, *Minimal surfaces spanning closed manifolds*, Proc. Nat. Acad. Sci. USA **26** (1940), 194–199.

193. R. Courant, *Dirichlet's principle, conformal mappings and minimal surfaces*, Interscience, New York, 1950.

194. ——, *The existence of minimal surfaces of given topological structure under prescribed boundary conditions*, Acta Math. **72** (1940), 51–98.

195. ——, *On a generalized form of Plateau's problem*, Trans. Amer. Math. Soc. **50** (1941), 40–47.

196. ——, *On Plateau's problem with free boundaries*, Proc. Nat. Acad. Sci. USA **31** (1945), 242–246.

197. D. S. Freed and K. K. Uhlenbeck, *Instantons and four-manifolds*, Springer-Verlag, Berlin, Heidelberg, and New York, 1984.

198. Đào Trọng Thi, *Isoperimetric inequalities for multivarifolds*, Acta Math. Vietnam. **6** (1981), 88–94.

199. G. Darboux, *Leçons sur la théorie générale des surfaces*, Première partie, Gauthier-Villars, Paris, 1941.

200. N. Davids, *Minimal surfaces spanning closed manifolds and having prescribed topological position*, Amer. J. Math. **64** (1942), 348–362.

201. J. Douglas, *Minimal surfaces of general topological structure*, J. Math. Phys. **15** (1936), 105–123.

202. ——, *The most general form of the problem of Plateau*, Proc. Nat. Acad. Sci. USA **24** (1938), 360–364.

203. ——, *Solutions to the problem of Plateau*, Trans. Amer. Math. Soc. **33** (1931), 263–321.

204. ——, *Minimal surfaces of higher topological structure*, Proc. Nat. Acad. Sci. USA **24** (1938), 343–353.

205. ——, *Minimal surfaces of higher topological structure*, Ann. of Math. (2) **40** (1939), 205–298.

206. ——, *The higher topological form of Plateau's problem*, Ann. Scuola Norm. Sup. Pisa (2) **8** (1939), 1–24.

207. E. H. Brown, Jr., *Cohomology theories*, Ann. of Math. (2) **75** (1962), 467–484.

208. J. Eells and J. Wood, *Maps of minimal energy*, J. London Math. Soc. (2) **23** (1981), 303–310.

209. J. Eells and J. H. Sampson, *Harmonic mappings of Riemannian manifolds*, Amer. J. Math. **86** (1964), 109–160.

210. J. Eells and L. Lemaire, *A report on harmonic maps*, Bull. London Math. Soc. **10** (1978), 1–68.

211. ——, *Selected topics in harmonic maps*, CBMS Regional Conf. Series in Math., 50, Amer. Math. Soc., Providence, R.I., 1983.

212. S. Eilenberg and N. Steenrod, *Foundations of algebraic topology*, Princeton Univ. Press, Princeton, N.J., 1952.

213. L. P. Eisenhart, *An introduction to differential geometry*, Princeton Univ. Press, Princeton, N.J., 1949.

214. *Encyklopädie der mathematischen Wissenschaften mit Einschluss ihrer Anwendungen. Physik*. Fünfter Band in drei Teilen, Erster Teil, Leipzig, 1903–1921.

215. H. Federer, *Currents and area*, Trans. Amer. Math. Soc. **98** (1961), 204–233.

216. ——, *Geometric measure theory*, Springer-Verlag, Berlin, Heidelberg, and New York, 1969.

217. H. Federer and W. H. Fleming, *Normal and integral currents*, Ann. of Math. (2) **72** (1960), 458–520.

218. H. Federer, *Hausdorff measure and Lebesgue area*, Proc. Nat. Acad. Sci. USA **37** (1951), 90–94.

219. ——, *Real flat chains, cochains and variational problems*, Indiana Univ. Math. J. **24** (1974), 351–407.

220. ——, *Measure and area*, Bull. Amer. Math. Soc. **58** (1952), 306–378.

221. ——, *The singular sets of area minimizing rectifiable currents with codimension one and of area minimizing flat chains modulo two with arbitrary codimensions*, Bull. Amer. Math. Soc. **76** (1970), 767–771.

222. ——, *Some theorems on integral currents*, Trans. Amer. Math. Soc. **117** (1965), 43–67.

223. R. Finn, *Isolated singularities of solutions of non-linear partial differential equations*, Trans. Amer. Math. Soc. **75** (1953), 383–404.

224. ——, *On equations of minimal surface type*, Ann. of Math. (2) **60** (1954), 397–416.

225. ——, *New estimates for equations of minimal surface type*, Arch. Rational Mech. Anal. **14** (1963), 337–375.

226. ——, *Remarks relevant to minimal surfaces and to surfaces of prescribed mean curvature*, J. Mech. Math. **14** (1965), 139–160.

227. R. Finn and R. Osserman, *On the Gauss curvature of non-parametric minimal surfaces*, J. Anal. Math. **12** (1964), 351–364.

228. W. H. Fleming, *On the oriented Plateau problem*, Rend. Circ. Mat. Palermo **11** (1962), 69–90.

229. ——, *Functions of several variables*, 2nd ed., Springer-Verlag, New York and Heidelberg, 1977.

230. ——, *An example in the problem of least area*, Proc. Amer. Math. Soc. **7** (1956), 1063–1074.

231. W. H. Fleming and L. C. Young, *A generalized notion of boundary*, Trans. Amer. Math. Soc. **76** (1954), 457–484.

232. A. T. Fomenko, *Multidimensional Plateau problem on Riemannian manifolds. On the problem of the algorithmic recognizability of the standard three-dimensional sphere*, Proc. Internat. Cong. Math., Vancouver, 1974, vol. 1, 515–525.

233. T. Frankel, *On the fundamental group of a compact minimal submanifold*, Ann. of Math. (2) **83** (1966), 68–73.

234. H. Whitney, *Geometric integration theory*, Princeton Univ. Press, Princeton, N.J., 1957.

235. M. Giaquinta and L. Pepe, *Esistenza e regolarità per il problema dell'area minima con ostacoli in n variabili*, Ann. Scuola Norm. Sup. Pisa (3) **25** (1971), 481–507.

236. E. De Giorgi, *Una estensione del teorema di Bernstein*, Ann. Scuola Norm. Sup. Pisa (3) **19** (1965), 79–85.

237. ——, *Nuovi teoremi relativi alle misure* $(r-1)$*-dimensionali in uno spazio ad r dimensioni*, Ricerche Mat. **4** (1955), 95–113.

238. ——, *Frontiere orientate di misura minima*, Sem. Mat. Scuola Norm. Sup. Pisa, 1960–61, 1–56.

239. ——, *Su una teoria generale della misur* $(r-1)$*-dimensionale in uno spazio ad r dimensioni*, Ann. di Mat. (4) **36** (1954), 191–213.

240. ——, *Complementi alla teoria della misura* $(n-1)$*-dimensionale in uno spazio n-dimensionale*, Sem. Mat. Scuola Norm. Sup. Pisa, 1960–61.

241. E. Giusti, *Superfici minime cartesiane con ostacoli discontinui*, Arch. Rational Mech. Anal. **40** (1971), 251–267.

242. A. Gray, *Minimal varieties and almost Hermitian submanifolds*, Michigan Math. J. **12** (1965), 273–287.

243. M. Gromov, *Volume and bounded cohomology*, Publ. Math. IHES **56** (1982), 5–99.

244. M. Grüter, S. Hildebrandt, and J. C. C. Nitsche, *On the boundary behaviour of minimal surfaces with a free boundary which are not minima of the area*, Manuscripta Math. **35** (1981), 387–410.

245. R. Gulliver and J. Spruck, *On embedded minimal surfaces*, Ann. of Math. (2) **103** (1976), 331–347.

246. R. D. Gulliver, R. Osserman, and H. L. Royden, *A theory of branched immersions of surfaces*, Amer. J. Math. **95** (1973), 750–812.

247. R. D. Gulliver, *Regularity of minimizing surfaces of prescribed mean curvature*, Ann. of Math. (2) **97** (1973), 275–305.

248. R. Gulliver and F. D. Lesley, *On boundary branch points of minimal surfaces*, Arch. Rational Mech. Anal. **52** (1973), 20–25.

249. A. Haar, *Über das Plateausche problem*, Math. Ann. **97** (1927), 127–158.

250. P. Hartman, *On homotopic harmonic maps*, Canad. J. Math. **19** (1967), 673–687.

251. R. Harvey and H. B. Lawson, *On boundaries of complex analytic varieties*. I, Ann. of Math. (2) **102** (1975), 233–290.

252. ——, *On boundaries of complex analytic varieties*. II, Anal. of Math. (2) **106** (1977), 213–238.

253. ——, *Calibrated geometries*, Acta Math. **148** (1982), 47–157.

254. ——, *Extending minimal varieties*, Invent. Math. **28** (1975), 209–226.

255. ——, *Calibrated foliations (foliations and mass-minimizing currents)*, Amer. J. Math. **104** (1982), 607–633.

256. E. Heinz and S. Hildebrandt, *Some remarks on minimal surfaces in Riemannian manifolds*, Comm. Pure Appl. Math. **23** (1970), 371–377.

257. S. Helgason, *Totally geodesic spheres in compact symmetric spaces*, Math. Ann. **165** (1966), 309–317.

258. ——, *Differential geometry and symmetric spaces*, Academic Press, New York and London, 1962.

259. H. H. Goldstine, *A history of the calculus of variations from the* 17*th through the* 19*th century*, Springer-Verlag, New York and Berlin, 1980.

260. S. Hildebrandt and J. C. C. Nitsche, *Optimal boundary regularity for minimal surfaces with a free boundary*, Manuscripta Math. **33** (1981), 357–364.

261. S. Hildebrandt, *Boundary behaviour of minimal surfaces*, Arch. Rational Mech. Anal. **35** (1969), 47–82.

262. S. Hildebrandt and J. C. C. Nitsche, *Minimal surfaces with free boundaries*, Acta Math. **143** (1979), 251–272.

263. S. Hildebrandt and H. Kaul, *Two-dimensional variational problems with obstructions, and Plateau's problem for H-surfaces in a Riemannian manifold*, Comm. Pure Appl. Math. **25** (1972), 187–223.

264. S. Hildebrandt and J. C. C. Nitsche, *A uniqueness theorem for surfaces of least area with partially free boundaries on obstacles*, Arch. Rational Mech. Anal. **79** (1982), 189–218.

265. D. A. Hoffman and R. Osserman, *The geometry of the generalized Gauss map*, Mem. Amer. Math. Soc. **28** (1980), No. 236.

266. ——, *The area of the generalized Gaussian image and the stability of minimal surfaces in S^n and R^n*, Math. Ann. **260** (1982), 437–452.

267. W.-Y. Hsiang and H. B. Lawson, *Minimal submanifolds of low cohomogeneity*, J. Diff. Geom. **5** (1971), 1–38.

268. W.-Y. Hsiang, *On the compact homogeneous minimal submanifolds*, Proc. Nat. Acad. Sci. USA **56** (1966), 5–6.

269. ——, *Remarks on closed minimal submanifolds in the standard Riemannian m-sphere*, J. Diff. Geom. **1** (1967), 257–267.

270. ——, *New examples of minimal embeddings of S^{n-1} into $S^n(1)$—the spherical Bernstein problem for $n = 4, 5, 6$*, Bull. Amer. Math. Soc. (N.S.) **7** (1982), 377–379.

271. W.-T. Hsiang, W.-Y. Hsiang, and P. Tomter, *On the construction of infinitely many mutually noncongruent examples of minimal imbeddings of S^{2n-1} into $\mathbf{C}P^n$, $n \geq 2$*, Bull. Amer. Math. Soc. (N.S.) **8** (1983), 463–465.

272. H. Jenkins and J. Serrin, *Variational problems of minimal surface type*. I, Arch. Rational Mech. Anal. **12** (1963), 185–212.

273. ——, *Variational problems of minimal surface type*. II: *Boundary value problems for the minimal surface equation*, Arch. Rational Mech. Anal. **21** (1965–66), 321–342.

274. ——, *Variational problems of minimal surface type*. III: *The Dirichlet problem with infinite data*, Arch. Rational Mech. Anal. **29** (1968), 304–322.

275. ——, *The Dirichlet problem for the minimal surface equation in higher dimensions*, J. Reine Angew. Math. **229** (1968), 170–187.

276. F. Klein, *Vorlesungen über die Entwicklung der Mathematik in* 19 *Jahrhundert*, Band I, Berlin, 1926.

277. K. T. Klotz and L. Sario, *Existence of complete minimal surfaces of arbitrary connectivity and genus*, Proc. Nat. Acad. Sci. USA **54** (1965), 42–44.

278. J. L. Lagrange, *Essai d'une nouvelle méthode pour determiner les maxima et les minima des formules intégrales indéfinies*, Miscellanea Taurinensia II, 1760–61, 173–195; Oeuvres, Vol. 1, 335–362.

279. H. B. Lawson and J. Simons, *On stable currents and their application to global problems in real and complex geometry*, Ann. of Math. (2) **98** (1973), 427–450.

280. H. B. Lawson and R. Osserman, *Non-existence, non-uniqueness and irregularity of solutions to the minimal surface system*, Acta Math. **139** (1977), 1–17.

281. H. B. Lawson, *The equivariant Plateau problem and interior regularity*, Trans. Amer. Math. Soc. **173** (1972), 231–249.

282. ——, *Some intrinsic characterizations of minimal surfaces*, J. Analyse Math. **24** (1971), 151–161.

283. ——, *Complete minimal surfaces in S^3*, Ann. of Math. (2) **92** (1970), 335–374.

284. ——, *The global behaviour of minimal surfaces in S^n*, Ann. of Math. (2) **92** (1970), 224–237.

285. ——, *The unknottedness of minimal embeddings*, Invent. Math. **11** (1970), 183–187.

286. ——, *The stable homology of a flat torus*, Math. Scand. **36** (1975), 49–73.

287. ——, *Local rigidity theorems for minimal hypersurfaces*, Ann. of Math. (2) **89** (1969), 187–197.

288. H. B. Lawson and S. T. Yau, *Compact manifolds of nonpositive curvature*, J. Diff. Geom. **7** (1972), 211–228.

289. A. M. Legendre, *L'intégration de quelques équations aux différences partielles*, Mem. Div. Sav., **1787**, 311–312.

290. H. Lebesgue, *Intégrale, longueur, aire*, Ann. Mat. Pura Appl. **7** (1902), 231–359.

291. P. Lelong, *Fonctions plurisousharmoniques et formes différentielles positives*, Gordon and Breach, London and New York, 1968.

292. L. Lemaire, *Applications harmoniques des surfaces Riemanniennes*, J. Diff. Geom. **13** (1978), 51–87.

293. P.-F. Leung, *On the stability of harmonic maps*, Lecture Notes in Math. **949** (1982), 122–129.

294. P. Lévy, *Leçons d'analyse fonctionelle*, Gauthier-Villars, Paris, 1922.

295. ——, *Le problème de Plateau*, Math. Z. **23** (1948), 1–45.

296. H. Lewy and G. Stampacchia, *On the regularity of the solution of a variational inequality*, Comm. Pure Appl. Math. **22** (1969), 153–188.

297. H. Lewy, *A priori limitations for solutions of Monge-Ampère equations*. II, Trans. Amer. Math. Soc. **41** (1937), 365–374.

298. A. Lichnerowicz, *Applications harmoniques et variétés kähleriennes*, Symposia Math. Vol. III, Academic Press, London, 1970, pp. 341–402.

299. Ü. G. Lumiste, *On the theory of two-dimensional minimal surfaces*, I–IV, Tartu Riikl. Ül. Toimetised No. 102 (1961), 3–15, 16–28; No. 129 (1962), 74–89, 90–102.

300. Matsumoto Makato, *Intrinsic character of minimal hypersurfaces in flat spaces*, J. Math. Soc. Japan **9** (1957), 146–157.

301. W. H. Meeks and S. T. Yau, *Topology of three-dimensional manifolds and the embedding problems in minimal surface theory*, Ann. of Math. (2) **112** (1980), 441–484.

302. ——, *The classical Plateau problem and the topology of three-dimensional manifolds. The embedding of the solution given by Dehn's lemma*, Topology **21** (1982), 409–442.

303. W. H. Meeks, L. Simon, and S. T. Yau, *Embedded minimal surfaces, exotic spheres, and manifolds with positive Ricci curvature*, Ann. of Math. (2) **116** (1982), 621–659.

304. W. H. Meeks and S. T. Yau, *The existence of embedded minimal surfaces and the problem of uniqueness*, Math. Z. **179** (1982), 151–168.

305. W. H. Meeks, *The classification of complete minimal surfaces in* $\mathbf{R}^3$ *with total curvature greater than* -8, Duke Math. J. **48** (1981), 523–535.

306. ——, *The topological uniqueness of minimal surfaces in three-dimensional Euclidean space*, Topology **20** (1981), 389–410.

307. ——, *Uniqueness theorems for minimal surfaces*, Illinois J. Math. **25** (1981), 318–336.

308. M. Miranda, *Sul minimo dell'integrale del gradiente di una funzione*, Ann. Scuola Norm. Sup. Pisa (3) **19** (1965), 626–665.

309. ——, *Un teorema di esistenza e unicità per il problema dell'area minima in n variabili*, Ann. Scuola Norm. Sup. Pisa (3) **19** (1965), 233–249.

310. ——, *Dirichlet problem with* L^1 *data for the nonhomogeneous minimal surface equation*, Indiana Univ. Math. J. **24** (1974), 227–241.

311. ——, *Nouveaux résultats pour les hypersurfaces minimales*, Actes Congrès Internat. Math., Nice, 1970, 853–858.

312. G. Monge, *Une méthode d'intégrer les équations aux différences ordinaires*, Mem. Div. Sav., Paris, 1784.

313. ——, *Application de l'analyse à la géometrie* (*A l'usage de l'École Impériale Polytechnique*), Paris, 1809.

314. F. Morgan, *A smooth curve in* $\mathbf{R}^4$ *bounding a continuum of area minimizing surfaces*, Duke Math. J. **43** (1976), 867–870.

315. ——, *A smooth curve in* $\mathbf{R}^3$ *bounding a continuum of minimal surfaces*, Arch. Rational Mech. Anal. **71** (1981), 193–197.

316. ——, *Almost every curve in* $\mathbf{R}^3$ *bounds a unique minimizing surface*, Invent. Math. **45** (1978), 253–297.

317. S. Mori, *Projective manifolds with ample tangent bundles*, Ann. of Math. (2) **110** (1979), 593–606.

318. C. B. Morrey, *Multiple integrals in the calculus of variations*, Springer-Verlag, Berlin, 1966.

319. ——, *The higher-dimensional Plateau problem on a Riemannian manifold*, Proc. Nat. Acad. Sci. USA **54** (1965), 1029–1035.

320. ——, *The problem of Plateau on a Riemannian manifold*, Ann. of Math. (2) **49** (1948), 807–851.

321. M. Morse, *The first variation in minimal surface theory*, Duke Math. J. **6** (1940), 263–289.

322. M. Morse and C. B. Tompkins, *Existence of minimal surfaces of general critical type*, Ann. of Math. (2) **40** (1939), 443–472.

323. ——, *Unstable minimal surfaces of higher topological type*, Proc. Nat. Acad. Sci. USA **26** (1940), 713–716.

324. ——, *Minimal surfaces not of minimum type by a new mode approximation*, Ann. of Math. (2) **42** (1941), 62–72.

325. ——, *Unstable minimal surfaces of higher topological structure*, Duke Math. J. **8** (1941), 350–375.

326. G. D. Mostow, *Strong rigidity of locally symmetric spaces*, Ann. of Math. Studies 78, Princeton Univ. Press, Princeton, N.J., 1973.

327. C. H. Müntz, *Die Lösung des Plateausche Problems über konvexen Bereichen*, Math. Ann. **94** (1925), 54–96.

328. J. C. C. Nitsche, *Vorlesungen über Minimalflächen*, Springer-Verlag, Berlin, Heidelberg, and New York, 1975.

329. ——, *On new results in the theory of minimal surfaces*, Bull. Amer. Math. Soc. **71** (1965), 195–270.

330. ——, *On the non-solvability of Dirichlet's problem for the minimal surface equation*, J. Math. Mech. **14** (1965), 779–788.

331. ——, *The boundary behaviour of minimal surfaces, Kellog's theorem and branch points on the boundary*, Invent. Math. **8** (1969), 313–333.

332. ——, *Minimal surfaces with partially free boundary*, Ann. Acad. Sci. Fennicae Ser. A, No. 483, 1971, 1–21.

333. ——, *A new uniqueness theorem for minimal surfaces*, Arch. Rational Mech. Anal. **52** (1973), 319–329.

334. ——, *Non-uniqueness for Plateau's problem. A bifurcation process*, Ann. Acad. Sci. Fennicae Ser. A **2** (1976), 361–373.

335. ——, *Variational problems with inequalities as boundary conditions or how to fashion a cheap hat for Giacometti's brother*, Arch. Rational Mech. Anal. **35** (1969), 83–113.

336. *Non-linear problems in geometry*, Proc. Conf. at Katata, Tokyo, 1979.

337. R. Osserman, *Global properties of minimal surfaces in* E^3 *and* E^n, Ann. of Math. (2) **80** (1964), 340–364.

338. ——, *Global properties of classical minimal surfaces*, Duke Math. J. **32** (1965), 565–573.

339. ——, *Minimal surfaces in the large*, Comment. Math. Helv. **35** (1961), 65–76.

340. ——, *On complete minimal surfaces*, Arch. Rational Mech. Anal. **13** (1963), 392–404.

341. ——, *A proof of the regularity everywhere of the classical solution to Plateau's problem*, Ann. of Math. (2) **91** (1970), 550–569.

342. R. Osserman and M. Schiffer, *Doubly-connected minimal surfaces*, Arch. Rational Mech. Anal. **58** (1975), 285–307.

343. R. Osserman, *The isoperimetric inequality*, Bull. Amer. Math. Soc. **84** (1978), 1182–1238.

344. ——, *Minimal varieties*, Bull. Amer. Math. Soc. **75** (1969), 1092–1120.

345. T. Otsuki, *Minimal hypersurfaces in a Riemannian manifold of constant curvature*, Amer. J. Math. **92** (1970), 145–173.

346. ——, *A construction of closed surfaces of negative curvature in* E^4, Math. J. Okayama Univ. **3** (1954), 95–108.

347. ——, *Minimal hypersurfaces with three principal curvature fields in* S^{n+1}, Kodai Math. J. **1** (1978), 1–29.

348. ——, *Minimal submanifolds with* M*-index* 2, J. Diff. Geom. **6** (1972), 193–211.

349. ——, *Minimal submanifolds with* m*-index* 2 *and generalized Veronese surfaces*, J. Math. Soc. Japan **24** (1972), 89–122.

350. R. Palais, *Morse theory on Hilbert manifolds*, Topology **2** (1963), 299–340.

351. ——, *Homotopy theory of infinite-dimensional manifolds*, Topology **5** (1966), 1–16.

352. R. Palais and S. Smale, *A generalized Morse theory*, Bull. Amer. Math. Soc. **70** (1964), 165–172.

353. M. Pini, *B-Kugelbilder reeler Minimalflächen in* $\mathbf{R}^4$, Math. Z. **59** (1953), 290–295.

354. ——, *Minimalflächen fester Gausscher Krümmung*, Math. Ann. **136** (1968), 34–40.

355. J. Plateau, Mém. de l'Acad. Roy. de Belgique, 1843–1868; Mém. de l'Acad. de Bruxelles **16** (1843).

356. ——, *Sur les figures d'équilibre d'une masse liquide sans pesanteur*, Mém. de l'Acad. Roy. de Belgique (N.S.), Vol. 23.

357. ——, *Statique expérimentale et théorique des liquides soumis aux seules forces moléculaires*, Gauthier-Villars, Paris, 1873.

358. N. Quien, *Über die endliche Lösbarkeit des Plateau-Problems in Riemannschen Mannigfaltigkeiten*, Manuscripta Math. **39** (1982), 313–338.

359. H. Poincaré, *Capillarité* (Leçons), Paris, 1895.

360. S. D. Poisson, *Nouvelle théorie de l'action capillaire*, Paris, 1831.

361. ——, *Mémoire sur l'équilibre des fluides*, Mém. de l'Acad. Sci. Paris, Vol. IX.

362. T. Poston, *The Plateau problem. An invitation to the whole of mathematics*, Preprint, Summer College on global analysis and its application, 4 July–25 August 1972, Internat. Centre for Theoretical Physics, Trieste, 1972.

363. T. Rado, *The problem of least area and the problem of Plateau*, Math. Z. **32** (1930), 763–796.

364. ——, *On the problem of Plateau*, Springer-Verlag, Berlin, 1933.

365. E. R. Reifenberg, *Solution of the Plateau problem for* m*-dimensional surfaces of varying topological type*, Acta Math. **104** (1960), 1–92.

366. ——, *An epiperimetric inequality related to the analyticity of minimal surfaces*, Ann. of Math. (2) **80** (1964), 1–14.

367. ——, *On the analyticity of minimal surfaces*, Ann. of Math. (2) **80** (1964), 15–21.

368. J. Sacks and K. K. Uhlenbeck, *The existence of minimal immersions of* 2*-spheres*, Ann. of Math. (2) **113** (1981), 1–24.

369. ——, *Minimal immersions of closed Riemann surfaces*, Trans. Amer. Math. Soc. **271** (1982), 639–652.

370. H. Samelson, *Beitrage zur Topologie der Gruppen-Mannigfaltigkeiten*, Ann. of Math. (2) **42** (1941), 1091–1137.

371. R. M. Schoen and K. K. Uhlenbeck, *A regularity theorem for harmonic maps*, J. Diff. Geom. **17** (1982), 307–335.

372. ——, *Boundary regularity and the Dirichlet problem for harmonic maps*, J. Diff. Geom. **18** (1983), 253–268.

373. ——, *Regularity of minimising harmonic maps into the sphere*, Invent. Math. **78** (1984), 89–100.

374. R. M. Schoen and S. T. Yau, *Complete three-dimensional manifolds with positive Ricci curvature and scalar curvature*, Ann. of Math. Studies 102, Princeton Univ. Press, Princeton, N.J., 1982, pp. 209–228.

375. ——, *On univalent harmonic maps between surfaces*, Invent. Math. **44** (1978), 265–278.

376. ——, *Existence of incompressible minimal surfaces and the topology of three-dimensional manifolds with nonnegative scalar curvature*, Ann. of Math. (2) **110** (1979), 127–142.

377. ——, *Compact group actions and the topology of manifolds with non-positive curvature*, Topology **18** (1979), 361–380.

378. E. M. Schröder, *Dürer: Kunst und Geometrie*, Birkhäuser, Basel and Boston, 1980.

379. H. A. Schwarz, *Gesammelte math. Abh.*, Vol. 1, Berlin, 1890.

380. Shing Tung Yau (ed.), *Seminar on Differential Geometry*, Ann. of Math. Studies 102, Princeton Univ. Press, Princeton, N.J., 1982.

381. J. Serrin, *On surfaces of constant mean curvature which span a given space curve*, Math. Z. **112** (1969), 77–88.

382. ——, *A priori estimates for solutions of the minimal surface equation*, Arch. Rational Mech. Anal. **14** (1963), 376–383.

383. L. Simon, *Boundary behaviour of solutions of the non-parametric least area problem*, Bull. Austral. Math. Soc. **26** (1982), 17–27.

384. ——, *Boundary regularity for solutions of the non-parametric least area problem*, Ann. of Math. (2) **103** (1976), 429–455.

385. L. Simon and R. Hardt, *Boundary regularity and embedded solutions for the oriented Plateau problem*, Ann. of Math. (2) **110** (1979), 439–486.

386. ——, *Boundary regularity and embedded solutions for the oriented Plateau problem*, Bull. Amer. Math. Soc. (N.S.) **1** (1979), 263–265.

387. J. Simons, *Minimal varieties in Riemannian manifolds*, Ann. of Math. (2) **88** (1968), 62–105.

388. ——, *Minimal cones, Plateau's problem and the Bernstein conjecture*, Proc. Nat. Acad. Sci. USA **58** (1967), 410–411.

389. ——, *A note on minimal varieties*, Bull. Amer. Math. Soc. **73** (1967), 491–495.

390. Y. T. Siu and S. T. Yau, *Compact Kähler manifolds of positive bisectional curvature*, Invent. Math. **59** (1980), 189–204.

391. Y. T. Siu, *The complex analyticity of harmonic maps and the strong rigidity of Kähler manifolds*, Ann. of Math. (2) **112** (1980), 73–111.

392. M. Shiffman, *The Plateau problem for minimal surfaces of arbitrary topological structure*, Amer. J. Math. **61** (1939), 853–882.

393. S. Smale, *On the Morse index theorem*, J. Math. Mech. **14** (1965), 1049–1056.

394. R. T. Smith, *The second variation formula for harmonic mappings*, Proc. Amer. Math. Soc. **47** (1975), 229–236.

395. ——, *Harmonic mappings of spheres*, Amer. J. Math. **97** (1975), 364–387.

396. D. J. Struik, *Outline of a history of differential geometry*, Isis **19** (1933), 92–120; **20** (1933), 161–191.

397. S. Suzuki, *On homeomorphisms of a 3-dimensional handlebody*, Canad. J. Math. **29** (1977), 111–124.

398. D'Arcy Thompson, *On growth and form*, London, 1917 (abridged edition edited by J. T. Bonner).

399. J. Todhunter, *A history of the progress of the calculus of variations during the nineteenth century*, Cambridge, 1861.

400. F. Tomi, *On the local uniqueness of the problem of least area*, Arch. Rational Mech. Anal. **52** (1973), 312–318.

401. F. Tomi and A. Tromba, *Extreme curves bound an embedded minimal surface of the type of a disc*, Math. Z. **158** (1978), 137–145.

402. D. Triscari, *Sulle singolarita delle frontiere orientate di misura minima*, Ann. Scuola Norm. Sup. Pisa Fiz. e Nat. Ser. (3) **17** (1963), 349–371.

403. A. J. Tromba, *On the number of simply-connected minimal surfaces spanning a curve*, Mem. Amer. Math. Soc. **12** (1977), No. 194.

404. ——, *A general approach to Morse theory*, J. Diff. Geom. **12** (1977), 47–85.

405. K. K. Uhlenbeck, *Minimal 2-spheres and tori in* S^k, Preprint, 1975.

406. ——, *Morse theory by perturbation methods with applications to harmonic maps*, Trans. Amer. Math. Soc. **267** (1981), 569–583.

407. ——, *Minimal spheres and other conformal variational problems*, in: Seminar on minimal submanifolds, Ann. of Math. Studies 103, Princeton Univ. Press, Princeton, N.J., 1983, pp. 169–176.

408. G. W. Whitehead, *Generalized homology theory*, Trans. Amer. Math. Soc. **102** (1962), 227–283.

409. J. A. Wolf, *Elliptic spaces in Grassmann manifolds*, Illinois J. Math. **7** (1963), 447–462.

410. ——, *Geodesic spheres in Grassmann manifolds*, Illinois J. Math. **7** (1963), 425–446.

411. Y. L. Xin, *Some results on stable harmonic maps*, Duke Math. J. **41** (1980), 609–613.

412. S. T. Yau, *Kohn-Rossi cohomology and its application to the complex Plateau problem*. I, Ann. of Math. (2) **113** (1981), 67–110.

413. ——, *On almost minimally elliptic singularities*, Bull. Amer. Math. Soc. **83** (1977), 362–364.

414. D. B. Fuks, *Maslov-Arnol′d characteristic classes*, Dokl. Akad. Nauk SSSR **178** (1968), 303–306; English transl. in Soviet Math. Dokl. **9** (1968).

415. Le Hong Van, *The growth of a two-dimensional minimal surface*, Uspekhi Mat. Nauk **40**, no. 3 (1985), 209–210; English transl. in Russian Math. Surveys **40**, no. 3 (1985).

416. ——, *Minimal surfaces and calibration forms in symmetric spaces*, Trudy Sem. Vektor. Tenzor. Anal. No. **22** Izdat. Moskov. Univ., Moscow, 1985, 107–118. (Russian)

417. B. C. White, *Mappings that minimize area in their homotopy classes*, J. Diff. Geom. **20** (1984), 433–446.

418. ——, *Homotopy classes in Sobolev spaces and energy minimizing maps*, Bull. Amer. Math. Soc. (N.S.) **13** (1985), 166–168.

419. A. T. Fomenko, *Symmetries of soap films*, Comput. Math. Appl. **128** (1986), 825–834.

420. Le Hong Van, *New Examples of globally minimal surfaces*, Geometry, differential equations and mechanics, Izdat. Moskov. Univ., Moscow, 1986, 102–105. (Russian)

421. A. A. Tuzhilin, *On the bifurcation of certain two-dimensional minimal surfaces under a two-parameter variation of the contour*, Geometry, differential equations and mechanics, Izdat. Moskov. Univ., Moscow, 1986, 140–145. (Russian)

422. ——, *Indices of two-dimensional minimal surfaces*, Global analysis, Izdat. Voronezh. Univ., Voronezh, 1987. (Russian)

423. A. O. Ivanov, *Globally minimal symmetric surfaces in Euclidean space*, Geometry, differential equations and mechanics, Izdat. Moskov. Univ., Moscow, 1986, 69–71. (Russian)

424. I. S. Balinskaya, *Volumes of orbits of smooth actions of Lie groups*, Geometry, differential equations and mechanics, Izdat. Moskov. Univ., Moscow, 1986, 49–51. (Russian)

425. ——, *Volumes of orbits of smooth actions of Lie groups*, Global analysis, Izdat. Voronezh. Univ., Voronezh, 1986. (Russian)

426. I. V. Shklyanko, *Minimal geodesics in symmetric spaces*, Geometry, differential equations and mechanics, Izdat. Moskov. Univ., Moscow, 1986, 159–161. (Russian)

427. T. N. Fomenko, *On an effective construction of a retraction of certain spaces to the sphere*, Analysis on manifolds and differential equations, Izdat. Voronezh. Univ., Voronezh, 1986, 164–173. (Russian)

428. S. A. K. El′makhi, *Volumes of orbits in certain homogeneous spaces*, Trudy Sem. Vektor. Tenzor. Anal. No. **22**, Izdat. Moskov. Univ., Moscow, 1985, 187–189. (Russian)

429. I. S. Balinskaya, *Minimal cones of the adjoint action of the classical Lie groups*, Uspekhi Mat. Nauk **41**, no. 6, (1986), 165–166; English transl. in Russian Math. Surveys **41** (1986).

430. F. A. Berezin and M. A. Shubin, *The Schrödinger equation*, Izdat. Moskov. Univ., Moscow, 1983; English transl., Reidel, Dordrecht.

431. M. do Carmo and M. Dajezer, *Rotation hypersurfaces in spaces of constant curvature*, Trans. Amer. Math. Soc. **277** (1983), 685–709.

432. D. Fischer-Colbrie, *On complete minimal surfaces with finite Morse index in three-manifolds*, Invent. Math. **82** (1985), 121–132.

433. P. H. Bérard and G. Besson, *On the number of bound states and estimates on some geometric invariants*, Lecture Notes in Math. **1324** (1988), 30–40.

434. J. Tysk, *Eigenvalue estimates with applications to minimal surfaces*, Pacific J. Math. **128** (1987), 361–366.

435. F. Lopez and A. Ros, *Complete minimal surfaces with index one and stable constant mean curvature surfaces*, Preprint, Granada, 1987.

436. R. Osserman, *A survey of minimal surfaces*, Van Nostrand, New York, London, and Melbourne, 1969.

437. T. M. Rassias, *Foundations of global non-linear analysis*, Teubner, Leipzig, 1986.

438. H. Mori, *Minimal surfaces of revolution in H^3 and their stability properties*, Indiana Math. J. **30** (1981), 787–794.

439. A. O. Ivanov, *A sufficient condition for minimality of a cone of arbitrary codimension with invariant boundary*, Trudy Sem. Vektor. Tenzor. Anal. (to appear).

440. G. Lawlor, *A sufficient criterion for a cone to be area-minimizing*, Ph. D. Thesis, Stanford Univ., 1988.

441. J. K. Moser, *On the volume elements on a manifold*, Trans. Amer. Math. Soc. **120** (1965), 286–294.

442. B. Cheng, *Area minimizing equivariant cones and coflat calibrations*, Ph. D. Thesis, MIT, 1987.

443. D. V. Alekseevskiĭ and A. M. Perelomov, *Invariant Kähler-Einstein metrics on compact homogeneous spaces*, Funktsional. Anal. i Prilozh. **20**, no. 3 (1986), 1–16; English transl. in Functional Anal. Appl. **20** (1986).

444. A. Borel, *Sur la cohomologie des espaces fibrés principaux et des espaces homogènes de groupes de Lie compacts*, Ann. of Math. (2) **57** (1953), 115–207.

445. N. Bourbaki, *Éléments de mathématique*, Fasc. XXXVIII, *Groupes et algèbres de Lie*, Ch. VII, VIII, Hermann, Paris, 1975.

446. E. B. Vinberg and A. L. Onishchik, *A seminar on Lie groups and algebraic groups*, "Nauka", Moscow, 1988. (Russian)

447. E. B. Dynkin, *Maximal subgroups of the classical groups*, Trudy Moskov. Mat. Obshch. **1** (1952), 39–166.

448. A. A. Kirillov, *Elements of the theory of representations*, "Nauka", Moscow, 1978; English transl., Springer-Verlag, Berlin and New York, 1976.

449. S. Kobayashi and K. Nomizu, *Foundations of differential geometry* (2 vols.), Interscience, New York and London, 1963, 1969.

450. Le Hong Van, *Absolutely minimal surfaces and calibrations on orbits of the adjoint representation of classical Lie groups*, Dokl. Akad. Nauk SSSR **298** (1988), 1308–1311; English transl. in Soviet Math. Dokl. **37** (1988).

451. ——, *Minimal surfaces in homogeneous spaces*, Izv. Akad. Nauk SSSR Ser. Mat. **52** (1988), 408–423; English transl. in Math. USSR-Izv. **32** (1989).

452. Le Hong Van and A. T. Fomenko, *Lagrangian manifolds and the Maslov index in the theory of minimal surfaces*, Dokl. Akad. Nauk SSSR **299** (1988), 42–45; English transl. in Soviet Math. Dokl. **37** (1988).

453. ——, *A criterion for the minimality of Lagrangian submanifolds of Kähler manifolds*, Mat. Zametki **42** (1987), 559–572; English transl. in Math. Notes **42** (1987).

454. Le Hong Van, *Minimal Φ-Lagrangian surfaces in almost-Hermitian manifolds*, Mat. Sb. **180** (1989), 924–936; English transl. in Math. USSR-Sb. **67** (1990).

455. Le Hong Van and A. T. Fomenko, *Volumes of minimal surfaces and the curvature tensor of Riemannian manifolds*, Dokl. Akad. Nauk SSSR **300** (1988), 1308–1312; English transl. in Soviet Math. Dokl. **37** (1988).

456. Le Hong Van, *Each globally minimal surface in a compact homogeneous space has an invariant calibration*, Dokl. Akad. Nauk SSSR **310** (1989), No. 2; English transl. in Soviet Math. Dokl. **41** (1990), No. 1.

457. ——, *Relative calibrations and the problem of stability of minimal surfaces*, New ideas in global analysis, Izdat. Voronezh. Univ., Voronezh, 1989. (Russian); English transl. in Lectures Notes in Math.

458. V. V. Trofimov, *New construction of characteristic classes of Lagrangian submanifolds* (to appear).

459. A. T. Fomenko, *Symplectic geometry*, Izdat. Moskov. Univ., Moscow, 1988; English transl.: Advanced Studies in Math., vol. 5, Gordon and Breach, 1988.

460. J. Brother, *Stability of minimal orbits*, Trans. Amer. Math. Soc. **294** (1986), 537–552.

461. R. L. Bryant, *Minimal Lagrangian submanifolds of Kähler-Einstein manifolds*, Lecture Notes in Math. **1255** (1987), 1–12.

462. J. Dadok, R. Harvey, and F. Morgan, *Calibrations on* R^8, Trans. Amer. Math. Soc. **305** (1988), 1–39.

463. Y. Ohnita, *On stability of minimal submanifolds in compact symmetric spaces*, Compositio Math. **64** (1987), 157–190.

464. J. Wolf, *The geometry and structure of isotropy irreducible homogeneous spaces*, Acta Math. **120** (1968), 59–148.

465. M. Wang and W. Ziller, *On normal Einstein homogeneous manifolds*, Ann. Sci. École Norm. Sup. (4) **18** (1985), 536–633.

466. I. S. Balinskaya, *Minimal cones of the adjoint action of the classical Lie groups*, Trudy Sem. Vektor. Tenzor. Anal. No. 23, Izdat. Moskov. Univ., Moscow, 1988, 11–17. (Russian)

467. E. N. Gilbert and H. O. Pollak, *Steiner minimal trees*, SIAM J. Appl. Math. **16** (1968), 1–29.

468. D. Z. Du and F. K. Hwang, *Steiner minimal tree for points on a circle*, Proc. Amer. Math. Soc. **95** (1985), 613–618.

469. ——, *Steiner minimal tree for points on a zig-zag line*, Trans. Amer. Math. Soc. **278** (1983), 149–156.

470. A. O. Ivanov and A. A. Tuzhilin, *A deformation of a manifold that decreases the volume with the greatest speed*, Vestnik Moskov. Univ. Ser. I Math. Mekh. **1989**, no. 3, 14–18; English transl. in Moscow Univ. Math. Bull.

471. A. O. Ivanov and A. A. Tuzhilin, *The solution of Steiner problem for convex boundaries*, Uspekhi Mat. Nauk **45** (1990), No. 2, 207–208; English transl. in Russian Math. Surveys **45** (1990).

Subject Index